Soil Management for Sustainability

Soil Management for Sustainability

Editors

Chiara Piccini
Rosa Francaviglia

MDPI • Basel • Beijing • Wuhan • Barcelona • Belgrade • Manchester • Tokyo • Cluj • Tianjin

Editors
Chiara Piccini
Council for Agricultural
Research and Economics
(CREA), Research Centre for
Agriculture and Environment
Italy

Rosa Francaviglia
Council for Agricultural
Research and Economics
(CREA), Research Centre for
Agriculture and Environment
Italy

Editorial Office
MDPI
St. Alban-Anlage 66
4052 Basel, Switzerland

This is a reprint of articles from the Special Issue published online in the open access journal *Land* (ISSN 2073-445X) (available at: http://www.mdpi.com).

For citation purposes, cite each article independently as indicated on the article page online and as indicated below:

LastName, A.A.; LastName, B.B.; LastName, C.C. Article Title. *Journal Name* **Year**, *Volume Number*, Page Range.

ISBN 978-3-0365-2323-1 (Hbk)
ISBN 978-3-0365-2324-8 (PDF)

Cover image courtesy of Claudia Di Bene

Contents

About the Editors

Chiara Piccini is a geologist, working at the Council for Agricultural Research and Economics, Research Centre for Agriculture and Environment (CREA-AA), in Rome, Italy, since 1998. Her main research interests are the spatialization of experimental data using geostatistics, digital soil mapping techniques and satellite imagery; the application of crop simulation models for soil nitrogen and carbon dynamics; soil quality and environmental concerns; the monitoring of nutrients leaching from soil; nitrate contamination; statistical data processing; the monitoring of soil chemical features; soil chemical, physical and hydrological characterization; soil chemical fertility; soil water dynamics; irrigation water quality; and irrigation with non-optimal-quality water.

Rosa Francaviglia, with expertise in agronomy, is a senior researcher at the Council for Agricultural Research and Economics, Research Centre for Agriculture and Environment (CREA-AA), in Rome, Italy, since 1981 (now retired). Her main research topics include the effects of climate change on agriculture, carbon sinks and agricultural soils, soil organic carbon simulation models, soil fertility, conservation agriculture, crop diversification, agro-environmental evaluations, soil quality indicators, and good agro-environmental conditions (GAEC) under the EU Common Agricultural Policy.

Preface to "Soil Management for Sustainability"

Soils are an essential and non-renewable natural resource, supplying goods and services fundamental to ecosystems and human life. Soils are necessary for producing crops, feed, fiber, and fuel, in addition to filtering and cleaning water. As a major carbon sink, soils also help to regulate the emissions of carbon dioxide and other greenhouse gases, which is fundamental for climate regulation. Nevertheless, evidence recently provided in the Status of the World's Soil Resources report and other studies show that about one-third of global soils are moderately or highly degraded due to unsustainable management practices, aggravated by the increased frequency of extreme weather events resulting in high soil erosion rates, with consequences for overall soil quality and ecosystem services. Particularly in agricultural soils, there is a continuing degradation trend arising from highly intensive agricultural systems, even though reduced tillage, crop rotations, and cover crops are spreading due to the adoption of conservation agriculture as a sustainable soil management system.

Soil management is sustainable if "the supporting, provisioning, regulating, and cultural services provided by soil are maintained or enhanced without significantly impairing either the soil functions that enable those services, or biodiversity". Thus, sustainable soil management (SSM) is crucial for effective soil functioning, also strongly contributing to climate change adaptation and mitigation, combating desertification, and promoting biodiversity. SSM is an integral part of sustainable land management; thus, a territorial perspective is important in such studies. This Special Issue welcomes research and review papers focusing on minimizing soil erosion; enhancing soil organic matter content; fostering soil nutrient balance and cycles; preventing, minimizing, and mitigating soil salinization and alkalinization; preventing and minimizing soil contamination; preventing and minimizing soil acidification; preserving and enhancing soil biodiversity; minimizing soil sealing; preventing and mitigating soil compaction; and improving soil water management.

However, the adoption of any sustainable soil management practice is highly dependent on the environmental–socioeconomic context, and therefore, guidance from land-use planners and advisory services should be made available from the local to regional scale. Monitoring and verifying that sustainable management practices are being applied correctly and assessing their impacts on the different ecosystem services is a major challenge. In this sense, geospatial modeling and the predictive mapping of selected soil properties, coupled with land-use and soil management information and other environmental covariates, is a valuable tool for assessing the improvement or degradation of soil quality and ecosystem services.

Chiara Piccini, Rosa Francaviglia
Editors

land

Article

Soil Carbon, Nitrogen and Phosphorus Contents along a Gradient of Agricultural Intensity in the Kilombero Valley, Tanzania

John Livsey [1,2,*], Edmond Alavaisha [1,3], Madaka Tumbo [3], Steve W. Lyon [1,2,4], Antonio Canale [5], Michele Cecotti [5], Regina Lindborg [1,2] and Stefano Manzoni [1,2]

1 Department of Physical Geography, Stockholm University, 106 91 Stockholm, Sweden; edmond.alavaisha@natgeo.su.se (E.A.); lyon.248@osu.edu (S.W.L.); regina.lindborg@natgeo.su.se (R.L.); stefano.manzoni@natgeo.su.se (S.M.)
2 Bolin Centre for Climate Research, Stockholm University, 106 91 Stockholm, Sweden
3 Institute of Resource Assessment, University of Dar es Salaam, Dar es Salaam 35097, Tanzania; tumbo.madaka@udsm.ac.tz
4 School of Environment and Natural Resources, Ohio State University, Columbus, OH 43210, USA
5 Department of Statistical Sciences, University of Padova, 35121 Padua, Italy; canale@stat.unipd.it (A.C.); michele.cecotti@studenti.unipd.it (M.C.)
* Correspondence: john.livsey@natgeo.su.se

Received: 12 March 2020; Accepted: 16 April 2020; Published: 18 April 2020

Abstract: The preservation of soils which provide many important services to society is a pressing global issue. This is particularly the case in countries like Tanzania, which will experience rapid population growth over coming decades. The country is also currently experiencing rapid land-use change and increasing intensification of its agricultural systems to ensure sufficient food production. However, little is known regarding what the long term effects of this land use change will be, especially concerning soil quality. Therefore, we assessed the effect of irrigation and fertilization in agricultural systems, going from low intensity smallholder to high intensity commercial production, on soil organic carbon (SOC), total nitrogen (TN), and total phosphorous (TP) concentrations and stocks. Soil sampling was conducted within Kilombero Plantations Ltd. (KPL), a high intensity commercial farm located in Kilombero, Tanzania, and also on surrounding smallholder farms, capturing a gradient of agricultural intensity. We found that irrigation had a positive effect on SOC concentrations and stocks while fertilization had a negative effect. Rain-fed non-fertilized production had no effect on soil properties when compared to native vegetation. No difference was found in concentrations of TN or TP across the intensity gradient. However, TN stocks were significantly larger in the surface soils (0–30 cm) of the most intensive production system when compared to native vegetation and smallholder production.

Keywords: soil organic carbon; agricultural intensity; nitrogen; phosphorous; irrigation; fertilization

1. Introduction

The preservation of soils is central to many of the challenges facing society. These challenges include ensuring food and energy security, climate change mitigation, biodiversity protection and the continued provision of numerous ecosystem services [1]. The most important role that soils play for society is in the provision of food. Large quantities of the earth's topsoil have been directly altered for anthropogenic land-use, with one third of all land cover (excluding Greenland and Antarctica) being dedicated to agriculture [2]. Therefore, sustaining global food production represents both a challenge for soils and an opportunity to ensure soil preservation through proper soil management.

With the global population expected to reach 9 billion by 2050, an increasing demand for food will place a significant pressure on land and soil. However, the expansion of global population will not be homogeneous. Due to regional differences in birth rates and mortality, population related pressures will be greater in some places than others. Sub-Saharan Africa will be the greatest contributor to the expansion of the world's population over coming decades, with the region becoming the world's most populous by 2100 [3]. Therefore, the importance of ensuring enough food is locally produced is one of the greatest challenges for nations within this region, placing pressure on the region's soils.

Tanzania is one such country, given that its population is expected to increase from 56 million in 2018 to 129 million by 2050 [3]. Food security is thus a critical challenge for the nation. This is further complicated by climate change, which will likely have negative effects on crop yields as a result of increased incidence of pest [4], changes in the availability of water resources [5], and heat stress [6]. While agriculture experiences the negative effects of climate change, it plays an important role as a contributor to greenhouse gas (GHG) emissions [7]. Agriculture is also increasingly being discussed as a potential mitigator of GHG emissions [8] through initiatives such as "4 per 1000" [9]. Due to the rapidly expanding population and increase in agricultural area, increases in GHG emissions from Tanzanian agriculture are among the fastest growing in the world [10].

Despite these challenges, Tanzania has great potential to succeed in meeting future production demands locally. Providing 78% of employment and contributing to 29% of the nation's gross domestic product [11], the country's agriculture sector is dominated by low intensity smallholder farming. Further, only 15% of nation's potential arable land is currently utilized for crop production, and the majority of this cropland is rain-fed. Therefore, the country has opportunities to both expand its production area and increase its production intensity. To meet future production needs, the country has embarked upon a modernization process to commercialize its smallholder dominated agricultural sector, via the Kilimo Kwanza (Agriculture First) initiative, and programs such as Southern Agricultural Growth Corridor of Tanzania [12]. But to ensure food security and improve livelihoods, this development of the sector needs to be achieved in a sustainable way.

The intensification of agriculture, through mechanization, crop selection, and the use of inorganic inputs has increased production [13], but often at the expense of other ecosystem services. Within Sub-Saharan Africa, where increased production has often been made possible through the conversion of forest to agricultural land, agricultural intensification has increased threats to biodiversity [14] and resulted in habitat loss and fragmentation [15]. This loss of biodiversity, coupled with the use of pesticides, may have indirect negative consequences for crop production through detrimentally effecting pollinators [16] and the loss of predators that feed on agricultural pests [17]. It may also lead to lower inherent soil fertility [18], potentially resulting in negative effects on yields. However, the significance of these effects is dependent on the original status of the soil and the land management practices (LMPs) that are implemented.

A loss of soil organic carbon (SOC) is often noted as a consequence of agricultural production. Winowiecki et al. [19] found that cultivation has a negative effect on soil organic carbon concentrations when compared to non-cultivated land in Tanzania. Similar effects have been seen globally, with the conversion of native forests to agriculture resulting, on average, in a 24% reduction in SOC stocks [20]. Yet, certain systems may also result in increased SOC. For example, the production of paddy rice has been responsible for SOC accumulation [21,22], although at the expense of relatively higher methane emissions, due to flooded fields limiting the degradation of organic matter [23]. However, the global trend in SOC, as a result of agriculture, has been a net loss [24]. Changes in SOC are also associated with changes in macronutrients, such as nitrogen (N) and phosphorus (P), which are chemically bound to carbon (C) in organic compounds [25]. Thus, SOC reductions caused by agricultural activities may result in loss of the soil nutrient capital. As well as being related to soil nutrient status, SOC also affects soil compaction, with bulk density (BD) commonly being inversely related to SOC content [26]. This makes changes in SOC a common indicator of soil fertility.

Considering that increased soil organic matter (including C, N, and P) may have yield benefits [27], it is important to both quantify and mitigate any loss that may occur as a result of agricultural intensification globally, particularly in countries like Tanzania that need to dramatically increase agricultural output to meet the needs of their rapidly expanding population. However, little focus has been put on assessing the effect on soils of land-use change for agriculture in Eastern Africa. In a recent review [28], Namirembe et al. identified only 15 studies which assessed changes in SOC as a result of bushland, woodland or forest conversion to agriculture in Ethiopia, Kenya, Rwanda, Tanzania, Uganda, or Burundi. The average sampling depth of the identified studies was 32 cm. Further, less focus has been put on comparing the effect of land-use conversion across a gradient of production intensities. Therefore, in this under-represented area, there is a need to assess changes in soil properties (and thus fertility) both along gradients of agricultural intensity, and at a depth in the soil profile where such changes have largely been neglected so far.

The long-term effects of land-use change and agricultural intensification should be urgently assessed in areas of Tanzania that are anticipated to experience a significant increase in both. In this study, we aim to identify the effects of agricultural management along a gradient of agricultural intensification within the Kilombero Valley, Tanzania. While the valley is dominated by smallholder production, it is also the location of Kilombero Plantations Ltd. (KPL), an industrial producer of rice and maize. Therefore, we consider the effect of agriculture management across a range of intensities, from unfertilized and rain-fed smallholder (low intensity) to fertilized and irrigated industrial (high intensity). We also compare agriculture to soil conditions under native vegetation. Specifically, we ask how does agricultural intensity affect concentrations and stocks of soil organic carbon, soil nitrogen and soil phosphorous?

2. Materials and Methods

2.1. Site Description

The Kilombero Valley is located in the Morogoro Region of southern central Tanzania. The valley covers approximately 39,000 km^2, with a complex network of streams moving down from the surrounding mountains and joining to form Kilombero River. The northern and western sides of the valley are bordered by the Udzungwa mountains, and the eastern side by the Mahenge highlands [29,30]. Annual precipitation ranges between 1200 and 1400 mm, falling mainly in the November to April wet season. Mean daily temperature is 22–23 °C, with a relative humidity between 70% and 87% in forest and highland areas, and 58% and 85% in the lowlands. Soils surrounding the study location, within the catchment's lowlands, are characterized as fluvisols [31].

The valley is home to both smallholder and commercial farms, with rice, maize and sugar cane being major produced crops. There are nine small-to-medium scale irrigation schemes with well-established irrigation infrastructure and two prominent commercial irrigation schemes in Msolwa (sugar plantation) and Mngeta (rice and maize plantation).

Sampling for this research was conducted on and around Mngeta farm, operated by KPL. The farm is an industrial rice and maize producer situated approximately 62 km west south west of Ifakara. Purchased in 2006, the farm sits on the site of a former Tanzania–North Korean agricultural operation which cleared the land for production during the 1980's but was then later abandoned. Covering over 5000 ha, the farm grows both center-pivot irrigated and rain-fed rice during the wet season and maize during the dry season. The farm fertilizes the rain-fed and irrigated areas using a combination of DAP, MOP, Urea and Nitrabor as basal and top dressings, with total application rates of approximately 90–125 kg N ha^{-1} and 15–25 kg P ha^{-1} per crop. Rice yields on the commercial farm are typically between 3 and 4 t ha^{-1}, but with a large degree of variability between years and fields.

The land surrounding KPL is resident to smallholder farmers who also produce rice and maize. Whilst the number of large scale production activities within Kilombero is increasing [32], agriculture within the valley is dominated by smallholder and subsistence farming [33]. Yields on the surrounding smallholder farms can be as low as 1 t ha^{-1} in unfertilized rain-fed areas, up to 6 t ha^{-1} in fields where System of Rice Intensification practices have been implemented. In both the KPL farm and surrounding smallholder farms, land is ploughed prior to crop planting. Ploughing is conducted using tractor drawn plough or, on smallholder land, using a power tiller. Although variable, plough depth is between 15 and 30 cm.

As a result of migration to the valley, its agricultural area has increased by 3430 km^2 (11.3%) between 1990 and 2016 [33] mainly through the conversion of bushland and forest. The increasing production area and intensity has led to increased crop production, but has resulted in changes in river chemistry and biota [29], and also species loss and habitat fragmentation [34].

2.2. Sampling Methods

Field sampling was conducted between June and July 2018, after the harvesting of the wet season rice crop and before the planting of maize, on and around the KPL farm. Four land management practices (LMPs), detailed in Figure 1 and Table 1, were selected for comparison of soil properties. We consider these LMPs to represent a gradient of production intensity, with mechanization and increasing use of chemical inputs being markers of production intensity. We considered fields in the commercial farm KPL (denoted by 'C-' in the LMP codes) and in the surrounding small holder farms (denoted by 'S-'). Moreover, we identified fields that were either fertilized or left unfertilized ('F' vs. 'U'), and either irrigated or rain-fed ('I' vs. 'R'). Soil samples were obtained at five sites within each LMP (Figure 1). All sites had been consistently managed in the same way for at least 10 years, excluding the C-FI sites where irrigation was introduced in the 2014–2015 growing season. Sites within C-FR were randomly selected from within the area, covering 120 ha, reported by the KPL farm management as being solely used for rain-fed rice production. Similarly, C-FI sites were randomly selected from within the pivot irrigated areas installed in 2014, which cover an area of 250 ha, within the KPL farm. The selection of smallholder fields that were neither irrigated or fertilized (S-UR) was limited to five small fields (between 0.5 and 1.5 ha) that were found to be appropriate after interviewing a local extension officer and farmers in the area. Therefore, for this LMP, a random site selection was not possible. However, the sampled sites were scattered over a wide area and are thus regarded as representative of this land management (Figure 1). Further, access to sites with native land cover (NAT) within the KPL farm was restricted by the density of the vegetation within these areas. This meant that NAT sites were selected by walking through the forest, on a path dictated by accessibility, and then digging at the location reached after thirty minutes, this time equated to a distance that was deemed sufficient to ensure that the sampled sites were representative of relatively undisturbed forest. The forest area within the farm had been set aside as a wildlife refuge, and no signs of anthropogenic disturbance were observed.

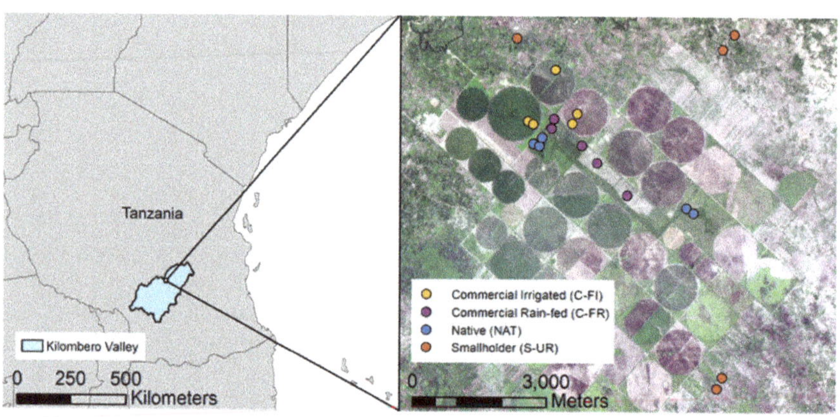

Figure 1. Soil sampling locations (sites) on and around the Kilombero Plantations Ltd. farm.

Table 1. Land management practices on the locations selected for soil sampling.

Land Management Practice	Description	Label	Figure Label Color	Agricultural Intensity
On-farm irrigated (C-FI)	Fields within the industrial farm on which irrigated and chemically fertilized rice is grown in the wet season. Irrigated maize is grown in the dry season. Fertilizer application is provided primarily via pivot irrigation systems	C-FI		High intensity
On-farm non irrigated (C-FR)	Fields within the industrial farm on which rain-fed and chemically fertilized rice is grown in the wet season. Fields are not utilized in the dry season.	C-FR		
Off-farm non irrigated (S-UR)	Smallholder fields outside, but located close to, the industrial farm. Fields are used for rice production, and are neither irrigated or fertilized.	S-UR		
Native (NAT)	Undisturbed forest and shrubland within the industrial farm which has been set aside as a green space	NAT		Low intensity

The sampling protocol was adapted from Alavaisha et al. (2019) [32] to allow integration of the current new dataset with previous soil analyses in the Kilombero Valley. For analysis of soil texture, pH, SOC, total nitrogen (TN), and total phosphorous (TP), soil samples were taken at five points for each site. One point was located at the center of the site and four at equidistant locations around a 10 m radius circle around the center point. Soils from each point were collected at six depths (0–20, 20–30, 30–40, 40–50, 50–60, and 60–80 cm) using a soil auger (Unoson Environment AB). The samples from each of the five points were then homogenized, with visible roots removed, and stored in sealed plastic bags. A single BD sample was taken, for each depth, at the center point of each site, with the bulk density ring driven horizontally into the soil profile at each depth's mid-point.

2.3. Soil Analysis

Laboratory analyses were conducted at the Department of Soil and Geological Science, Sokoine University of Agriculture, Morogoro. Samples were air dried and passed through a 2 mm sieve, with large particles not passing through being crushed and then passed through the sieve again [35]. Texture was measured using the hydrometer bouyoucos method, pH with a Oakton Ion 700 bench meter using a 1:2.5 soil-distilled water solution, and TN with a Foss Tecato Kjeltec™ 2100 Auto Distillation Digestion system using the Kjeldahl method, as described by Klute (1986). SOC was measured via the Walkley–Black method [36], and TP using the dry combustion method [37] measured with a Biomate 6 UV Spectrophotometer.

2.4. Statistical Procedure

The effects of land management practices, on concentrations of SOC, TN and TP, and on C:N and C:P ratios, were analyzed with a mixed-effect model [38] accounting for the dependence of the measurements taken in the same field. For each soil element, we determined the value of the logarithm of the concentration (y_{ij}) in crop $j = 1,\ldots,20$ at the i-th depth with $i = 1,\ldots,6$ and then fit a model such that:

$$y_{ij} = \alpha_0 + X\beta + Z\theta + \gamma_j + \delta_j d_{ij} + \epsilon_{ij}, \quad (1)$$

where α represents an intercept term, X is a matrix of dummy variables labelling the crop type as reported in Table S1 with β the related vector of regression coefficients measuring the additive effect (on a log scale) of the land-use. The matrix Z contains additional confounders (the percentage of silt and depth) with θ the related regression coefficients. Finally, γ_j and δ_j are crop-specific random effect parameters representing crop-specific intercept and regression coefficient associated to the depth d_{ij} of measurement i in crop j. The model specification is completed assuming ε_{ij} represents a random Gaussian noise. The mixed effect model was fitted in R utilizing the nlme package and using the restricted maximum likelihood approach.

Stocks of SOC, TN and TP (Mg ha^{-1}) were calculated for each soil sampling depth from the percent concentrations at that depth (x):

$$Stock_{mn} = C_{mn} \times BD_{mn} \times D_{mn} \quad (2)$$

where m is the element (SOC, TN, TP), n the sampling depth, C and BD are the element concentration (expressed as percentages on a dry weight basis) and soil bulk density (g cm^{-3}) derived from the laboratory analysis, respectively, and D the thickness of the sampling depth (cm). Stocks through the profile were summed to calculate the stock of each element in the total soil profile, as well as for the root zone (0–30 cm) and the subsoil (30–80 cm).

One-way analysis of variance (ANOVA) with Tukey's honestly significant difference (HSD) post hoc test was used to assess for significant difference ($p < 0.05$) in element stocks in root zone, subsoil and full profile, and also in concentrations of SOC, TN, TP and C:N and C:P ratios at each sampling depth.

3. Results

We first present the observed concentrations of SOC and soil nutrients along vertical soil profiles, comparing trends seen between land-use types; second, nutrient stocks are presented, and finally soil C:N and C:P ratios. Variability in soil texture can also be found in the Supplementary Figure S1. The complete original data file (Data S1) can also be found in the Supplementary Materials.

3.1. *Observed Soil Organic Carbon and Nutrient Concentrations*

3.1.1. Vertical Profiles of Soil Organic Carbon and Nutrient Concentrations

Depth had a significantly negative effect on SOC, TN and TP ($p < 0.01$) (Table 2). Percent SOC and TN consistently decreased with increasing depth below the land surface within all LMPs (Table 3). Reductions in TP concentrations were milder than for SOC and TP, and concentrations between 20 and 60 cm were more variable between sites within the two commercial LMPs, C-FR and C-FI, compared to the two non-commercial LMPs (Table 3). In all LMPs, bulk density consistently increased with depth (Table 3).

Table 2. Estimates and regression using the mixed effect model for soil organic carbon (SOC), total nitrogen (TN), total phosphorus (TP), and carbon to nitrogen ratio (C:N) and carbon to phosphorus ratio (C:P) ratios.

Parameter	SOC			TN			TP			C:N			C:P		
	Estimate	SE	*p*-Value	Estimate	SE	*p*-Value	Estimate	SE	*p*-Value	Estimate	SE	*p*-Value	Estimate	SE	*p*-Value
α(*Intercept*)	1.1633	0.2990	0.0002	−1.6329	0.1313	0.0000	−2.1360	0.2265	0.0000	2.7517	0.2682	0.0000	3.1979	0.3520	0.0000
β_1(*Irrigated*)	0.5752	0.2691	0.0483	−0.0283	0.1158	0.8102	0.1270	0.2076	0.5495	0.6296	0.2130	0.0093	0.3075	0.2954	0.3135
β_2(*Fertilized*)	−0.4949	0.2694	0.0848	0.0151	0.1159	0.8978	0.3724	0.2076	0.0918	−0.5205	0.2132	0.0267	−0.7821	0.2956	0.0176
β_3(*Non−natural*)	−0.2807	0.2763	0.3249	0.1006	0.1171	0.4027	0.0529	0.2076	0.8021	−0.3340	0.2196	0.1478	−0.1902	0.2983	0.5328
θ_1(*Percent Silt*)	0.0279	0.0089	0.0021	0.0128	0.0044	0.0044	−0.0003	0.0091	0.9758	0.0155	0.0086	0.0733	0.0274	0.0123	0.0277
θ_2(*Depth*)	−0.0217	0.0025	0.0000	−0.0147	0.0013	0.0000	−0.0086	0.0027	0.0022	−0.0070	0.0026	0.0085	−0.0132	0.0039	0.0011

Note: soil organic carbon (SOC); total nitrogen (TN); total phosphorous (TP); carbon to nitrogen ratio (C:N) and carbon to phosphorus ratio (C:P); standard error (SE).

Table 3. Mean percentages of: soil organic carbon (SOC; top rows); total nitrogen (TN; middle top); total phosphorous (TP; middle bottom); and bulk density (BD, bottom) at each sampling depth for the four land management practices (LMPs). Bracketed numbers are standard errors (n = 5). Different letters in the same column indicate significant difference ($p < 0.05$) between LMPs determined using one-way analysis of variance and Tukey's honestly significant difference post hoc test.

LMP	0–20 cm	20–30 cm	30–40 cm	40–50 cm	50–60 cm	60–80 cm
			% SOC			
NAT	2.54(0.3)a	2.24(0.22)a	2.07(0.24)a	2.18(0.44) b	1.41(0.10) b	1.10(0.06)a
S-UR	3.64(0.53)a	2.34(0.62)a	1.67(0.39)a	1.34(0.25)ab	0.90(0.26)ab	0.74(0.24)a
C-FR	2.89(0.19)a	2.48(0.39)a	1.25(0.16)a	0.50(0.11) a	0.53(0.15) a	0.54(0.10)a
C-FI	3.81(0.27)a	2.69(0.27)a	1.46(0.19)a	1.00(0.16) a	0.83(0.12)ab	1.06(0.27)a
			% TN			
NAT	0.19(0.02)a	0.14(0.01)a	0.12(0.01)a	0.12(0.01)a	0.11(0.01)a	0.09(<0.00)a
S-UR	0.23(0.03)a	0.19(0.03)a	0.14(0.02)a	0.12(0.02)a	0.10(0.02)a	0.08(0.01)a
C-FR	0.20(0.02)a	0.21(0.05)a	0.13(0.01)a	0.13(0.02)a	0.11(0.02)a	0.09(0.02)a
C-FI	0.23(0.01)a	0.18(0.02)a	0.15(0.01)a	0.11(0.02)a	0.09(0.01)a	0.09(0.01)a
			% TP			
NAT	0.12(0.02)a	0.10(0.02)ab	0.09(0.02)a	0.08(0.01)a	0.07(0.01)a	0.09(0.02)a
S-UR	0.13(0.01)a	0.09(0.02) b	0.10(0.02)a	0.09(0.02)a	0.08(0.02)a	0.08(0.03)a
C-FR	0.16(0.02)a	0.14(0.03)ab	0.18(0.05)a	0.12(0.06)a	0.13(0.03)a	0.07(0.02)a
C-FI	0.19(0.03)a	0.22(0.04) a	0.14(0.03)a	0.11(0.03)a	0.25(0.12)a	0.17(0.05)a
			BD (g cm^{-3})			
NAT	0.99(0.05)a	1.07(0.05)a	1.16(0.04)a	1.11(0.06)a	1.16(0.04)a	1.17(0.03)a
S-UR	0.84(0.06)a	0.94(0.08)a	1.01(0.1)a	1.10(0.09)a	1.13(0.05)a	1.21(0.08)a
C-FR	0.96(0.04)a	1.03(0.06)a	1.09(0.08)a	1.23(0.07)a	1.24(0.08)a	1.31(0.07)a
C-FI	0.93(0.06)a	0.98(0.09)a	1.10(0.09)a	1.14(0.06)a	1.17(0.05)a	1.26(0.05)a

While SOC concentrations decreased with increasing depth below the soil surface, the decrease was more pronounced in the agricultural LMPs than the NAT sites. This can be seen in the larger absolute values of the slope coefficients for the three agricultural LMPs compared against that of the NAT sites in Figure 2a. TN concentration profiles were also less steep in NAT sites compared to two of the agricultural LMPs (Figure 2b). Compared to SOC and TN, the concentration profiles for TP were more variable between sites and LMPs, demonstrated by the larger standard errors in Figure 2c.

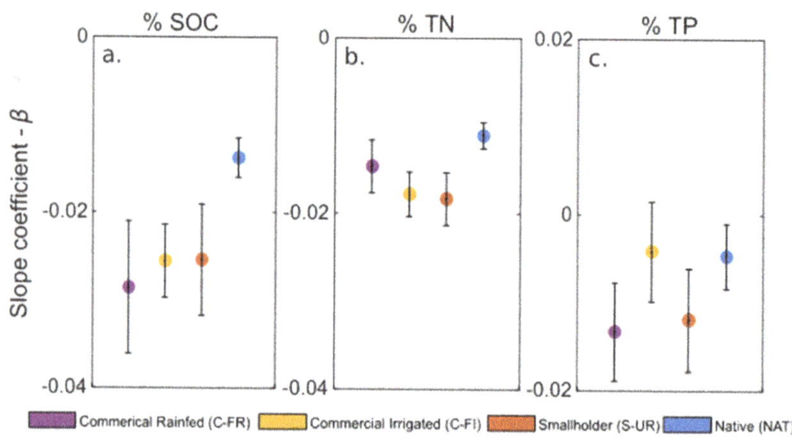

Figure 2. Slope coefficients (β) for a fitted single term exponential model ($y = ae^{\beta x}$) for: (**a**), % SOC; (**b**), % TN; (**c**), % TP. More negative values indicate steeper vertical declines in element concentrations. Colored circles along the x-axis of each pane denote the results for each LMP, circles represent the estimated β values for each LMP, black whiskers are the related standard errors (n = 5).

3.1.2. Differences in Soil Organic Carbon and Nutrient Concentrations between LMPs

We found some statistical differences between SOC concentrations in the four LMPs, seen in the estimates of β and their marginal significance t-tests, reported in Table 2. Land management within the irrigated C-FI sites had a mild but significantly positive effect on log % SOC ($p = 0.048$), while fertilization had a significantly negative effect only at the 10% level ($p = 0.085$). Also, no difference was detected between SOC concentrations between the native vegetation and the smallholder producers. Analysis of SOC concentrations within individual soil layers found that between 40 and 60 cm, concentrations of C-FR sites were significantly smaller ($p < 0.05$) than those of NAT sites (Table 3).

While fertilization had a significant effect on TP concentrations at the 10% level ($p = 0.092$), no significant difference was found in concentrations of TN and TP between any of the four LMPs. Silt content was also found to be significantly related to SOC and nutrient concentrations. Clay content (Figure S1) and pH, which ranged from 5.05 and 6.49 between all sites and depths (mean 5.72), had no significant effect and did not improve the model fit.

3.2. Carbon and Nutrient Stocks

Observed stocks of SOC across the entire sampled soil profile were highest within the NAT sites with a mean value of 177 Mg ha^{-1} (Figure 3c). Similarly, SOC stocks were also higher in the NAT sites between 30 and 80 cm than they were in the other LMPs (Figure 3b). However, only C-FR was significantly different ($p < 0.05$) from NAT between 30–80 cm (Figure 3b). Conversely, while not significantly different, NAT sites had the lowest mean SOC stocks in the surface soils (Figure 3a). Stocks of TN were similar between all LMPs, with the mean values being marginally, but not significantly, higher in the two fertilized LMPs (C-FR and C-FI) than in the non-fertilized ones (S-UR and NAT) at all depths (Figure 3d–f). For TP, mean observed stocks were greater in the two fertilized LMPs than in the non-fertilized LMPs when comparing stocks in the surface soils (Figure 3g), the subsurface (Figure 3h), and the total soil profile (Figure 3i). The difference in TP stocks was significant in C-FI compared to S-UR and NAT at 0–30 cm (Figure 3g).

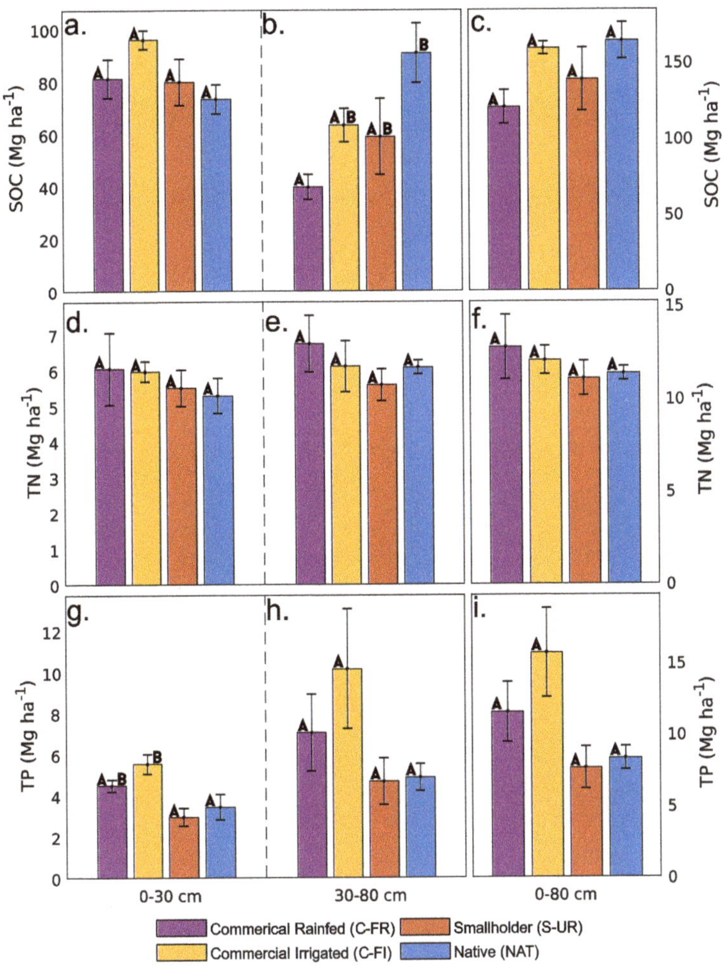

Figure 3. Stocks of SOC (**a–c**), TN (**d–f**) and TP (**g–i**) for: left, 0–30 cm, center 30–80 cm from the soil surface; right, 0–80 cm from the soil surface. Note the different vertical scales between the two left columns and the right one; black whiskers are standard errors (n = 5). Different letters (A or B) above bars represent significant difference ($p < 0.05$) between LMPs determined using one-way analysis of variance and Tukey's honestly significant difference post hoc test.

3.3. C:N and C:P Ratios

There was a diverging trend in C:N and C:P ratios with increasing depth (Table 4), the difference in mean values was more constrained in the surface layer (0–20 cm) than in any other soil layer. The most pronounced difference in C:N ratios occurred between the C-FR and NAT, which were significantly different ($p < 0.05$) between 30–60 cm (Table 4). Below 30 cm, mean C:N markedly decreases in C-FR. For NAT, despite an increase at intermediate soil depths, the ratio was relatively consistent between soil layers. Depth had a mild but significantly negative effect on both the C:N ($p = 0.009$) and C:P ratio

($p = 0.001$) (Table 2). With regard to management practices, irrigation had a significantly positive effect on the C:N ratio ($p < 0.009$), and fertilization had a significantly negative effect on the C:N ($p = 0.026$) and C:P ratios ($p = 0.018$) (Table 2).

Table 4. Means of the observed soil C:N (top) and C:P (bottom) ratios through the sampled depths for each LMP. Bracketed numbers are standard error (n = 5). Different letters in the same column indicate significant difference ($p < 0.05$) between LMPs determined using one-way analysis of variance and Tukey's honestly significant difference post hoc test.

	0–20 cm	20–30 cm	30–40 cm	40–50 cm	50–60 cm	60–80 cm
			C:N			
NAT	14.6(3.24)a	15.9(1.34)a	16.6(1.35)b	18.2(2.96)b	13.3(0.61)b	12.4(1.02)a
S-UR	15.7(0.88)a	11.8(1.09)a	11.5(1.63)ab	11.4(1.79)ab	9.48(2.49)ab	9.37(2.64)a
C-FR	14.5(0.94)a	14.7(3.91)a	9.68(0.77)a	3.77(0.65)a	4.54(0.86)a	6.09(1.08)a
C-FI	16.7(0.64)a	14.8(0.73)a	10.1(1.13)a	9.39(1.50)a	9.09(0.72)ab	14.6(6.13)a
			C:P			
NAT	24.4(6.51)a	24.6(2.59)a	28.4(6.79)b	34.2(12.0)b	19.9(1.34)a	13.3(1.84)a
S-UR	30.0(5.70)a	32.4(11.41)a	19.9(4.69)ab	15.7(3.41)ab	19.7(11.3)a	26.7(19.5)a
C-FR	18.5(1.17)a	22.7(6.50)a	8.68(1.67)a	6.53(2.13)a	4.83(1.65)a	8.10(2.01)a
C-FI	21.5(1.99)a	13.9(2.34)a	13.5(4.46)ab	12.1(3.26)ab	5.90(1.89)a	9.89(3.54)a

4. Discussion

The conversion of native vegetation to agricultural land affects the physical, chemical, and biological properties of soil, with the extent of these effects being controlled by the agricultural practices implemented [39]. In turn, changes in soil properties may have positive or negative effects on crop yields. For example, SOC has been shown to be positively correlated with yields of wheat and maize [40]. However, little work has been done to investigate the effect of land-use change on soil properties and nutrient stocks within Africa [41]. Given that the Sub-Saharan Africa population will rapidly increase over coming decades, and this will likely lead to increasing conversion of native vegetation to make room for agriculture, the effects of land-use change on soil properties and their related ecosystem services should not be ignored. As such, this study provides valuable insights into changes in soil properties in a data limited area which is currently undergoing significant land-use conversion and agricultural intensification.

SOC concentrations seen across all sampling locations within this study agree with both in situ measurements [32] and SOC values interpolated from a broad-scale survey [42] elsewhere in the Kilombero Valley. Also, the concentrations of SOC and TN consistently decrease with soil depth—a well-known pattern occurring in most ecosystems and attributed to the higher inputs of C and nutrients at the soil surface compared to deeper soil layers [43,44]. As concentrations decrease with depth, organic matter becomes progressively enriched in N and P, leading to lower C:N and C:P ratios at depth, as also found in previous studies [45]. The increase in soil bulk density seen with depth (Table 3) is not surprising. The opposing trends of SOC and bulk density are a common signal in soil profiles due to the importance of SOC in regulating soil compressibility [46]. Changes among vertical profiles and carbon or nutrient stocks along our agricultural intensity gradient are more subtle, as discussed in the following sections.

4.1. Comparison of Soil Organic Carbon and Nutrient Concentrations between LMPs

We found that agricultural intensification has a significant positive effect on SOC concentrations at our study location. No difference was detected between NAT and S-UR, suggesting that low intensity smallholder production had little effect on SOC when compared to native vegetation. However, irrigation was found to have a mild but significantly positive effect on SOC concentrations, and fertilization had a significantly negative effect. The positive effect of irrigation on SOC concentrations is likely due to the use of pivot irrigation systems allowing for the production of two crops (rice and maize) on the C-FI sites. Irrigation in arid and semi-arid areas commonly increases SOC concentrations, but irrigation's effect in humid or sub-humid areas is not consistently positive [47]. Here, the growth of two crops increases the production of below ground biomass compared to single crop systems which, when not balanced by higher C mineralization, will lead to increasing SOC concentrations. Elsewhere in the Kilombero Valley, maize production has been found to increase SOC concentrations when compared to rice production [32]. This result is consistent with our findings, because maize produces more biomass than rice, and thus more residues that provide organic matter to the soil.

The negative effect of fertilization on SOC concentrations is likely not a direct consequence of chemical fertilizer use. Fertilization generally increases SOC concentrations when compared to unfertilized agriculture [48], because fertilization promotes plant growth. However, in the present study, the use of fertilizer is noted as an indicator of production intensity, and only implemented on the commercial farm sites. Therefore, the negative effect of fertilization on SOC concentrations may actually be an indication of other land management practices, such as higher intensity tillage on the commercial farm sampling locations (C-FI and C-FR) compared to the smallholder locations (S-UR). Both no-till and intermediate intensity tillage may promote SOC retention compared to high intensity tillage [49]. Previous studies in the Kilombero Valley found that fertilization had no effect on SOC [32]. However, their study looked solely at smallholder production systems. Here, due to a lack of intermediate LMPs between S-UR (non-irrigated and unfertilized smallholder) and C-FR (commercial fertilized and non-irrigated), it is not possible to state whether this negative effect on SOC is actually due to fertilization, or due to some other factor which is also associated with production on the commercial farm.

No significant difference was seen in concentrations of TN or TP between any of the land-uses. While this lack of difference may partially be due to the small number of plots sampled for each land-use, due to the explorative nature of the study, it is also clear that there is little difference in the TN and TP concentration profiles (Figure 2b,c). The three agricultural LMPs appear to have elevated SOC concentrations in the surface soils and reduced concentrations in the deep soil layers when compared to the native vegetation, resulting in a more negative slope coefficient for the agricultural soils (Figure 2a). However, TN and TP do not follow a similar trend. Therefore, agriculture may have little effect on soil nitrogen and phosphorous between the LMPs sampled within this study.

4.2. Comparison of Soil Organic Carbon and Nutrient Stocks between LMPs

Soil organic matter is a principal regulator of bulk density [26]. Our results are consistent with this known relationship, with native vegetation having the lowest mean SOC concentrations and highest bulk density in the upper sampling depths, and the highest SOC and lowest bulk density in the deeper sampling depths, when compared to the agricultural LMPs (Table 3). Changes in SOC concentrations affect soil quality, but it is also important to consider changes in SOC stocks to understand the net C flux between soils and the atmosphere. While no significant difference in total SOC stocks was found between the four LMPs, the higher mean SOC stocks in the sub-surface soils of the NAT sites compared to the agricultural LMPs (Figure 3b) warrants further investigation to understand the effect of land conversion on SOC storage at depth.

SOC in subsoils is often assumed to be more stable than within topsoils [50]—this has resulted in the dynamics of subsoil SOC often being ignored [51]. However, knowledge of land conversion impacts on subsoil SOC is increasing. Loss of SOC has been seen up to 1 m as a result of forest conversion for crop production [52]. Loss of deep SOC is probably larger where native vegetation has deeper roots than crops, such as in seasonally-dry forests like the Miombo woodland of Kilombero Valley. Woodland species with deep roots might have contributed C to deep layers that are not reached by crop roots after conversion—without fresh inputs, this deep SOC can be lost. The conversion of tropical forest to cropland has, on average, been reported to reduce SOC stocks by 25% [41], with soil sampling conducted to a mean depth of 36 cm (mostly overlapping with the plough layer). Such shallow sampling depth likely skews attempts to quantify the effects of agriculture on nutrient stocks.

Therefore, the effect of agriculture on soils is not limited to only those depths which are subject to direct physical disturbance. In the Kilombero Valley, where additional native land is not only likely to be converted to agriculture given current trends, but also where this "new" agriculture will become increasingly intensive, soil surveys should consider the role of deep soil layers in carbon stock accounting.

Agricultural intensity has been positively correlated with phosphorous content [53], and is seen here (Figure 3) with mean TP stocks increasing with increasing management intensity (S-UR < C-FR < C-FI). However, the trend seen within this study was not significant. For TP, as well as SOC and TN, this lack of significance may be due to the time scale on which intensified production has taken place within the industrial farm. Sampling within the C-FI sites was conducted on the three oldest pivot irrigation fields found within the farm. These pivots became operational during the 2014–2015 growing season, compared to other LMPs being implemented for at least 10 years. Changes in soil properties are generally slow, and respond to land-use change over periods of decades [54]. Given the short time since the initiation of the pivot irrigation system, and the growing of two crops on the C-FI land, a more marked effect may be visible in future sampling on and around the farm.

4.3. Comparison of C:N and C:P Ratios between LMPs

Even though we do not see significant effects of irrigation and fertilization on TN and TP, the effect of irrigation on C:N, and fertilization on both the C:N and C:P ratio (Table 2), as well as differences at specific depths between LMPs (Table 4), may be evidence of changes that are not clear in the assessment of individual elements. Conversion of grasslands to agricultural fields is known to lower the soil organic matter C:N and C:P ratios, at least in the topsoil [55]. This nutrient enrichment of soil organic matter can be explained by the accelerated decomposition occurring in agricultural fields, which tends to release more C compared to N and P, especially in the plough layer. Here, we find larger changes in C:N and C:P ratios at depth, which could be due to the removal of C-rich inputs from the deep roots of the native vegetation. In particular, fertilization negatively affected the C:N and C:P ratios, indicating a relatively larger enrichment in N and P in fertilized fields. This result is not unexpected, as fertilization is likely increasing N and P concentrations in the crop biomass, which in turn promotes the formation of organic matter with correspondingly higher N and P concentrations and lower C:nutrient ratios. Moreover, the negative effect of fertilization on SOC (Table 2) suggests a second mechanism—C might be preferentially removed via respiration, while nutrients are immobilized by soil microorganisms and retained in the organic matter.

5. Conclusions

We studied the effect of agricultural production intensity on the concentration and stocks of soil carbon and nutrients. Our results suggest that the concentrations of SOC, and stocks of SOC and TP, are mildly affected by the conversion of native forest to agriculture and by the intensity of production systems. While no consistent trend was seen across the whole gradient of intensity, we did find a significant negative effect of fertilization and a significant positive effect of irrigation on SOC concentrations. Despite the small effects on C and nutrient stocks, fertilization significantly decreased organic matter C:N and C:P ratios, and irrigation increased the C:N ratio, suggesting that soil properties are changing along the agricultural intensity gradient.

Concentrations of SOC in surface soils are presently greater than the thresholds below which crop production is negatively impacted as reported in other studies. For example, maize yields can be negatively affected as SOC concentrations decline below 2% in surface soils [40]. Our results also suggest that it is unlikely that surface SOC concentrations would rapidly decrease below this threshold in these relatively organic matter-rich soils. However, to further assess the effect of agricultural intensity on yields, it would be valuable to look at changes in plant available forms of macro and micro nutrients. This would require repeat sampling, as these forms are prone to large temporal variation, which is a potential way forward for research within the Kilombero Valley.

Whilst not statistically significant, all three agricultural LMPs point in the direction of increased SOC stocks in surface soils and reduced SOC stocks in the subsoils, compared to native vegetation. While often unaccounted for, the potential reduction in subsoil SOC stocks is an important factor to consider with regard to climate change, as SOC loss at depth may reduce benefits gained from increasing SOC stocks in surface soil.

Supplementary Materials: The following are available online at http://www.mdpi.com/2073-445X/9/4/121/s1, Figure S1: Soil texture ternary plot for each soil layer, for: (a) 0–20 cm; (b) 20–30; (c) 30–40 cm; (d) 40–50 cm; (e) 50–60 cm; (f) 60–80 cm. Points consisting of an outer ring of a different color to its center represent points with identical results between two LMPs. Table S1: Dummy variable values. A 1 denotes the presences of a specific property, Data S1; the original data file containing all sampling results.

Author Contributions: Conceptualization, J.L. and S.M.; methodology, J.L., S.M. and E.A.; formal analysis, J.L., A.C., M.C.; writing—original draft preparation, J.L.; writing—review and editing, J.L., E.A., M.T., S.W.L., A.C., M.C., R.L., and S.M.; visualization, J.L. All authors have read and agreed to the published version of the manuscript.

Funding: This research was funded by Vetenskapsrådet, grant number VR 2016-06313; the Bolin Centre for Climate Research; and Swedish International Development Agency (Sida), grant number SWE-2011-066 and Sida Decision 2015-000032 Contribution 51170071 Sub-project 2239.

Acknowledgments: The authors would like to thank all the farmers who allowed access to their land, the management of Kilombero Plantation Ltd. for local assistance and logistics, and all support workers who provided assistance during sampling. The authors also thank three reviewers for their constructive comments.

Conflicts of Interest: The authors declare no conflict of interest.

References

1. McBratney, A.; Field, D.J.; Koch, A. The dimensions of soil security. *Geoderma* **2014**, *213*, 203–213. [CrossRef]
2. UNCCD. *The Global Land Outlook*, 1st ed.; UNCCD: Bonn, Germany, 2017.
3. United Nations. *Department of Economic and Social Affairs, Population Division. World Population Prospects 2019: Data Booklet*; United Nations: New York, NY, USA, 2019.
4. Deutsch, C.A.; Tewksbury, J.J.; Tigchelaar, M.; Battisti, D.S.; Merrill, S.C.; Huey, R.B.; Naylor, R.L. Increase in crop losses to insect pests in a warming climate. *Science (80-.)* **2018**, *361*, 916–919. [CrossRef] [PubMed]
5. Piemontese, L.; Fetzer, I.; Rockström, J.; Jaramillo, F. Future hydroclimatic impacts on Africa: Beyond the Paris Agreement. *Earth's Future* **2019**, *7*, 748–761. [CrossRef]
6. Teixeira, E.I.; Fischer, G.; Van Velthuizen, H.; Walter, C.; Ewert, F. Global hot-spots of heat stress on agricultural crops due to climate change. *Agric. For. Meteorol.* **2013**, *170*, 206–215. [CrossRef]
7. Tubiello, F.N.; Salvatore, M.; Ferrara, A.F.; House, J.; Federici, S.; Rossi, S.; Biancalani, R.; Condor Golec, R.D.; Jacobs, H.; Flammini, A.; et al. The Contribution of Agriculture, Forestry and other Land Use activities to Global Warming, 1990–2012. *Glob. Chang. Biol.* **2015**, *21*, 2655–2660. [CrossRef]
8. Paustian, K.; Lehmann, J.; Ogle, S.; Reay, D.; Robertson, G.P.; Smith, P. Climate-smart soils. *Nature* **2016**, *532*, 49–57. [CrossRef]
9. Minasny, B.; Malone, B.P.; McBratney, A.B.; Angers, D.A.; Arrouays, D.; Chambers, A.; Chaplot, V.; Chen, Z.; Cheng, K.; Das, B.S.; et al. Soil carbon 4 per mille. *Geoderma* **2017**, *292*, 59–86. [CrossRef]
10. Tongwane, M.I.; Moeletsi, M.E. A review of greenhouse gas emissions from the agriculture sector in Africa. *Agric. Syst.* **2018**, *166*, 124–134. [CrossRef]
11. URT. *Agricultural Sector Development Programme Phase II (ASDP-II)*; The United Republic of Tanzania Ministry of Agriculture: Dodoma, Tanzania, 2017.
12. World Bank. *Tanzania-Southern Agriculture Growth Corridor Investment Project: Environmental Assessment (Vol. 3): Strategic Regional Environmental and Social Assessment: Final Report (English)*; World Bank: Tanzania, South Africa, 2013.
13. Evans, L.T. *Feeding the Ten Billion: Plants and Population Growth*; Cambridge University Press: Cambridge, UK, 1998; ISBN 0-521-64685-5.
14. Perrings, C.; Halkos, G. Agriculture and the threat to biodiversity in sub-saharan Africa. *Environ. Res. Lett.* **2015**, *10*, 095015. [CrossRef]

15. Zabel, F.; Delzeit, R.; Schneider, J.M.; Seppelt, R.; Mauser, W.; Václavík, T. Global impacts of future cropland expansion and intensification on agricultural markets and biodiversity. *Nat. Commun.* **2019**, *10*, 1–10. [CrossRef]
16. Potts, S.G.; Imperatriz-Fonseca, V.; Ngo, H.T.; Aizen, M.A.; Biesmeijer, J.C.; Breeze, T.D.; Dicks, L.V.; Garibaldi, L.A.; Hill, R.; Settele, J.; et al. Safeguarding pollinators and their values to human well-being. *Nature* **2016**, *540*, 220–229. [CrossRef] [PubMed]
17. Zhao, Z.H.; Hui, C.; He, D.H.; Li, B.L. Effects of agricultural intensification on ability of natural enemies to control aphids. *Sci. Rep.* **2015**, *5*, 8024. [CrossRef] [PubMed]
18. Matson, P.A.; Parton, W.J.; Power, A.G.; Swift, M.J. Agricultural intensification and ecosystem properties. *Science (80-.)* **1997**, *277*, 504–509. [CrossRef] [PubMed]
19. Winowiecki, L.; Vågen, T.G.; Massawe, B.; Jelinski, N.A.; Lyamchai, C.; Sayula, G.; Msoka, E. Landscape-scale variability of soil health indicators: Effects of cultivation on soil organic carbon in the Usambara Mountains of Tanzania. *Nutr. Cycl. Agroecosystems* **2016**, *105*, 263–274. [CrossRef]
20. Murty, D.; Kirschbaum, M.U.F.; Mcmurtrie, R.E.; Mcgilvray, H. Does conversion of forest to agricultural land change soil carbon and nitrogen? A review of the literature. *Glob. Chang. Biol.* **2002**, *8*, 105–123. [CrossRef]
21. Huang, S.; Sun, Y.; Zhang, W. Changes in soil organic carbon stocks as affected by cropping systems and cropping duration in China's paddy fields: A meta-analysis. *Clim. Chang.* **2012**, *112*, 847–858. [CrossRef]
22. Chiti, T.; Gardin, L.; Perugini, L.; Quaratino, R.; Vaccari, F.P.; Miglietta, F.; Valentini, R. Soil organic carbon stock assessment for the different cropland land uses in Italy. *Biol. Fertil. Soils* **2012**, *48*, 9–17. [CrossRef]
23. Livsey, J.; Kätterer, T.; Vico, G.; Lyon, S.W.; Lindborg, R.; Scaini, A.; Da, C.T.; Manzoni, S. Do alternative irrigation strategies for rice cultivation decrease water footprints at the cost of long-term soil health? *Environ. Res. Lett.* **2019**, *14*, 74011. [CrossRef]
24. Sanderman, J.; Hengl, T.; Fiske, G.J. Soil carbon debt of 12,000 years of human land use. *Proc. Natl. Acad. Sci. USA* **2017**, *114*, 9575–9580. [CrossRef]
25. Cleveland, C.C.; Liptzin, D. C:N:P stoichiometry in soil: Is there a "Redfield ratio" for the microbial biomass? *Biogeochemistry* **2007**, *85*, 235–252. [CrossRef]
26. Williams, R.J.B. *Relationships Between the Composition of Soils and Physical Measurements Made on Them. Rothamsted Experimental Station Report for 1970*; Rothamsted Research: Harpenden, UK, 1971; Volume 2.
27. Lal, R. Enhancing crop yields in the developing countries through restoration of the soil organic carbon pool in agricultural lands. *Land Degrad. Dev.* **2006**, *17*, 197–209. [CrossRef]
28. Namirembe, S.; Piikki, K.; Sommer, R.; Söderström, M.; Tessema, B.; Nyawira, S.S. Soil organic carbon in agricultural systems of six countries in East Africa—A literature review of status and carbon sequestration potential. *S. Afr. J. Plant Soil* **2020**, *1862*, 1–15. [CrossRef]
29. Alavaisha, E.; Lyon, S.W.; Lindborg, R. Assessment of water quality across irrigation schemes: A case study of wetland agriculture impacts in Kilombero Valley, Tanzania. *Water* **2019**, *11*, 671. [CrossRef]
30. RAMSAR Information Sheet on Ramsar Wetland: The Kilombero Valley Floodplain. Available online: https://rsis.ramsar.org/RISapp/files/RISrep/TZ1173RIS.pdf (accessed on 20 September 2019).
31. Dewitte, O.; Jones, A.; Spaargaren, O.; Breuning-Madsen, H.; Brossard, M.; Dampha, A.; Deckers, J.; Gallali, T.; Hallett, S.; Jones, R.; et al. Harmonisation of the soil map of africa at the continental scale. *Geoderma* **2013**, *211–212*, 138–153. [CrossRef]
32. Alavaisha, E.; Manzoni, S.; Lindborg, R. Different agricultural practices affect soil carbon, nitrogen and phosphorous in Kilombero -Tanzania. *J. Environ. Manag.* **2019**, *234*, 159–166. [CrossRef]
33. Msofe, N.K.; Sheng, L.; Lyimo, J. Land use change trends and their driving forces in the Kilombero Valley Floodplain, Southeastern Tanzania. *Sustainability* **2019**, *11*, 505. [CrossRef]
34. Seki, H.A.; Shirima, D.D.; Courtney Mustaphi, C.J.; Marchant, R.; Munishi, P.K.T. The impact of land use and land cover change on biodiversity within and adjacent to Kibasira Swamp in Kilombero Valley, Tanzania. *Afr. J. Ecol.* **2018**, *56*, 518–527. [CrossRef]
35. *Methods of Soil Analysis: Part. One-Physical and Minerological Methods*, 2nd ed.; Klute, A. Ed.; American Society of Agronomy, Inc. and Soil Science Society of America, Inc.: Madison, WI, USA, 1986; ISBN 0-89118-811-8.
36. Walkey, A.; Black, I.A. An examination of the Degtjareff method for determining soil organic matter and a proposed modification of the chromic acid titration method. *Soil Sci.* **1934**, *37*, 29–38. [CrossRef]
37. Bray, R.; Kurtz, L. Determination of total, organic, and available forms of phosphorus in soils. *Soil Sci.* **1945**, *59*, 39–46. [CrossRef]

38. Davidson, A.C. *Statistical Models*; Cambridge University Press: Cambridge, UK, 2003; ISBN 9780511672996.
39. Tully, K.; Sullivan, C.; Weil, R.; Sanchez, P. The State of soil degradation in sub-Saharan Africa: Baselines, trajectories, and solutions. *Sustainability* **2015**, *7*, 6523–6552. [CrossRef]
40. Oldfield, E.E.; Bradford, M.A.; Wood, S.A. Global meta-analysis of the relationship between soil organic matter and crop yields. *Soil* **2019**, *5*, 15–32. [CrossRef]
41. Don, A.; Schumacher, J.; Freibauer, A. Impact of tropical land-use change on soil organic carbon stocks—A meta-analysis. *Glob. Chang. Biol.* **2011**, *17*, 1658–1670. [CrossRef]
42. Kempen, B.; Dalsgaard, S.; Kaaya, A.K.; Chamuya, N.; Ruipérez-González, M.; Pekkarinen, A.; Walsh, M.G. Mapping topsoil organic carbon concentrations and stocks for Tanzania. *Geoderma* **2019**, *337*, 164–180. [CrossRef]
43. Jobbágy, E.G.; Jackson, R.B.; Biogeochemistry, S.; Mar, N. The Distribution of Soil Nutrients with Depth: Global Patterns and the Imprint of Plants. *Biogeochemistry* **2001**, *53*, 51–77. [CrossRef]
44. Jobbágy, E.G.; Jackson, R.B. The vertical distribution of soil organic carbon and its relation to climate and vegetation. *Ecol. Appl.* **2000**, *10*, 423–436. [CrossRef]
45. Tipping, E.; Somerville, C.J.; Luster, J. The C:N:P:S stoichiometry of soil organic matter. *Biogeochemistry* **2016**, *130*, 117–131. [CrossRef]
46. Ruehlmann, J.; Körschens, M. Calculating the effect of soil organic matter concentration on soil bulk density. *Soil Sci. Soc. Am. J.* **2009**, *73*, 876–885. [CrossRef]
47. Trost, B.; Prochnow, A.; Drastig, K.; Meyer-Aurich, A.; Ellmer, F.; Baumecker, M. Irrigation, soil organic carbon and N2O emissions. A review. *Agron. Sustain. Dev.* **2013**, *33*, 733–749. [CrossRef]
48. Han, P.; Zhang, W.; Wang, G.; Sun, W.; Huang, Y. Changes in soil organic carbon in croplands subjected to fertilizer management: A global meta-analysis. *Sci. Rep.* **2016**, *6*, 1–13.
49. Haddaway, N.R.; Hedlund, K.; Jackson, L.E.; Kätterer, T.; Lugato, E.; Thomsen, I.K.; Jørgensen, H.B.; Isberg, P.-E. How does tillage intensity affect soil organic carbon? A systematic review. *Environ. Evid.* **2017**, *6*, 30. [CrossRef]
50. Hobley, E.; Baldock, J.; Hua, Q.; Wilson, B. Land-use contrasts reveal instability of subsoil organic carbon. *Glob. Chang. Biol.* **2017**, *23*, 955–965. [CrossRef] [PubMed]
51. Rumpel, C.; Kögel-Knabner, I. Deep soil organic matter—A key but poorly understood component of terrestrial C cycle. *Plant Soil* **2011**, *338*, 143–158. [CrossRef]
52. Osinaga, N.A.; Álvarez, C.R.; Taboada, A.M. Effect of deforestation and subsequent land use management on soil carbon stocks in the South American Chaco. *Soil* **2018**, *4*, 251–257. [CrossRef]
53. De Neve, S.; Van Den Bossche, A.; Sleutel, S.; Hofman, G. Soil nutrient status of organic farms in flanders: An overview and a comparison with the conventional situation. *Biol. Agric. Hortic.* **2006**, *24*, 217–235. [CrossRef]
54. Hartemink, A.E.; McSweeney, K. (Eds.) *Soil Carbon*; Springer International Publishing: Cham, Switzerland, 2014; ISBN 978-3-319-04083-7.
55. Li, S.; Xu, J.; Tang, S.; Zhan, Q.; Gao, Q.; Ren, L.; Shao, Q.; Chen, L.; Du, J.; Hao, B. A meta-analysis of carbon, nitrogen and phosphorus change in response to conversion of grassland to agricultural land. *Geoderma* **2020**, *363*, 114149. [CrossRef]

land

MDPI

Article

A Comparative Analysis of a Detailed and Semi-Detailed Soil Mapping for Sustainable Land Management Using Conventional and Currently Applied Methodologies in Greece

Orestis Kairis *, Vassiliki Dimitriou, Chrysoula Aratzioglou, Dionisios Gasparatos, Nicholas Yassoglou, Constantinos Kosmas and Nikolaos Moustakas

Laboratory of Soil Science and Agricultural Chemistry, Agricultural University of Athens, 75 Iera Odos Street, 118 55 Athens, Greece; vassiliki_demetriou@hotmail.com (V.D.); aratzioglou@aua.gr (C.A.); gasparatos@aua.gr (D.G.); nick.yasoglou@gmail.com (N.Y.); ckosm@aua.gr (C.K.); nmoustakas@aua.gr (N.M.)

* Correspondence: kairis@aua.gr

Received: 14 April 2020; Accepted: 12 May 2020; Published: 15 May 2020

Abstract: Two soil mapping methodologies at different scales applied in the same area were compared in order to investigate the potential of their combined use to achieve an integrated and more accurate soil description for sustainable land use management. The two methodologies represent the main types of soil mapping systems used and still applied in soil surveys in Greece. Diomedes Botanical Garden (DBG) (Athens, Greece) was used as a study area because past cartographic data of soil survey were available. The older soil survey data were obtained via the conventional methodology extensively used over time since the beginnings of soil mapping in Greece (1977). The second mapping methodology constitutes the current soil mapping system in Greece recently used for compilation of the national soil map. The obtained cartographic and soil data resulting from the application of the two methodologies were analyzed and compared using appropriate geospatial techniques. Even though the two mapping methodologies have been performed at different mapping scales, using partially different mapping symbols and different soil classification systems, the description of the soils based on the cartographic symbols of the two methodologies presented an agreement of 63.7% while the soil classification by the two taxonomic systems namely Soil Taxonomy and World Reference Base for Soil Resources had an average coincidence of 69.5%.

Keywords: soil survey; soil classification; soil mapping; botanical garden

1. Introduction

Soil surveys provide a source of information and an inventory of soil parameters of an area of interest assisting land users to make accurate predictions for the response of a specific land to a certain use [1]. An integrated soil survey delineates the groups of soils of a region and describes their characteristics by using a specific mapping and classification system. Taking into consideration this information, the behavior of soils and their interaction with various land uses can be foreseen [2]. Therefore, there is an interdependent and interactive relationship between soil surveys and soil mapping [3] since the information of soil properties and their spatial distribution, given by detailed and accurate maps, are necessary for evaluation and land suitability analysis [4]. The landscape-soil relationship is reflected and emphatically imprinted in soil surveys and soil maps [5] and therefore together they consist an important driver for sustainable land management [6].

There are generally two approaches to mapping the soils, the modern and recently increasingly used digital soil mapping (DSM) [7,8] and the traditional soil mapping. Traditional soil mapping is conducted on the basis of Soil Mapping Unit (SMU), which is concerned as a distinguishable spatial object that delineates areas on the earth surface with similar physical and chemical properties [9]. This approach indicates a certain degree of subjectivity in the delineations of SMUs [10] since their nature is also transitional [11]. Actually, the physical soil in the landscape is segregated into discrete entities via the SMUs [12], consisting of one or several Soil Typological Units (STUs) [13], that represent soil volumes having the same arrangement of soil horizons (soil pedon) [14]. As the scale becomes more detailed the number of STUs in an SMU diminishes up to the level of very detailed mapping, where the SMU boundaries are identical to STU boundaries [14]. At the beginnings of soil mapping in Greece, due to a lack of technology, the initial delineation of SMUs was carried out on topographic backgrounds during field crossings of the mapped area and related observations of the environment [15]. With the progress of technology, new computer-based techniques of preliminary delineation of SMUs were developed [16]. SMU's delineations are based on the principle that the same factors of soil genesis create repeated geomorphological structures that can be identified both on a combination of various cartographic backgrounds and on the Earth's surface [2]. Those repetitive patterns are reflected in soils under the effect of soil-forming factors and can be identified at scales from continental to microscopic consisting the cornerstone for soil identification and mapping at various scales [2]. So, a SMU corresponds to a specific area in the map as well as to a specific part of the landscape in the physical environment [17].

The size of SMUs is determined from the mapping scale and the specific purposes of soil survey. Specifically, the minimum legible delineation (MLD) is defined as the smallest distinguishable area of any map that can be legibly delineated. MLD conventionally represents an area of 0.4 cm^2 independently of the mapping scale [1]. According to the mapping scale, the minimum legible area (MLA) that can be defined on a map, is totally connected and emerges from MLD. In terms of investigating which mapping scale is the most appropriate for the composition of a map for a specific soil/land use, the MLA must be equal or smaller than the minimum area of mapping interest for this specific soil/land use [1]. Taking into consideration the above-mentioned concepts, soil mapping of the same area in semi-detailed and detailed scales will result in the creation of few and extended SMUs in the first case and to more and less extended SMUs in the latter case, respectively. According to Food and Agriculture Organization of the United Nations (FAO) guidelines technical paper [18] semi-detailed soil surveys are typically at scales from 1:25,000 to 1:50,000, while detailed soil map's scales ranging from 1:10,000 to 1:25,000. The mapping scale largely influences the accuracy of soils grouping in SMUs. Detailed and semi-detailed soil maps are widely used for agricultural applications such as land resources assessment and land use planning [19].

The soil map of a country constitutes a basic national and infrastructural project for agricultural development and for the sustainable management of the primary sector. [20]. In Greece, the first actual integrated efforts for soil mapping have been carried out since 1977 through the implementation of a relevant law (Government Gazette Issue 186/A'/30-6-1977) with economic assistance from United Nations and Europe. Nowadays and due to the previous efforts, a plethora of soil surveys exist, mainly in detailed scale (1:5000–1:20,000) [21], which cover approximately the 15% of the agricultural areas in a fragmented pattern. Quite recently [22] the Greek Ministry of Rural Development and Food (Greek Payment and Control Agency for Guidance and Guarantee Community Aid, 2014) has funded the compilation of the national soil map in a semi-detailed scale (1:30,000) which includes approximately the 85% of the agricultural areas of the country. The results of the old (before 2014) and the new (2014) soil mapping of the agricultural areas in Greece present some spatial overlaps but mainly the two efforts complement each other as far as their spatial distribution is concerned. Summarizing, today, despite the numerous soil surveys in agricultural areas for which have been utilized considerable financial resources and a lot of working time there is not available a united, normalized and integrated soil map of the agricultural areas of Greece.

The objective of this study was the comparison of the two soil mapping systems (old and new) in order to investigate the potential of their combined use utilizing the already existing soil maps in Greece to complete the national soil map of the country. Particularly, the comparison focused on—(a) the volume of soil information that the two systems can record; (b) the reliability of the conclusions drawn from them on soil's characteristics and land utilization and (c) the potential for combining them to create final and integrated thematic maps; combining the detailed spatial information of conventional soil mapping (old) and the smaller-scale spatial information of current soil mapping (new) in the study area of the DBG.

2. Materials and Methods

2.1. The Study Area

The Diomedes Botanical Garden (DBG) is located in Attica (Greece) west of Athens, covering an area of approximately 175 ha (Figure 1). It is mostly a sloping area with slopes ranging from 2%–65% and is crossed by few small gorges and waterways. The DBG is a social welfare area including a private legal entity under the administration of National and Kapodistrian University of Athens. This garden is covered by natural vegetation, which consists mainly of *Pinus halepensis*, *Pinus brutia*, *Cupressus sempervirens*, *Quercus coccifera*, *Pistacia lentiscus* whereas a significant part of the garden is occupied by "phrygana" with the most representative species to be *Sarcopoterium spinosum*, *Cistus spp.*, *Phlomis fruticosa*, *Euphorbia acanthothamnos*, *Coridothymus capitatus* and *Satureja thymbra*. The chasmophytic vegetation of the garden is also remarkable consisting of *Campanula celsii subsp. celsii*, *Inula verbascifolia subsp. methanaea* and so forth [23]. The garden was established at 1951, in the west region of Athens, north of Aigaleo mountain in a hilly area where the dominant parent material is limestone, followed by schist [24]. The study area is characterized by a Mediterranean climate, with long hot and dry summers and moderately wet and cold winters. The annual precipitation is ranging between 259 and 576 mm, whereas the average annual temperature is approximately 19.8 °C [25]. This study area was selected due to the availability of old mapping data and to its proximity to the Agricultural University of Athens (AUA) facilities where most of the co-authors work. The lack of financial assistance in completing this study had also a decisive influence in the selection of the study area.

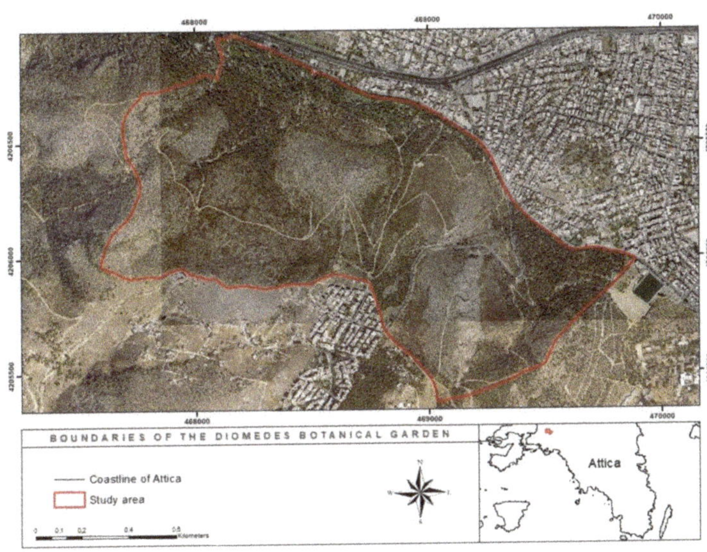

Figure 1. The boundaries of the study area located west of Athens.

2.2. Conventional (Old) and Currently Applied (New) Soil Mapping Methodologies in Greece.

According to the conventional soil mapping system, SMUs were delineated in the field by using topographical and geological maps of the interested area and the field work was partially confirmed afterwards by laboratory soil analyses. Under the principles of the conventional soil mapping system, topographic maps in a detailed scale (1:5000–1:10,000) [26] were the basic backgrounds for the initial delineation of SMUs in the field. Actually, in this mapping methodology the preliminary delineations of SMUs boundaries were carried out based on the detailed cartographic background of the topographic map and then were identified on the basis of geological maps, macroscopical field characteristics and measured soil properties. This kind of delineation method is called physiographic [27] since it identifies the repeated geomorphological patterns depicted from the combination of the topographical, geological and vegetative characteristics which consist a physiographic region. The final obtained soil data originated from field observations including both description of representative soil profiles accompanied by soil sampling for laboratory analysis and drilling holes by augers at distances depending on the scale of soil mapping [28–30]. The soil survey system was developed after extensive studies and significant experience in the countryside, taking into consideration the needs of cultivation practices and the evaluation for various land uses [31]. In this context, soils characterized according to their taxonomical category under the principles of United States Department of Agriculture (USDA) Keys to Soil Taxonomy [32]. The specific key consists of 6 taxonomic categories with increasing detail from the level of Order to the level of Series (Order, Suborder, Great group, Subgroup, Family, Series). The majority of the soil surveys compiled via the conventional methodology (old) in Greece were published at a scale ranging from 1:5000 to 1:20,000 [20]. According to this methodology, developed by Yassoglou et al. [21,31], the soils were initially classified to the taxonomic level of Great group [33] or, in some cases, to the level of Subgroup and afterwards were subdivided using a set of soil parameters (depth, texture and so forth) which determine soil productivity and management (Families and Series characteristics, in the broad context), in order the soil map to be published in a semi-detailed scale. Finally, SMUs were coded using a mapping symbol in which soil properties were designated by alphanumeric characters [34]. The alphanumeric expressions of the conventional method map symbol for the alluvial plains or the lowlands of Greece correspond to eight (8) different descriptive soil parameters, which are representative of the SMU properties and referred to: the degree of drainage of the soil profile, soil texture at depths 0–25 cm, 25–75 cm and 75–150 cm, slope gradient, degree of erosion, presence of carbonates in the soil profile and taxonomic characterization, through symbols of Soil Taxonomy, referring to soil Order, Suborder and Great group/Subgroup (Appendix A, Table A1).

The criteria for the different soil Orders include properties that reflect major differences in the genesis of soils such as the presence or absence of diagnostic horizons. The soil Suborders within an Order are discerned on the basis of any soil property which can influence the absence of horizon differentiation as for example the soil moisture regime. Great group category is a subdivision of a suborder in which all the principal soil properties of the soil solum are considered collectively such as the number and the kind of soil horizons [30], the moisture and the temperature regimes [35]. Subgroup category identifies distinctive soil features among various soils within a soil Great group [36].

The previous described methodology for detailed soil mapping introduced by Yassoglou et al. [21,31] used a different, more limited in extent and information, mapping symbol for the hilly or mountainous residual soils of Greece. This symbol records seven (7) descriptive soil parameters (the degree of drainage of the soil profile, parent material, soil depth, slope gradient, degree of erosion, type of vegetation and taxonomic characterization). The two cartographic symbols of the conventional method have four (4) soil properties in common and differ in five (5) properties those of soil texture and inorganic carbonates for the lowlands and parent material, soil depth and vegetation type for the hilly soils. (Appendices A and A.1, Tables A6 and A7). The mapping process for the hilly soils symbol follows the principles outlined above.

In the framework of the new soil mapping system (currently applied), SMUs are preliminarily delineated in a satellite orthoimage background or in an ortho-rectified photomap, usually of a semi-detailed scale, using Geographic Information System (GIS) software. The delineation of SMUs in this method is conducted on the basis of the image tone analysis macroscopically or by using image analysis techniques. Topographic, geological and vegetation maps are also used auxiliary for the preliminary delineation of SMUs. As in the case of the conventional method this kind of SMU delineation is in the context of physiographic method [27]. This soil mapping system uses geometrically corrected satellite images and geological as well as vegetation maps in semi-detailed scale (1:30,000–1:50,000) as its main background for the preliminary draw of SMUs. As in the conventional method, the finalization of SMU limits is carried out by certain morphological soil properties identified macroscopically in the field accompanied by both detailed description of the soil profiles and by laboratory analyses of selected soil samples. The mapping symbol of the currently applied methodology consists of coded letters and numbers presenting the following fourteen (14) soil properties: drainage conditions, soil texture at depths of 0–25 cm, 25–75 cm and 75–150 cm, slope gradient of soil surface, soil depth, rock fragments on soil surface, parent material, degree of soil erosion, presence of inorganic carbonates, limiting layers, electrical conductivity, soil alkalization and soil taxonomic unit [22] (Appendix A, Table A1). The taxonomic classification of soils in the context of the new soil mapping method is carried out in accordance with the rules of World Reference Base for Soil Resources (WRB) Taxonomy System [37]. The WRB system consists of two taxonomical categories in increasing detail from Reference Soil Groups (RSGs) to Principal and Supplementary Qualifiers. RSGs are defined according to primary soil-forming processes and the subsequent diagnostic soil features, excepting the case where the parent material is of prominent significance. At the second level, soil units are differentiated according to any secondary pedogenetic process that has a great influence on primary soil features. Qualifiers are subdivided in Principal Qualifiers (PQs), describing typical characteristics of RSGs and Supplementary Qualifiers (SQs), which describe additional characteristics of them [37]. The WRB classification system is recommended to be used only in soil mapping at scales from 1:250,000 to 1:1,000,000 indicating the Reference Soil Group name plus the first three PQs ranked in an order of importance with the most significant PQ placed closest to the name of the RSG [37,38]. However, as in the conventional method, new soil mapping approach also uses a combination of specific soil parameters in its mapping symbol that eventually lead the delineation of SMUs to a semi-detailed scale.

2.3. Soil Mapping of the Study Area with the Two Methods

Prior to the official soil mapping of the country (1977) the first attempt for soil mapping conducted on the study area of DBG at 1976 via the conventional soil mapping method [26]. Following the prompts of the old mapping method SMUs were delineated locally in the field using a detailed (1:5000) topographic map (Geographical Military Service of Greece—GMS) as the basic cartographic background and utilizing observations from the natural landscape (geology, soils, vegetation) in combination with geological information provided at 1:50,000 scale [24]. SMUs were delineated on the basis of the attributes of the two mapping symbols (lowlands and hilly areas) of the conventional method, mentioned in the previous section and soils were classified to the level of Subgroup [33,39]. The 1976 soil survey report [26] also provided descriptions of the five (5) representative soil profiles and the results of the laboratory analyses of fourteen (14) soil samples, which were sampled from the soil profiles obtained to verify the taxonomical units of the lowland soils (Appendix C–C1, Table A13, Figure 2). These older analog (hard copies) cartographic and soil data, that were obtained through the old methodology, were digitized and corrected geometrically in the ArcGIS v.10.4 software (Environmental Systems Research Institute—ESRI, Redlands, California, United States of America). Geometrical correction of the digitized old map grid was conducted based on a satellite orthoimage (Greek Cadaster, year 2007) of the area pre-corrected in the national coordinate system (Greek Geodetic Reference System 87—GGRS87). The geo-reference of the old map grid was achieved by identifying five characteristic and unmodified points over the years, that were recognizable both on the satellite image and on the map. In this way, the old and geographically uncorrected existing spatial soil information was digitized and connected to a specific geographical coordinate system acting as a reference base background for the spatial concurrence and comparison of the results of the two methodologies in the next phase. Specifically, a geodatabase file was created in order to receive the cartographic and analytical soil data of the 1976 mapping [25]. The overall digitization of the old mapping was carried out in order all the necessary soil data to be electronically available facilitating the comparison of the two soil mapping systems (conventional and current). Each digitized polygon corresponded to a particular SMU and to a specific soil group with similar soil properties which is different from the rest SMUs. A weakness of the 1976 mapping was the limiting number of soil profiles due to lack of adequate financial support. Additionally, no soil samples were taken to confirm the SMUs boundaries of the hilly regions of the DBG. Those shortcomings were adequately addressed during the 2019 soil mapping procedure.

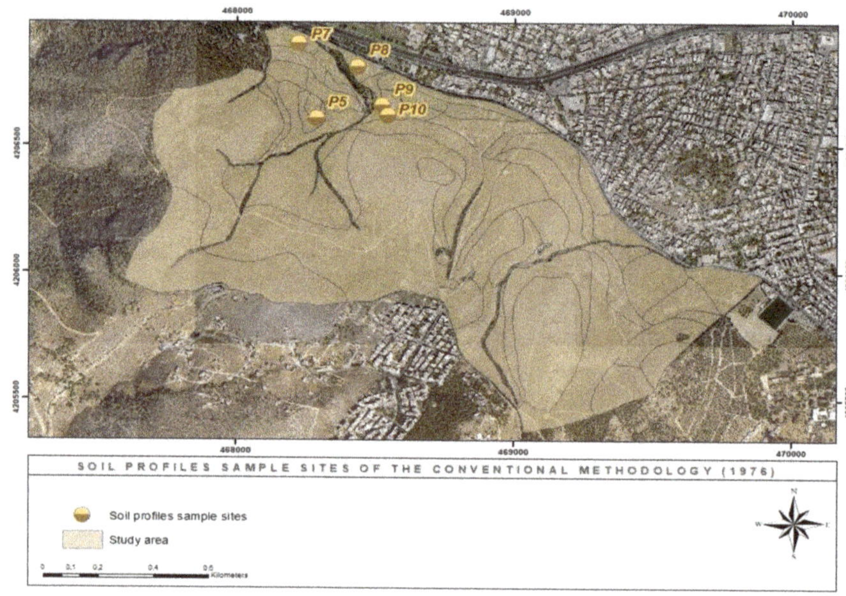

Figure 2. Soil profiles sample sites of the conventional soil mapping method.

The second mapping of the study area was carried out in 2019 according to the currently used soil mapping system following the physiographic methodology. In the context of the new soil mapping system, the preliminary delineation of SMUs was achieved by incorporating three digitized backgrounds in the GIS software, those of topography (GMS map at scale 1:50,000), geology [24] and the geometrically corrected ortho-photo map of the interested area from the Greek Cadaster (2007) as the main cartographic background at the scale 1:30,000. Finalization of SMUs boundaries was achieved by complementary on-site visual observations in the field in order to confirm or correct the initial delineated SMUs boundaries, using the Collector for ArcGIS software. Taking into consideration that the laboratory soil sampling analyses as well as the descriptions of the lowlands soil profiles of the 1976 study remained unchanged, they were used in the new mapping system to confirm the delineated SMUs of the lower parts of the Garden and to verify the corresponding taxonomical soil units. Based on the mapping scale and the size of the mapped area twelve soil sampling sites were selected and two soil samples were taken for laboratory analyses from the surface and subsurface horizons or layers of each sampling site, where possible (Figure 3).

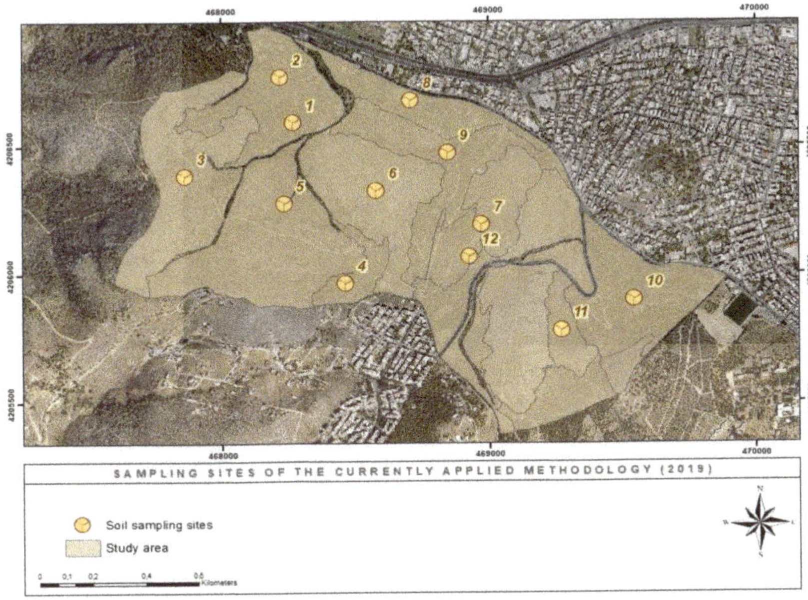

Figure 3. Soil sampling sites of the currently applied mapping method.

The obtained results from the soil sample analyses, along with field measurements and observations, were used to confirm delineations and mapping symbols of SMUs of the 2019 and 1976 mapping methodologies. In order to also cover the hilly areas of the DBG and to increase the number of observations four (4) additional profiles were prepared and described in detail in existing soil cuts confirming the delineations and classifications of both methods (Figures 4 and 5). The data were finally introduced in a specific geodatabase, in the ArcGIS v.10.4 environment.

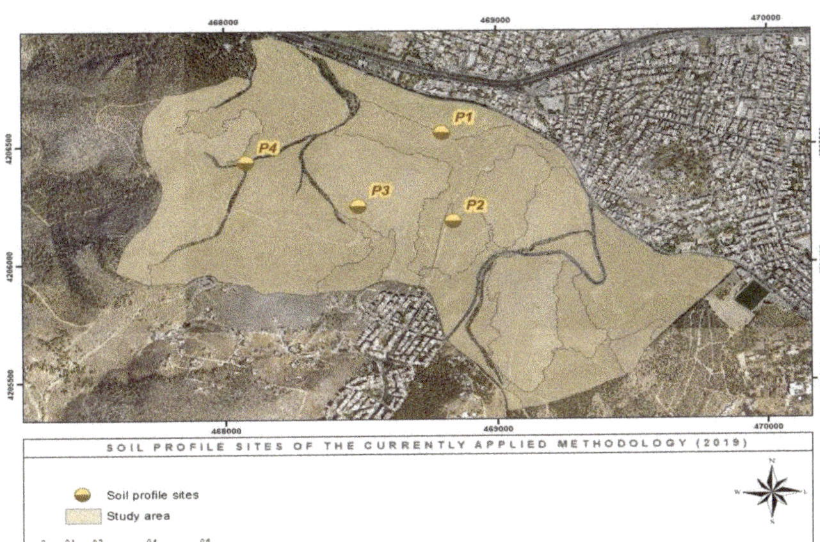

Figure 4. Profile sites of the new soil mapping method.

Figure 5. Representative soil profiles described in 2019: (**a**) Cambisol formed on limestone parent material with calcic horizon (Profile P1), (**b**) Cambisol formed on limestone parent material with a petrocalcic horizon at the depth >80 cm (Profile P2), (**c**) Cambisol formed on limestone parent material rich in rock fragments (Profile P3), (**d**) Leptosol formed on limestone parent material with bedrock at depth ≥ 20 cm (Profile P4).

2.4. Comparison of the Results of the Two Methodologies

The comparison of the two methodologies was made on the basis of the spatial coincidence between the two classification systems and the successful or not common description of the soils achieved via the two cartographic symbols. Specifically, with the use of geospatial techniques, in the ArcGIS v.10.4. environment, the eight taxonomic categories of the 2019 mapping system were used as clipping surfaces for the extraction of the 1976 SMUs that spatially coincided with them. Subsequently, the spatial correspondences between each taxonomic category of the 2019 mapping, that was used as the base reference for the analysis and the taxonomic categories of the 1976 SMUs were emerged. Then, an analysis of the participation rates (%) of the 1976 SMUs areas that presented the same classification with each taxonomic category of the 2019 mapping was performed inside the eight clipped common areas (Section 4.2, Appendix B, Table A11). Based on the areas, resulted from the participation rates and grossed up to the total area of each taxonomic category of 2019 mapping, the percentages of agreement between the classifications of the two methods were calculated (Section 4.2, Appendix B, Table A12). The overall percentage of taxonomic coincidence between the two systems, presented in conclusions (Section 5), emerged as the average of the individual percentages of agreement (Figure 6), (Appendix B, Table A12). Afterwards, the properties of the cartographic symbols of the two methodologies were compared in order to evaluate the description of the soils achieved by the two systems. Regarding the comparison of the two cartographic symbols, it was considered that the area corresponded to each taxonomic category of 2019, multiplied by its individual percentage of agreement with the 1976 classification, defined the new area for the comparison of the two symbols concerning all of the other soil properties except classification as it was already being evaluated. Then, depending on whether it was observed inside the spatial boundaries of the eight new areas a coincidence or not of the soil properties of the two symbols greater than 80% based on the participation rates, explained previously, the areas corresponding to a common description of soil properties (except classification) between the two systems were calculated. These areas were considered as the rates of success for the common description of the soils by the symbols of the two methods. The overall percentage (Section 5) of success for the common description of the soils by the mapping symbols of the two systems, presented in conclusions (Section 5), emerged as the summation of the individual success rates respectively, grossed up to the total area of DBG (Figure 6), (Appendix B, Table A12).

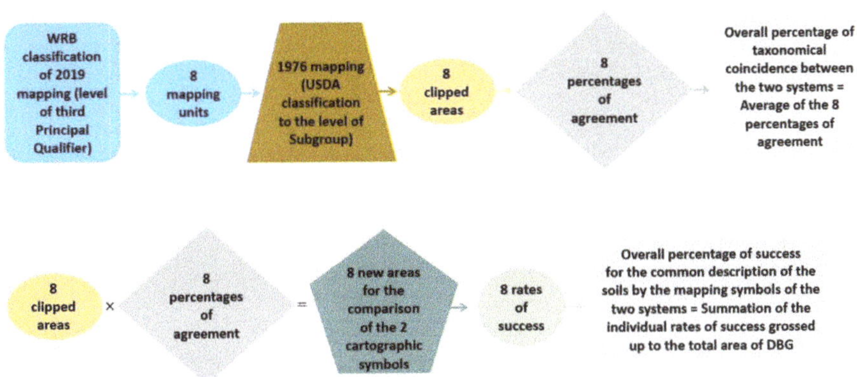

Figure 6. Schematic presentation showing the procedures for comparison of the two mapping systems. The spatial object of WRB classification fragmented into 8 mapping units according to the 8 taxonomical categories. The 8 mapping units also included all the information of the cartographic symbol of 2019 mapping. Each mapping unit was used as a clipping surface for the extraction of SMUs of 1976 mapping that contained all the information of the cartographic symbol of 1976 mapping including the USDA classification.

3. Results

3.1. Soil Map and Soil Groups Derived by the Conventional Mapping System

According to the 1976 soil mapping, fifty-five (55) SMUs were identified [32], delineated and described in the scale of 1:5,000 (Figure 7, Appendix A-A1, Table A1, Appendix B, Table A8). The main identified soil order was Entisols covering 111.9 ha or 64.1% of the mapped area, while the next important soil order was Inceptisols covering 62.7 ha or the 35.9% of the study area. The subgroup of Lithic Xerorthents prevailed in Entisols covering 110 ha or the 98.3% of these soils. A small area of 1.1 ha or 1.0% of Entisols were characterized as Typic Xerofluvents. In addition, Typic Xerorthents also covered a small area of 0.8 ha or the 0.7% of Entisols (Table 1., Figure 8). Typic Xerofluvents included mainly deep, calcaric and medium textured allochthonous soils formed on Holocene alluviums characterized by a xeric soil moisture regime and located in the lower part of the DBG. Typic or Lithic Xerorthents are located in the hilly part of DBG, strongly sloping, formed mainly on limestone parent material. These two Subgroups of Xerorthents are distinguished by the presence or not of a lithic contact within 50 cm of the mineral soil surface. In Inceptisols the majority of the mapped soils were characterized as Typic Calcixerepts (54.7 ha or the 87.2% of Inceptisols) while Petrocalcic Calcixerepts covered 8 ha or the 12.8% of these soils. Typic Calcixerepts were soils mainly freely drained with presence of a calcic horizon within 100 cm of the mineral soil surface, while Petrocalcic Calcixerepts were characterized by a petrocalcic horizon within 100 cm of the mineral soil surface (Figure 8).

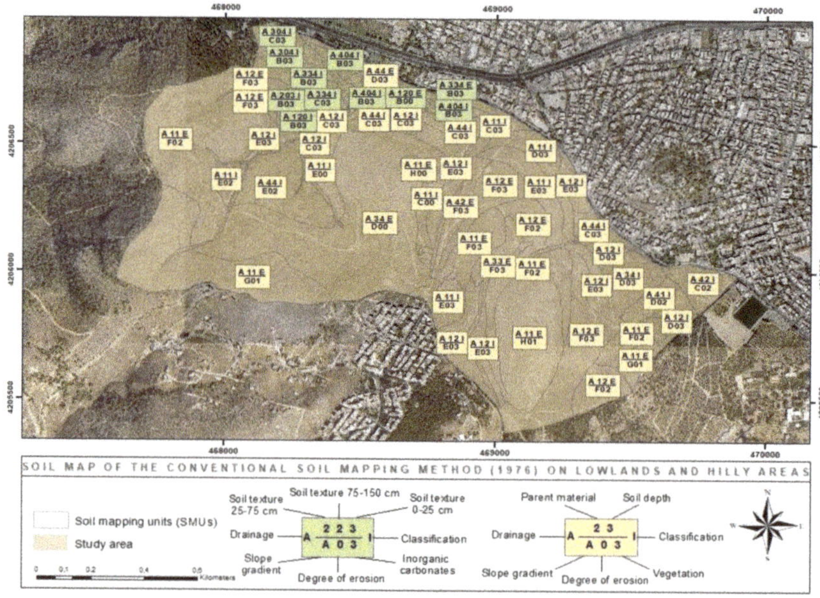

Figure 7. Soil Mapping Units (SMUs) delineations and mapping symbols according to the conventional mapping methodology (1976).

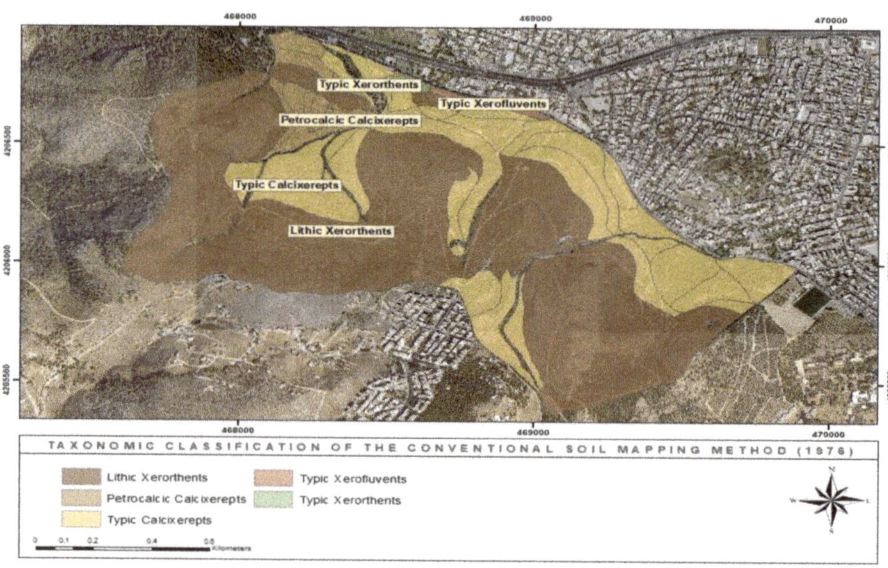

Figure 8. Soil taxonomic units (Subgroups) according to the conventional mapping methodology (1976).

Table 1. Grouping of soils according to the United States Department of Agriculture (USDA) taxonomical system to the level of Subgroup.

Classification	Number of Soil Mapping Units (SMUs)	Area (ha)
Lithic Xerorthents	19	110
Typic Xerorthents	1	0.8
Typic Xerofluvents	1	1.1
Petrocalcic Calcixerepts	7	8
Typic Calcixerepts	27	54.7
Total	55	174.6

3.2. Soil Map and Soil Groups Derived From the Currently Applied (2019) Mapping Methodology

In the context of the 2019 mapping system, nineteen (19) SMUs were emerged, delineated and described in the scale of 1:30,000 (Figure 9, Appendices A and A.1, Table A1, Appendix B, Table A9). The prevailing RSG was Leptosols covering 106.9 ha or 61.4% of the area, while 52.3 ha were identified as Cambisols corresponding to 30.1% of the mapped area. Calcisols occupied a small part of the DBG covering 14.7 ha or 8.5% of the studied area. Calcaric Skeletic Nudilithic Leptosols (LP-nt.sk.ca) [1] were the major soil group of Leptosols covering 38.1 ha or 35.6% of the RSGs followed by Skeletic Calcaric Cambic Leptosols (LP-cm.ca.sk) occupying an area of 25.7 ha or 24.1% of the RSG. Skeletic Calcaric Nudilithic Leptosols (LP-nt.ca.sk) and Cambic Calcaric Skeletic Leptosols (LP-sk.ca.cm) shared almost the same percentages (20.4% and 19.9%) of Leptosols covering 21.8 and 21.3 ha respectively. LP-nt.sk.ca were shallow soils (depth ≤ 25 cm) primarily with bedrock exposed on the soil surface, having 40% (by volume) rock fragments and evidences of calcaric material. LP-cm.ca.sk characterized as shallow soils (depth ≤ 25 cm) primarily having a cambic horizon (thickness ≥ 15 cm) with evidences of calcaric material and presenting 40% (by volume) rock fragments. LP-nt.ca.sk were also shallow soils (depth ≤ 25 cm) with bedrock partially exposed on the soil surface having evidences of calcaric material as well as a considerable amount of rock fragments 40% (by volume). LP-sk.ca.cm were shallow soils (depth ≤ 25 cm) primarily presenting 40% (by volume) rock fragments and also having evidence of calcaric material and a cambic horizon (thickness ≥ 15 cm). As far as Cambisols is concerned, Leptic Skeletic Calcaric Cambisols (CM-ca.sk.le) were the most significant group mapped covering 25.2 ha or the 48.2% of the RSG. The second most important soil group of Cambisols was Skeletic Calcaric Leptic Cambisols (CM-le.ca.sk) covering 18.6 ha or 35.5% of this area. Chromic Leptic Calcaric Cambisols (CM-ca.le.cr) occupied 8.5 ha or 16.3% of Cambisols. Calcisols were grouped as Skeletic Cambic Petric Calcisols (CL-pt.cm.sk) covering only 14.7 ha (Table 2., Figure 10). CM-ca.sk.le were soils with a cambic horizon (thickness ≥ 15 cm) mainly presenting evidence of calcaric material, 40% (by volume) rock fragments and bedrock at a depth ≤ 100 cm from the soil surface. CM-le.ca.sk were soils with a cambic horizon (thickness ≥ 15 cm) mainly with bedrock at a depth ≤ 100 cm from the soil surface, having also evidences of calcaric material and an amount of 40% (by volume) rock fragments. CM-ca.le.cr were soils with a cambic horizon (thickness ≥ 15 cm) mainly with evidences of calcaric material having bedrock at a depth ≤ 100 cm from the soil surface and a layer between 25 and 150 cm from the soil surface (thickness ≥ 30 cm) having a Munsell color hue redder than 7.5 YR. CL-pt.cm.sk included soils with a calcic or petrocalcic horizon at a depth ≤ 100 cm from the soil surface mainly presenting a cemented or indurated layer starting at a depth ≤ 100 cm from the soil surface, evidences of a cambic horizon (thickness ≥ 15 cm) and 40% (by volume) rock fragments (Figure 10). Fluvisols were not possible to map, whereas Fluvents were found in the conventional method, because the MLA of the currently applied methodology (due to the scale of 1:30,000) was 3.6 ha and Fluvents covered an area of 1.1 ha.

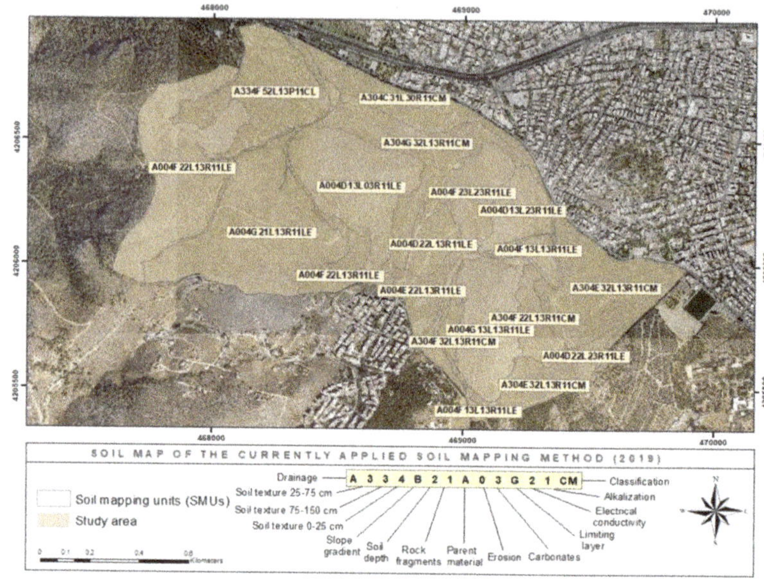

Figure 9. SMUs delineations and mapping symbols according to the currently applied methodology (2019).

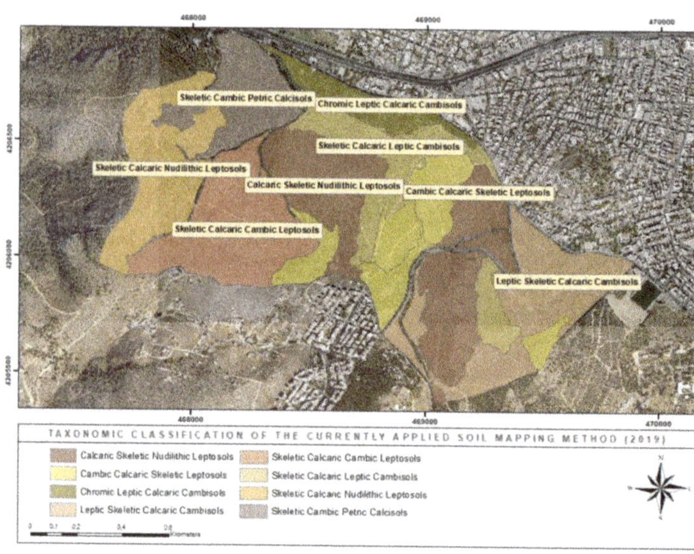

Figure 10. World Reference Base for Soil Resources (WRB) soil classification according to the currently applied mapping methodology.

Table 2. Grouping of soils according to the World Reference Base for Soil Resources (WRB) taxonomical system to the level of third Principal Qualifier.

Classification	Number of Soil Mapping Units (SMUs)	Area (ha)
LP-nt.sk.ca	5	38.1
LP-sk.ca.cm	5	21.3
LP-nt.ca.sk	1	21.8
LP-cm.ca.sk	1	25.7
CM-le.ca.sk	2	18.6
CM-ca.sk.le	3	25.2
CM-ca.le.cr	1	8.5
CL-pt.cm.sk	1	14.7
Total	19	173.9

4. Discussion

As it was highlighted above, the numerical alteration between the fifty-five (55) and the nineteen (19) SMUs of the conventional (old) and the current (new) soil mapping system, respectively, was attributed to the different used scale of the two methodologies. As it had been thoroughly reported in the case of the present work, the delineation of 1976 was mainly conducted on a detailed topographical background of 1:5,000 scale under a detailed SMU delineation method, while the delineation of 2019 was mainly conducted on semi-detailed ortho-imagery backgrounds of 1:30,000, under a semi-detailed SMU delineation methodology.

4.1. Data Provided by the Two Soil Mapping Systems

In the Appendix A (Table A1) the soil properties with the corresponding classes of the two mapping symbols used by the conventional (1976) and currently applied (2019) systems are shown. The two symbols and thus the two soil mapping systems recorded and transferred largely the same amount of soil information since nine (9) of the fifteen (15) mapped soil properties were common between the two systems (an example of a common soil property is given in Figures 11 and 12). The soil classification was identified by using different taxonomical systems in the two mapping systems (USDA and WRB) but as it will be discussed, the two classification systems were largely compatible concerning the types of soil information recorded per taxonomic category. The cartographic symbol of the conventional method included one (1) additional property concerning the prevailing type of vegetation. The cartographic symbol of the new method included an additional set of five (5) soil properties, over that of the old method, related to soil depth, presence of rock fragments (gravels and cobbles) on the soil surface, limiting layers, electrical conductivity and alkalization. All the above-mentioned additional properties are of great value for agricultural production and soil protection. Soil depth can be extracted inductively and indirectly from the texture of the three soil layers of the lowlands old mapping symbol up to a depth of 150 cm. However, it is very important the soil depth to be measured more accurate due to its great importance on soil water storage capacity and plant growth. Parent material was presented in the mapping symbol of the conventional system only for hilly soils. However, parent material is a significant soil parameter even in transported allochthonous lowland soils affecting chemical and physical properties. For example, soils formed on recent alluvial deposits are usually more fertile compared to soils formed on alluvial terraces. Additionally, the percentage of rock fragments in the soil surface affects soil moisture conservation and soil erosion susceptibility [40]. The conventional mapping system for the alluvial soils (lowlands) recorded the presence of gravels in the three textural layers, depending on the depth that they were observed (Appendix A, Table A1 and Appendix B, Table A8). In fact, gravels affect effective soil water storage capacity and rooting depth considered this recording as an advantage in relation to the new mapping system. The type and the kind of a limiting layer is an important property to be recorded affecting soil water movement and penetration of plant roots. Finally, electrical conductivity and alkalization are key soil parameters especially for the characterization of the salt-affected soils of the lowlands and coastal areas. In conclusion, the new mapping symbol can be considered as an effort of unifying the two symbols used to cover the plain and hilly soils according to the conventional mapping under one mapping symbol. Additionally, the new mapping symbol included more soil characteristics, easily identified in the field, of great importance for plant growth and soil protection. The use of one mapping system for all soils gives the opportunity to the user of having a uniform soil database independently of the origin of the soils.

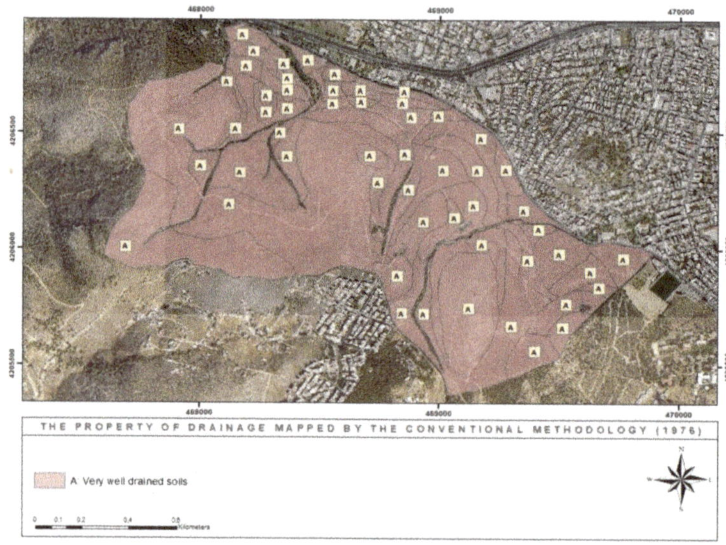

Figure 11. Drainage map of the conventional mapping system.

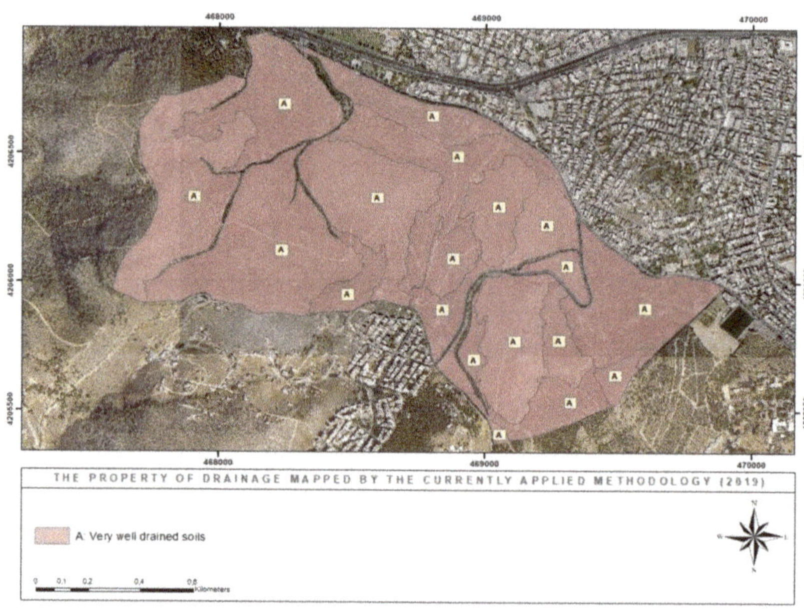

Figure 12. Drainage map of the currently applied mapping system.

4.2. Reliability and Compatibility of the wo Mapping Systems

According to USDA [39] and WRB [37] soil taxonomical systems it was evident that there was an identification between the general taxonomical classes of Order and Reference Soil Group (RSG) as far as Inceptisols and Cambisols were concerned. Regarding the Order of Entisols on the basis of the previously mentioned taxonomical systems they can only partially agree with the RSGs of Leptosols and Fluvisols.

Based on the results of the present work, Skeletic Calcaric Nudilithic Leptosols (SMU 18 from the 2019 mapping including 82% of SMU 11 and 41% of SMU 12 of the 1976 mapping) corresponded to Lithic Xerorthents in 93.6% of their area, highlighting an excellent agreement between the classifications of the two systems (Table 3). In addition, the remaining soil parameters of the cartographic symbols of SMUs of both mapping systems were largely coincided and characterized the soils mainly as shallow, freely drained, strongly inclined, formed on limestone and subjected to none or weak erosion (Appendix B, Table A10). The other 6.4% of Skeletic Calcaric Nudilithic Leptosols corresponded to Typic Calcixerepts in 1976 mapping (SMU 18 from the 2019 mapping including 100% of SMU 43). However, apart from the negligible mismatch in classification the soils of this group were similarly characterized by the two methodologies as very well drained, shallow, strongly inclined, formed on limestone and subjected to none or weak erosion (Appendix B, Table A10).

Table 3. Table of differences between the classifications of the two systems on a spatial basis.

WRB, 2019	USDA, 1976	Percentage (%) of 2019 Classification Corresponding to 1976 Classification
CL-pt.cm.sk	Petrocalcic Calcixerepts	54.0
	Typic Calcixerepts	6.0
	Lithic Xerorthents	40.0
CM-ca.le.cr	Typic Calcixerepts	65.0
	Typic Xerorthents	9.4
	Lithic Xerorthents	9.4
	Typic Xerofluvents	16.1
CM-le.ca.sk	Typic Calcixerepts	55.0
	Lithic Xerorthents	45.0
CM-ca.sk.le	Typic Calcixerepts	58.0
	Lithic Xerorthents	42.0
LP-nt.ca.sk	Lithic Xerorthents	93.6
	Typic Calcixerepts	6.4
LP-nt.sk.ca	Lithic Xerorthents	75.9
	Typic Calcixerepts	24.1
LP-sk.ca.cm	Lithic Xerorthents	75.6
	Typic Calcixerepts	24.4
LP-cm.ca.sk	Lithic Xerorthents	73.2
	Typic Calcixerepts	26.8

Calcaric Skeletic Nudilithic Leptosols (SMUs 2,4,6,11,16 from the 2019 mapping including 43% of SMU 13, 92% of SMU 16, 12% of SMU 18, 65% of SMU 20, 100% of SMU 27,62% of SMU 31 of the 1976 mapping) characterized as Lithic Xerorthents in 75.9% of their area, noting a very good match between the two classification systems (Table 3). The recorded soil properties by the two mapping methods in this taxonomic group were largely coincided characterizing the soils mainly as shallow, freely drained, strongly inclined, formed on limestone and subjected to no or weak erosion (Appendix B, Table A10). The remaining 24.1% of Calcaric Skeletic Nudilithic Leptosols characterized as Typic Calcixerepts in 1976 mapping (SMUs 2,4,6,11,16 from the 2019 mapping including 48% of SMU 19, 15% of SMU 25, 52% of SMU 28, 42% of SMU 32, 68% of SMU 33, 31% of SMU 34, 43% of SMU 35, 97% of SMU 44, 100% of SMU 45, 30% of SMU 47) and recorded a small mismatch in the classification property. However, as far as the rest soil properties of this group are concerned the two mapping systems described the soils in a similar way mostly as very well drained, formed on limestone, moderately to strongly inclined and subjected to none or weak erosion (Appendix B, Table A10).

Cambic Calcaric Skeletic Leptosols (SMUs 1,8,9,12,17 from the 2019 mapping including 3.8% of SMU 12, 13% of SMU 13, 98% of SMU 26, 7% of SMU 29, 14% of SMU 30, 37% of SMU 31,90% of SMU 36, 100% of SMU 37, 29% of SMU 39, 82% of SMU 49 of the 1976 mapping) corresponded to Lithic Xerorthents in 75.6% of their area, presenting a very good match between the two classification systems (Table 3). The two soil mapping methods characterized the soils of the above-mentioned group partly in a similar manner mainly as very well drained, strongly inclined, formed on limestone and subjected to none or weak erosion. There were some discrepancies in parent material and soil depth in SMUs 13 and 26 of the 1976 mapping characterized the soils as deep and formed on shale. The majority of the soils of this group (1976 SMUs 30,31,36,39) according to the 1976 mapping were mainly recorded as deep in contradiction to the 2019 mapping method (Appendix B, Table A10). The remaining 24.4% to complete the class of Cambic Calcaric Skeletic Leptosols was characterized as Typic Calcixerepts in 1976 mapping (SMUs 1,8,9,12,17 from the 2019 mapping including 39% of SMU 14, 95% of SMU 15, 40% of SMU 32, 13% of SMU 34). However, there was a close match with the rest soil properties between the two systems that characterized the soils of this group as shallow, very well drained, moderately to strongly inclined, formed on limestone and subjected to none or weak erosion (Appendix B, Table A10).

Skeletic Calcaric Cambic Leptosols (SMU 19 from the 2019 mapping including 54% of SMU 12 and 38% of SMU 13 of the 1976 mapping) characterized as Lithic Xerorthents in 73.2% of their area and showed a very good match between the two classification methods (Table 3). The soil properties mapped by the two methods for these soils were not similar because of the discrepancies of 1976 SMU 13 that characterized the soils mainly as deep and formed on shale (Appendix B, Table A10). The rest 26.8% of Skeletic Calcaric Cambic Leptosols corresponded to Typic Calcixerepts in 1976 mapping (SMU 19 from the 2019 mapping including 100% of SMU 51) and noted a mismatch in the classification and in basic soils properties between the two systems since 1976 classification characterized soils of this group mainly as deep and strongly inclined (Appendix B, Table A10).

Chromic Leptic Calcaric Cambisols (SMU13 from the 2019 mapping including 13% of SMU 34, 79% of SMU 40, 76% of SMU 41, 38% of SMU 42, 98% of SMU 46, 5% of SMU 47, 97% of SMU 50, 55% of SMU 53 of the 1976 mapping) matched in 65.0% of their area to Typic Calcixerepts, demonstrating a good agreement of the two classification systems (Table 3). The majority of these soils had the same characteristics between the two mapping methods recorded mainly as deep, moderately fine textured in the surface layer (0–25 cm), very well drained, slightly or moderately inclined with strong reaction on the soil surface (Appendix B, Table A10). The remaining 35% of Chromic Leptic Calcaric Cambisols was differentiated into three taxonomic categories of 1976 mapping those of Typic Xerorthents (SMU 13 from the 2019 mapping including 100% of SMU 55), Lithic Xerorthents (SMU 13 from the 2019 mapping including 100% of SMU 54) and Typic Xerofluvents (SMU 13 from the 2019 mapping including 100% of SMU 52). Although the taxonomic class of 2019 mapping was divided into 3 classes in 1976 mapping, it was observed a great overlap in the description of the soils of this group which were mainly characterized as deep, very well drained, slightly or moderately inclined, with strong reaction on the soil surface and subjected to none or weak erosion soils (Appendix B, Table A10).

Skeletic Cambic Petric Calcisols (SMU 15 from the 2019 mapping including 98% of SMU 1, 100% of SMU 2, 100% of SMU 4, 100% of SMU 6, 100% of SMU 7, 100% of SMU 8, 100% of SMU 10, 100% of SMU 48 of the 1976 mapping) characterized as Typic and Petrocalcic Calcixerepts in 60% of their area, presenting a moderate good match between the classifications of the two systems (Table 3). Over one half of Typic and Petrocalcic Calcixerepts classes (1976 SMUs 1,2,4,6,48) corresponding to the lowland's mapping system, were mostly characterized as very well drained soils, moderately fine textured in the surface layer (0–25 cm), inclined, with strong reaction on the soil surface and subjected to none or weak erosion as Skeletic Cambic Petric Calcisols (Appendix B, Table A10). The rest of the Typic and Petrocalcic Calcixerepts class (1976 SMUs 7,8,10) presented the same soil characteristics as Skeletic

Cambic Petric Calcisols except soil texture, which did not appear on the mapping symbol because they were mapped as hilly soils. The remaining 40% of Skeletic Cambic Petric Calcisols corresponded to Lithic Xerorthents (SMU 15 from the 2019 mapping including 100% of SMU 3, 100% of SMU 5, 100% of SMU 9, 18% of SMU 11). However, apart from the considerable mismatch in classification the soils of this group were similarly characterized by the two methodologies mainly as deep, very well drained, strongly inclined, formed on limestone, with strong reaction on the soil surface and subjected to none or weak erosion (Appendix B, Table A10).

Leptic Skeletic Calcaric Cambisols (SMUs 3,5,7 from the 2019 mapping including 55% of SMU 14, 93% of SMU 17, 46% of SMU 19, 98% of SMU 21, 98% of SMU 22, 97% of SMU 23, 97% of SMU 24, 82% of SMU 25, 56% of SMU 35 of the 1976 mapping) corresponded to Typic Calcixerepts in 58% of their area indicating a moderate good match between the classifications of the two methodologies (Table 3). The two mapping methodologies had also the same results as far as soil parameters of the mapping symbols are concerned and characterized the soils mainly as deep, very well drained, moderately to strongly inclined, formed on limestone and subjected to none or weak erosion, except of a minor mismatch of SMU 24 that characterized the soils as deep and formed on shale (Appendix B, Table A10). The remaining 42% of Leptic Skeletic Calcaric Cambisols corresponded to Lithic Xerorthents (SMUs 3,5,7 from the 2019 mapping including 59% of SMU 18, 14% of SMU 20, 86% of SMU 29, 80% of SMU 30, 18% of SMU 49). This soil group had a moderate mismatch in classification but regarding the soil properties of the mapping symbols of the two methods the results were similar and the soils were characterized mainly as very well drained, moderately deep or deep, strongly inclined, formed on limestone and subjected to none or weak erosion (Appendix B, Table A10).

Skeletic Calcaric Leptic Cambisols (SMUs 10,14 from the 2019 mapping including 48% of SMU 28, 18% of SMU 32, 30% of SMU 33, 44% of SMU 34, 97% of SMU 38, 19% of SMU 40, 24% of SMU 41, 63% of SMU 42, 3% of SMU 44, 64% of SMU 47, 45% of SMU 53 of the 1976 mapping) characterized as Typic Calcixerepts in 55.0% of their area, presenting a moderate to good agreement between the classifications of the two methodologies (Table 3) and a good match between the soil properties of the two different mapping symbols that characterized the soils mainly as freely drained, deep, strongly inclined and subjected to none or weak erosion (Appendix B, Table A10). The other 45.0% of the Skeletic Calcaric Leptic Cambisols area was characterized as Lithic Xerorthents (SMUs 10,14 from the 2019 mapping including 7% of SMU 13, 7% of SMU 16, 28% of SMU 18, 19% of SMU 20, 4% of SMU 29, 5% of SMU 30, 10% of SMU 36, 71% of SMU 39) and a moderate mismatch was noted in the classification property and a moderate mismatch in soil depth and parent material since SMU 13 was differentiated from 2019 mapping characterized the soils as deep and formed on shale parent material (Appendix B, Table A10).

5. Conclusions

Our study showed that the conventional mapping system was based on more detailed mapping backgrounds and thorough crossings within the SMUs for their delineation. Following the conventional soil mapping system, the delineated SMUs were smaller and more detailed than the SMUs of the currently applied mapping system. Due to the technological shortcomings of that time and the absence of a global or national geodetic reference system, the detailed delineations of SMUs of the conventional system could not be placed in their exact positions due to errors generated during spatial information transferring from the digitized and uncorrected old map grids to georeferenced backgrounds in modern reference systems.

The comparison of the cartographic symbols of the two mapping systems (conventional and currently applied) showed that the two mapping symbols convey a common critical mass of information since nine over fifteen properties are common between them (Section 4.1.). However, the conventional soil mapping system have used two cartographic symbols one for the hilly residual soils and one for the lowland alluvial soils with the risk of creating confusion in organizing a database. In the opposite, in the currently applied soil mapping system, one cartographic symbol was been assigned for both hilly residual soils and lowland alluvial soils providing the advantage of an easily established database. (Section 4.1.). A major weakness of the new mapping symbol is that two soil parameters, namely electrical conductivity and alkalization, require measurement in the field with scientific instruments or laboratory soil analysis. This disadvantage differentiates the new symbol from the general philosophy of creating soil cartographic symbols using parameters easily recognizable in the field.

The comparison of the taxonomic systems (WRB and USDA) of the two mapping methodologies have shown an average coincidence of 69.5% (Section 4.2., Appendix B, Table A12). Regarding the descriptions of the soils (except classification) based on the cartographic symbols of the two methodologies (conventional and currently applied), the percentage of agreement between the two methods reached 63.7% (Section 4.2., Appendix B, Table A12).

According to the results of this work, the two mapping systems (conventional and currently applied) can be creatively combined and can function complementary to each other for a better mapping of soils. In many cases the more detailed but uncorrected SMUs of the conventional mapping could be used to highlight some important areas of specific agricultural management within the coarser and georeferenced SMUs of the currently applied mapping system. Considering that the main soil information recorded by the two systems is common, the two cartographic symbols could be combined satisfactory for a more detailed and accurate description of soil parameters.

We finally consider the results of this work as a starting point in an effort of utilizing and integrating the existing old soil mapping data in the national soil map of Greece. Of course, towards to this direction several other assessments of the compatibility of the two systems, especially in extended and lowland agricultural areas, should follow given that our work has been limited mainly to hilly forest soils of a small area. At a country level, given that this effort will be massive and laborious, it is very important to be supported as much as possible with the products and techniques of DSM [7,8]. In the present study, digital mapping was not used supportively because the subject was the comparison of two classical soil mapping methods and the study area was restricted.

Author Contributions: Conceptualization, O.K., C.K. and N.M.; methodology, O.K., N.Y., C.K. and N.M.; software, O.K. and V.D.; field supervision, C.K. and N.M. data analysis, O.K., V.D. and C.A.; writing—original draft preparation, O.K., V.D. and C.A.; writing—review and editing, O.K., D.G., C.K. and N.M. All authors have read and agreed to the published version of the manuscript.

Funding: This research received no external funding.

Acknowledgments: The authors must acknowledge and thank Soulios Stelios who allowed all the necessary field work for soil mapping in Diomedes Botanical Garden to be carried out appropriately.

Conflicts of Interest: The authors declare no conflict of interest.

Abbreviations

AUA	Agricultural University of Athens
DBG	Diomedes Botanical Garden
DSM	Digital Soil Mapping
ESRI	Environmental Systems Research Institute
FAO	Food and Agriculture Organization of the United Nations
GGRS87	Greek Geodetic Reference System 87
GIS	Geographic Information System
MLA	Minimum Legible Area
MLD	Minimum Legible Delineation
PQ	Principal Qualifier
RSG	Reference Soil Group
SMU	Soil Mapping Unit
STU	Soil Typological Unit
SQ	Supplementary Qualifier
USDA	United Stated Department of Agriculture
WRB	World Reference Base for Soil Resources

Appendix A

Table A1. Parameters of the conventional and currently applied soil mapping symbol and their corresponding classes.

Parameters and Their Corresponding Classes						
Drainage conditions *						
Very well drained soils (>150 cm) A	Well drained soils 100–150 cm B	Moderately well drained soils 50–100 cm C	Imperfectly drained soils 30–50 cm D	Poorly drained soils < 30 cm E	Very poorly drained soils F Gley (75–150 cm), G Gley < 75 cm F	G
Soil texture (0–25 cm) *,1,2						
Very coarse (S, LS) 1	Coarse (SL) 2	Medium (L, SiL, Si) 3	Moderately fine (CL, SCL, SICL) 4	Fine (C, SC, SiC) 5		
Soil texture (25–75 cm) *,1,2						
Very coarse, coarse (S, LS, SL) 1	Medium (L, SiL, Si) 2	Moderately fine (CL, SCL, SICL) 3	Fine (C, SC, SiC) 4			
Soil texture (75–150 cm) *,1,2						
Very coarse, coarse (S, LS, SL) 1	Medium (L, SiL, Si) 2	Moderately fine (CL, SCL, SiCL), fine C, SC, SiC) 3				
Soil depth (cm) *						
0–15 1	15–30 2	30–60 3	60–100 4	100–150 5	>150 6	
Slope gradient (%) *						
0–2 A	2–6 B	6–12 C	12–18 D	18–25 E	25–35 F	>35 G
Rock fragments (%) **						
<20 1	20–60 2	>60 3				
Parent material *						
Marl M	Conglomerates C	Limestone, marbles L	Alluvial deposits A	Schist S	Flysch P	
Acid igneous O	Basic igneous B	Clay deposits G	Volcanic ash H	Magmatic conglomerates K	Alluvial fan R	
Lake deposits Y	Sand dunes D	Organic deposits X	Alluvial terraces T	Sandstone I		

Table A1. *Cont.*

Parameters and Their Corresponding Classes					
Inorganic carbonates *					
Strong reaction on the soil surface	Weak reaction on the soil surface	Reaction in the subsurface horizon or in substratum	No reaction		
3	2	1	0		
Degree of erosion *					
No erosion	Weak erosion (< 25% A horizon)	Moderate erosion (25–75% A horizon)	Severe erosion (no A horizon)	Very severe erosion (gullies)	
0	1	2	3	4	
Limiting Layers **					
No	Bedrock	Gravels or sand	Compact horizon		
0	R	G	F4		
Electrical conductivity (dS/m) **					
0–4	4–8	8–15	>15		
1	2	3	4		
Alkalization (Exchangeable Sodium Percentage—ESP) **					
ESP < 6	ESP = 6.1–15	ESP > 15			
1	2	3			
Soil classification (WRB)					
Cambisols	Calcisols	Regosols	Fluvisols	Luvisols	Leptosols
CM	CL	RG	FL	LV	LP
Soil classification (USDA)					
Entisols	Inceptisols	Alfisols	Histosols		
E	I	A	Hs		

*—Common properties of conventional and currently applied mapping symbols; **—Additional properties of currently applied mapping symbol; [1]—In soil texture symbols of the three textural layers of the alluvial soils, according to the conventional method, the presence of gravels was recorded depending on the depth that they were observed by noting their presence with the symbol x (>60%) or o (<60%) as exponents above the number denoting the texture of each depth; [2]—In cases where there is no soil in a particular layer the symbol 0 is assigned.

Appendix A

Appendix A.1. Guidelines for the Characterization of the Soil Properties Used in the Conventional and Currently Applied Soil Mapping Symbols

Drainage Conditions

The characterization of the drainage conditions is based on the presence of iron or manganese mottles and the subsoil colorings. The six (6) hydromorphic classes which are used in the soil mapping system are the following:

Class A—Very well drained soils

They are characterized by the absence of iron and manganese mottles throughout the whole soil profile. The brownish colors prevail, the soil usually has a high hydraulic conductivity and the water is infiltrated into the soil's deepest layers. The soil remains wet only during the wet period of the year (wet months). Drainage is not required.

Class B—Well drained soils

They are characterized by the presence of iron and manganese mottles or gray mottles at a depth between 100 and 150 cm from the soil surface. The brown colors prevail throughout the whole soil profile. During the growing season, these soils are not sufficiently wet for a long period of time to adversely affect the growth of the plants. Drainage is not required.

Class C—Moderately well drained soils

They are characterized by the presence of iron and manganese mottles or gray mottles at a depth between 50 and 100 cm from the soil surface. In some soils of this class, there may be mottles at depths of less than 50 cm but its percentage is less than 2%. The underground aquifer in the wet months rises and may adversely affect perennial crops. These soils require drainage for sensitive crops.

Class D—Imperfectly drained soils

They are characterized by the presence of iron and manganese mottles or some reductive spots at a depth between 30 and 50 cm from the soil surface. The percentage of mottles in this layer is less than 20%. These soils are characterized by high moisture for a long period of the year close to the soil surface, resulting adverse consequences to the cultivations during the spring. Drainage is required for the perennial crops.

Class E—Poorly drained soils

They are characterized by the presence of iron and manganese mottles at a depth less than 30 cm from the soil surface while the presence of iron and manganese mottles or reductive spots covers a percentage of 20–50% at a depth between 30 and 50 cm from the soil surface. These soils have a high level of ground water table during the wet months of the year. The cultivation of perennial crops or early spring crops requires drainage.

Class F, G—Very poorly drained soils

Soils with a permanent ground water table at a depth commonly higher than 75 cm from the soil surface. If reducing conditions prevail at a percentage higher than 50 % at the depth of 75–150 cm, the soil is characterized by F drainage class. If the reductive conditions prevail at a depth less than 75 cm, the soil is characterized by G drainage class. If there is a seasonal fluctuation of the aquifer, the drainage class may be characterized by combining two of the previous classes (e.g., E/F, E/G andso forth). These soils are wet to the surface for the longest period of the year and therefore prevent the normal growth of most cultivations. Drainage is absolutely required.

Soil Texture

The soil texture is determining in the field using the sense of touch. The soil sample is moistened between the fingers index and thumb and the soil aggregates are broken due to the pressure and friction applied to the soil. The class of soil texture is determined accordingly to the sense of sticking (clay), gliding (silt) or gritting (sand). Soils that have a high percentage of sand have a gritty sense. Soils that have a high percentage of silt feel smooth while soils that have a high percentage of clay have a sticky feel. The corresponding symbols and sub-classes of soil texture are shown in Table A2.

Table A2. Classes of soil texture with the corresponding symbols for the three parts of soil profile.

Map Symbol	Part A (0–25 cm)	Part B (25–75 cm)	Part C (75–150 cm)
1	Coarse-textured or layers with predominant coarse-textured materials	Coarse-textured, moderately coarse-textured or predominant coarse-textured materials	Coarse-textured, moderately coarse-textured or predominant coarse-textured materials
2	Moderately coarse-textured or predominant moderately coarse-textured materials	Medium-textured or predominant medium-textured materials	Medium-textured or predominant medium-textured materials
3	Medium-textured or predominant medium-textured materials	Moderately fine-textured or predominant moderately fine-textured materials	Moderately fine-textured or predominant moderately fine-textured materials
4	Moderately fine-textured or predominant moderately fine-textured materials	Fine-textured or predominant fine-textured materials	
5	Fine-textured or predominant fine-textured materials		
6	Muck	Muck	Muck
Coarse-textured		Sandy (S), Loamy-sand (LS)	
Moderately coarse-textured		Sandy-loam (SL)	
Medium-textured		Loamy (L), Silty-loam (SiL), Silty (Si) andfine Sandy-loam (fSL)	
Moderately fine-textured		Sandy-Clay-Loam (SCL), Clay-Loam (CL) and Silty-Clay-Loam (SiCL)	
Fine-textured		Silty-Clay (SiC), Clay(C) and Sandy-Clay (SC)	

Slope Gradient

The slope of the soil surface is determined on the field by the usage of a clysimeter, a topographical map or through estimation after acquisition of relevant experience. The slope gradient is distinguished in the following six (6) classes: almost flat (slope 0–2%), slightly inclined (slope 2–6%), moderately inclined (slope 6–12%), strongly inclined (slope 12–18%), moderately steep (slope 18 –25%), steep (slope 25–35%) and very steep (slope > 35%).

Rock Fragments

The content of rock fragments (gravels and cobbles) on the soil surface is estimated in the field using the classes which are shown in Table A3. Gravel is defined as the part of rock fragments with a diameter ranging from 2 mm to 7.5 cm. The cobbles include the rock fragments with a diameter > 7.5 cm.

Table A3. Classes of rock fragments with the corresponding symbols.

Map Symbol	Class Description
1	Gravels (diameter 2 mm–7.5 cm) and cobbles (diameter > 7.5 cm) on the soil surface in a percentage lower than 20 %
2	Gravels (diameter 2 mm–7.5 cm) and cobbles (diameter >7.5 cm) on the soil surface in a percentage ranging from 20 % to 40 %
3	Gravels (diameter 2 mm–7.5 cm) and cobbles (diameter > 7.5 cm) on the soil surface in a percentage higher than 60 %

Parent Material

The parent material in each SMU is determined by field observations and the use of geological maps. As parent material is defined the upper geological layer on which the soil was formed. In each parent material a letter of the alphabet was given as shown in Table A1 (Appendix A).

Degree of Erosion

Erosion classes are defined by the presence or absence of diagnostic horizons and rills or gullies as follows (Table A4):

Table A4. Erosion classes and their characteristics.

Map Symbol	Class Description
0	No Erosion
1	Soils which have lost part of the surface horizon A but on average less than 25% of the initial horizon A. Indications for erosion class 1 are (a) few rills, (b) concentration of soil sediments at the base of the slope or in a cavity, (c) scattered spots where the horizon of cultivation contains materials from the underlying horizon.
2	Soils which have lost an average of 25–75% of the initial A horizon. In erosion class 2, the surface layer is consisted of a mixture of horizon A materials and the underlying subsurface horizon. In some areas there may be a mixed state of spots without any erosion signs and spots where all the A horizon has been removed. Where the horizon A is thick enough, minimum or no mixing of horizon A materials with materials of the underlying horizon has taken place.
3	Soils that have lost the whole A horizon and some of the deeper horizons to their greatest extent. The initial soil can be identified only on individual spots.
4	Soils that have lost the whole horizon A and some or all of the deeper horizons to their greatest extent. The initial soil can be identified only on individual spots. A complex system of rills and gullies is observed on the soil surface.

Inorganic Carbonates

The inorganic carbonates are determined accordingly to their concentration and the depth where are detected indirectly by the reaction in dilute hydrochloric acid as follows (Table A5):

Table A5. Classes of inorganic carbonates with the corresponding symbols.

Map Symbol	Class Description
0	No reaction throughout the whole soil profile
1	No reaction at the surface horizon of 0–30 cm (part A) while there is reaction at the subsurface horizon of 30–75 cm (part B) and/or at the substratum of 75–150 cm (part C).
2	Weak reaction on the surface horizon (part A) while the reaction at the deeper layers is not taken into account.
3	Strong reaction on the soil surface, while the reaction at the deeper layers is not taken into account.

Limiting Layers

The presence of limiting layers that affects the growth of plant roots was observed through opening holes using soil-drills or in existing exposed soil profiles. As limiting layers are considered (a) solid rock, (b) cobbles or sand, (c) solid horizons impervious to water and roots such as a fragipan horizon. The corresponding symbols are given in Table A1 (Appendix A).

Electrical Conductivity

Electrical conductivity was recorded in the field according to the presence or absence of water-soluble salts and the formation on the soil surface of a white crust. The recording of the presence of salts or not was confirmed by laboratory measurement of the electrical conductivity of the soil. The following classes were distinguished 0–4 ds/m, 4–8 ds/m, 8–15 ds/m and> 15 ds/m with the corresponding symbols given in Table A1 (Appendix A).

Alkalization

The alkalization of the soil was determined according to the percentage of exchangeable sodium (ESP) and the cation exchange capacity. In alkaline soils, the amount of sodium retained by clay minerals is 15% or greater than the total cation exchange capacity (CEC). Alkalization was assessed based on previous soil studies and analyses. The used ESP classes were: ESP < 6, ESP = 6–15, ESP > 15% with the corresponding symbols given in Table A1 (Appendix A).

Soil Classification

The classification of the soils in the context of the two mapping methodologies (1976 and 2019) was carried out as described in the text of the present work based on the principles of USDA and WRB taxonomical keys.

Parent material of the hilly and mountainous soils

Table A6. Soil mapping symbols for the hilly and mountainous areas according to the parent material.

Parent Material/Soil Depth (cm)	0–30	30–100	100–150	>150
Granite	01	02	03	04
Limestone	11	12	13	14
Peridotite	21	22	23	24
Shale	31	32	33	34
Conglomerate Limestone Rock	41	42	43	44
Hornstone	51	52	53	54
Colluvial	61	62	63	64
Alluvial by diagenesis	71	72	73	74
Sandstone	81	82	83	84
Clayey Marls by diagenesis	91	92	93	94
Sandstone-Marls by diagenesis	891	892	893	894
Marls-Sandstone by diagenesis	981	982	983	984
Talc	771	772	773	774

Vegetation type of the hilly and mountainous soils

Table A7. Soil mapping symbols for the hilly and mountainous areas according to the type of vegetation.

Symbol	Type of Vegetation
0	No vegetation
1	Shrubs
2	Not dense forests
3	Dense forests
4	Olives orchards
5	Vineyards

Appendix B

Table A8. SMU numbering, symbolization, occupied area and soil's taxonomy of the conventional methodology (1976 mapping system).

SMU_No	Conventional Soil Mapping Symbol	Order	Suborder	Great Group	Subgroup	Area (ha)
1	A3$^{\times}$04$^{\times}$I/C03	Inceptisols	Xerepts	Calcixerepts	Petrocalcic	1.3
2	A304I/B03	Inceptisols	Xerepts	Calcixerepts	Petrocalcic	1.2
3	A12E/F03	Entisols	Orthents	Xerorthents	Lithic	1.1
4	A334I/C03	Inceptisols	Xerepts	Calcixerepts	Petrocalcic	0.4
5	A12E/F03	Entisols	Orthents	Xerorthents	Lithic	1.3
6	A203I/B03	Inceptisols	Xerepts	Calcixerepts	Petrocalcic	0.4
7	A12I/B03	Inceptisols	Xerepts	Calcixerepts	Petrocalcic	0.8
8	A12I/C03	Inceptisols	Xerepts	Calcixerepts	Petrocalcic	3.0
9	A12E/F03	Entisols	Orthents	Xerorthents	Lithic	0.2
10	A12I/E03	Inceptisols	Xerepts	Calcixerepts	Petrocalcic	0.9
11	A11E/F02	Entisols	Orthents	Xerorthents	Lithic	17.6
12	A11E/G01	Entisols	Orthents	Xerorthents	Lithic	14.7
13	A34E/D00	Entisols	Orthents	Xerorthents	Lithic	28.7
14	A12I/E03	Inceptisols	Xerepts	Calcixerepts	Typic	3.0
15	A11I/E03	Inceptisols	Xerepts	Calcixerepts	Typic	2.3
16	A11E/H01	Entisols	Orthents	Xerorthents	Lithic	10.6
17	A12E/E03	Inceptisols	Xerepts	Calcixerepts	Typic	1.5
18	A12E/F03	Entisols	Orthents	Xerorthents	Lithic	8.5
19	A12I/E03	Inceptisols	Xerepts	Calcixerepts	Typic	2.0
20	A11E/F02	Entisols	Orthents	Xerorthents	Lithic	4.9
21	A41I/D02	Inceptisols	Xerepts	Calcixerepts	Typic	2.3
22	A12I/D03	Inceptisols	Xerepts	Calcixerepts	Typic	0.6
23	A42I/C02	Inceptisols	Xerepts	Calcixerepts	Typic	2.7
24	A34I/D03	Inceptisols	Xerepts	Calcixerepts	Typic	3.7
25	A12I/D03	Inceptisols	Xerepts	Calcixerepts	Typic	0.8
26	A33E/F03	Entisols	Orthents	Xerorthents	Lithic	1.7
27	A11E/H00	Entisols	Orthents	Xerorthents	Lithic	1.0
28	A11I/C00	Inceptisols	Xerepts	Calcixerepts	Typic	1.7
29	A11E/F02	Entisols	Orthents	Xerorthents	Lithic	2.7
30	A12E/F02	Entisols	Orthents	Xerorthents	Lithic	2.8
31	A12E/F02	Entisols	Orthents	Xerorthents	Lithic	2.6
32	A11I/E03	Inceptisols	Xerepts	Calcixerepts	Typic	3.5
33	A12I/E03	Inceptisols	Xerepts	Calcixerepts	Typic	1.4
34	A11I/D03	Inceptisols	Xerepts	Calcixerepts	Typic	3.4
35	A44I/C03	Inceptisols	Xerepts	Calcixerepts	Typic	1.7
36	A12E/F03	Entisols	Orthents	Xerorthents	Lithic	3.6
37	A11E/F03	Entisols	Orthents	Xerorthents	Lithic	3.5
38	A12I/E03	Inceptisols	Xerepts	Calcixerepts	Typic	4.5
39	A42E/F03	Entisols	Orthents	Xerorthents	Lithic	2.2
40	A12I/C03	Inceptisols	Xerepts	Calcixerepts	Typic	0.5
41	A44I/C03	Inceptisols	Xerepts	Calcixerepts	Typic	2.1
42	A11I/C03	Inceptisols	Xerepts	Calcixerepts	Typic	1.2
43	A11I/E02	Inceptisols	Xerepts	Calcixerepts	Typic	1.4
44	A11I/E00	Inceptisols	Xerepts	Calcixerepts	Typic	2.2
45	A12I/C03	Inceptisols	Xerepts	Calcixerepts	Typic	0.5
46	A4$^{\times}$04$^{\times}$I/B03	Inceptisols	Xerepts	Calcixerepts	Typic	1.2
47	A44I/C03	Inceptisols	Xerepts	Calcixerepts	Typic	1.1
48	A334I/B03	Inceptisols	Xerepts	Calcixerepts	Typic	0.8
49	A11E/G01	Entisols	Orthents	Xerorthents	Lithic	1.5
50	A404I/B03	Inceptisols	Xerepts	Calcixerepts	Typic	1.1
51	A44I/E02	Inceptisols	Xerepts	Calcixerepts	Typic	6.9
52	A334E/B03	Entisols	Fluvents	Xerofluvents	Typic	1.1
53	A4$^{\times}$04$^{\times}$I/B03	Inceptisols	Xerepts	Calcixerepts	Typic	0.6
54	A12E/B00	Entisols	Orthents	Xerorthents	Lithic	0.8
55	A44E/D03	Entisols	Orthents	Xerorthents	Typic	0.8

Table A9. SMU numbering, symbolization, occupied area and soil's taxonomy of the currently applied methodology (2019 mapping system).

SMU_No	Current Soil Mapping Symbol	Principal Qualifier 3	Principal Qualifier 2	Principal Qualifier 1	RSG	Area (ha)
1	A004D22L23R11LE	Cambic	Calcaric	Skeletic	Leptosols	1.9
2	A004F13L13R11LE	Calcaric	Skeletic	Nudilithic	Leptosols	2.2
3	A304E32L13R11CM	Leptic	Skeletic	Calcaric	Cambisols	5.6
4	A004F13L13R11LE	Calcaric	Skeletic	Nudilithic	Leptosols	0.4
5	A304E32L13R11CM	Leptic	Skeletic	Calcaric	Cambisols	16.2
6	A004D13L23R11LE	Calcaric	Skeletic	Nudilithic	Leptosols	5.9
7	A304F32L13R11CM	Leptic	Skeletic	Calcaric	Cambisols	3.4
8	A004E22L13R11LE	Cambic	Calcaric	Skeletic	Leptosols	4.2
9	A004F23L23R11LE	Cambic	Calcaric	Skeletic	Leptosols	7.1
10	A304F22L13R11CM	Skeletic	Calcaric	Leptic	Cambisols	4.2
11	A004G13L13R11LE	Calcaric	Skeletic	Nudilithic	Leptosols	12.5
12	A004D22L13R11LE	Cambic	Calcaric	Skeletic	Leptosols	4.2
13	A304C31L03R11CM	Chromic	Leptic	Calcaric	Cambisols	8.5
14	A304G32L13R11CM	Skeletic	Calcaric	Leptic	Cambisols	14.4
15	A334E52L13P11CL	Skeletic	Cambic	Petric	Calcisols	14.7
16	A004D13L03R11LE	Calcaric	Skeletic	Nudilithic	Leptosols	17.1
17	A004F22L13R11LE	Cambic	Calcaric	Skeletic	Leptosols	3.9
18	A004F22L13R11LE	Skeletic	Calcaric	Nudilithic	Leptosols	21.8
19	A004G21L13R11LE	Skeletic	Calcaric	Cambic	Leptosols	25.7

Table A10. Correlation table of the cartographic symbols and soil classification of the two mapping systems.

2019 SMUs Numbers	2019 SMU Symbol	WRB Classification	1976 SMUs Numbers	1976 SMU Symbols	USDA Classification	Percentage (%) of the Total Area
15	A334E52L13P11CL	CL-pt.cm.sk	1, 2, 4, 6, 7, 8, 10	1: A304I/C03 2: A304I/B03 4: A334I/C03 6: A203I/B03 7: A120I/B03 8: A12I/C03 10: A12I/E03	Petrocalcic Calcixerepts	4.6%
			48	48: A334I/B03	Typic Calcixerepts	0.5%
			3, 5, 9, 11	3: A12E/F03 9: A12E/F03 11: A11E/F02 5: A12E/F03	Lithic Xerorthents	3.5%
13	A304C31L03R11CM	CM-ca.le.cr	40,47,41,46,53,50, 42,34	40: A12I/C03 47: A44I/C03 41: A44I/C03 46: A404I/B03 53: A404I/B03 50: A404I/B03 42: A11I/C03 34: A11I/D03	Typic Calcixerepts	3.3%
			55,54,52	55: A44E/D03 54: A120E/B00 52: A334E/B03	Typic Xerorthent Lithic Xerorthent Typic Xerofluvent	0.5% 0.5% 0.7%
18	A004F22L13R11LE	LP-nt.ca.sk	11,12	11: A11E/F02 12: A11E/G01	Lithic Xerorthents	11.7%
			43	43: A11I/E02	Typic Calcixerepts	0.8%

Table A10. *Cont.*

2019 SMUs Numbers	2019 SMU Symbol	WRB Classification	1976 SMUs Numbers	1976 SMU Symbols	USDA Classification	Percentage (%) of the Total Area
14	A304G32L13R11CM	CM-le.ca.sk	38,28,33,40,41,42, 44,47,53,32,34	38: A12I/E03 28: A11I/C00 33: A12I/E03 40: A12I/C03 41: A44I/C03 42: A11I/C03 44: A11I/E00 47: A44I/C03 53: A404I/B03 32: A11I/E03 34: A11I/D03	Typic Calcixerepts	5.9%
			13,39	13: A34E/D00 39: A42E/F03	Lithic Xerorthents	4,7%
10	A304F22L13R11CM		20,16,18,29,30,36	20: A11E/F02 16: A11E/H01 29: A11E/F02 30: A12E/F02 36: A12E/F03 18: A12E/F03		
16	A004D13L03R11LE	LP-nt.sk.ca	45,47,44,28	45: A12I/C03 47: A44I/C03 44: A11I/E00 28: A11I/C00	Typic Calcixerepts	Typic Calcixerepts corresponding to Calcaric Skeletic Nudilithic Leptosols 5.3%
			13,27	13: A34E/D00 27: A11E/H00	Lithic Xerorthents	
6	A004D13L23R11LE		33,34, 35	33: A12I/E03 34: A11I/D03 32: A11I/E03 35: A44I/C03	Typic Calcixerepts	
			31	31: A12E/F02	Lithic Xerorthents	

Table A10. *Cont.*

2019 SMUs Numbers	2019 SMU Symbol	WRB Classification	1976 SMUs Numbers	1976 SMU Symbols	USDA Classification	Percentage (%) of the Total Area
2	A004F13L13R11LE		25,19	25: A12I/D03 19: A12I/E03	Typic Calcixerepts	Lithic Xerorthents corresponding to Calcaric Skeletic Nudilithic Leptosols 16.6%
			18,20	18: A12E/F03 20: A11E/F02	Lithic Xerorthents	
11	A004G13L13R11LE		16,20	16: A11E/H01 20: A11E/F02	Lithic Xerorthents	
4	A004F13L13R11LE		18	18: A12E/F03	Lithic Xerorthents	
17	A004F22L13R11LE	LP-sk.ca.cm	13	13: A34E/D00	Lithic Xerorthents	Typic Calcixerepts corresponding to Cambic Calcaric Skeletic Leptosols 3.0%
1	A004D22L23R11LE		29,30,49	29: A11E/F02 30: A12E/F02 49: A11E/G01	Lithic Xerorthents	
9	A004F23L23R11LE		36,31,37, 26	36: A12E/F03 31: A12E/F02 37: A11E/F03 26: A33E/F03	Lithic Xerorthents	
			34,32	34: A11I/D03 32: A11I/E03	Typic Calcixerepts	Lithic Xerorthents corresponding to Cambic Calcaric Skeletic Leptosols 9.2%
12	A004D22L13R11LE		39,36,37	39: A42E/F03 36: A12E/F03 37: A11E/F03	Lithic Xerorthents	
8	A004E22L13R11LE		13	13: A34E/D00	Lithic Xerorthents	
			15,14	15: A11I/E03 14: A12I/E03	Typic Calcixerepts	
19	A–4G21L13R11LE	LP-cm.ca.sk	13,12	13: A34E/D00 12: A11E/G01	Lithic Xerorthents	10.8%
			51	51: A44I/E02	Typic Calcixerepts	4.0%

Table A10. *Cont.*

2019 SMUs Numbers	2019 SMU Symbol	WRB Classification	1976 SMUs Numbers	1976 SMU Symbols	USDA Classification	Percentage (%) of the Total Area
5	A304E32L13R11CM	CM-ca.sk.le	29,18,49	29: A11E/F02 18: A12E/F03 49: A11E/G01	Lithic Xerorthents	Lithic Xerorthents corresponding to Leptic Skeletic Calcaric Cambisols 6.1%
			35,25,24, 21,23,22, 19,14	35: A44I/C03 25: A12I/D03 24: A34I/D03 21: A41I/D02 23: A42I/C02 22: A12I/D03 14: A12I/E03 19: A12I/E03	Typic Calcixerepts	
3	A304E32L13R11CM		29,30,18	29: A11E/F02 30: A12E/F02 18: A12E/F03	Lithic Xerorthents	Typic Calcixerepts corresponding to Leptic Skeletic Calcaric Cambisols 8.4%
7	A304F32L13R11CM		20 17	20: A11E/F02 17: A12E/E03	Lithic Xerorthents Typic Calcixerepts	

Table A11. Table of the participation rates of 1976 SMUs area to the classification of 2019 mapping.

Classification of 2019 Mapping (WRB)	2019 SMUs Numbers	Classification of 1976 Mapping (USDA)	1976 SMUs Numbers	Participation Rates (%)
Skeletic Calcaric Nudilithic Leptosols	18	Lithic Xerorthents	11 12	82 41
		Typic Calcixerepts	43	100
Calcaric Skeletic Nudilithic Leptosols	2,4,6, 11,16	Lithic Xerorthents	13 16 18 20 27 31	43 92 12 65 100 62

Table A11. *Cont.*

Classification of 2019 Mapping (WRB)	2019 SMUs Numbers	Classification of 1976 Mapping (USDA)	1976 SMUs Numbers	Participation Rates (%)
Cambic Calcaric Skeletic Leptosols	1,8,9,12,17	Typic Calcixerepts	19	48
			25	15
			28	52
			32	42
			33	68
			34	31
			35	43
			44	97
			45	100
			47	30
		Lithic Xerorthents	13	13
			26	98
			29	7
			30	14
			31	37
			36	90
			37	100
			39	29
			49	82
			12	3.8
		Typic Calcixerepts	14	39
			15	95
			32	40
			34	14
Skeletic Calcaric Cambic Leptosols	19	Lithic Xerorthents	12	54
			13	38
		Typic Calcixerepts	51	100

Table A11. *Cont.*

Classification of 2019 Mapping (WRB)	2019 SMUs Numbers	Classification of 1976 Mapping (USDA)	1976 SMUs Numbers	Participation Rates (%)
Chromic Leptic Calcaric Cambisols	13	Typic Calcixerepts	34	12
			40	79
			41	76
			42	38
			46	98
			47	5
			50	97
			53	55
		Typic Xerorthents	55	100
		Lithic Xerorthents	54	100
		Typic Xerofluvents	52	100
Skeletic Cambic Petric Calcisols	15	Typic&Petrocalcic Calcixerepts	1	98
			2	100
			4	100
			6	100
			7	100
			8	100
			10	100
			48	100
		Lithic Xerorthents	3	100
			5	100
			9	100
			11	18
Leptic Skeletic Calcaric Cambisols	3,5,7	Typic Calcixerepts	14	55
			17	93
			19	46
			21	98
			22	98
			24	97
			23	97
			25	82
			35	56

Table A11. *Cont.*

Classification of 2019 Mapping (WRB)	2019 SMUs Numbers	Classification of 1976 Mapping (USDA)	1976 SMUs Numbers	Participation Rates (%)
		Lithic Xerorthents	18	59
			20	14
			29	86
			30	80
			49	18
Skeletic Calcaric Leptic Cambisols	10,14	Typic Calcixerepts	28	48
			32	18
			33	30
			34	44
			38	97
			40	19
			41	24
			42	63
			47	64
			53	45
			44	3
		Lithic Xerorthents	13	7
			16	7
			18	28
			20	19
			29	4
			30	5
			36	10
			39	71

Table A12. Explanatory table of the methodology for comparison the results of the two methods.

2019 Classification (WRB)	Area (ha)	Percentage of Agreement (%) Between the Classifications of the Two Systems	Overall Percentage of Taxonomical Coincidence of the Two Systems	Algorithm for the Calculation of the Area Corresponding to a Common Description of Soil Properties Except Classification Between the Two Systems—Rates of Success (ha)	Summation of the Rates of Success Corresponding to a Common Description of Soil Properties Except Classification Between the two Systems	Overall Percentage of Success Corresponding to a Common Description of Soil Properties Except Classification Between the Two Systems
LP-nt.ca.sk LP-nt.sk.ca	21.8 38.1	93.6 75.9	69.5	93.6% × 21.8 = 20.4 75.9% × 38.1 = 28.9	111.3	63.7
LP-sk.ca.cm	21.3	75.6		75.6% × 75% (due to the participation rates of 1976 SMUs 13 and 26 resulting to a differentiated description) × 21.3 = 12		
LP-cm.ca.sk	25.7	73.2		73.2% × 58% (due to the participation rates of 1976 SMUs 13 resulting to a differentiated description) × 25.7 = 10.9		
CM-ca.le.cr CL-pt.cm.sk CM-ca.sk.le CM-le.ca.sk	8.5 14.7 25.2 18.6	65.0 60.0 58.0 55.0		65% × 8.5 = 5.5 60% × 14.7 = 8.8 58% × 25.2 = 14.6 55% × 18.6 = 10.2		

Appendix C

Table A13. Soil analyses of the 14 soil samples of the 5 soil profiles of the 1976 mapping system.

Soil Profiles Sites	Soil Depth (cm)	Soil Texture [a] Sand (%)	Silt (%)	Clay (%)	Characterization	[b] CaCO3 (%)	[c] pH	[d] Organic Matter (%)	CEC [e] (cmole/kg Soil)	[e] Exchangeable K (cmole/kg Soil)	[f] Moisture Equivalent (%)	[g] P (mg P/kg Soil)	[h] N (%)
P5	0–10	69.3	15.8	14.8	SL	49.6	7.7	3.5	8	0.8	12.8	>45	2.1
	10–22	64.2	17.7	18.1	SL	55.2	7.8	1.9	4	0.5	14.8	>45	1.2
	22–52	57.5	25.4	17.1	SL	71.2	7.9	0.1	9	0.02	13.2	23	0.4
	>52	53.1	26.2	20.6	SCL	66.4	7.9	0.1	9	0.02	15.0	30	0.4
P7	0–10	59.9	16.3	26.7	SCL	56	7.9	0.7	9	0.5	13.4	22	0.5
	10–60	64.3	15.5	20.2	SCL	55.2	8.2	0.7	8	0.2	12.5	20	0.6
	60–80	66.3	13.5	20.2	SCL	54.4	8.4	0.3	7	0.2	13.5	16	0.6
P8	0–10	71.2	17.4	11.1	SL	43.2	7.8	3.6	12	0.7	13.4	44	2.2
	10–60	67.1	14.0	18.8	SL	41.6	7.8	1.2	13	0.4	13.5	15	0.8
	60–80	66.2	19.5	21.3	SCL	52	7.8	0.7	6	0.3	15.3	17	0.6
P9	0–5	72.5	17.2	10.2	SL	34	7.9	8.7	14	0.3	16.3	>45	3.5
	5–10	71.6	18.4	10.3	SCL	29.2	8.2	4	8	0.8	14.2	>45	2.1
P10	0–10	68.5	16.2	15.2	SL	27.6	7.9	6.7	13	0.8	15.8	>45	2.6
	10–60	61.2	16.6	21.7	SCL	28.8	8.4	0.3	9	0.2	15.2	17	0.5

Note: [a]—Hydrometer method; [b]—Bernard Method; [c]—Soil: distilled water (w-v) suspension (1: 1); [d]—Walkley-Black wet digestion; [e]—Ammonium acetate extraction; [f]—Gravimetric method; [g]—Olsen method; [h]—Kjeldahl digestion.

Appendix C.1 Soil Profiles Description (1976)

Profile P5

Location: Diomedes Botanical Garden
Coordinates: x: 468,283, y: 4,206,607
Elevation: 145.0 m
Parent material: limestone
Landform: hilly
Hydrological group: well drained
Slope: 2–6%
Erosion: none
Vegetation: Pinus halepensis
Groundwater: deep
Classification: Inceptisol

Table A14. Profile P5 horizons description.

Horizon	Depth (cm)	Description
A1	0–10	Dark reddish brown (5YR 4/8) moist; sandy loam (SL); moderate, medium, subangular blocky (msbk); dry hard; strongly effervescent; clear boundary
B1	10–22	Dark reddish brown (5YR 4/8) moist; sandy loam (SL); moderate, medium, subangular blocky (msbk); dry hard; strongly effervescent; clear boundary
C	22–52	Dark reddish brown (5YR 4/8) moist; sandy loam (SL); moderate, medium, subangular blocky structure (msbk); dry hard; strongly effervescent; smooth boundary
C	>52	Dark reddish brown (5YR 4/8) moist; sandy clay loam (SCL); moderate, medium, subangular blocky (msbk); slightly hard; strongly effervescent

Profile P7

Location: Diomedes Botanical Garden
Coordinates: x: 468,220, y: 4,206,902
Elevation: 128.0 m
Parent material: limestone
Landform: hilly
Hydrological group: well drained
Slope: 6–12%
Erosion: none
Vegetation: Pinus halepensis
Groundwater: deep
Classification: Inceptisol

Table A15. Profile P7 horizons description.

Horizon	Depth (cm)	Description
A1	0–10	Reddish yellow (5YR 6/8) moist; sandy clay loam (SCL); moderate, medium, subangular blocky (msbk); dry very hard; gravels 40%strongly effervescent; clear boundary
B1	10–50	Yellowish red (5YR 6/8) moist; sandy clay loam (SCL); moderate, medium, subangular blocky (msbk); dry very hard; gravels 30%; strongly effervescent; clear boundary
B2	50–95	Yellowish red (5YR 6/8) moist; sandy clay loam (SCL); moderate, medium, subangular blocky (msbk); dry very hard; gravels 40%; strongly effervescent

Profile P8

Location: Diomedes Botanical Garden
Coordinates: x: 468,432, y: 4,206,812
Elevation: 133.0 m
Parent material: limestone
Landform: hilly
Hydrological group: well drained
Slope: 2–6%
Erosion: none
Vegetation: Pinus halepensis
Groundwater: deep
Classification: Inceptisol

Table A16. Profile P8 horizons description.

Horizon	Depth (cm)	Description
A1	0–10	Yellowish red (5YR 5/8) moist; sandy loam (SL); moderate, fine, subangular blocky (fsbk); dry very hard; gravels 45%; strongly effervescent; clear boundary
B1	10–60	Yellowish red (5YR 4/8) moist; sandy loam (SL); moderate, coarse, subangular blocky (csbk); dry very hard; gravels 30%; strongly effervescent; clear boundary
B2	60–80	Yellowish red (5YR 6/8) moist; sandy clay loam (SCL); moderate, fine, subangular blocky (fsbk); dry very hard; gravels 40%; strongly effervescent

Profile P9

Location: Diomedes Botanical Garden
Coordinates: x: 468,521, y: 4,206,660
Elevation: 138.4 m
Parent material: limestone
Landform: hilly
Hydrological group: well drained
Slope: 2–6%
Erosion: none
Vegetation: Pinus halepensis
Groundwater: deep
Classification: Inceptisol

Table A17. Profile P9 horizons description.

Horizon	Depth (cm)	Description
A1	0–5	Dark reddish brown (5YR 3/4) moist; sandy loam (SL); moderate, fine, subangular blocky (fsbk); dry slightly hard; strongly effervescent; clear boundary
B1	5–20	Yellowish red (5YR 4/6) moist; sandy loam (SL); moderate, coarse, subangular blocky (csbk); dry very hard; gravels 30%; strongly effervescent

Profile P10

Location: Diomedes Botanical Garden
Coordinates: x: 468,540, y: 4,206,621
Elevation: 142.8 m
Parent material: limestone
Landform: hilly
Hydrological group: well drained
Slope: 6–12%
Erosion: none
Vegetation: Pinus halepensis
Groundwater: deep
Classification: Inceptisol

Table A18. Profile P10 horizons description.

Horizon	Depth (cm)	Description
A1	0–10	Dark reddish brown (5YR 3/4) moist; sandy loam (SL); moderate, medium, subangular blocky (msbk); dry slightly hard; gravels 30%; strongly effervescent; clear boundary
B1	10–60	Yellowish red (5YR 4/6) moist; sandy loam (SL); moderate, medium, subangular blocky structure (msbk); dry very hard; gravels 35%; strongly effervescent

Note: Petrocalcic horizon is not formed in T5 soil profile but at a depth of 20–30 cm deeper from the soil surface there is marl. In the rest of the soil profiles, the petrocalcic horizon is formed 100 cm deeper from the soil surface.

Table A19. Soil analyses of the 20 soil samples of the 12 soil sampling sites of the 2019 mapping system.

Sample	Depth (cm)	Soil Texture				Carbonates (%)
		Sand (%)	Silt (%)	Clay (%)	Characterization	
1	(0–30)	42	32.9	25.1	L	20.2
	(30–60)	38	29.7	32.3	CL	28.0
2	(0–30)	45.7	30	24.3	L	30.1
	(30–60)	37.7	25.7	36.6	CL	33.0
3	(0–30)	28.6	42	29.4	CL	0.78
	(30–60)	22.6	34	43.4	C	0.24
4	(0–30)	46.9	28	25.1	SCL	61.7
	(30–60)	50.9	24	25.1	SCL	65.1
5	(0–30)	18.6	40	41.4	SiCL	1.63
	(30–60)	18.6	32	49.4	C	1.80
6	(0–30)	30.6	30	39.4	CL	1.59
7	(0–30)	19.9	29	51.1	C	12.28
	(30–60)	20.9	26	53.1	C	13.51
8	(0–30)	48.7	21	30.3	SCL	25.67
	(30–60)	40	23.7	36.3	CL	30.10
9	(0–30)	25.2	31.4	43.4	C	32.15
	(30–60)	20.1	35.1	44.8	C	38.08
12	(0–30)	25.2	36.7	38.1	CL	26.52
	(30–60)	38.5	26.5	35	CL	25.44

Appendix C.2 Description of the 2019 roadside profiles.

Profile P1

Location: Diomedes Botanical Garden
Coordinates: x: 468,801, y: 4,206,553
Elevation: 146.6 m
Parent material: limestone
Landform: hilly
Hydrological group: well drained
Slope: 6–12%
Erosion: none
Vegetation: Pinus halepensis
Groundwater: deep
Classification: Cambisol (Inceptisol)

Table A20. Profile P1 horizons description.

Horizon	Depth (cm)	Description
A	0–30	Yellowish red (5YR 4/6) moist; clay (C); fine, subangular blocky (fsbk); dry hard; gravels 45%; strongly effervescent; clear boundary
B	30–60	Reddish brown (5YR 4/4) moist; clay (C); fine, subangular blocky (fsbk); dry hard; gravels 30%; strongly effervescent;

Profile P2

Location: Diomedes Botanical Garden
Coordinates: x: 468,839, y: 4,206,180
Elevation: 200.0 m
Parent material: limestone
Landform: hilly
Hydrological group: well drained
Slope: 18–25%
Erosion: none
Vegetation: Pinus halepensis
Groundwater: deep
Classification: Cambisol (Inceptisol)
Remark: A petrocalcic horizon observed at depth >80 cm

Table A21. Profile P2 horizons description.

Horizon	Depth (cm)	Description
A	0–30	Yellowish red (5YR 4/6) moist; clay (C); fine, subangular blocky (fsbk); dry hard; gravels 45%; strongly effervescent; clear boundary
B	30–60	Yellowish red (5YR 5/8) moist; clay (C); fine, subangular blocky (fsbk); dry hard; gravels 30%; strongly effervescent;
C	>60	

Profile P3

Location: Diomedes Botanical Garden
Coordinates: x: 468,492, y: 4,206,244
Elevation: 198.4 m
Parent material: limestone
Landform: hilly
Hydrological group: well drained
Slope: 18–25%
Erosion: none
Vegetation: Pinus halepensis
Groundwater: deep
Classification: Cambisol (Inceptisol)

Table A22. Profile P3 horizons description.

Horizon	Depth (cm)	Description
A	0–10	Yellowish red (5YR 4/8) moist; clay loam (CL); moderate fine, subangular blocky (fsbk); dry hard; gravels 30–45%; strongly effervescent; clear boundary
B	10–30	Yellowish red (5YR 4/6) moist; clay loam (CL); moderate fine, subangular blocky (fsbk); dry hard; gravels 30–45%; strongly effervescent;

Profile P4

Location: Diomedes Botanical Garden
Coordinates: x: 468,080, y: 4,206,429
Elevation: 156.8 m
Parent material: limestone
Landform: hilly
Hydrological group: well drained
Slope: 25–35%
Erosion: none
Vegetation: Pinus halepensis
Groundwater: deep
Classification: Leptosol (Entisol)

Table A23. Profile P4 horizons description.

Horizon	Depth (cm)	Description
A	0–10	Yellowish red (5YR 4/8) moist; loam (L); medium, subangular blocky (fsbk); dry hard; gravels 45%; strongly effervescent; clear boundary
B	10–20	Yellowish red (5YR 4/6) moist; loam (L); medium; moderate fine, subangular blocky (fsbk); dry hard; gravels 30%; strongly effervescent;

References

1. Forbes, T.; Rossiter, D.; Van Wambeke, A. Guidelines for Evaluating the Adequacy of Soil Resources Inventories. In *Soil Management support series*; Technical Monograph; Cornell University, Department of Agronomy, New York State College of Agriculture and Life Sciences; U.S. Department of Agriculture, Soil Conservation Service, Soil Management Support Service (SMSS): Ithaca, NY, USA; Washington, DC, USA, 1987; 4, p. 10.
2. United States Department of Agriculture (USDA)—Soil Science Division Staff. *Soil Survey Manual*; Ditzler, C., Scheffe, K., Monger, H.C., Eds.; USDA Handbook 18; Government Printing Office: Washington, DC, USA, 2017.
3. Food and Agriculture Organization (FAO). *Soil Survey Interpretation and its Use*; FAO Soils Bulletin, No. 8; FAO: Rome, Italy, 1967; p. 70.
4. Gobin, A.; Campling, P.; Feyen, J. Soil-landscape modeling to quantify spatial variability of soil texture. *Phys. Chem. Earth Part B Hydrol. Oceans Atmos.* **2001**, *26*, 41–45. [CrossRef]
5. Zhu, A.X.; Hudson, B.; Burt, J.; Lubich, K.; Simonson, D. Soil Mapping Using GIS, Expert Knowledge, and Fuzzy Logic. *Soil Sci. Soc. Am. J.* **2001**, *65*, 1463–1464. [CrossRef]
6. Castrignanò, A.; Buttafuoco, G.; Comolli, R.; Castrignanò, A. Using digital elevation model to improve soil pH prediction in an alpine doline. *Pedosphere* **2011**, *21*, 259–270. [CrossRef]
7. McBratney, A.; Mendonça, S.M.; Minasny, B. On Digital Soil Mapping. *Geoderma* **2003**, *117*, 3–52. [CrossRef]
8. Arrouays, D.; McBratney, A.; Bouma, J.; Libohova, Z.; Richer-de-Forges, A.C.; Morgan, C.L.C.; Roudier, P.; Poggio, L.; Mulder, V.L. Impressions of digital soil maps: The good, the not so good, and making them ever better. *Geoderma Reg.* **2020**, *20*, e00255. [CrossRef]
9. Dobos, E.; Carré, F.; Hengl, T.; Reuter, H.I.; Tóth, G. *Digital Soil Mapping as a Support to Production of Functional Maps*; EUR 22123 EN; Office for Official Publications of the European Communities: Luxemburg, 2006; p. 30.
10. Legros, J.-P. *Mapping of the soil*, 1st ed.; Science Publishers: Enfield, NH, USA, 2006; p. 411.
11. Campbell, J. Spatial Variation of Sand Content and pH Within Single Contiguous Delineations of Two Soil Mapping Units. *Soil Sci. Soc. Am. J.* **1978**, *42*, 460–464. [CrossRef]
12. Lin, H.; Wheeler, D.; Bell, J.; Wilding, L. Assessment of soil spatial variability at multiple scales. *Ecol. Model.* **2005**, *182*, 271–290. [CrossRef]
13. Constantini, E.A.C.; Barbetti, R.; Fantapie, M.; Abate, G.L.; Lorenzetti, R.; Napoli, R.; Marchetti, A.; Riveccio, R. The soil map of Italy: A hierarchy of geodatabases from soil regions to sub-systems. In *Global Soil Map*, 1st ed.; Arrouays, D., McKenzie, N., Hempel, J., de Forges, A.R., McBratney, A., Eds.; CRC Press: London, UK, 2014; pp. 109–112.

14. Bouma, J. Back to the Old Paradigms of Soil Classification. In *Soil Classification*; Eswaran, H., Ahrens, R., Rice, T., Stewart, B., Eds.; CRC Press: Boca Raton, FL, USA, 2002; pp. 51–56. [CrossRef]
15. Miller, B.A.; Schaetzl, R.J. The historical role of base maps in soil geography. *Geoderma* **2014**, *230–231*, 329–339. [CrossRef]
16. Lagacherie, P. Digital soil mapping: A state of the art. In *Digital Soil Mapping with Limited Data*, 1st ed.; Hartemink, A.E., McBratney, A., Mendonça-Santos, M.L., Eds.; Springer: Dordrecht, The Netherlands, 2008; pp. 3–14.
17. Valentine, K.W.G. How soil map units and delineations change with survey intensity and map scale. *Can. J. Soil Sci.* **1981**, *61*, 535–551. [CrossRef]
18. Food and Agriculture Organization (FAO). *Guidelines: Land Evaluation for Irrigated Agriculture*; FAO Soils Bulletin, No. 55; FAO: Rome, Italy, 1985.
19. Basayigit, L.; Senol, S. Comparison of Soil Maps with Different Scales and Details Belonging to the Same Area. *Soil. Water Res.* **2008**, *3*, 31–39. [CrossRef]
20. Yang, L.; Zhu, A.X.; Zhao, Y.G.; Li, D.C.; Zhang, G.L.; Zhang, S.J.; Band, L.E. Regional Soil Mapping Using Multi-Grade Representative Sampling and a Fuzzy Membership-Based Mapping Approach. *Pedosphere* **2017**, *27*, 344–357. [CrossRef]
21. Yassoglou, N. Soil Survey in Greece. In *Soil Resources of Europe*, 2nd ed.; European Soil Bureau Research Report No. 9, EUR 20559 EN; Jones, R.J.A., Houšková, B., Bullock, P., Montanarella, L., Eds.; Office for Official Publications of the European Communities: Luxembourg, 2005.
22. Greek Payment and Control Agency for Guidance and Guarantee Community Aid. 2014. Available online: https://iris.gov.gr/SoilServices/js/pdf/SOIL%20MAP%20OF%20GREECE%20e-SOILBOOK.pdf (accessed on 14 May 2020).
23. Botanical Garden Julia & Alexander N. Diomedes. Available online: http://www.diomedes-bg.uoa.gr/endemic-en.html (accessed on 10 February 2020).
24. Gaitanakis, P. *Geological Map of Greece at Map Scale 1:50000*; Hellenic Survey of Geology and Mineral Exploration—H.S.G.M.E.: Athens, Greece, 1976.
25. Data Station's Information from National Observatory of Athens. Available online: http://www.meteo.gr/climatic.cfm (accessed on 10 February 2020).
26. Moustakas, N. Soil mapping of the Diomedes Botanical garden. Undergraduate Dissertation, Library of the Agricultural University of Athens, Athens, Greece, 1976; p. 42.
27. Shalaby, A.; AbdelRahman, M.A.E.; Belal, A.A. A GIS Based Model for Land Evaluation Mapping: A Case Study North Delta Egypt. *Egypt. J. Soil Sci.* **2017**, *57*, 341–342.
28. Dent, D.; Young, A. *Soil Survey and Land Evaluation*; George Allen & Unwin: London, UK, 1981.
29. Avery, B.W. *Soil survey methods: A review*; Soil Survey Technical Monograph No. 18; Soil Survey and Land Research Centre: Silsoe, UK, 1987.
30. Grunwald, S. *Environmental Soil-Landscape Modeling, Geographic Information Technologies and Pedometrics*, 1st ed.; CRC Press: Boca Raton, FL, USA, 2005; p. 37. [CrossRef]
31. Yassoglou, N.; Nychas, A.; Kosmas, C. Parametric designation of mapping units for soil surveys and land evaluation in Greece based on Soil Taxonomy. In *Agronomy Abstracts, Proceedings of the Diamond Jubilee of the American Society of Agronomy, 74th Annual Meeting, American Society of Agronomy–Crop Science Society of America–Soil Science Society of America, Society of Agronomy–Crop Science Society of America–, Anaheim, CA, USA November 28–December 3, 1982*; Soil Science Society of America: Anaheim, CA, USA; p. 212.
32. Soil Survey Staff. *Soil Taxonomy: A Basic System of Soil Classification for Making and Interpreting Soil Surveys*; USDA–NRCS Agricultural Handbook No. 436; US Government Printing Office: Washington, DC, USA, 1975; p. 774.
33. Soil Survey Staff. *Soil taxonomy: A Basic System of Soil Classification for Making and Interpreting Soil Surveys*, 2nd ed.; USDA–NRCS Agricultural Handbook No. 436; US Government Printing Office: Washington, DC, USA, 1999; p. 886.
34. Yassoglou, N.; Tsadilas, C.; Kosmas, C. Soil Classification. In *The soils of Greece*; Springer International Publishing: New York, NY, USA, 2017; pp. 19–25.
35. Ditzler, C.A.; Hempel, J. *Soil Taxonomy and Soil Classification*; USDA-NRCS National Soil Survey Center: Lincoln, NE, USA, 2017; p. 6.

36. Rice, R.W.; Gilbert, R.A.; Daroub, S.H. *Application of the Soil Taxonomy Key to the Organic Soils of the Everglades Agricultural Area*; SS-AGR-246; Institute of Food and Agricultural Sciences, University of Florida: Gainesville, FL, USA, 2005; p. 3.
37. IUSS Working Group WRB. *World Reference Base or Soil Resources 2014, Update 2015 International Soil Classification System for Naming Soils and Creating Legends for Soil Maps*; World Soil Resources Reports No. 106; FAO: Rome, Italy, 2005.
38. Spaargaren, O.; Schad, P.; Micheli, E.; Jones, A.R. Soil Solutions for a Changing World. In Proceedings of the World Congress of Soil Science, Brisbane, Australia, 1–6 August 2010.
39. Soil Survey Staff. *Soil classification: A comprehensive system, 7th Approximation*; US 1544; Government Printing Office: Washington, DC, USA, 1960.
40. Moustakas, N.; Kosmas, C.; Danalatos, N.; Yassoglou, N. Rock fragments I. Their effect on runoff, erosion and soil properties under field conditions. *Soil Use Manag.* **1995**, *11*, 115–120. [CrossRef]

land

MDPI

Article

A Spatial-Temporal Analysis of the Effects of Households' Land-Use Behaviors on Soil Available Potassium in Cropland: A Case Study from Urban Peripheral Region in Northeast China

Hongbin Liu [1,2], Zhanli Sun [2], Xiaojuan Luo [3,4,*], Xiuru Dong [1] and Mengyao Wu [1]

1 College of Land and Environment, Shenyang Agricultural University, Shenyang 110866, China; liuhongbinsy@syau.edu.cn (H.L.); dongxiuru@syau.edu.cn (X.D.); 2019220430@stu.syau.edu.cn (M.W.)
2 Leibniz Institute of Agricultural Development in Transition Economies (IAMO), 06120 Halle (Saale), Germany; sun@iamo.de
3 Jiangxi Economic Development Research Institute, College of Finance, Jiangxi Normal University, Nanchang 330022, China
4 College of Public Administration, Nanjing Agricultural University, Nanjing 210095, China
* Correspondence: lxj918@126.com; Tel.: +86-185-070-07012

Received: 17 April 2020; Accepted: 19 May 2020; Published: 20 May 2020

Abstract: Available potassium (AVK) in the soil of cropland is one of the most important factors determining soil quality and agricultural productivity. Thus, it is crucial to understand the variation of AVK and its influencing factors for sustaining soil fertility and mitigating land degradation. Farm households are the ultimate land users, and their land-use behaviors inevitably play an important role in the variation of AVK. This paper, therefore, aims to explore the effects of households' land-use behaviors on soil AVK from spatial and temporal perspectives. Taking an urban peripheral region in Northeast China as the study area, we firstly use geostatistics (Kriging interpolation) and GIS tools to map out the spatial AVK distributions in 1980, 2000, and 2010, based on soil sampling data points, and then assess the impacts of land-use behaviors on AVK using econometric models. The results show that, although the AVK content in the study area has a largely downward trend over the 30 years, there are distinct trends in different stages. The disparity of trends can be attributed to the changes in households' land-use behaviors over time. The spatial variation of AVK is also substantial and intriguing: the closer to the urban area, the greater the decline of soil AVK content, while the farther away from the urban area, the greater the rise of soil AVK content. This spatial disparity can too be largely explained by the obvious differences in households' land-use behaviors in various regions.

Keywords: soil available potassium; land-use behavior; spatial-temporal analysis; soil quality; Kriging interpolation

1. Introduction

China faces the great challenge of meeting the ever-increasing food demand, due to its large and ever-growing population and dietary shifts accompanying rapidly increasing income, with scarce agricultural land. Land conservation is not only essential for the realization of sustainable social and economic development in China, but also of great strategic significance for ensuring world food security and stabilizing international food prices [1–3]. Chinese governments have prioritized maintaining the soil quality of arable lands to safeguard national food security. Therefore, researches on how and where improvements can be made to enhance the sustainability of agricultural production are of paramount importance to policymakers. While much attention has been focused on maintaining the arable land area—as reflected by the so-called "cropland red line" policy, aiming to maintain at least 1.8 billion mu (i.e., 0.12 billion ha) area of quality arable land, the deterioration of soil quality and land degradation has long been overlooked. Recently, soil preservation has been recognized as one of the most important factors restricting food security and agricultural product safety in China [4,5]. Potassium is one of the four essential macronutrients (N, P, K, and S) needed by plants, which directly and indirectly affect soil fertility. Soil available potassium (AVK) determined by ammonium acetate leaching-flame photometric method is an effective method that determines soil availability of potassium, which can be can directly absorbed and utilized by plants [6,7]. It is thus a good indicator of the supply capacity of potassium nutrient and soil fertility of cropland, which in turn affects the yield and quality of crops [8,9]. The critical soil test values or the AVK values at which relative yield to potassium fertilizer application is equal to 90% for each crop in the study. Therefore, the study on the spatial and temporal variation of AVK in soil and its influencing factors are critical for monitoring the dynamic evolution characteristics of cropland soil fertility and scientific planning and management of croplands.

There is a growing body of literature on the spatial and temporal variations of soil fertility and its driving factors [10,11]. The current literature mainly focused on two aspects. One is on the spatial variation of soil physical properties and soil salinity from the perspective of natural science [12–14]. In past years, geostatistical methods have been increasingly used to analyze the spatial variation of soil nutrients [15–17], for example, soil organic matter [18], NO^{-3}-N [19], available phosphorus [5], AVK [20,21], nugget effect and correlation degree [22], coefficient of variation [23]. At the same time, studies on the influencing factors of soil fertility have gained much attention. Researches showed that soil fertility changes are directly or indirectly affected by soil erosion, farming systems, land-use modes, and fertilizer inputs [24–28]. However, the existing researches still have the following limitations: firstly, the change of soil nutrient content is affected not only by natural conditions, but also by human activities. In intensive land use areas, for example, in urban peripheral regions, human factors play an important role. However, existing researches focus more on the natural perspective, and the discussions on human production behavior and activities are relatively rare; convincing empirical analyses are still lacking. Secondly, most studies have been either on the spatial variation or temporal changes of soil nutrient content in a certain region. There is a lack of spatial-temporal analysis of continuous monitoring data, especially in the area of intensive land use, where land-use behaviors significantly influence soil nutrients changes. Thirdly, there are few researches on the mechanism of the change of soil AVK based on multivariate data—especially on the data of soil fertility, the role of farm households' behaviors, and socio-economic statistics.

To bridge the existing research gaps, this study aims to investigate the link between soil AVK and farm households' behaviors from a theoretical and empirical perspective. For that, we take a suburban area of Shenyang, a megacity in Northeast China, as the study area. This region is at the interface between urban expansion and cropland conservation, and forms intricate regional complexes [29–31]; substantial changes have taken place in both households' land-use behaviors and management measures ever since the implementation of the household contract responsibility system (HRS) in China in the early 1980s. Small households, with diverse characteristics are the main land users and the ultimate decision-makers on the ground. This provides a perfect setting to study how land-use behaviors influence the change of soil AVK content.

Specifically, we try to answer two research questions: first, what changes have taken place in the content of AVK in cropland soil in the urban peripheral region, where land-use changes most dramatically? Second, what is the relationship between these changes and the households' land-use behaviors? In this research, first of all, we embark on a multidisciplinary approach, by integrating theories and methods from soil sciences, land-use change science, and household economics of social sciences. Secondly, by revealing the temporal and spatial variation of soil AVK content, we hope to provide technical guidance for the sustainable utilization of regional cropland by preserving the soil quality of cropland, and at the same time, provide scientific bases for policymakers in designing policy measurements in regulating households' land-use behaviors to maintain and enhance cropland quality and health, while improving agricultural productivity in intensive agricultural areas.

2. Conceptual Framework

Before the analytical study, here, we construct a conceptual framework to illustrate the causal relationship between land-use behaviors and soil fertility change. This framework shows how households' decisions and behaviors affect the AVK content in soil from temporal and spatial perspectives. Households are the ultimate decision-makers in the family-based farming system in China; they are both producers and consumers; they make decisions on land use, capital, and labor inputs in agricultural production to partially fulfil demand from consumers. According to household behavior theory in neoclassical economics, the goal of household production is to maximize utility [32,33]. However, the utility function of households may vary in different stages of economic development, and so does the goal of agricultural farming of households. The level of development has been constantly changing. During this process, households generate and measure their own needs, namely food demand and monetary demand. The former can meet the household consumption of food, and the latter can meet the family's money expenditure needs. Under the strong constraints of household consumption, households have a priority to meet the needs of household food demand, and the decision-making basis is the satisfaction of household food consumption capacity. After household consumption is met, farmer households then focus on attaining the maximum economic value from the land. The process would eventually match the value and function of the cropland, according to the needs of the household.

However, the difference in households' land-use targets in different periods results in the temporal and spatial variability of households' land-use behaviors. Generally speaking, since the reform and opening up in the late 1970s, the main demands of households have shifted from food self-consumption, to pursuing monetary returns from the land to meet the basic money expenditure needs on diversifying food consumption, to the purchase of basic needs like clothing, and then to profit maximization demand, based on the market-oriented production system. The land use target of households is consistent with the realization of the goals of their economic activities. Households' land-use behaviors vary in different stages, accordingly. Specifically, it passed from the stage of maximizing grain production, to the stage of grain production and profit optimization, and then to profit maximization. The differentiation of households has also presented the stage of farm-oriented households, to part-time households, and then to the off-farm oriented households, in different spatial extents at the same time. The change of land-use pattern, degree, and input intensity would inevitably lead to different evolutionary laws of soil AVK in time and space (see Figure 1).

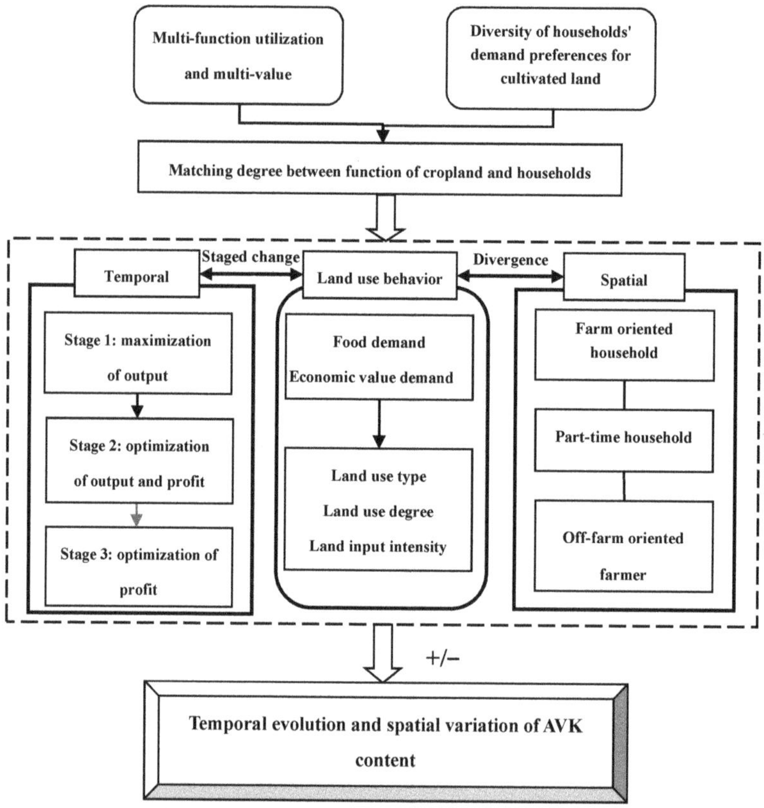

Figure 1. The theoretical analysis framework of the effects of households' land-use behaviors on soil available potassium (AVK).

3. Materials and Methods

3.1. Study Areas

Sujiatun District, our case study area, is located in the south of Shenyang City, 15 km away from the center of Shenyang, which is the only mega-city in the three northeastern provinces of China. The total area is 782 km^2, located at east longitude ranging from 123°09′ to 123°47′ and north latitude ranging from 41°27′ to 41°43′. The study area belongs to a continental semi-humid monsoon climate of the warm temperate zone with four distinct seasons, abundant sunshine, and concentrated rainfall. The annual average temperature is about 8 degrees Celsius, The average frost-free period is 150.5 days, the longest is 175 days (1975). The shortest is 128 days (1972, 1974). The earliest final frost date was April 20 (1967). The latest date is May 18 (1960). The annual average precipitation is 659.6 mm, the highest year is 1055.3 mm (1953), and the lowest is 445 mm (1965). The annual sunshine hours are 2527 h on average. Annual average evaporation 1430.3 mm. Sujiatun is a national commodity grain base and demonstration area of grain self-sufficiency. It is also a demonstration area of agricultural standardization production in Liaoning Province, a base of high-quality rice in Shenyang City, and a suburban agricultural demonstration area [2]. In this study, we select Linhu Street (near suburb), Yongle Township (outside suburb), and Wanggangbao Township (middle area) as sample areas, which are located in the western plain area (see Figure 2). The soil type in this area is cultivated loamy meadow soil, and the land use type is dry land. Under the condition of relatively homogeneous natural conditions, the change of AVK content in soil is mainly driven by human factors, which determine the potassium balance of the agricultural production system. Thus this region makes a suitable study area for our study.

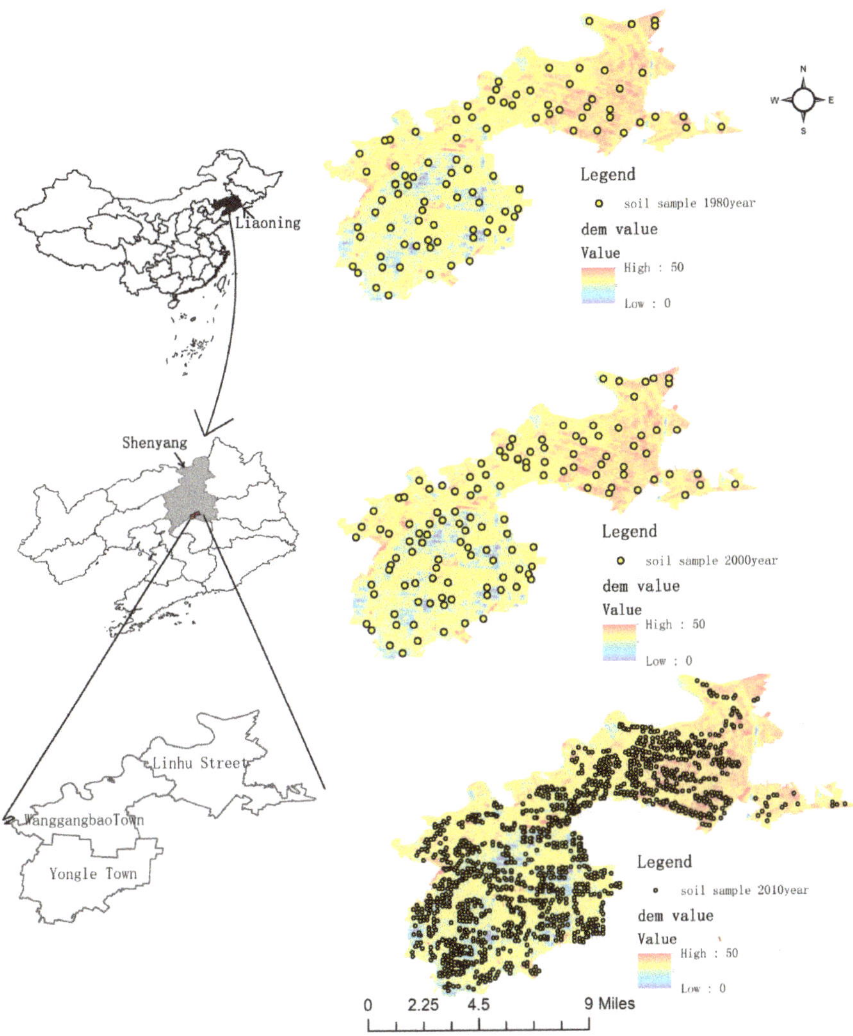

Figure 2. Study area map and spatial distribution of soil sampling points in 1980, 2000, and 2010.

3.2. Data Collection and Processing

3.2.1. Soil Observations

To ensure the continuity and comparability of soil sample collection, the number and location of sampling points were determined according to the difference of plot size, planting system, crop species, and yield level of cropland. The sample points used in this study include soil observation data from three periods, 1980, 2000, and 2010. The number of sampling points has increased over time, with 119 sampling points collected in 1980, 141 sampling points collected in 2000, and 1437 sampling points collected in 2010. The initial soil data and sample distribution map in the year 1980 comprehensively used the data and soil map of the second national soil survey in Sujiatun District. The soil map at a scale of 1:50,000 was used to facilitate the collection of 119 sampling points—evenly distributed across the study area—between May and June in 1980. Based on the main section of the second national soil survey, 141 soil observations were collected from May to June in 2000. In 2010, as soil tests had become more widespread in the region over time, the number of sampling points in the cropland concentrated area increased to 1437 according to the geographical coordinates of the soil sample points in 2000. Approximately 15~20 sampling points of the 0–20 cm soil layer of cropland were taken in a chessboard spatial sampling pattern. GPS was used to obtain the geographic coordinates of soil sample points in 2000 and 2010. As the 2000 sampling points include the 1980 sampling points, the GPS locations of the 1980 sampling points can be determined. Removing the plant roots and debris, stones, insect corpses, and other debris from the soil samples is a standard practice in preparing soil samples for analysis using a soil sieve. Subsequently, soil samples were air-dried and passed through a 0.15–1.0 mm sieve to remove plant roots and debris, stones, insect corpses, and other debris. The chemical analysis methods of soil AVK in three periods were the same, which is the ammonium acetate leaching-flame photometric method [6,7].

3.2.2. Socioeconomic Data

The socioeconomic data used in this study mainly comes from two sources, one is the statistical data provided by the local statistical bureau and the other is the survey data of the households in the study area.

The socioeconomic statistics are mainly derived from Shenyang Statistical Yearbook (1995–2010), Shenyang City Economic Statistics Yearbook (1985–1991), Compilation of National Economic Data of Sujiatun District (1984–1997) and Compilation of Statistical Materials of Sujiatun District (1998–2010). The household-level data of this study comes from the sample survey of the households in Sujiatun District from March to June 2010. To support this research, the soil sampling data and the households' survey data were designed to be matched with each other: while taking soil samples in households' land, the households were interviewed to record their land use, planting systems, fertilization status, water conservancy facilities, irrigation water sources, irrigation systems, and average yields. The sampling approach is based on the average distribution, according to the principle of representativeness and variability. The clustering, stratification, and random sampling methods were carried out to ensure the reliability of data collection. Overall 240 households were interviewed; excluding non-representative invalid questionnaires, 238 valid questionnaires were obtained: 79, 78, and 81 in Linhu Street, Wanggangbao Township, and Yongle Township, accounting for 33.2%, 32.8%, and 34% of the total sample size, respectively.

3.3. *Methodology*

We first used geostatistical analysis, the Kriging interpolation method, to generate the soil AVK surface maps, based on the soil survey data at sampling points. Then, we further related the AVK content value to human activities, namely, the households' land-use behaviors, by a linear regression model.

3.3.1. Geostatistical Analysis Methodology

We firstly conducted spatial superposition analysis using the semi-variance function [34–37] and the Kriging interpolation on the soil AVK in 1980, 2000, and 2010 [38–40]. The process is described as follows: The coordinate transformation had to be carried out, because the reference ellipsoids used in the two spatial coordinate systems are different. The land and topographic survey was based on the Beijing 54 Coordinate System in China, while the GPS survey data was based on the WGS84 geocentric coordinate system. In China, the 1:50,000 soil map is projected to the two-dimensional plane rectangular coordinate system, according to the three-degree band Gauss-Kruger, therefore the GPS survey data needs to be reprojected to match with the GIS data. In this study, we used MapGIS®

software to read in GPS data, complete precise projection conversion and data format conversion, and then convert GPS data in the WGS84 coordinate system to Beijing 54 Coordinate system. Finally, data in ESRI shapefile format was exported, which can be read by ESRI ArcView®. Thus, we obtained the soil sampling point bitmaps of 1980, 2000, and 2010 at the plot scale. Then, we calculate the theoretical model of the semivariance function, draw the graph, and count the area of each level by the ESRI ArcGIS® v10.

3.3.2. Econometric Model Construction

Based on the theoretical analysis framework presented earlier, we constructed the following econometric analysis model:

$$LUB = f(GCC, MCI, LII) \quad (1)$$

The model indicates that households' land-use behaviors (LUB), including land-use patterns, land use levels, and land input intensity, can be quantified by three quantifiable dependent variables. GCC indicates whether households plant economic crops, indicating the difference in land-use patterns; MCI means the land multiple cropping index, indicating the difference in land use level; while LII indicates the amount of land capital investment per unit area of household showing the difference in household's land input intensity.

$$AVK = g(LUB) \quad (2)$$

Model (1) is brought into Model (2) to obtain Model (3) to model the effects of households' land-use behaviors on the change of AVK content in cropland.

$$AVK = g(LUB) = g[f(GCC, MCI.LII)] = h\left(GCC^{+/-}, MCI^{+/-}, LII^{+/-}\right) \quad (3)$$

The model represents the theoretical model of the relationship between households' land-use behavior and AVK content.

In addition, it should be noted that the above model is only a general model of the interaction mechanism between the land-use behavior of households and the change of AVK in cropland. In specific applications, we need to choose a specific model form according to the specific characteristics of each study area, as well as the availability of data.

4. Results

4.1. Temporal and Spatial Evolution Characteristics of AVK

4.1.1. Temporal Evolution Characteristics of Soil AVK

Using the semi-variance model parameters fitted in GS+win9® (a software platform developed by Tetoc scientific instrument in China). The optimal models are exponential, spherical, and exponential in 1980, 2000, and 2010, respectively. The coefficients of determination R^2 are 0.878, 0.775, and 0.838, respectively. It indicates that the interpolation models have good accuracy, and the interpolated values are consistent with the spatial distribution characteristics of soil AVK. Based on the result of the semi-variance model, the Ordinary Kriging interpolation method was chosen in Geostatistical module of ArcGIS® 9.3 to interpolate the soil AVK levels [18] in the three periods, respectively (Table 1). The root-mean-square errors (RMSE) are 2.176 in 1980, 3.173 in 2000, and 2.749 in 2010, respectively. This meets the standard requirements for the accuracy of spatial interpolation. As a result, the soil AVK grade area tables and spatial interpolation maps of AVK1980, AVK2000, and AVK2010 were obtained (Figure 3).

Table 1. Statistics on the area of AVK in various levels from 1980 to 2010.

Region	Level	Content	In 1980		In 2000		In 2010	
			Area (ha)	Ratio (%)	Area (ha)	Ratio (%)	Area (ha)	Ratio (%)
Linhu Street	I	≥180	183	1.6	——	——	——	——
	II	160–180	1923	17.1	——	——	——	——
	III	140–160	684	6.1	——	——	——	——
	IV	120–140	147	1.3	34	0.3	559	5.0
	V	100–120	164	1.5	1037	9.2	2314	20.6
	VI	<100	66	0.6	2097	18.7	295	2.6
Wanggangbao Township	I	≥180	343	3.1	——	——	——	——
	II	160–180	1976	17.6	——	——	35	0.3
	III	140–160	1146	10.2	——	——	778	6.9
	IV	120–140	117	1.0	123	1.1	2135	19.0
	V	100–120	——	——	1673	14.9	634	5.7
	VI	<100	——	——	1785	15.9	——	——
Yongle Township	I	≥180	——	——	——	——	——	——
	II	160–180	——	——	——	——	492	4.4
	III	140–160	1930	17.2	——	——	2879	25.7
	IV	120–140	2542	22.7	699	6.2	976	8.7
	V	100–120	——	——	2619	23.3	124	1.1
	VI	<100	——	——	1154	10.3	——	——
Total	I	≥180	526	4.7	——	——	——	——
	II	160–180	3899	34.7	——	——	522	4.7
	III	140–160	3760	33.5	——	——	3663	32.6
	IV	120–140	2806	25.0	856	7.6	3670	32.7
	V	100–120	164	1.5	5329	47.5	3070	27.4
	VI	<100	66	0.6	5036	44.9	295	2.6

AVK levels were grouped according to national soil survey levels—Level 1: > 180 mg/kg, Level 2: 160–180 mg/kg, Level 3: 140–160 mg/kg, Level 4: 120–140 mg/kg, Level 5: 100–120 mg/kg, and Level 6: < 100 mg/kg. In general, the soil AVK positively correlate soil quality and crop productivity. Thus, soil in Level 1 is the best, while Level 5 is the worst in terms of soil quality. From the temporal evolution, it can be seen that the average content of AVK in the soil in 1980 was 149.58 mg/kg, which was mainly distributed in Level II (160–180 mg/kg), III (140–160 mg/kg), and IV (120–140 mg/kg). Among them, the area of cropland in Level II was the largest (3899 ha), accounting for 34.7%, mainly distributed in Linhu Street and Wanggangbao Township. Followed by Level III, the cropland area was 3760 ha, accounting for 33.5%, mainly distributed in the north of Yongle Township and the south of Wanggangbao Township. Soil AVK was high on the whole, and specifically, the content in the north was higher than that in the south. By 2000, the average content of AVK in soil was 103.62 mg/kg, which was mainly distributed in Level IV (120–140 mg/kg), V (100–120 mg/kg), and VI (less than 100 mg/kg). Among them, the area of cropland in Level V was the largest, accounting for 47.5% of the total area of cropland in the study area, which mainly located in the middle south of Wanggangbao Township and middle west in Yongle Township. Then followed by Level VI, the area is 5036 ha, accounting for 47.5%, which mainly distributed in Linhu Street and Wanggangbao Township. In short, from 1980 to 2000, the content of AVK in soil decreased significantly. By 2010, the average content of AVK in soil was 134.29 mg/kg, mainly distributed in Level III, IV, and V, and the area was 3657 ha, 3670 ha, and 3072 ha, respectively. During the 30 years from 1980 to 2010, the average content of AVK in soil decreased, but the trends vary spatially among places. AVK decreased in the first 20 years and increased in the next 10 years, but the ranges of change were different. The average annual decrease was 0.51 mg/kg. In the first two decades (1980–2000), the average content of AVK decreased by 46.04 mg/kg, with an average annual decrease of 2.30 mg/kg, while the average annual increase in the next 10 years (2000–2010) was 1.5 mg/kg.

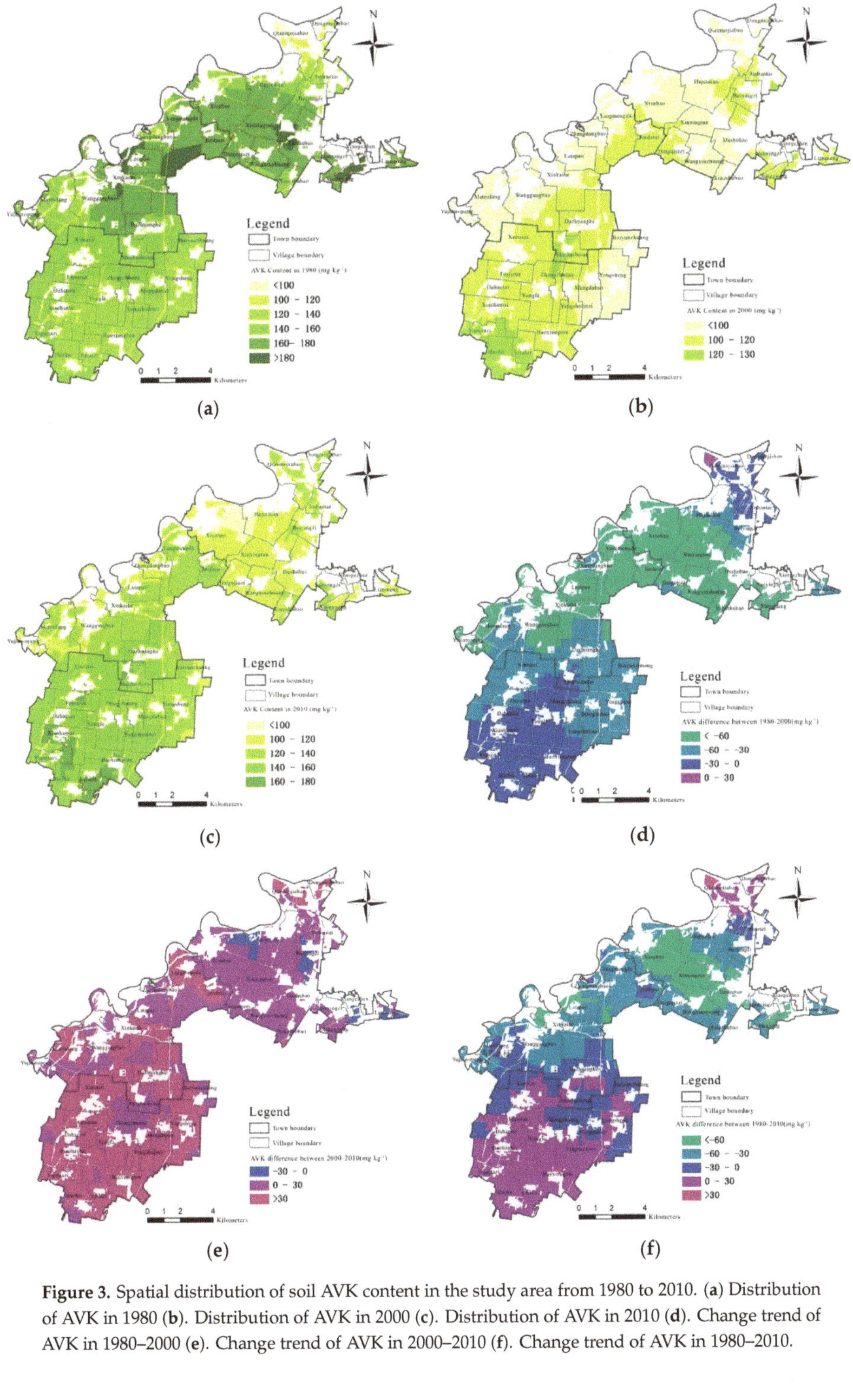

Figure 3. Spatial distribution of soil AVK content in the study area from 1980 to 2010. (**a**) Distribution of AVK in 1980 (**b**). Distribution of AVK in 2000 (**c**). Distribution of AVK in 2010 (**d**). Change trend of AVK in 1980–2000 (**e**). Change trend of AVK in 2000–2010 (**f**). Change trend of AVK in 1980–2010.

4.1.2. Spatial Distribution Characteristics of Soil AVK

As shown in Table 2 and Figure 3, the changes of AVK varied over time and space from 1980 to 2010. The magnitude of the AVK increase was between 0 and 30 mg/kg. Overall, an area of 3512 ha, or 31.30% of the total area of the study area, showed an increase of AVK over that study period. It was mainly distributed in the south of Yongle Township. The range of the AVK decline was between −60 to −30 mg/kg to 0 mg/kg, accounting for 53.97% of the total area of the study area. It was mainly distributed in Linhu Street and Wanggangbao Township. The results showed that the content of AVK in the soil near the peripheral region of the city showed a downward trend, and the extent of the decline was gradually increasing, while it showed a slightly increasing trend far away from the city. From 1980 to 2000, the soil AVK content showed a downward trend. The largest decline was more than 60 mg/kg, covering an area of 4508 ha, accounting for 40.17% of the total area of the study area. It mainly distributed in Linhu Stree and Wanggangbao Township. While the declining trend was more moderate in Yongle Township, and the range was 0~30mg/kg, indicating a trend that the nearer the city is, the larger the decline is. From 2000 to 2010, the soil AVK in the whole study area showed an upward trend. The largest increase was more than 30 mg/kg, covering an area of 5788 ha, accounting for 51.59% of the total area of the study area, mainly concentrated in the southern part of Wanggangbao Township and Yongle Township. That is, the farther away from the urban area is, the greater the increase is.

Table 2. Distribution of AVK content in cropland soil.

Period	Area and Proportion	Range of Content Various (mg/kg)				
		<−60	−60–−30	−30–0	0–30	>30
2010–1980	Area (ha)	1595	3334	2721	3512	57
	Proportion (%)	14.22%	29.72%	24.25%	31.30%	0.51%
2000–1980	Area (ha)	4508	3843	2826	43	-
	proportion (%)	40.17%	34.25%	25.19%	0.39%	-
2010–2000	Area (ha)	-	-	340	5092	5788
	proportion (%)	-	-	3.03%	45.38%	51.59%

4.2. Households' Land-Use Behaviors Change over Time and Its Effect on Soil AVK

4.2.1. Households' Land-Use Behaviors Change over Time

Based on the above theoretical analysis and socioeconomic and environmental changes in the study area, this paper mainly selects some specific indicators to represent land-use behaviors. The areas of grain crops and cash crops are used to measure land-use type, the multiple cropping index is regarded as land use degree (or land use intensity), while chemical fertilizer input is used as land input intensity.

Land-Use Type Change over Time

Figure 4 displays the change in planting area of grain crops (corn and rice) and cash crops (vegetables and melons) from 1980 to 2010 in the Sujiatun District. It indicates that grain crops planting area has declined overall, from 32,457 ha in 1980 down to 25,812 ha in 2010. The total decrease amounts to 6645 ha in 30 years, with an average annual reduction of 221.5 ha. In particular, the reduction in grain crop area has become more pronounced since 2000. By 2003, the area of grain crops was essentially equal to those of cash crops.

The change in the vegetable planting area has gone through three stages: slow growth, rapid growth, and stabilization (Figure 5). In the first 10 years, the vegetable planting area increased slowly from 2271 ha in 1983 to 3271 ha in 1990, with an average annual growth of 100 ha. In the second 10 years, from 1991 to 1999, the vegetable planting area showed a rapid growth trend, from 4030 ha in 1992 to 8230 ha in 2000, with an average annual growth of 420 ha. In the last 10 years, from 2001 to 2010, the area remained at a high level—about 8000 ha overall.

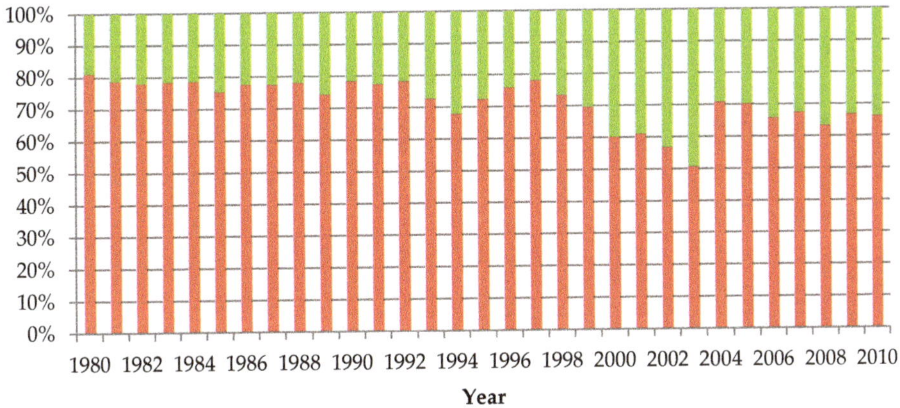

Figure 4. Changes in planting area of grain crops and cash crops in the study area from 1980 to 2010.

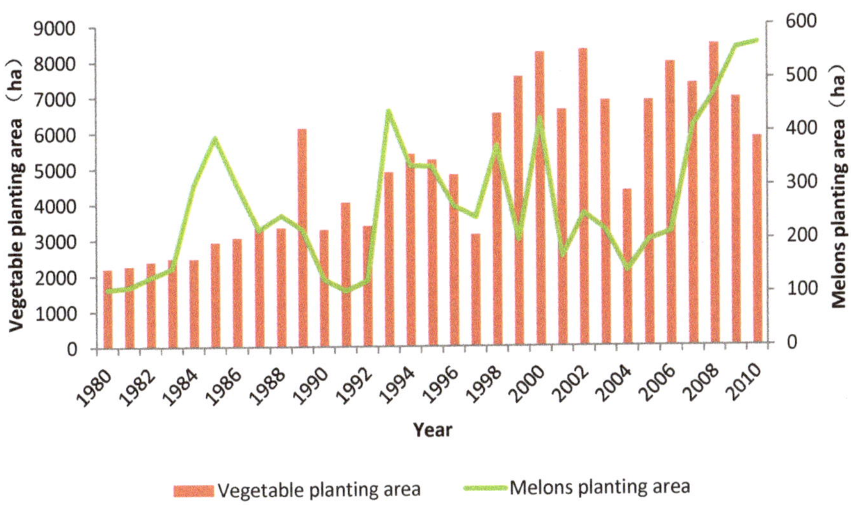

Figure 5. Change of economic crop planting area from 1980 to 2010.

The change in the planting area of melon and fruit mainly experienced two stages. In the first two decades, the area of melon and fruit cultivation stayed at a low level with an average area of 257 ha. However, during the last 10 years, from 2001 to 2010, the area of the melon and fruit area began to rise rapidly to 565 ha in 2010, which increased nearly four times in ten years.

Land Use Degree Change over Time

As an indicator to measure land-use degree, the multiple cropping index can reflect the final impact of population pressure on the cropping system under certain natural resource conditions. The multiple cropping index of households in Sujiatun District from 1980 to 2010 shows that households' land-use degree is gradually increasing, which can be roughly divided into three stages (shown in Figure 6). In the first stage from 1980 to 1990, the multiple cropping index increased from 100.23% to 102.96%, with a peak of 105.04% in 1986 and a trough of 100.16% in 1989. In the second stage between 1991 and 2000, the change of multiple cropping index is in a relatively stable stage with an annual growth rate of

0.026%. However, the multiple cropping index first rose sharply and then grew steadily in the third stage from 2001 to 2010. The multiple cropping index rose from 103.42% in 2000 to 111.88% in 2010, with a trough of 101.27% in 2003 and a peak of 112.37% in 2006.

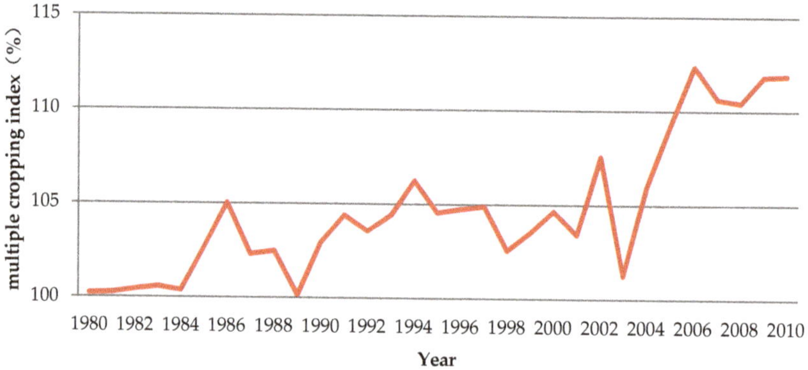

Figure 6. Change of the multiple cropping index of Sujiatun District from 1980 to 2010.

Land Input Intensity Change over Time

It can be seen from Figure 7 that the application intensity of potash fertilizer and compound fertilizer increased from 1980 to 2010, increasing from 0 kg/ha to 96 kg/ha and 20 kg/ha to 257 kg/ha, respectively. The temporal trends of these two fertilizers differed significantly. For the amount of potash fertilizer applied, the changes can be divided into three phases: slow rise (1980–1995), rapid rise (1996–2005), and slow decline (2006–2010). In contrast, the compound fertilizer side, the application amount experienced a slow increase stage (1980–2000), and a significant increase stage (2000–2010). The average annual increase was 3.9 kg/ha in the first stage, while the application amount was 15.9 kg/ha in the second stage, nearly four times as much as the former.

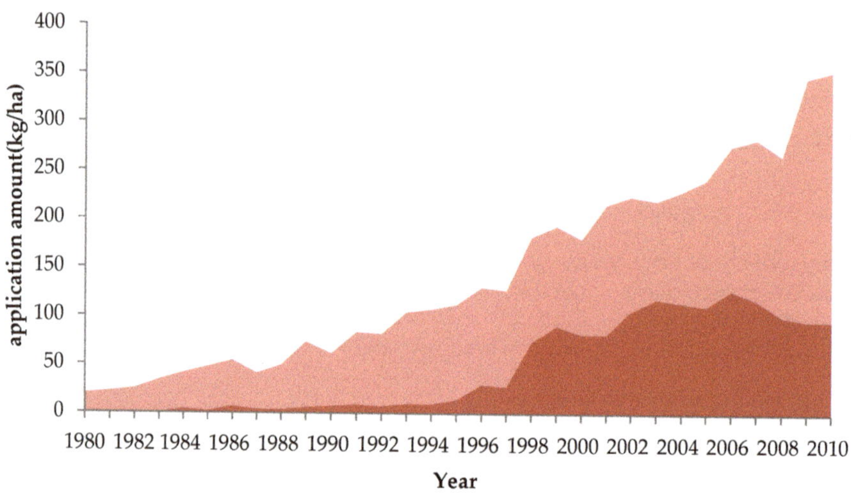

Figure 7. Change of application of potash fertilizer and compound fertilizer from 1980 to 2010.

4.2.2. Impact of Households' Land-Use Behaviors on Temporal Evolution on AVK

In the first stage (1980–2000), the soil AVK content in the whole study area decreased, and the average content decreased from 149.56 mg/kg in 1980 to 103.52 mg/kg in 2000. This time corresponds to the early stage of reform and opening up in China, and GDP per capita has gradually increased from a low level at RMB 1013 (USD 680) in 1980 (shown in Figure 8), and then grew to RMB 10,125 (USD 1212) in 1995 and RMB 15,666 (USD 1982) in 2000[1]. Under such a socioeconomic background, households' land-use behaviors are mainly characterized by the cultivation of corn and rice, and the area of cash crops remained low, but started to slowly increase. As a result of the cultivation of corn and rice, the multiple cropping index also grew slowly. In terms of agricultural inputs, the input of compound fertilizer and potash fertilizer is at a low level, and this leads to a significant decline in soil AVK content in this area.

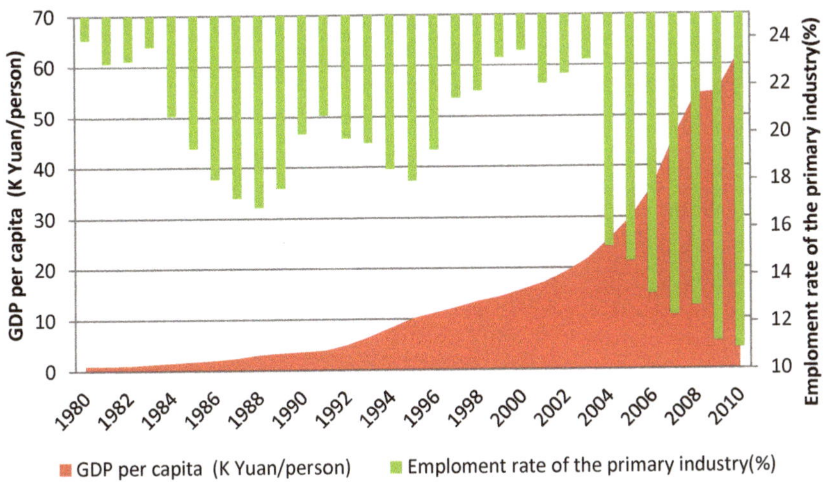

Figure 8. Trends of GDP per capita and primary industry employment ratio from 1980 to 2010.

In the second stage (2000–2010), the soil AVK content in the whole study area tended to increase, and the average content increased from 103.52 mg/kg in 2000 to 134.27 mg/kg in 2010. At this stage, GDP per capita increased from RMB 15,666 (USD 1892) in 2000 to RMB 62,357 (USD 9211) in 2010—nearly an four-fold increase. The labor force engaged in agriculture also began to shift and differentiate. The proportion of employment in agriculture dropped from 23.5% in 2000 to 10.9% in 2010. Under the influence of the external socioeconomic environment, the land use behavior of households had also undergone major changes. The proportion of cash crops increased rapidly, and so did multiple cropping index and fertilizer application; all of them reached a high level in 2010. Negative potassium balance can happen when potassium removal from agricultural cultivation is greater than the application of potassium fertilizer. The deficiency of potassium fertilizer has become the main constrain for the increase in crop yield. Realizing the problem, farm households have increased the amount of potassium fertilizer since 1998. While after 2005, local soil and fertilizer stations began to promote soil testing and customized fertilization technology, so that households gradually learned that an optimal ratio of nitrogen, phosphorus, and potassium in the soil could increase the yield more effectively. As a result of these changes in the land-use behaviors of the households, the AVK

1 USD/RMB exchange rate was USD 1 = RMB1.49 in 1980; USD 1 = RMB 8.351 in 1995; USD 1 = RMB 8.2784 in 2000; USD 1 = RMB 6.7695 in 2010.

content increased significantly and households began to practice intensive farming to obtain the maximal benefit.

4.3. Temperal Evolution of Households' Land-Use Behaviors and Its Effect on Soil AVK

4.3.1. Spatial Variation of Households' Land-Use Behaviors

Spatial Variations of Land Use Type

Through the investigation on the three study areas, it was found that there were obvious spatial differences in households' crop selection behaviors, and the crop selection options were divided into three types of households including grain crops only, both grain crops and cash crops, and cash crops only. Moreover, as the distance from the city center increases, the grain crop area gradually decreases, while the cash crops area gradually increase. As shown in Table 3, Linhu Street, the closest to Shenyang City, 93.7% of the respondents in this area only planted grain crops, while the proportion in Yongle Township, the farthest from the city center, is just 2.5%. Nearly 45% of the respondents in Wanggangbao Township selected both grain crops and cash crops, where is in the middle of the three study area.

Table 3. Land-use behavior of households in the study area.

Land-Use Behavior		Classification	Linhu Street		Wanggangbao Township		Yongle Township		Total	
			Obs	Proportion %	Obs	Proportion %	Obs	Proportion %	Obs	Proportion %
GCC		Grain crops only	74	93.7	23	29.5	2	2.5	115	48.3
		Both grain crops and cash crops	1	1.3	35	44.9	31	38.3	50	21.0
		Cash crops only	4	5.10	20	25.6	48	59.2	73	30.7
MCI		1.0	74	93.7	39	50.0	2	2.5	115	48.3
		1.0–1.5	3	3.8	26	33.3	22	27.2	51	21.4
		1.5–2.0	0	0	5	6.4	32	39.5	37	15.6
		2.0–2.5	0	0	3	3.9	12	14.8	15	6.3
		2.5–3.0	2	2.5	5	6.4	13	16.0	20	8.4
LII	self-employed (day per mu)	0–10	56	70.9	29	37.2	19	23.5	104	43.7
		10–20	19	24.1	40	51.3	26	32.1	85	35.7
		20–30	2	2.5	5	6.4	11	13.6	18	7.6
		>30	2	2.5	4	5.1	25	30.8	31	13.0
	employment (day per mu)	0	47	59.9	63	80.8	9	11.1	119	50.0
		1–10	30	38.0	6	7.7	21	25.9	57	23.9
		11–20	2	2.5	5	6.4	23	28.4	30	12.6
		>20	0	0.0	4	5.1	28	34.6	32	13.5
	capital investment (Yuan per mu) [a]	¥0–500 ($0–74)	72	91.1	38	48.7	9	11.1	119	50.0
		¥500–1000 ($74–128)	6	7.6	17	21.8	7	8.6	30	12.6
		¥1000–1500 ($128–222)	1	1.3	16	20.5	9	11.1	26	10.9
		>¥1500(>$222)	0	0	7	9.0	56	69.1	63	26.5

Note: [a] The value in brackets is US dollars. USD/RMB exchange rate was USD 1 = RMB 6.7695 in 2010.

Spatial Evolution of Land Use Degree

Regarding the multiple cropping index (MCI), we divided the respondents into five groups between 1.0 and 3.0, with 0.5 as intervals. The results show that the farther from the city center is, the higher of MCI is. The MCI values of 93.7% of the respondents in Linhu Street were 1.0—as they mainly grew corn once per year, while half of the respondents in Wanggangbao Township have MCI values above 1.0. However, the proportion of respondents, whose MCI is above 1.0 in Yongle Township, is about 97.5%; specifically, 39.5% of them have MCI between 1.5 and 2.0, and more than 30% of them have MCI over 2.0.

Spatial Evolution of Land Input Intensity

Land investment mainly includes labor input and capital investment, and the labor force is divided into self-employed labor and employment. Therefore, this study is mainly focused on the differences in household's self-employment input, employee input, and household's capital investment. The spatial evolution between self-employment and employment is similar, that is, as the distance from the city increases, the number of self-employed and employment inputs increased gradually. In terms of the capital investment intensity, the proportion of households in Linhu Street decreased with the increase of land input intensity, which are 91.1%, 7.6%, 1.3%, and 0.0%, respectively. In the opposite direction of Yongle Township, the proportion increased by 11.1%, 8.6%, 11.1%, and 69.2%, respectively, meaning the land input intensity was higher. Wanggangbao Township in the outer peripheral region is located between the first two regions, and the change range is relatively small.

These results indicate that with the accelerating process of industrialization and urbanization, households' land-use targets and land use types in the study area have significantly diverged. Specifically, the households in the suburban Linhu Street mainly engage in off-farm work or business, and agriculture becomes less important in livelihood. Farmers plant grain crops such as corn, which are time-saving and labor-saving, with low land-use and labor intensity. While the households who are in the outer suburb of Yongle Township mainly involved in agriculture. They planted vegetables and greenhouse vegetables, meaning the land use degree and land input intensity are the highest. The households in Wanggangbao Township located in the middle place, are mainly part-time households, and they plant both corn and vegetables. As a result, the land use degree and land input intensity are also in the middle of Linhu Street and Yongle Township. The differences in land-use behaviors of various types of households in different regions are bound to have different effects on the changes of AVK in cropland soil.

4.3.2. Empirical Results of the Impact of Households' Land-Use Behavior in Spatial Evolution on AVK

To quantitively analyze the impact of households' land-use behaviors on the AVK content, multiple linear regression models were used by testing and comparing different model estimation forms. As shown in Table 4, land use types, land use degree and input intensity have different effects on the change of AVK content in cropland soil. Specifically, households' crop selection behavior has a significant positive impact on the soil AVK content in Wanggangbao Township, with a significant level of 1%. It indicates that when other conditions remain unchanged, the switch from grain to cash crops will lead to an increase of soil AVK by an average of 61.205 mg/kg. The multiple cropping index had a significant negative impact on AVK in Linhu Street. While holding other variables constant, soil AVK in the study area decreases by an average of 15.469 mg/kg for each additional unit of the multiple cropping index. While land input intensity has a significant positive impact on the soil AVKcontent in Wanggangbao Township and Yongle Township, indicating that land input intensity increases by 1 yuan per mu unit, the average AVK content increased by 0.009 mg/kg and 0.037 mg/kg, respectively. That is, the impact on Yongle Township was greater than that on Wanggangbao Township.

Table 4. Estimation results.

	Linhu Street			Wanggangbao Township			Yongle Township		
	B	*t*-Value	Beta	B	*t*-Value	Beta	B	*t*-Value	Beta
GCC				61.205 ***	2.702	0.399			
MCI	−15.469 *	−1.817	−0.282						
LII				0.009 *	1.509	0.073	0.037 *	1.884	0.225

Note: *, **, *** denote statistical significance at 10%, 5%, and 1%. The factor corresponding to the parameter is blank, which means that the model has not reached a significant level. B value is the coefficient of the regression equation. The positive value of the coefficient indicates that the explained variable increases correspondingly when the explanatory variable increases by one unit value, while the negative value indicates that the explained variable decreases correspondingly when the unit value is increased. Beta value is expressed as the relative weight of each explanatory variable in the model. The larger the absolute value, the greater the effect of the factor.

Due to the different effects of land-use behaviors of households in various regions, the soil AVK appears obvious spatial differences. The main reason is that households in the suburban Linhu street are affected by urban expansion and with more off-farm employment opportunities. The households there mainly plant corn and other field crops, and land use degree and land input intensity are both low, due to the relatively low efficiency of agricultural production. While the households in the remote suburb of Yongle Township, they mainly plant land vegetables and greenhouse vegetables to meet the need of urban residents, with the highest level of land use degree and land input intensity. Households in Wanggangbao Township in the middle of the study area can choose both grain crops and cash crops, and they mainly grow corn and terrestrial vegetables, with higher land-use degrees and greater land input intensity. The resulting land-use intensity forms a ring structure surrounding the central city and resembles a distribution pattern of "anti Thunen circle" in space, which is a special manifestation of classical agricultural location theory.

5. Discussion

Compared with previous studies [5,19,41], this study reveals the temporal and spatial evolution characteristics of soil AVK in the marginal zone of large cities based on continuous, high-density soil sampling data, and discusses the underlying reasons for this evolutionary feature from the micro perspective of households, according to long-term sequence statistics and households' survey data. It answers what kind of change takes place in the AVK content in time and space at the plot scale, and its relationship with households' land-use behavior in the peripheral region of the big cities. It is of great significance to explore the regular pattern of soil fertility change in cropland. On one hand, this study establishes links between soil science and economics, and reveals the influence mechanism of households' land-use behaviors on soil AVK content from the temporal and spatial dimensions, which may inspire future multi-disciplinary researches. On the other hand, this study has important policy implications. Generally, the fringe area of large cities is the most sensitive area for developing urban modern agriculture. The understanding of how households' land-use behaviors affect the AVK content of the soil helps the government and policymakers to propose corresponding systems and policies to regulate households' agricultural production activities and promote the sustainability of soil production capacity and environment. Finally, it realizes the dynamic analysis using multivariate data, including the soil survey data, households' survey data, and socio-economic statistics.

Although we have obtained important and interesting research results, there is no denying that this study still has the following limitations: firstly, regarding the data, we only obtained three phases data of soil AVK content, therefore, it is impossible to analyze the impact of households' land-use behavior on soil AVK content in the time dimension, by constructing an econometric time series analysis—which still lacks in this research, as we only have a households survey at a single point in time. Secondly, biophysical factors in this study, given the relatively short-time period, are assumed to be constant, despite the fact that biophysical factors and human factors are the two most important factors for the quality of soil quality change. The biophysical factors, such as topography, climate, parent material, and organisms, usually change slowly compared to human factors, so the effects need to be analyzed on a medium and long term basis. Additionally, we selected the area that is least affected by biophysical factors and greatest influenced by human factors as the study area. Nevertheless, future researches should examine how biophysical factors, especially climate change affect long-term soil AVK content.

With the increasing influences of human activities on land quality, social, economic, and human factors have become increasingly prominent in affecting the quality of cropland. Therefore, multidisciplinary and multi-dimensional pattern detection and analysis are important to investigate the temporal and spatial dynamics of cropland quality, such as the connecting spatial information data with socialeconomic data of microeconomic subjects. The multi-model coupling process and mechanism analysis are therefore promising and important fields for future researches in this regard, for example, the simulation of households' land-use behaviors affecting the quality of cultivated land under different incentive policies and measurements. At the same time, the soil quality variation of croplands is a dynamic and complex process. Further researches are thus needed on the optimization of spatial soil sampling to facilitate the continuous soil quality monitoring. In addition, panel data on households' land-use behaviors, which is lacking in this research, would help greatly in deepening the research.

6. Conclusions

This paper constructs a theoretical analysis framework for the impact of households' land-use behaviors on soil AVK content from the spatial and temporal dimensions. The empirical research was conducted based on the matching data of high-density soil sampling data, long-term sequence statistics, and cross-sectional household survey data in the Sujiatun area of Shenyang City, Liaoning Province, China, comprehensively using the approach of geostatistics, econometric methods, and GIS. The results of this study can be summarized as:

(1) Although the AVK content of the soil in the study area has a largely downward trend in the past 30 years, there are different trends in different stages. This variation can be attributed to to the gradual evolution of households' land-use behaviors. From 1980 to 2000, the average value of AVK decreased from 149.56 mg/kg to 103.52 mg/kg, due to the underdeveloped economy, limited investment capacity, technology level, and management level, as well as the plunder of cropland. After the year 2000, with the acceleration of the urbanization process and driven by economic interests, households gradually seek to maximize profits by increasing agricultural production. This was achieved by improving potassium use management: the use of potassium soil test, potassium nutrient budgeting, and the increasing use of potassium fertilizers. As a result, the average of AVK rose to 134.27 mg/kg in 2010 with an increased rate of 29.70% compared to the year 2000.

(2) The spatial variation of AVK is also substantial and intriguing. The closer to the urban area, the greater the decline of soil AVK content, while the farther away from the urban area, the greater the rise. This can be attributed to the differences in households' land-use behaviors in different areas. The households in the near peripheral region mainly engage in off-farm work and only cultivate time and labor-saving corns. This leads to a low land-use degree and low input intensity, while the households further away from the city center mainly rely on agriculture as the major income source and pursue profit maximization resulting in the highest land-use degree and land input intensity.

This research provides a glimpse of how the land-use behaviors of small farm households influence the soil AVK through an empirical study in Northeastern China. It reveals the complex spatial dynamics of soil AVK driven by socioeconomic development. The results also shed light on how policy measurements can be designed to steer farmer's behaviors and preserve soil for sustainable agricultural production.

Author Contributions: Conceptualization, H.L., Z.S. and X.L.; Formal Analysis, H.L. and X.D.; Funding Acquisition, X.L. and H.L.; Data Curation, H.L. and M.W.; Methodology, H.L. and X.D.; Writing-Review & Editing, H.L., Z.S. and X.L. All authors have read and agreed to the published version of the manuscript.

Funding: This research was funded by National Natural Science Foundation of China, under Grant: 71503174; 71503113; 71373127; Liaoning Province Social Science Planning Foundation, under Grant: L18BJY006; Foundation for Young Scientific and Innovative Talents in Shenyang City (RC170180); Jiangxi Province Social Science Planning Foundation, under Grant: 18GL08, and Research Project of Social Science and Economic Development in Liaoning Province, under Grant: 20201s1ktyb-077.

Conflicts of Interest: The authors declare no conflict of interest.

References

1. Jin, J.J.; He, R.; Wang, W.Y.; Gong, H. Valuing cultivated land protection: A contingent valuation and choice experiment study in China. *Land Use Policy* **2018**, *74*, 214–219. [CrossRef]
2. Jiang, G.; Zhang, R.; Ma, W.; Zhou, D.; Wang, X.; He, X. Cultivated land productivity potential improvement in land consolidation schemes in Shenyang, China: Assessment and policy implications. *Land Use Policy* **2017**, *68*, 80–88. [CrossRef]
3. Liu, J.; Guo, Q. A spatial panel statistical analysis on cultivated land conversion and Chinese economic growth. *Ecol. Indic.* **2015**, *51*, 20–24. [CrossRef]
4. Wu, Y.; Shan, L.; Guo, Z.; Peng, Y. Cultivated land protection policies in china facing 2030: Dynamic balance system versus basic farmland zoning. *Habitat Int.* **2017**, *69*, 126–138. [CrossRef]
5. Xia, M.; Zhao, B.; Hao, X. Soil quality in relation to agricultural production in the North China Plain. *Pedosphere* **2015**, *25*, 592–604. [CrossRef]
6. Evangelou, V.P.; Wang, J.; Phillips, R.E. New developments and perspectives on soil potassium quantity/intensity Relationships. *Acad. Press* **1994**, *52*, 173–227.
7. Zanati, M.R.; Guirguis, M.A.; Saber, M.S.M. Biological and chemical determination of available potassium in soil. Zentralblatt Für Bakteriologie, Parasitenkunde, Infektionskrankheiten Und Hygiene. Zweite Naturwissenschaftliche Abteilung: Allgemeine. *Landwirtsch. Techn. Mikrobiol.* **1973**, *128*, 572–577.
8. Bilias, F.; Barbayiannis, N. Potassium availability: An approach using thermodynamic parameters derived from quantity-intensity relationships. *Geoderma* **2019**, *338*, 355–364. [CrossRef]
9. Zhang, S.; Zhang, X.; Liu, X. Spatial distribution of soil available potassium in different slope profiles of typical black soil region. *Soil* **2014**, *46*, 218–224. (In Chinese)
10. John, V.S. *Precision Agriculture '19*, 1st ed.; Wageningen Academic Publishers, Wageningen University & Research: Wageningen, The Netherlands, 2019; pp. 1–1030.
11. Cammarano, D.; Holland, J.; Ronga, D. Spatial and temporal variability of spring barley yield and quality quantified by crop simulation model. *Agronomy* **2020**, *10*, 393. [CrossRef]
12. Milkevych, V.; Munkholm, L.J.; Chen, Y.; Tavs, N. Modelling approach for soil displacement in tillage using discrete element method. *Soil Tillage Res.* **2018**, *183*, 60–71. [CrossRef]
13. Lin, T.; Sun, C.; Li, X.; Zhao, Q.; Zhang, G.; Rubing, G.; Ye, H.; Huang, N.; Yin, K. Spatial pattern of urban functional landscapes along an urban–rural gradient: A case study in Xiamen City, China. *Int. J. Appl. Earth Obs. Geoinf.* **2016**, *46*, 22–30. [CrossRef]
14. Ho Leung, I.P.R.; Li, W.K. Matérn cross-covariance functions for bivariate spatio-temporal random fields. *Spat. Stat.* **2016**, *17*, 22–37.
15. Abdelrahman, M.A.E.; Natarajan, A.; Hegde, R.; Prakash, S.S. Assessment of land degradation using comprehensive geostatistical approach and remote sensing data in GIS-model builder. *Egypt. J. Remote Sens. Space Sci.* **2018**, *12*, 1–12. [CrossRef]
16. Hou, D.; O'Connor, D.; Nathanail, P.; Tian, L.; Ma, Y. Integrated GIS and multivariate statistical analysis for regional scale assessment of heavy metal soil contamination: A critical review. *Environ. Pollut.* **2017**, *231*, 1188–1200. [CrossRef] [PubMed]
17. Abdelrahman, M.A.E.; Natarajan, A.; Srinivasamurthy, C.A. Estimating soil fertility status in physically degraded land using GIS and Remote Sensing techniques in Chamarajanagar District, Karnataka, India. *Egypt. J. Remote Sens. Space Sci.* **2016**, *19*, 95–108. [CrossRef]
18. Wang, B.; Waters, C.; Orgill, S. Estimating soil organic carbon stocks using different modelling techniques in the semi-arid rangelands of Eastern Australia. *Ecol. Indic.* **2018**, *88*, 425–438. [CrossRef]
19. Behm, S.; Haupt, H.; Schmid, A. Spatial detrending revisited: Modelling local trend patterns in NO_2-concentration in Belgium and Germany. *Spat. Stat.* **2018**, *28*, 331–351. [CrossRef]
20. He, P.; Yang, L.; Xu, X. Temporal and spatial variation of soil available potassium in China (1990–2012). *Field Crops Res.* **2015**, *173*, 49–56. [CrossRef]
21. Zhang, L.; Shuang, W.; Yun, A. Spatial and temporal variability of available potassium in soils and its influencing factors in Quzhou County, Hebei Province during the past 30 years. *Chin. Agric. Sci.* **2014**, *47*, 923–933. (In Chinese)
22. Dai, W.; Zhao, K.; Fu, W.; Jiang, P.; Li, Y.; Zhang, C. Spatial variation of organic carbon density in topsoils of a typical subtropical forest, Southeastern China. *Catena* **2018**, *167*, 181–189. [CrossRef]

23. Shi, S.; Cao, Q.; Yao, Y.; Tang, H.; Yang, P.; Wu, W. Influence of climate and socio-economic factors on the spatio-temporal variability of soil organic matter: A case study of central Heilongjiang Province, China. *J. Integr. Agric.* **2014**, *13*, 1486–1500. [CrossRef]
24. Zimmerman, E.K.; Tyndall, J.C.; Schulte, L.A.; Larsen, G.L.D. Farmer and farmland owner views on spatial targeting for soil conservation and water quality. *Water Resour. Res.* **2019**, *55*, 3796–3814. [CrossRef]
25. Kuria, A.W.; Barrios, E.; Pagella, T.; Muthuri, C. Farmers' knowledge of soil quality indicators along a land degradation gradient in Rwanda. *Geoderma Reg.* **2019**, *16*, 1–14. [CrossRef]
26. Bai, Z.; Caspari, T.; Gonzalez, M.R.; Niels, H.B.; Paul, M.; Else, K.B.; Ron de, G.; Lijbert, B.; Minggang, X.; Carla, S.S.F.; et al. Effects of agricultural management practices on soil quality: A review of long-term experiments for Europe and China. *Agric. Ecosyst. Environ.* **2018**, *265*, 1–7. [CrossRef]
27. Ebanyat, P.; de Ridder, N.; de Jager, A.; Robert, J.D.; Mateete, A.B.; Ken, E.G. Drivers of land use change and household determinants of sustainability in smallholder farming systems of Eastern Uganda. *Popul. Environ.* **2010**, *31*, 474–506. [CrossRef]
28. Marenya, P.P.; Barrett, C.B. Soil quality and fertilizer use rates among smallholder farmers in Western Kenya. *Agric. Econ.* **2009**, *40*, 561–572. [CrossRef]
29. Su, M.; Guo, R.; Hong, W. Institutional transition and implementation path for cultivated land protection in highly urbanized regions: A case study of Shenzhen, China. *Land Use Policy* **2019**, *81*, 493–501. [CrossRef]
30. Li, W.; Wang, D.; Li, H. Urbanization-induced site condition changes of peri-urban cultivated land in the black soil region of Northeast China. *Ecol. Indic.* **2017**, *80*, 215–223. [CrossRef]
31. Jin, J.J.; Chong, J.; Lun, L. The economic valuation of cultivated land protection: A contingent valuation study in Wenling City, China. *Landsc. Urban Plan.* **2013**, *119*, 158–164.
32. Kong, X.B.; Li, C.; Liang, Y.; Wang, H. Arable Land Productivity and Its Elastic Loss on the Basis of Farm Household Land Use Behavior. *Prog. Geogr.* **2010**, *29*, 869–877. (In Chinese)
33. Chen, M.; Wu, Y.; Liu, T. Research and prospect of farmland protection in China based on farmer's behavior. *J. Nanjing Agric. Univ. (Soc. Sci. Ed.)* **2012**, *12*, 66–72. (In Chinese)
34. Pfarrhofer, M.; Piribauer, P. Flexible shrinkage in high-dimensional bayesian spatial autoregressive models. *Spat. Stat.* **2019**, *29*, 109–128. [CrossRef]
35. Mattos, C.L.C.; Barreto, G.A. A stochastic variational framework for recurrent gaussian processes models. *Neural Netw.* **2019**, *112*, 54–72. [CrossRef]
36. Kerebel, A.; Gélinas, N.; Déry, S. Landscape aesthetic modelling using bayesian networks: Conceptual framework and participatory indicator weighting. *Landsc. Urban Plan.* **2019**, *185*, 258–271. [CrossRef]
37. Hallin, C.; Huisman, B.J.A.; Larson, M.R.; Walstra, D.; Hanson, H. Impact of sediment supply on decadal-scale dune evolution—Analysis and modelling of the kennemer dunes in the Netherlands. *Geomorphology* **2019**, *337*, 94–110. [CrossRef]
38. Soler, R.; Soler, J.R.; Araya, I. Subjects in the blended learning model design: Theoretical—Methodological elements. *Procedia Soci. Behav. Sci.* **2017**, *237*, 771–777. [CrossRef]
39. Alegría, A.; Caro, S.; Bevilacqua, M.; Porcu, E.; Cerda, J.C.D.L. Estimating covariance functions of multivariate skew-gaussian random fields on the Sphere. *Spat. Stat.* **2017**, *22*, 388–402. [CrossRef]
40. Goovaerts, P. Geostatistical modelling of uncertainty in soil science. *Geoderma* **2001**, *103*, 3–26. [CrossRef]
41. Lyu, K.; Chen, K.; Zhang, H. Relationship between land tenure and soil quality: Evidence from China's soil fertility analysis. *Land Use Policy* **2019**, *80*, 345–361. [CrossRef]

land

MDPI

Article

Predicted Maps for Soil Organic Matter Evaluation: The Case of Abruzzo Region (Italy)

Chiara Piccini *, Rosa Francaviglia and Alessandro Marchetti

Council for Agricultural Research and Economics, Research Centre for Agriculture and Environment, Via della Navicella 2–4, 00184 Rome, Italy; rosa.francaviglia@crea.gov.it (R.F.); alessandro.marchetti@crea.gov.it (A.M.)

* Correspondence: chiara.piccini@crea.gov.it

Received: 25 August 2020; Accepted: 22 September 2020; Published: 24 September 2020

Abstract: Organic matter, an important component of healthy soils, may be used as an indicator in sustainability assessments. Managing soil carbon storage can foster agricultural productivity and environmental quality, reducing the severity and costs of natural phenomena. Thus, accurately estimating the spatial variability of soil organic matter (SOM) is crucial for sustainable soil management when planning agro-environmental measures at the regional level. SOM variability is very large in Italy, and soil organic carbon (SOC) surveys considering such variability are difficult and onerous. The study concerns the Abruzzo Region (about 10,800 km^2), in Central Italy, where data from 1753 soil profiles were available, together with a Digital Elevation Model (DEM) and Landsat images. Some morphometric parameters and spectral indices with a significant degree of correlation with measured data were used as predictors for regression-kriging (RK) application. Estimated map of SOC stocks, and of SOM related to USDA (United States Department of Agriculture) texture—an additional indicator of soil quality—were produced with a satisfactory level of accuracy. Results showed that SOC stocks and SOM concentrations in relation to texture were lower in the hilly area along the shoreline, pointing out the need to improve soil management to guarantee agricultural land sustainability.

Keywords: soil organic carbon; digital soil mapping; regression-kriging; central Italy

1. Introduction

One of the main challenges for the future is to maintain soil functions, but despite many efforts to promote more sustainable land management, soil degradation in the European Union (EU) is increasing [1], with a severe impact on food production and the supply of ecosystem services. Among the soil properties impacting soil quality, soil organic carbon (SOC)—and soil organic matter (SOM) derived from its determination—deserves special attention, representing a key indicator for evaluating soil quality [2], but also impacting the chemical and physical properties of the soil and its overall health. Properties affected by organic matter include soil structure, water holding capacity, diversity and activity of soil organisms, buffering capacity, and nutrient availability. SOM also regulates the efficiency of soil amendments, fertilizers, pesticides, and herbicides. According to the Food and Agriculture Organization of the United Nations (FAO) [3], one of the characteristics of sustainable soil management (SSM) is a stable or increasing storage of SOM, ideally close to the optimal level for the local environment, for all land uses. Thus, the most efficient way to improve soil quality is to stimulate better SOM management.

SOC is also an essential component of the global carbon cycle [4], representing one of the largest reservoirs of terrestrial carbon that may influence global warming [5,6]. The global carbon budget, and the CO_2 emissions associated with its major components (i.e., atmosphere, ocean, fossil fuels, soil, and biosphere) is vital to support the environmental policy addressing climate change [7].

SOC sequestration, resulting from increased soil C inputs and/or reduced C losses [1,8,9], has been recognized as an important process to mitigate the rise of the atmospheric greenhouse gas (GHG) concentration. SSM measures such as increasing SOC stocks can thus mitigate the climate change, at least for several years after their adoption [10].

Soil carbon losses are related to changes in land use, soil management, and climate change issues. Across Europe, most soils have unbalanced SOC/SOM contents, resulting from non-conservative land management practices and land use. The consequence is an acceleration of SOM decomposition [11], and opposing this process is highly advisable to ensure sustainability in European arable soils [12].

The decrease of SOC/SOM is particularly relevant within the Mediterranean area, in consequence of the decreased soil fertility and the increased risk of soil erosion and desertification [13]. It was estimated that 74% of the territory in southern Europe has soils with less than 2% of organic carbon (i.e., 3.4% organic matter in the shallow layers) [14]). In Italy, SOC/SOM variability is very high due to the peculiar geological and geomorphological situation, and soil surveys taking into consideration a similar variability are difficult and onerous. In areas with Mediterranean climates, like central and southern Italy, SOC/SOM degradation is higher due to the coupled effect of high temperatures and low soil moisture, increasing the mineralization rate. Moreover, lower SOC/SOM accumulation often results from intensive and non-conservative agronomic practices (e.g., deep tillage), usually adopted in clayey soils of these areas to enhance soil structure, permeability, and aeration, and to assist crop growth, especially in hilly lands [15,16]. A similar soil management causes higher aeration, accelerating the SOC/SOM degradation rate, and the mixing with underlying horizons with lower SOC/SOM, diluting SOC/SOM in the arable layer. The soil is thus more exposed to wind and water erosion [17]. Therefore, ongoing interest in ensuring a sustainable use and management of European soil resources give rise to a priority need for reliable quantitative information on the present state of SOC/SOM.

SOC/SOM distribution is controlled by many factors (e.g., climate, hydrology, soil type, land use, etc. [18]), whose spatial variation is often wide and not linear. Thus, an accurate estimate of such spatial variability is crucial in soil quality evaluation and in assessing the carbon sequestration potential, providing an operative tool for land use planners and decision makers. Mapping SOM is a common task in site-specific crop management, whereas mapping SOC and its changes over time is an important issue in research and in quantifying and monitoring changes in soil carbon stocks. Both objectives can be achieved at minimum cost and high accuracy and precision [19].

Up-to-date and accurate information on SOC/SOM is essential for tailoring site-specific management, but traditional mapping is labor-intensive, time-consuming, and requires expensive sample collection. Point samples at the local scale are usually more easily available, but such soil information needs to be interpolated in space from a limited number of observations, estimating soil properties over the whole area of interest. Studies have revealed that there is a significant correlation among terrain variables and SOC/SOM [20,21], meaning that spatial estimates can be improved by using auxiliary data—exhaustive and spatially extensive—that can provide relevant information at unsampled locations. It has been largely demonstrated that spatial prediction methods based on ancillary information usually produce maps with higher accuracy, given that the primary and secondary variables are significantly correlated [22,23]. Digital soil mapping (DSM) techniques have gained more increasing appeal in recent years, providing rapid and cost-efficient tools for mapping soil properties across large areas. These methods integrate measured data and auxiliary information, usually available at finer spatial resolution than the point values of a primary target sampled variable [24]. Several authors have used DSM techniques for SOC/SOM mapping, for example, Adhikari et al. [25] adopted regression-kriging (RK) in Denmark, using 18 environmental variables as predictors for mapping the spatial distribution of SOC. Guo et al. [26] applied random forest plus residuals kriging in an area of China to estimate the spatial arrangement of SOM. Song et al. [27] combined estimation methods with local terrain features to enhance their SOC prediction performance. Wang et al. [28] tested three machine learning techniques for mapping SOC stocks in Australia, using several environmental covariates from remote sensing. Lamichhane et al. [29] reviewed the applications of different DSM techniques reported in the scientific literature from 2013 until 2019 for mapping SOC concentration and stocks.

Spatial interpolation by RK is a method that merges a regression of the primary dependent variable on secondary ancillary variables with simple kriging of the regression residuals [30–33]. Ancillary information is often derived from the digital elevation model (DEM), which provides topographic information that allows for the calculation of terrain parameters, and from satellite imagery, both easily available at relatively low cost. Since maintaining or increasing SOM is one of the main targets of SSM, our research question was whether such a technique, already tested in limited areas in Italy [15,34], could be suitable for SOM assessment at the regional level. We hypothesized that RK could yield accurate estimates for site-specific analyses without further sampling expenses. The Abruzzo region in central Italy was chosen as the study area due to the complexity of its territory, and for the availability of suitable data for RK application.

In this framework, the study aimed to provide a spatial evaluation—at a regional level—of SOC, soil USDA (United States Department of Agriculture) texture, and SOM levels based on the soil texture from point data, estimating values in non-sampled locations by applying RK. Then, by transferring them into a GIS software, we can produce a reliable estimate and a valid evaluation tool for a SSM.

2. Materials and Methods

2.1. Study Area

The study area, located in central Italy, consists of about 10,800 km^2 corresponding to the territory of the Abruzzo region. Among the 20 administrative regions of Italy, Abruzzo is one of the most mountainous, located in the central peninsular part of the country. Bordered by the Adriatic Sea in the east and by the Apennines in the west, its territory is very complex and heterogeneous. Within a few kilometers, the environment changes from high mountains (Gran Sasso and Maiella) to the seashore, passing through all the intermediate landscapes: mountain grasslands and woods, hills, plains and river basins. While the mountain ranges lie along a NW–SE direction, the rivers cross them toward the sea along a SE–NW direction. About 40% of the total surface is represented by utilized agricultural area. Along the coastline, the climate is Mediterranean, warm and dry, gradually becoming continental moving inlands. Mean annual air temperature goes from 6 °C in the mountains to 15 °C near the sea. Mean annual rainfall ranges from 600–800 mm in the plains and in the river basins to 1000–1200 mm in the hills, reaching 1600 mm in the mountains. Summer is everywhere the dry season.

Soil regions (SR) are the largest units of soil description, depicting areas with similar soil-forming conditions. These are usually defined as typical associations of dominant soils, occurring in areas limited by a specific climate and/or a characteristic association of parent material. In our study area, soils belong to SRs 61.3, 61.1, and 16.4, as defined by the European Soil Bureau [35]:

- SR61.3: Hills of central and southern Italy on Pliocene and Pleistocene marine deposits and Holocene alluvial sediments along the Adriatic Sea;
- SR 61.1: Apennine and anti-Apennine relieves on sedimentary rocks (Tertiary arenaceous marly flysch) of central and southern Italy; and
- SR16.4: Apennine relieves on Mesozoic and Tertiary limestone, dolomite, and marl, and intra-mountain plains [36].

In Figure 1, the digital elevation model of the region is reported, together with the location of the sampling points and the boundaries of the SRs.

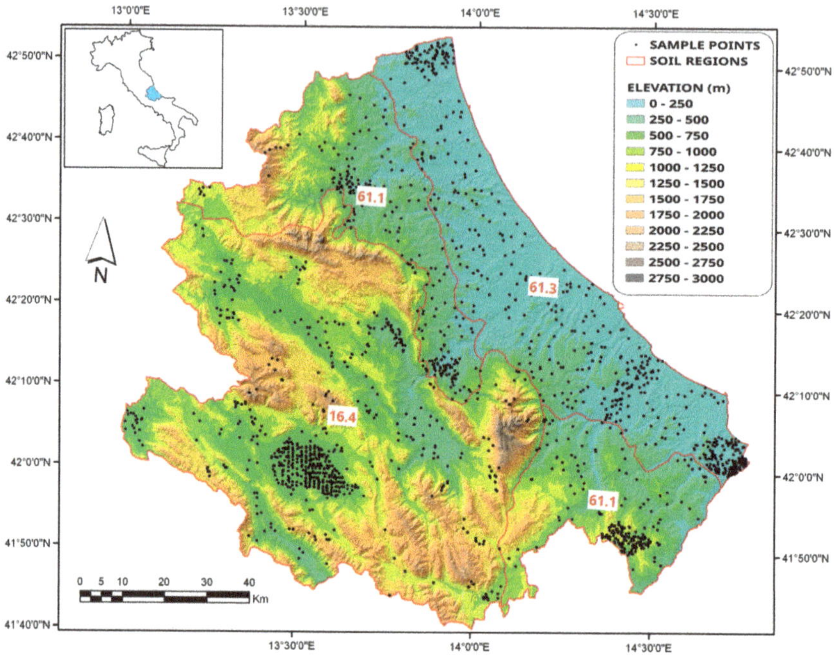

Figure 1. Digital elevation model draped on a hillshade, with the location of the sampling points and boundaries of the soil regions.

A land use map of the Abruzzo Region is reported as Supplementary Material (Figure S1).

2.2. Data Collection

The study dataset, provided by the Regional Agency for Agricultural Extension Services of Abruzzo Region (ARSSA), consists of 1753 georeferenced soil samples collected by an auger (0–25 and 25–50 cm) in accessible agricultural and forest land. The physical and chemical routine analyses included the measurement of particle size distribution according to the pipette method [37] and of SOC content according to the modified Walkley–Black method [38]. The SOC stock in kg m^{-2} for each soil profile was calculated as follows:

$$\text{SOCstock} = \sum_{1}^{n} soc \times bd \times th \times \frac{(100 - gr)}{100} \times 10^{-1} \tag{1}$$

where *soc* is the organic carbon concentration in % for each soil horizon; *bd* is the bulk density of the soil horizon in g cm^{-3} estimated by suitable pedofunctions, different for each type of horizon [39]; *th* is the thickness of the horizon in cm; *gr* is the gravel content in %; and *n* is the number of horizons in the soil profile.

The SOM content was then evaluated from SOC [40] by means of the following formula:

$$\text{SOM} = \text{SOC} \times 1.724 \tag{2}$$

From an agronomic standpoint, the pedo-climatic context cannot be neglected in the evaluation of SOM, because in different soil types, the same amount of SOM can differently impact soil functions.

Thus, the SOM content of the study area was classified into four different levels (i.e., very low, low, medium, and high) based on the USDA textural classes [41], as detailed in Table 1.

Table 1. Levels defined for Soil Organic Matter (SOM) evaluation based on USDA (United States Department of Agriculture) textural classes [41].

USDA Texture Class	SOM Content			
	Very Low	Low	Medium	High
		%		
Sand, Loamy Sand, Sandy Loam	<0.8	0.8–1.4	1.5–2.0	>2.0
Loam, Sandy Clay, Sandy Clay Loam, Silty Loam, Silt	<1.0	1.0–1.8	1.9–2.5	>2.5
Clay, Clay Loam, Silty Clay, Silty Clay Loam	<1.2	1.2–2.2	2.3–3.0	>3.0

Land use, vegetation, climate, and terrain features are the main factors affecting soil properties at the landscape scale, particularly in hills, hence DEM-derived terrain attributes can be used for the prediction of the spatial distribution of soil features. Ancillary data for the area were derived from: (i) a 30 m degraded version of the 20 m DEM provided by the Land Information Service of the Abruzzo Region; (ii) from Landsat 7 TM imagery (three visible bands and four infrared bands), and (iii) the 1:250,000 Soil Subsystems Map of Abruzzo available from ARSSA [42].

From the DEM, the following morphometric attributes were derived:

- Elevation (ELEV);
- slope gradient (SLOPE);
- curvature plan and profile (PLANC and PROFC), obtained from the second derivative of the maximum slope direction and the perpendicular one respectively [43];
- solar radiation (SOLAR);
- Topographic Wetness Index (TWI);
- flow accumulation (FLOWACC), which represents the contributing area (i.e., the surface over which water from rainfall, snowfall, etc. can be aggregated) [44]; and
- Stream Power Index (SPI).

TWI is a parameter correlating topography and the water movement in slopes, used to display the spatial distribution of soil moisture and the shallow saturation degree:

$$\mathrm{TWI} = \ln(A_s/\tan\beta) \tag{3}$$

where A_s is the specific catchment area and β is the slope [45].

SPI is used to describe potential flow erosion and related landscape processes. When specific catchment area and slope steepness increase, both the amount of water contributed by upslope areas and the velocity of water flow also increase, hence stream power and potential erosion increase:

$$\mathrm{SPI} = A_s \times \tan\beta \tag{4}$$

where A_s is the specific catchment area and β is the slope [46].

From the Landsat 7 TM imagery (July 2016, cloud cover 0%), Clay Index (CI) and Normalized Difference Vegetation Index (NDVI) were calculated. CI is correlated with the clay content of the soil:

$$\mathrm{CI} = \mathrm{MIR}/\mathrm{MIR2} \tag{5}$$

where MIR is Mid Infra Red (band 6) and MIR2 is Mid Infra Red (band 7) [47]. NDVI gives a quantitative and qualitative estimation of the vegetation:

$$\mathrm{NDVI} = (\mathrm{NIR} - \mathrm{R})/(\mathrm{NIR} + \mathrm{R}) \tag{6}$$

where NIR is Near Infra Red (band 5) and R is Red (band 4) [48].

From the 1:250,000 Soil Subsystems Map of Abruzzo Region by ARSSA [42], an additional variable—SST86—was derived, defined by 29 different soil systems and 87 soil subsystems. Converting the map into a raster, all these soil units were grouped in a single categorical variable.

Finally, a multivariate correlation analysis allowed us to consider as suitable covariates for prediction only those auxiliary data with a relatively stronger spatial correlation with the target soil variables.

2.3. Data Processing and Validation of Results

RK, a sort of BLUP (Best Linear Unbiased Prediction) method for spatial data, assumes that the local mean varies continuously into each neighborhood, and can be estimated by combining both directly measured data and correlated ancillary information [28,29]. The technique uses multiple regression to depict the relationship linking the field primary variable and the secondary data. Kriging is then applied to the regression residuals, and the results from both regression and kriging are joined to obtain the estimation [49,50].

The available measured dataset was randomly divided in a training dataset (75% of total samples), and a test dataset (25% of total samples), using only the training part for prediction and the test one to validate the results.

For estimating the target variables (SOC, sand and clay) by RK, the computational steps reported below were followed [34]:

1. set up and import predictor data layers (land-surface parameters and soil subsystems map);
2. match soil samples in the training dataset with land-surface parameters and build the regression matrix. Since using the calculated parameters and indices directly as predictors may cause multicollinearity and redundancy effects, the covariates were transformed in principal components (PCs) [51]. Eleven orthogonal and independent components were defined, and the choice of predictors for each variable was then performed by a stepwise regression, considering only the components significant at $p < 0.001$;
3. linear regression analysis and derivation of the regression residuals, resolving the regression coefficients by means of a maximum likelihood algorithm [52];
4. analysis of residuals for detecting spatial autocorrelation, and fitting of the theoretical variogram models;
5. run the interpolation; and
6. visualization and validation of the results using the test dataset.

A flowchart of the procedure is reported in Figure 2. The software SAGA 7.6.3 [53] and ILWIS 3.8.6 Open [54] were used for the derivation of parameters and indices, and for PCs definition. The statistical software R 3.6.3, with the packages *sp* and *rgdal* for spatial data preparation and *gstat* for geostatistical modeling and prediction, was used to perform the regression analysis [55]. The software ArcGIS 10.2® was used for drawing the estimated maps of SOC in kg m^{-2}, soil texture (USDA classification), and SOM levels based on the USDA texture as specified in Table 1.

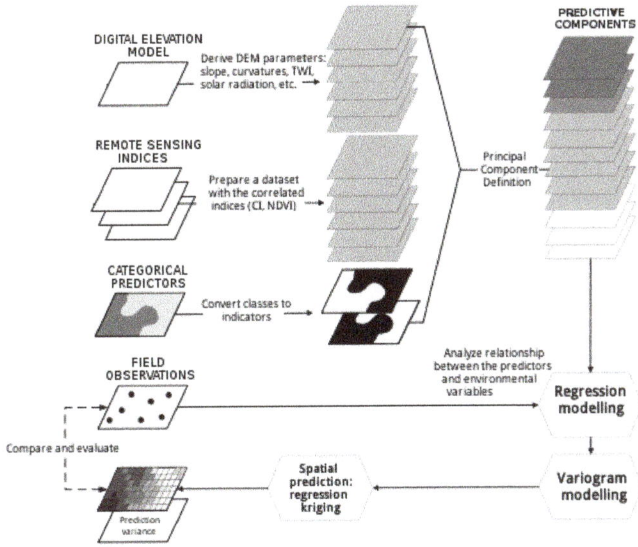

Figure 2. Flowchart showing the computational procedure for Digital Soil Mapping (DSM) based on Regression Kriging (RK) [47].

To assess the precision of prediction, estimated values from the training dataset were compared with the correspondent values from the test dataset and not used in the estimation procedure. Such validation allowed us to evaluate the accuracy of the prediction model by measuring the root mean square prediction error (RMSE):

$$\text{RMSE} = \sqrt{\frac{\sum_{i=1}^{N}\left(\hat{Z}(x_i) - Z(x_i)\right)^2}{N}} \quad (7)$$

where $\hat{Z}(x_i)$ and $Z(x_i)$ are the estimated values and actual observations, respectively, and N is the number of validation points. The RMSE expresses the difference between the model estimations and the observed values, presented in the same unit of measurement. If the value of RMSE is close (lower) to the standard deviation of the data, then the model is a good fit [56]. The *Geostatistical Analyst* extension of the software ArcGIS 10.2® was used to validate the estimation results.

3. Results and Discussion

Pre-processing of data included basic statistics calculation and frequency distribution analysis. Two variables needed to be transformed to approach a normal distribution: sand (square root) and SOC (cube root). The 11 parameters and indices derived from ancillary data (ELEVATION, SLOPE, PROFC, PLANC, TWI, SOLAR, FLOWACC, SPI, CI, NDVI, SST86) were converted in PCs. Table 2 shows the matrix of transformation coefficients, calculated from the covariance matrix. The PC1 and PC2 components were the most significant ones for the prediction of all the target variables. The last component—PC11—was excluded a priori to avoid any rounding effect in the PCs computation with ILWIS [46].

Table 2. Matrix of principal components coefficients and their significance for prediction.

	Elevation	Slope	Profc	Planc	Twi	Solar	Flowacc	SPI	CI	NDVI	SST86	Sand	Clay	SOC
PC1	0.277	0.236	0.334	0.334	0.256	0.419	0.026	0.055	0.358	0.404	0.330	***	***	***
PC2	−0.344	−0.514	0.108	0.146	0.522	0.417	0.103	−0.051	−0.257	−0.238	0.036	***	***	***
PC3	−0.371	−0.104	−0.022	−0.010	0.148	−0.040	0.042	0.026	0.433	0.443	−0.664	**	***	***
PC4	−0.065	0.223	0.424	0.422	−0.398	0.258	−0.183	−0.224	−0.228	−0.206	−0.429		***	
PC5	0.474	0.305	−0.184	−0.099	0.370	0.193	0.157	0.366	−0.228	−0.174	−0.474	***	**	***
PC6	0.326	−0.331	−0.175	−0.441	−0.293	0.570	−0.168	−0.295	0.098	0.117	−0.097	***	***	
PC7	−0.574	0.525	−0.154	−0.326	−0.084	0.422	−0.065	0.212	−0.011	−0.052	0.166		*	***
PC8	0.028	0.009	−0.013	−0.020	0.309	−0.110	−0.941	0.071	−0.009	−0.011	−0.010	*	*	
PC9	0.023	0.104	0.711	−0.612	0.204	−0.162	0.079	−0.174	−0.055	−0.021	−0.039		*	
PC10	−0.013	0.361	−0.320	0.056	0.336	−0.056	0.063	−0.801	−0.045	0.013	−0.009	***	***	
PC11	0.040	0.013	0.018	0.007	0.041	0.020	0.011	−0.035	0.707	−0.703	−0.026			
						R^2						0.72	0.81	0.93

Significance codes: *** = 0.001; ** = 0.01; * = 0.05.

The target variables were strongly correlated with the principal components, as can be seen from the R^2 values from the linear regression. As explained in the previous section, the choice of the components to be used as predictors was performed by the stepwise regression, considering a significance level of 0.001. For SOC estimation, the sum of PC1—PC2—PC3—PC5—PC7 was used; for sand estimation, the predictor was the sum of PC1—PC2—PC5—PC6—PC10 components; for clay estimation, the sum of PC1—PC2—PC3—PC4—PC6—PC10 components was chosen. Notably, all the components selected as predictors were related to the soil type and position.

RK was applied to the SOC content of the training dataset as well as to the sand and clay data to estimate soil texture. The prediction accuracy was evaluated comparing the values estimated using the training dataset with the measured values from the test dataset that did not enter in the estimation. The calculated RMSE was 0.31 for SOC, 1.76 for sand, and 11.56 for clay, lower than the standard deviation of the measured data of 0.38, 2.09, and 15.02, respectively, and thus considered a very good result, notwithstanding the irregular distribution of the sampling points.

In Figure 3, the estimated map of SOC is reported, showing that about 88% of the area has a SOC content below 3.0 kg m^{-2}.

In Figure 4, the map of USDA soil texture, estimated from sand and clay data using the training dataset, is reported, showing that the prevailing textures are loam and sandy loam.

Finally, SOM in g kg^{-1} was evaluated from SOC (Equation (2)), then SOM values were ranked in four classes (very low, low, medium, high) based on the estimated USDA texture (according to Table 1). The obtained map is reported in Figure 5.

The observed SOM levels were related to soil morphology: in the hilly belt along the coast, the map shows essentially a very low content, while in the interior, where soils are mainly under forest, the SOM content reached a high class in most of the territory (about 53%). In the coastal area, the SOM depletion can be considered both as a cause and effect of the active erosion on hills, usually exacerbated by intensive agricultural practices. This area is indeed intensely cropped, as can be seen from the land use map (Figure S1). Previous studies of the same type carried out in the northern part of the region, concerning a vineyard district, showed that an intervention to enhance SOM content is necessary [15,16,23,34]. Such a result is in line with the findings by Berhongaray et al. [57], who reported that cultivation caused a reduction of 16% in SOC content at 50 cm depth in the Argentine Pampas.

In Abruzzo, an improvement in agro-environmental planning is required to ensure an appropriate SOC/SOM content to soils, so that they can maintain their ecological and socio-economical functions, and crop yield sustainability in agricultural lands. Stimulating the adoption of SSM practices through the dissemination activity of agricultural extension services is fundamental to preserve soil resources. Such practices can enhance soil quality (water holding capacity, infiltration, soil structure, and soil fertility), improve the SOM cycle, and effectively contribute to climate change mitigation. A rational soil management can also help to reduce GHG emissions, especially carbon dioxide emissions, favoring a decrease in soil organic carbon losses, increasing the organic matter input, or combining both of them [15].

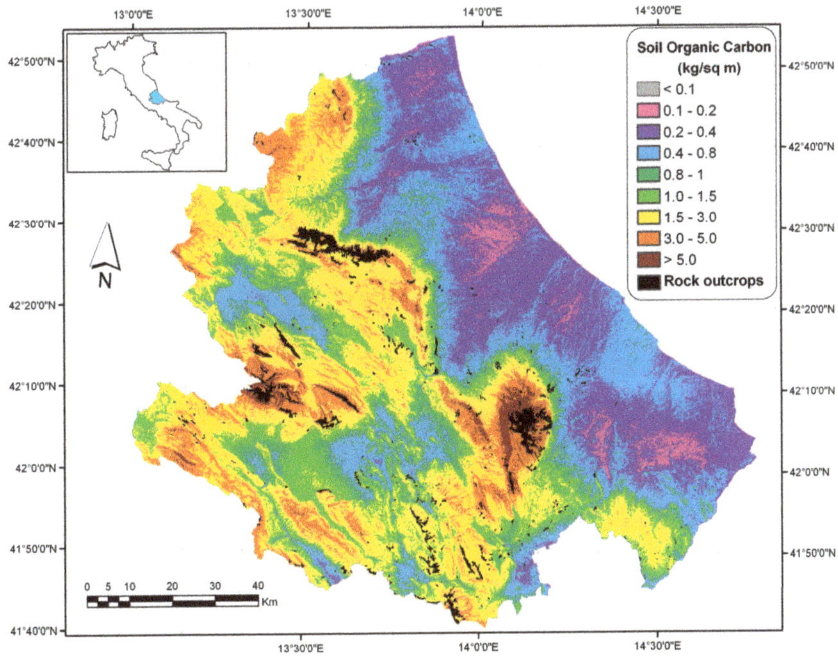

Figure 3. Soil organic carbon content predicted by RK.

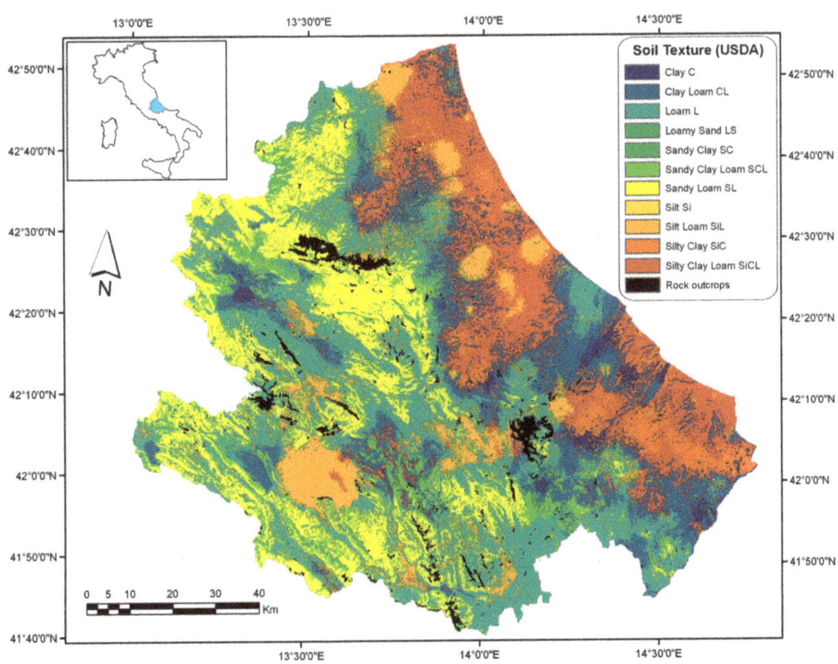

Figure 4. Soil texture predicted by RK (USDA).

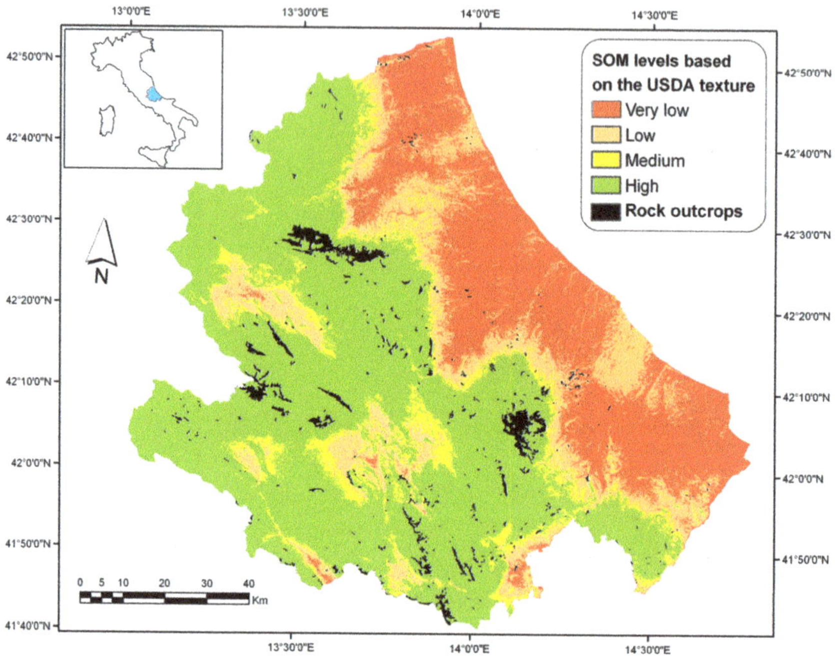

Figure 5. SOM levels based on the USDA texture predicted by RK.

This study, defining the current SOC/SOM status in a region of central Italy, provides evidence that SOM management in its agricultural areas should be improved. The results of our study show that in Abruzzo, most arable lands are susceptible to soil degradation, sped-up by dry conditions and high temperatures causing a rapid mineralization of SOM. These areas need appropriate management to guarantee agricultural land sustainability. Adopting conservative practices such as conservation tillage or no-tillage (e.g., direct seeding), improving rotations with forage crops, returning crop residues to soil, growing green manure crops, and supplying the soil with proper exogenous organic matter could lead to an appropriate SOM restoration [16]. Permanent grasslands are effective for soil carbon accumulation in mineral soils and the adoption of agroforestry—the integration of trees and shrubs on agricultural land—and crop diversification might contribute to SSM [1].

An accurate state-of-the-art of SOC/SOM distribution would allow us to foresee future trends, and to evaluate the effectiveness of soil conservation practices stabilizing and increasing carbon stock in soils, which should be adopted for a more sustainable soil management. As in our case, mapping SOC over a large area in Bosnia and Herzegovina by classical geostatistical methods [58] showed that the spatial distribution of SOC concentration was strongly influenced by the intense farming practices. In this case, however, the environmental variables had a reduced capacity to explain the spatial variability of SOC. In Brazil, Bonfatti et al. [59] applied RK for mapping SOC, finding again a similar situation: soils under arable crops and vineyard showed the lowest concentration.

In the northern part of Abruzzo, similar mapping work was carried out comparing ordinary kriging and RK as spatial interpolators, and their performance resulted in being approximately the same [34], depending on the available auxiliary information. Mapping targeted soil parameters such as SOC/SOM applying RK, both in limited areas [15] and at the regional level, can represent a valid tool, providing useful and accurate information to assess soil degradation in a cost-effective and little time-consuming way. Such maps can also improve the process of carbon budgeting and reporting—in line with global initiatives of minimizing greenhouse gas emission—and thus the impacts of climate change [60]. Moreover, they can really help farmers in managing soils, can influence and support local land-use planners, and can also be easily updated whenever new data become available.

4. Conclusions

A spatial representation of SOM is essential to supply a useful and proper reference tool to decision makers, facilitating and optimizing the regional planning of agro-environmental measures for a SSM. Soil surveys taking into account the extremely large variability of the Italian territory are very difficult and onerous. However, most soil attributes are spatially correlated with ancillary variables derived from DEM, Landsat imagery. and existing soil subsystem maps.

In the Abruzzo region, analyzing the available data and estimating values in non-sampled locations by means of RK—integrating measured data and ancillary variables—allowed us to map soil texture, SOC, and SOM levels based on the USDA texture with an acceptable precision. These maps, obtained at relatively low costs, could also be less accurate than the traditional ones, but providing the associated estimation error may anyway bring added value. The observed SOC/SOM distribution appears to be linked to the soil morphology and to the land use: low or very low in intensively cropped hills near the shore, and high in mountains and forest land.

RK proved to be a rapid and cost-efficient tool for mapping soil properties across large areas, allowing us to easily monitor their changes over time. Maps can be updated every time new information and/or new data become available. In the next future, it is hoped that this technique will be successfully coupled and associated with traditional soil surveying and mapping procedures, and then applied at the national level.

Supplementary Materials: The following are available online at http://www.mdpi.com/2073-445X/9/10/349/s1, Figure S1: Land use map (from CORINE Land Cover, https://land.copernicus.eu/pan-european/corine-land-cover/clc2018).

Author Contributions: Conceptualization, C.P., R.F., and A.M.; Investigation, C.P., R.F., and A.M.; Methodology, C.P., R.F. and A.M.; Validation, C.P., R.F. and A.M.; Writing—original draft, C.P.; Writing—review & editing, R.F. and A.M. All authors have read and agreed to the published version of the manuscript.

Funding: This research received no external funding.

Acknowledgments: The authors wish to thank Sergio Santucci and Igino Chiuchiarelli from ARRSA—Agricultural Extension Service of Abruzzo Region—for providing the soil dataset.

Conflicts of Interest: The authors declare no conflict of interest.

References

1. Hagemann, N.; Álvaro-Fuentes, J.; Siebielec, G.; Castañeda, C.; Bartke, S.; Dietze, V.; Maring, L.; Arrúe, J.L.; Playán, E.; Plaza-Bonilla, D.; et al. *Review of Economic, Social and Environmental Impacts and Implementation Barriers for Soil Protection and Sustainable Management Measures for Arable Land across the EU*; European Commission, DG Environment, Technical Report, 2019; Contract no. ENV.D.1/SER/2016/0041; European Commission: Luxembourg, 2016.
2. Gregorich, E.G.; Carter, M.R.; Angers, D.A.; Monreal, C.M.; Ellert, B.H. Toward minimum data set to assess soil organic matter quality in agricultural soils. *Can. J. Soil Sci.* **1994**, *74*, 885–901. [CrossRef]
3. Food and Agriculture Organization of the United Nations. *FAO Voluntary Guidelines for Sustainable Soil Management*; FAO: Rome, Italy, 2017; p. 16. Available online: http://www.fao.org/3/a-bl813e.pdf (accessed on 18 August 2020).
4. Smith, P.; Fang, C.M.; Dawson, J.J.C.; Moncrieff, J.B. Impact of global warming on soil organic carbon. *Adv. Agron.* **2008**, *97*, 1–43.
5. Lützow, M.V.; Kögel-Knabner, I.; Ekschmitt, K.; Matzner, E.; Guggenberger, G.; Marschner, B.; Flessa, H. Stabilization of soil organic matter in temperate soils: Mechanisms and their relevance under different soil conditions—A review. *Eur. J. Soil Sci.* **2006**, *57*, 426–445. [CrossRef]
6. Batjes, N.H. Total carbon and nitrogen in the soils of the world. *Eur. J. Soil Sci.* **1996**, *47*, 151–163. [CrossRef]

7. Friedlingstein, P.; Jones, M.W.; O'Sullivan, M.; Andrew, R.M.; Hauck, J.; Peters, G.P.; Peters, W.; Pongratz, J.; Sitch, S.; Le Quéré, C.; et al. Global Carbon Budget 2019. *Earth Syst. Sci. Data* **2019**, *11*, 1783–1838. [CrossRef]
8. Freibauer, A.; Rounsevell, M.D.A.; Smith, P.; Verhagen, J. Carbon sequestration in the agricultural soils of Europe. *Geoderma* **2004**, *122*, 1–23. [CrossRef]
9. Baker, J.M.; Ochsner, T.E.; Venterea, R.T.; Griffis, T.J. Tillage and soil carbon sequestration—What do we really know? *Agric. Ecosyst. Environ.* **2007**, *118*, 1–5. [CrossRef]
10. Lal, R. Soil carbon sequestration to mitigate climate change. *Geoderma* **2004**, *123*, 1–22. [CrossRef]
11. Smith, P.; Andren, O.; Karlsson, T.; Perala, P.; Regina, K.; Rounsevell, M.; Van Wesemael, B. Carbon sequestration potential in European croplands has been overestimated. *Glob. Chang. Biol.* **2005**, *11*, 2153–2163. [CrossRef]
12. Lal, R. Challenges and opportunities in soil organic matter research. *Eur. J. Soil Sci.* **2009**, *60*, 158–169. [CrossRef]
13. Francaviglia, R.; Renzi, G.; Rivieccio, R.; Marchetti, A.; Piccini, C. Spatial analysis and prediction of soil organic carbon in Friuli Venezia Giulia Region (Northern Italy). *Geoinfor. Geostat. Overv.* **2014**, *2*. [CrossRef]
14. Zdruli, P.; Jones, R.J.A.; Montanarella, L. *Organic Matter in the Soils of Southern Europe. European Soil Bureau Technical Report*; EUR 21083 EN; Office for Official Publications of the European Communities: Luxembourg, 2004.
15. Marchetti, A.; Piccini, C.; Francaviglia, R.; Santucci, S.; Chiuchiarelli, I. Estimating soil organic matter content by regression kriging. In *Digital Soil Mapping: Bridging Research, Environmental Application, and Operation*; Boettinger, J.L., Howell, D.W., Moore, A.C., Hartemink, A.E., Kienast Brown, S., Eds.; Springer: Dordrecht, The Netherlands, 2010; pp. 241–254.
16. Marchetti, A.; Piccini, C.; Francaviglia, R.; Mabit, L. Spatial distribution of soil organic matter using geostatistics: A key indicator to assess soil degradation status in central Italy. *Pedosphere* **2012**, *22*, 230–242. [CrossRef]
17. Bot, A.; Benites, J. *The Importance of Soil Organic Matter. Key to Drought-Resistant Soil and Sustained Food Production*; FAO Soil Bulletin: Rome, Italy, 2005; Available online: http://www.fao.org/docrep/009/a0100e/a0100e00.htm (accessed on 18 August 2020).
18. Fang, X.; Xue, Z.; Li, B.; An, S. Soil organic carbon distribution in relation to land use and its storage in a small watershed of the Loess Plateau, China. *Catena* **2012**, *88*, 6–13. [CrossRef]
19. Ping, J.L.; Dobermann, A. Variation in the precision of soil organic carbon maps due to different laboratory and spatial prediction methods. *Soil Sci.* **2006**, *171*, 374–387.
20. Mueller, T.G.; Pierce, F.J. Soil carbon maps: Enhancing spatial estimates with simple terrain attributes at multiple scales. *Soil Sci. Soc. Am. J.* **2003**, *67*, 258–267. [CrossRef]
21. Dobos, E.; Carré, F.; Hengl, T.; Reuter, H.I.; Tóth, G. *Digital Soil Mapping As a Support to Production of Functional Maps*; EUR 22123 EN; Office for Official Publications of the European Communities: Luxemburg, 2006.
22. McBratney, A.B.; Mendonça Santos, M.L.; Minasny, B. On digital soil mapping. *Geoderma* **2003**, *117*, 3–52. [CrossRef]
23. Marchetti, A.; Piccini, C.; Francaviglia, R.; Santucci, S.; Chiuchiarelli, I. Evaluation of soil organic matter content in Teramo province, Central Italy. *Adv. GeoEcol.* **2008**, *39*, 345–356.
24. McBratney, A.; Odeh, I.; Bishop, T.; Dunbar, M.; Shatar, T. An overview of pedometric techniques for use in soil survey. *Geoderma* **2000**, *97*, 293–327. [CrossRef]
25. Adhikari, K.; Hartemink, A.E.; Minasny, B.; Bou Kheir, R.; Greve, M.B.; Greve, M.H. Digital Mapping of Soil Organic Carbon Contents and Stocks in Denmark. *PLoS ONE* **2014**, *9*, e105519. [CrossRef]
26. Guo, P.T.; Li, M.F.; Luo, W.; Tang, Q.F.; Liu, Z.W.; Lin, Z.M. Digital mapping of soil organic matter for rubber plantation at regional scale: An application of random forest plus residuals kriging approach. *Geoderma* **2015**, *237–238*, 49–59. [CrossRef]
27. Song, X.; Liu, F.; Zhang, G.; Li, D.; Zhao, Y.; Yang, J. Mapping soil organic carbon using local terrain attributes: A comparison of different polynomial models. *Pedosphere* **2017**, *27*, 681–693. [CrossRef]
28. Wang, B.; Waters, C.; Orgill, S.; Gray, J.; Cowie, A.; Clark, A.; Liu, D.L. High resolution mapping of soil organic carbon stocks using remote sensing variables in the semi-arid rangelands of eastern Australia. *Sci. Total Environ.* **2018**, *630*, 367–378. [CrossRef] [PubMed]
29. Lamichhane, S.; Kumar, L.; Wilson, B. Digital soil mapping algorithms and covariates for soil organic carbon mapping and their implications: A review. *Geoderma* **2019**, *352*, 395–413. [CrossRef]

30. Odeh, I.O.A.; McBratney, A.B.; Chittleborough, D.J. Spatial prediction of soil properties from landform attributes derived from a digital elevation model. *Geoderma* **1994**, *63*, 197–214. [CrossRef]
31. Goovaerts, P. *Geostatistics for Natural Resource Evaluation*; Oxford University Press: New York, NY, USA, 1997.
32. Hengl, T.; Heuvelink, G.B.M.; Rossiter, D.G. About regression kriging: From equations to case studies. *Comput. Geosci.* **2007**, *33*, 1301–1315. [CrossRef]
33. Sun, W.; Minasny, B.; McBratney, A. Analysis and prediction of soil properties using local regression-kriging. *Geoderma* **2012**, *171–172*, 16–23. [CrossRef]
34. Piccini, C.; Marchetti, A.; Francaviglia, R. Estimation of soil organic matter by geostatistical methods: Use of auxiliary information in agricultural and environmental assessment. *Ecol. Indic.* **2014**, *36*, 301–314. [CrossRef]
35. European Communities. *European Soil Bureau Georeferenced Soil Database for Europe: Manual of Procedures, Version 1.1*; EUR 18092 EN; Office for Official Publications of the European Communities: Luxemburg, 2001; p. 178.
36. Costantini, E.A.C.; Urbano, F.; L'Abate, G. Soil Regions of Italy. 2004. Available online: http://www.soilmaps.it (accessed on 18 August 2020).
37. Gee, G.W.; Bauder, J.W. Particle-size analysis. In *Methods of Soil Analysis: Part I. Physical and Mineralogical Methods*, 2nd ed.; Soil Science Society of America Inc.: Madison, WI, USA, 1986; pp. 383–411.
38. Nelson, D.W.; Sommer, L.E. Total carbon, organic carbon, and organic matter. In *Methods of Soil Analysis*, 2nd ed.; Page, A.L., Ed.; ASA Monograph 9(2) American Society of Agronomy: Madison, WI, USA, 1982; pp. 539–579.
39. Hollis, J.M.; Hannam, J.; Bellamy, P.H. Empirically derived pedotransfer functions for predicting bulk density in European soils. *Eur. J. Soil Sci.* **2012**, *63*, 96–109. [CrossRef]
40. Jackson, M.L. *Soil Chemical Analysis*; Prentice-Hall: Englewood Cliffs, NJ, USA, 1965.
41. SILPA. *Società Italiana dei Laboratori Pubblici di Agrochimica Dall'analisi del Terreno al Consiglio di Concimazione*; ASSAM Regione Marche: Ancona, Italy, 1999. (In Italian)
42. Chiuchiarelli, I.; Paolanti, M.; Rivieccio, R.; Santucci, S. *Suoli e Paesaggi d'Abruzzo—Carta dei Suoli Della Regione Abruzzo*; ARSSA Regione Abruzzo: L'Aquila, Italy, 2006. (In Italian)
43. Olaya, V. Basic Land-Surface Parameters. In *Geomorphometry: Concepts, Software, Applications*; Hengl, T., Reuter, H.I., Eds.; Developments in Soil Science; Elsevier: Amsterdam, The Netherlands, 2009.
44. Gruber, S.; Peckham, S. Land-Surface Parameters and Objects in Hydrology. In *Geomorphometry: Concepts, Software, Applications*; Hengl, T., Reuter, H.I., Eds.; Developments in Soil Science; Elsevier: Amsterdam, The Netherlands, 2009.
45. Beven, K.; Kirkby, N. A physically based variable contributing area model of basin hydrology. *Hydrol. Sci. Bull.* **1979**, *24*, 43–69. [CrossRef]
46. Moore, I.D.; Burch, G.J.; Mackenzie, D.H. Topographic effects on the distribution of surface soil water and the location of ephemeral gullies. *Trans. ASAE* **1988**, *31*, 1098–1107. [CrossRef]
47. Hengl, T. *A Practical Guide to Geostatistical Mapping*, 2nd ed.; University of Amsterdam: Amsterdam, The Netherlands, 2009; Available online: http://spatial-analyst.net/book/system/files/Hengl_2009_GEOSTATe2c0w.pdf (accessed on 19 August 2020).
48. Colwell, B.J. Vegetation canopy reflectance. *Remote Sens. Environ.* **1974**, *3*, 175–183. [CrossRef]
49. Odeh, I.O.A.; McBratney, A.B.; Chittleborough, D.J. Further results on prediction of soil properties from terrain attributes: Heterotopic cokriging and regression-kriging. *Geoderma* **1995**, *67*, 215–226. [CrossRef]
50. Hengl, T.; Heuvelink, G.B.M.; Stein, A. A generic framework for spatial prediction of soil variables based on regression kriging. *Geoderma* **2004**, *120*, 75–93. [CrossRef]
51. Hengl, T.; MacMillan, R.A. *Predictive Soil Mapping with R*; OpenGeoHub Foundation: Wageningen, The Netherlands, 2019; Available online: www.soilmapper.org (accessed on 10 September 2020).
52. Bailey, M.; Clements, T.; Lee, J.T.; Thompson, S. Modelling soil series data to facilitate targeted habitat restoration: A polytomous logistic regression approach. *J. Environ. Manag.* **2003**, *67*, 395–407. [CrossRef]
53. SAGA User Group Association. SAGA 7.6.3 System for Automated Geoscientific Analyses. 2020. Available online: http://www.saga-gis.org (accessed on 19 March 2020).
54. 52 North Initiative ILWIS 3.8.6 Open Integrated Land and Water Information System. 2020. Available online: http://52north.org (accessed on 11 March 2020).

55. R Development Core Team. *R: A Language and Environment for Statistical Computing*; R Development Core Team: Vienna, Austria, 2020; ISBN 3-900051-07-0. Available online: http://www.R-project.org (accessed on 4 March 2020).
56. Smith, J.; Smith, P. *Introduction to Environmental Modelling*; Oxford University Press: New York, NY, USA, 2007.
57. Berhongaray, G.; Alvarez, R.; De Paepe, J.; Caride, C.; Cantet, R. Land use effects on soil carbon in the Argentine Pampas. *Geoderma* **2013**, *192*, 97–110. [CrossRef]
58. Bogunovic, I.; Pereira, P.; Husnjak, S.; Coric, R.; Brevik, E.C. Spatial distribution of soil organic carbon and total nitrogen stocks in a karst polje located in Bosnia and Herzegovina. *Environ. Earth Sci.* **2017**. [CrossRef]
59. Bonfatti, B.R.; Hartemink, A.E.; Giasson, E.; Tornquist, C.G.; Adhikari, K. Digital mapping of soil carbon in a viticultural region of Southern Brazil. *Geoderma* **2016**, *261*, 204–221. [CrossRef]
60. Vitharana, U.W.A.; Mishra, U.; Mapa, R.B. National soil organic carbon estimates can improve global estimates. *Geoderma* **2019**, *337*, 55–64. [CrossRef]

land

MDPI

Article

Using Machine Learning Algorithms to Estimate Soil Organic Carbon Variability with Environmental Variables and Soil Nutrient Indicators in an Alluvial Soil

Kingsley JOHN [1,*], Isong Abraham Isong [2], Ndiye Michael Kebonye [1], Esther Okon Ayito [2], Prince Chapman Agyeman [1] and Sunday Marcus Afu [2]

1 Department of Soil Science and Soil Protection, Faculty of Agrobiology, Food, and Natural Resources, Czech University of Life Sciences, Kamýcká 129, 16500 Prague, Czech Republic; kebonye@af.czu.cz (N.M.K.); agyeman@af.czu.cz (P.C.A.)

2 Department of Soil Science, Faculty of Agriculture, University of Calabar, Calabar P.M.B. 1115, Nigeria; eneaki1@unical.edu.ng (I.A.I.); irenotobong@unical.edu.ng (E.O.A.); sunnymarcus@unical.edu.ng (S.M.A.)

* Correspondence: johnk@af.czu.cz

Received: 4 November 2020; Accepted: 30 November 2020; Published: 2 December 2020

Abstract: Soil organic carbon (SOC) is an important indicator of soil quality and directly determines soil fertility. Hence, understanding its spatial distribution and controlling factors is necessary for efficient and sustainable soil nutrient management. In this study, machine learning algorithms including artificial neural network (ANN), support vector machine (SVM), cubist regression, random forests (RF), and multiple linear regression (MLR) were chosen for advancing the prediction of SOC. A total of sixty (n = 60) soil samples were collected within the research area at 30 cm soil depth and measured for SOC content using the Walkley–Black method. From these samples, 80% were used for model training and 21 auxiliary data were included as predictors. The predictors include effective cation exchange capacity (ECEC), base saturation (BS), calcium to magnesium ratio (Ca_Mg), potassium to magnesium ratio (K_Mg), potassium to calcium ratio (K_Ca), elevation, plan curvature, total catchment area, channel network base level, topographic wetness index, clay index, iron index, normalized difference build-up index (NDBI), ratio vegetation index (RVI), soil adjusted vegetation index (SAVI), normalized difference vegetation index (NDVI), normalized difference moisture index (NDMI) and land surface temperature (LST). Mean absolute error (MAE), root-mean-square error (RMSE) and R^2 were used to determine the model performance. The result showed the mean SOC to be 1.62% with a coefficient of variation (CV) of 47%. The best performing model was RF (R^2 = 0.68) followed by the cubist model (R^2 = 0.51), SVM (R^2 = 0.36), ANN (R^2 = 0.36) and MLR (R^2 = 0.17). The soil nutrient indicators, topographic wetness index and total catchment area were considered an indicator for spatial prediction of SOC in flat homogenous topography. Future studies should include other auxiliary predictors (e.g., soil physical and chemical properties, and lithological data) as well as cover a broader range of soil types to improve model performance.

Keywords: geostatistic; machine learning; geospatial modeling; predictive mapping; soil fertility indices; environmental covariates

1. Introduction

Globally, soils of the humid tropics have received overwhelming acceptance for agriculture. However, these soils in southeastern Nigeria have the potential that could be exploited for crop production. Unfortunately, they are both highly weathered and leached soils formed on alluvial deposits under excessive rainfall and high-temperature conditions [1,2]. This soil like other soils weathers through the actions of environmental conditions (i.e., topography, and other soil-forming factors) to give the soil their genetic properties (e.g., soil pH, texture, clay, CEC, exchangeable cations) [3]. Soil texture, nutrient status, and mineralogical properties of alluvial deposits bear the imprints of quartz oxides, which are not rich in most plant growth nutrients [4]. This status gives low crop yield if there is no application of appropriate nutrient amendments. For example, the yield of fresh fruit bunches (FFB) is estimated at 3–5 t·ha^{-1} from University of Calabar Teaching and Research Farm; under alluvial deposits soil is far less than the national average of 8–12 FFB t·ha^{-1} and world-record yields of 25–35 t·ha^{-1} in Malaysia [5].

Soil organic carbon (SOC) is an essential indicator of soil quality, and directly determines soil fertility and plant productivity [5]; it plays a significant role in supplying nutrients to the soil and in the formation of improved soil structure. In previous years, several soil researchers have reported variability of SOC in different ecological zones of the world [6–8]. These studies are in line with the different assumptions, including the fact that variation in crop yield within a given field reflects variation in SOC [9]. Their studies further explained that in order to achieve appropriate soil nutrient management for uniform crop yield, it is necessary to know where the low SOC, as well as soil nutrients, reside within a given field, and how much carbon or soil nutrient is present. This is essentially the importance of quantitative soil mapping. The accurate and up-to-date information obtained in the process ensures the application of site-specific nutrient management to match spatially variable conditions.

Variability of soil nutrients is a significant constraint for sustainable crop production due to the resulting non-uniformity of output across different sections of the field. One way of minimizing heterogeneity in the soil resulting in different crop yields is through digital soil mapping (DSM), but it is often constrained by within-site variability [10]. These issues became the target of a site-specific cropping system, otherwise known as precision agriculture. The technique of precision agriculture can delineate sites for specific management. Precision agriculture has now been developed to spatially varied nutrients and soil properties within a field relying on geospatial technologies and utilizing soil properties, remote sensing data, digital elevation model (DEM), micro-climatic data, and geology [11]. Precision agriculture allows farm managers to manage within-field variability to maximize the cost–benefit ratio of the proposed crop enterprise. Besides that, specific landscape attributes control the spatial distribution of SOC coupled with the interactive action of soil-forming factors [3,12].

In agro-ecosystems, the spatial distribution of soil properties is affected by natural ecological processes influenced by many factors, including climate, soil type, topography, and land use. It thus becomes a challenge to accurately model SOC at farm scales [10,13] over a broader area that spans several kilometres without taking into consideration these factors. Before the advent of geospatial technologies, the spatial distribution of soil properties including SOC was assessed from conventional soil surveys and laboratory analyses of collected soil samples utilizing classical statistics; an approach that is tedious, time-consuming, and expensive. The traditional soil survey method could not provide detailed information about soil variation required for many environmental applications. Thus, alternative approaches are needed. As an alternative, the digital soil mapping (DSM) technique was developed and became one focus of soil and environmental science. Under the framework of the DSM, several geostatistics prediction methods, as found in John et al. [9], have been developed to predict the spatial distribution of soil properties.

Through the advances in technology, there is a comprehensive application of machine learning algorithms such as multiple linear regression (MLR), artificial neural network (ANN), support vector machine (SVM), decision tree, cubist regression, and random forests in soil studies using auxiliary environmental data [8,14].

Environmental auxiliary data such as digital elevation models (DEM), remote sensing, climatic data, and geology have been combined via predictive models to estimate soil properties. A large number of existing DEM data sets (e.g., SRTM DEM and Aster GDEM) [6,15,16] has been used to extract terrain attributes (e.g., elevation, slope, aspect, topography wetness index) as predictors for predicting soil properties. Remote sensing images, on the other hand, have also served as excellent data for both qualitative and quantitative study of soil properties, including SOC [7,8,15]. Previous studies on predicting SOC primarily utilized multi-spectral optical sensors, including Landsat [6,8], MODIS [17], SPOT [18], RapidEye [19], Landsat and MODIS [15], and Landsat and ALOS PALSAR [20]. Remote sensing data provides a cost-effective, reproducible, and spontaneous approach to quantifying SOC variability [21]. This technique is achieved through the correlation between soil reflectance and SOC. In [22] it is reported that the increase in SOC is inversely proportional to an overall decrease in reflectance in the visible (Vis, 400–700 nm), near-infrared (NIR, 700–1400 nm), and shortwave infrared (SWIR, 1400–2500 nm) regions of the electromagnetic spectrum (McMorrow et al. [23,24]).

Fathololoumi et al., [25] worked on improved digital soil mapping with multitemporal remotely sensed satellite data fusion in Iran using random forest (RF) and cubist models. Their results showed that the cubist model exhibited greater accuracy than RF in the modeling of SOC. While in the high-resolution mapping of soil properties using remote sensing variables in southwestern Burkina Faso (studies conducted by Forkuor et al. [19]), RF performed better in the prediction of SOC. In addition, in the prediction and mapping of soil organic carbon using MLA in Northern Iran by Emadi et al. [15], the deep neural network (DNN) model was reported as a superior algorithm with the lowest prediction error and uncertainty. Bian et al. [7] utilizes multiple stepwise regression (MSR), boosted regression trees (BRT) model, and boosted regression trees hybrid residuals kriging (BRTRK) to model SOC in northeastern coastal areas of China. Similarly, Taghizadeh-Mehrjardi et al. [6] use the artificial neural network (ANN), support vector regression (SVR), k-nearest neighbour (kNN), random forest (RF), regression tree model (RT), and genetic programming (GP) to predict SOC. Their study recommended the combination of ANN and equal-area spline functions for predicting SOC spatial distribution in the Baneh region of Iran.

Despite the acceptability of MLA in DSM, few or no studies have considered the incorporation of soil nutrient indicators and environmental data in modeling SOC in southeastern Nigeria and the world at large. Additionally, the Nigeria environment is yet to get acquainted with the modeling program involving MLA in soil mapping, and no feasible study has been carried out elucidating this approach, despite the region's active engagement in agriculture production. Consequently, a fundamental knowledge gap remains, hindering the ability of farm managers and agronomists to improve the land and soil quality. Furthermore, we hypothesize that in flat terrain configuration, soil nutrient indicators play many roles in explaining SOC distribution to ancillary environmental data. Therefore, in this study, we applied five machine learning algorithms (RF, Cubist, ANN, MLR, and SVM) to estimate the SOC variability in a flat alluvial terrain condition with environmental variables and soil nutrient indicators known to influence SOC variability in the alluvial deposit of Calabar, Nigeria.

2. Materials and Methods

2.1. Description of the Study Area

The study was conducted in Calabar, Cross River State. The study area extends from latitudes 4°57′ N–5°00′ N and longitude 8°19′ E–8°24′ E (Figure 1), and spreads over an area of approximately 60 km^2 with an elevation range of 1 to 102 m above sea level. The area is characterized by a humid tropical climate with distinct wet and dry seasons. This area receives average annual rainfall exceeding 2500 mm per annum; and the average minimum and maximum temperatures of this area are about 22 °C and 30 °C, respectively, with a mean relative humidity of 83% [26]. The principal crops grown in the area include maize, sugar cane, cassava, groundnut, oil palm and vegetable crops (okra, *Telfairia occidentalis*, pepper, waterleaf, *Amaranthus cruentus*, etc.). The soils of the study area are developed on coastal plain sand parent material [27]. They are characterized by udic moisture regime and isohyperthemic temperature regimes, respectively [28]. Furthermore, according to USDA soil taxonomic classification, the soil order of the region is overwhelmingly Ultisols, and the soil is classified as Typic kandiudults [29].

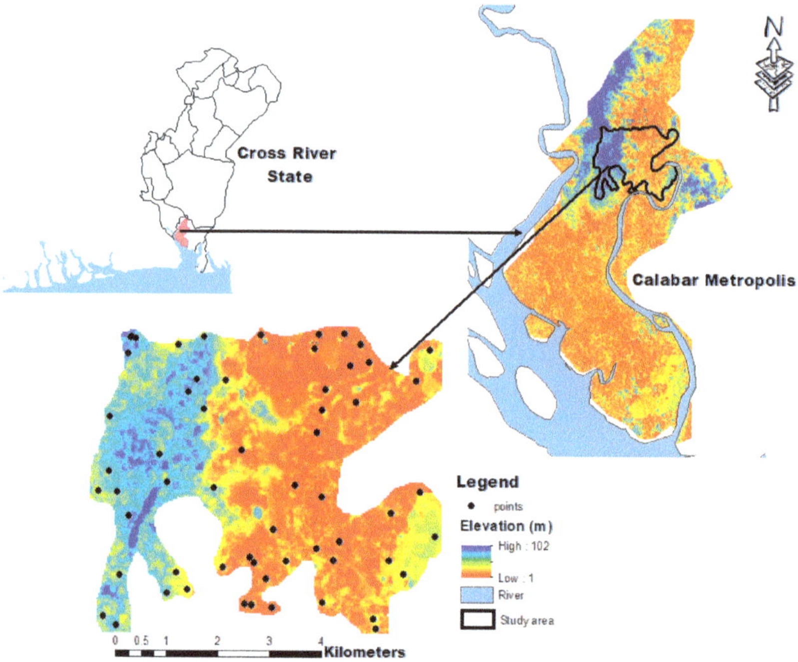

Figure 1. Geographical position of the study area in Cross River State.

2.2. Soil Sampling Regime and Laboratory Analysis

A total of sixty (n = 60) composite soil samples were collected at a depth of 0–30 cm with the aid of a soil auger at a sampling density of one sample per 3.3 m^2 and were thoroughly mixed in a Ziploc bag to obtain a homogenized sample. The soil sampling at 0–30 cm is the depth of the tillage zone. We sampled to this depth because there is no significant accumulation of SOC beyond 30 cm in the alluvial deposit. The sampling was aided by a hand-held global positioning system (GPS) (Garmin eTrex 10). These samples were adequately labeled and transported to the laboratory for analysis.

The samples were air-dried, ground, and passed through a 0.5 mm sieve. The SOC was determined by the standard Walkley–Black wet oxidation method using acid dichromate (K_2Cr_2O7) solution, as outlined in Udo et al. [30]. At the same time, effective cation exchange capacity (ECEC), base saturation, calcium (Ca), magnesium (Mg), and potassium (K) were obtained by standard laboratory procedure prescribed by Udo et al. [30]. These analyses were carried out at the University of Calabar Soil Science Department Laboratory. The soil nutrient indicators used as part of the explanatory variables was estimated from the already laboratory-measured soil properties, for example, Ca^{2+} to Mg^{2+}, K^+ to Mg^{2+} and K^+ to Ca^{2+} ratios were calculated using their representative basic cations; furthermore, in this study, they are represented as Ca_Mg, K_Mg, K_Ca, respectively.

2.3. Environmental Covariates

Environmental covariates were derived from both the digital elevation model (DEM), obtained at the spatial resolution of 30 m from ASTER GDEM, and Landsat 8 operational land imager (OLI) and a thermal infrared sensor (TIRS) acquired at https://earthexplorer.usgs.gov. DEM was processed using System for Automated Geoscientific Geographical Information System (SAGA-GIS) software terrain analysis toolbox.

The Landsat 8 Operation Land Imager (OLI) remote sensing data Path 187/Row 57 was acquired 2 January 2018 (growing season) with a cloud cover of 6.31% and SCENE_ID

"LC81880562018361LGN00", and used to derived spectral indices and land surface temperature (LST) (Table 1). The images contain nine spectral bands with a resolution of 30 m (multi-spectral), 15 m (panchromatic), and 100 m (TIRS bands 10 and 11), resampled to 30 m. The Landsat images were geometrically corrected and projected to a World Geodetic System 1984 (WGS 84) into a Universal Transverse Mercator (UTM) Zone 32N coordinate system. Detailed specifications and preprocessing method of the Landsat 8 OLI images to obtain surface reflectance images can be found in Roy et al. [31]. The area of interest (AOI) was demarcated in the satellite images with the help of the polygon feature using the ArcGIS 10.8 software (ESRI, Redlands, USA) environment.

Table 1. Environmental covariates for soil organic carbon prediction.

Environmental Covariates	Variable	Description
Landsat 8 OLI	b3	Green, 0.525–0.600 μm
	b4	Red, 0.630–0.680 μm
	b5	NIR, 0.845–0.885 μm
	Clay index (CI)	$CI = \frac{SWIR1}{SWIR2}$
	Iron index	$\text{Iron index} = \frac{Red}{Blue}$
	Normalized Difference build-up Index (NDBI)	$NDBI = \frac{(SWIR - NIR)}{(SWIR + NIR)}$
	Ratio Vegetation Index (RVI)	$RVI = \frac{NIR}{RED}$
	Soil Adjusted Vegetation Index (SAVI)	$SAVI = \frac{(NIR - RED)}{(NIR + RED + L)} \times (1 + L)$
	Normalized Difference Vegetation Index (NDVI)	$NDVI = \frac{(\text{Band 5} - \text{Band 4})}{(\text{Band 5} + \text{Band 4})}$
	Normalized Difference Moisture Index (NDMI)	$NDMI = \frac{(NIR - SWIR)}{(NIR + SWIR)}$
	Land surface temperature (LST)	$LST = \frac{BT}{\{1 + [(\lambda BT/\rho) ln\varepsilon]\}}$
ASTER GDEM	Elev	Elevation
	PCurv	Plan curvature
	TCA	Total catchment area
	CNBL	Channel Network base level
	TWI	Topographic wetness index

Retrieval of land surface temperature (LST) from thermal infrared sensor (TIRS) band 10 was carried out according to the following sequence of steps. The first step involves the conversion of the Digital number (DN) of the thermal infrared band into spectral radiance ($L\lambda$) as presented in Equation (1):

$$L\lambda = M_L \times Q_{cal} + A_L \tag{1}$$

where, $L\lambda$ = atmospheric spectral radiance (SR) in watts/(m^2 · srad · μm), M_L = band-specific multiplicative rescaling factor from the metadata, Q_{cal} = corresponds to band 10, A_L = band-specific additive rescaling factor from the metadata.

The second step involves the conversion of spectral radiance to brightness temperature in Celsius.

$$BT = \frac{K_2}{\ln\left(\frac{K_1}{L\lambda} + 1\right)} - 273.15 \tag{2}$$

where, BT is the satellite brightness temperature in Celsius, and K_1 and K_2 represent thermal conversion from the metadata.

$$L\lambda = \text{spectral radiance at the sensor's aperture} \left[W/\left(m^2 \cdot sr \cdot \mu m\right)\right]$$

where, W = Atmospheric water vapor content.

The next step was the calculation of the normalized difference vegetation index (NDVI), the proportion of vegetation (P_V), which is highly related to the NDVI, and emissivity (ε), which is related to the P_V.

$$\text{NDVI} = \frac{(\text{Band 5} - \text{Band 4})}{(\text{Band 5} + \text{Band 4})} \tag{3}$$

Estimation of the proportion of vegetation P_V

$$P_V = \left[\frac{\text{NDVI} - \text{NDVI}_{\text{min}}}{\text{NDVI}_{\text{max}} - \text{NDVI}_{\text{min}}}\right]^2 \tag{4}$$

Estimation of land surface emissivity (LSE)

$$\varepsilon = 0.004 \times P_V + 0.986 \tag{5}$$

Calculation of land surface temperature

$$LST = \frac{\text{BT}}{\{1 + [(\lambda BT/\rho) ln\varepsilon]\}} \tag{6}$$

where LST is Celsius, BT is the at-sensor brightness temperature in Celsius, λ (10.8 μm) is the wavelength of the emitted radiance: $\rho = h \times c/\sigma = 1.438 \times 10^{-2}$ mK, σ is the Stefan–Boltzmann constant, h is Planck's constant, c is the velocity of light, and ε is the land surface emissivity (LSE). The computation of other covariates from Landsat 8 OLI is shown in Table 1.

2.4. *Machine Learning Techniques*

In this study, five ML algorithms, including random forest (RF), cubist regression, artificial neural networks (ANN), support vector machine (SVM), and multiple linear regression, were chosen. A brief description of the ML techniques used in this study are presented as follows:

2.4.1. Random forest

Random forests (RF) is an ensemble of classification and regression trees (CART). This MLA was developed by Breiman [32] and is said to be as accurate as or better than adaptive boosting, yet computationally faster [33,34]. RF algorithm can handle both continuous and categorical variables. The RF algorithm is quite robust to noise in predictors and thus does not require a pre-selection of variables [35]. In RF, two hyperparameters are usually modified by users to regulate the complexity of the models, including (a) the number of trees (or iterations) (ntree), which also corresponds to the numbers of decision trees; random forests will overfit if the number is too large; (b) and mtry depicts the number of indicators that are randomly sampled as candidates at each split. In this case study, we will tune two parameters, namely the ntree and the mtry parameters that have the following effect on our random forest model.

In this present study, the model performance is obtained from each combination of the hyperparameters tuning with the grid search method [36] with cross-validation (CV) methods. K-fold CV is one of the extensively employed CV methods in machine learning and there is no definite rule for selecting the value of k. However, a value of k = 5 or 10 is ubiquitous in the field of applied machine learning and in this present study, we adopted this k = 10 in five repetitions. This was executed to avoid bias in data selection during RF hyperparameters tuning. According to Rodriguez [37], the bias of an accurate estimate will be smaller when the number of folds is either five or ten.

2.4.2. Cubist Regression

The cubist model was developed by Quinlan [38] as a rule-based model which is an extension of the M5 tree model. According to Kuhn [39], the model structure consists of a conditional component—or piecewise function acting as a decision tree, coupled with multiple linear regression models. The trees are reduced to a set of rules which are eliminated via pruning or combined for simplification. The main benefit of the cubist method is to add multiple training committees and boosting to make the weights more balanced [38–40]. The cubist model adds boosting with training committees (usually greater than one) which is similar to the method of "boosting" by sequentially developing a series of trees with adjusted weights. The number of neighbours in the cubist model is applied to amend the rule-based prediction [39]. This model was implemented in R with tuning two hyper-parameters: neighbors (Instances) and committees (Committees). These two parameters are the most likely parameters to have the largest effect on the final performance of the cubist model. Cubist followed a similar approach in RF

2.4.3. Artificial Neural Network

In predictive modeling and forecasting, as well as nonlinear and impermanent time series of processes where there is no exact solution and clear relationship to recognize and describe them, artificial neural networks have shown good performance. The frequently used ANN model is referred to as the multilayer perceptron (MLP). This model is occasionally used as a substitute for a feed-forward network. The MLP requires a well-known output so that to learn and train the network; this type of neural network is referred to as a supervised network. MLP produces a model that plots the input to the output using training data so that subsequently, the model is applied to predict the output when it is unknown. In the present study, and after some preliminary tests to choose the model, multilayer feed-forward back-propagation ANN was applied [41]. The ANN models are well adapted for modeling nonlinear behaviour. They have the capacity of learning for complex relationships between multiple inputs and output variables. The ANN model was run in R using the package "nnet." The best structure for the ANN model was obtained by changing the size (number of units in the hidden layer).

2.4.4. Multiple Linear Regression

Multiple linear regression (MLR) is a machine learning algorithm applied to regress a target variable that is SOC in this study against some selected covariates (e.g., environmental variables and soil nutrient indicators). In soil spatial prediction functions, MLR is a least-squares model where a targeted soil property is predicted from selected explanatory variables. So, in this present research, a linear relationship was established for SOC (response variable) using the explanatory variables. A simple MLR equation is presented in Equation (1).

$$y = a + \sum_{i-1}^{n} b_i \times x_i \pm \varepsilon_i \tag{7}$$

where n = number of predictors; y = response variable (SOC); x_i = explanatory variables or predictors (environmental and soil nutrient indicators variables); a = intercept (constant term); b_i = partial regression coefficients; ε_i = the model's error term (also known as the residuals).

This was automatically implemented in R using the k = 10 folds CV in five repetitions. In addition, the tuning parameter "intercept" was held constant at a value of true

2.4.5. Support Vector Machine (SVM)

Support vector machine (SVM) is a machine learning algorithm that produces an optimal separating hyperplane to differentiate classes that overlap and are not separable in a linear way. It was originally developed for classification purposes; however, it can also be used for regression problems [42]. In this study, SVM for regression (SVR) was implemented. SVR is a kernel-based learning regression method that was proposed by Cherkassky [43]. It is based on the computation of a linear regression function in a multidimensional feature space. Hence, modeling a linear regression hyperplane for nonlinear relationships is possible with the feature space. Two forms of SVM regression, namely, "epsilon (ε)-SVR" and "nu (v)-SVR," are commonly used in the SVM model. The original SVM formulations for regression (SVR) use parameter cost (c) and epsilon (ε) to apply a penalty to the optimization for points that are incorrectly predicted. Several studies, including Siewert [44] and Zhang et al. [45] have utilized SVR in environmental monitoring studies to predict SOC. In SVM regression, the Gaussian Radial Basis Function (RBF) kernel was applied. We employed the RBF kernel to obtain an optimal SVM regression model which is important to obtain the best set of penalty parameters C and kernel parameters gamma (γ) for the SOC training datasets. In the present study, we evaluated the training set and then tested the model performance on the validation set.

2.5. Data Scaling and Partitioning

The dataset used for modeling (n = 60) was scaled to a range between 0 and 1, indicating the lowest and the highest value, respectively. To evaluate the suitability of the different models for SOC prediction, a completely random technique was applied to divide the dataset into training (80%), and test (20%) datasets. Each model was fitted using the train data while the test data was used for validation. A 10-fold cross-validation was applied to the training dataset for each of the models used in the study and repeated five times. This and all modeling were performed in R software [46].

2.6. Model Validation and Accuracy Assessment

From the pool of twenty (22) SOC predictors, only the significant predictors (p-value < 0.1) were selected to build a prediction model. This was established using a simple correlation matrix. The models selected for this study were evaluated for their performance. The models were trained with 80% of the dataset (i.e., 48 observation points) and the validation set was tested by the remaining 20% of the dataset (i.e., 12 observation points). Mean absolute error (MAE), root-mean-square error (RMSE) and R^2 were used to determine the model performance according to the following equations:

$$MAE = \frac{1}{n}\sum_{i=1}^{n}\left|SOC(X_i) - SOC(\hat{X}_i)\right| \quad (8)$$

$$RMSE(\%) = \sqrt{\frac{1}{n}\sum_{i=1}^{n}\left[SOC(\hat{X}_i) - SOC(X_i)\right]^2} \quad (9)$$

$$R^2(\%) = 1 - \frac{\sum_i\left[SOC(X_i) - SOC(\hat{X}_i)\right]^2}{\sum_i\left[SOC(X_i) - SOC(\hat{X}_i)\right]^2} \quad (10)$$

where n = the size of the observations, $SOC(X_i)$ = measured response and $SOC(\hat{X}_i)$ = predicted response values, respectively, for the i-th term observation, $SOC(\overline{X_i})$ being the average of the response variable. Furthermore, a good model prediction was expected to have low MAE and RMSE as well as an R^2 value close to 1. Li et al. [47] proposes a classification criterion for R^2 values: $R^2 < 0.50$ (unacceptable prediction), $0.50 \leq R^2 < 0.75$ (acceptable prediction) and $R^2 \geq 0.75$ (good prediction). The same criterion was applied in the current study.

3. Results and Discussion

3.1. Descriptive Statistics

The descriptive statistics of the SOC of the study site are shown in Table 2. SOC value ranged from 0.32 to 3.10% with the mean of 1.62% and coefficient of variation (CV) of 47%. According to the classification proposed by Wilding and Drees [48], SOC samples indicated high variability (CV > 35%) which may be attributed to random factors such as environmental factors and measurement errors [49,50]. Using Landon [51] rating for tropical soils, the SOC of the study was generally low. The low SOC in the soil is consistent with the findings by Akpan-Idiok and Ogbaji [29] and Taghizadeh-Mehrjardi et al. [6] in Cross River State, and also with that of Bednář and Šarapatka [52] in the Czech Republic. The low SOC content may be attributed to the disturbance of the topsoil (0–30 cm) during tillage activities in preparation of the site for planting, in addition to high temperature, and high erodibility of the soils resulting from high rainfall intensity experienced in the area [1].

Table 2. Descriptive statistics of soil organic carbon (SOC).

	n	Mean	Median	SD	Min	Max	1st Quartile	3rd Quartile	CV
					→% ←				
SOC	60	1.62	1.38	0.76	0.32	3.10	1.0	2.24	47

Furthermore, intensive cultivation depletes soil organic matter accumulation, and in turn lowers SOC content through the increase in decomposition rate generated by the change in the aggregate structure of the soil due to the cultivation and mixing effect of tillage [53]. The current study is supported by the plausible reasons that intensive cultivated systems reduce SOC contents due to increased mineralization created through soil surface disturbance [54–57].

3.2. Correlation between SOC and Environmental Variables and Soil Indicators

Figure 2 shows the correlation between SOC and environmental variables and soil indicators. SOC was weakly correlated with b5 (r = 0.2), clay_index (r = 0.2), LST(r = −0.2), RVI (r = 0.2), SAVI (r = 0.2) and NDVI (r = 0.2) obtained from Landsat satellite imagery. Similarly, SOC was weak but significantly correlated with elevation (r = −0.2), total catchment area (r = 0.2), topographic wetness index (r = 0.2) and channel network base level (r = 0.2) derived from digital elevation model (DEM). The result obtained here showed that environmental variables obtained from Landsat imagery gave a poor relationship with SOC in a flat topographical system. Environmental variations in areas with a small range of topography, such as plains, are usually very small [20]. This factor including but not limited to the time of acquiring spaceborne data and intensive crop cultivation utilizing chemical fertilizers in the area, could be responsible for the low correlation between SOC and NDVI in the studied soil. Furthermore, NDVI may only show a high contribution to SOC when the crops are producing more crop biomass. The result is supported by the findings of Florinsky et al. [58] and Mosleh et al. [10].

Additionally, the effect of environmental variables for SOC in this low-relief area was weakly correlated, and the spatial variability of SOC cannot be obtained by total dependence on both terrain and remote sensing parameters. On the other hand, a good relationship was obtained between SOC and soil nutrient indicators. That is, SOC was strongly correlated with ECEC (r = 0.50), base saturation (BS) (r = 0.60), K_Ca (r = −0.60) and moderately correlated with Ca_Mg (r = 0.40). ECEC and BS increase with an increase in organic matter accumulation [56,59]. The observation between soil nutrient indicators (Ca_Mg and K_Ca) and SOC represents that the accumulation of organic materials in the soil surface may increase or decrease Ca, Mg and K in the soil.

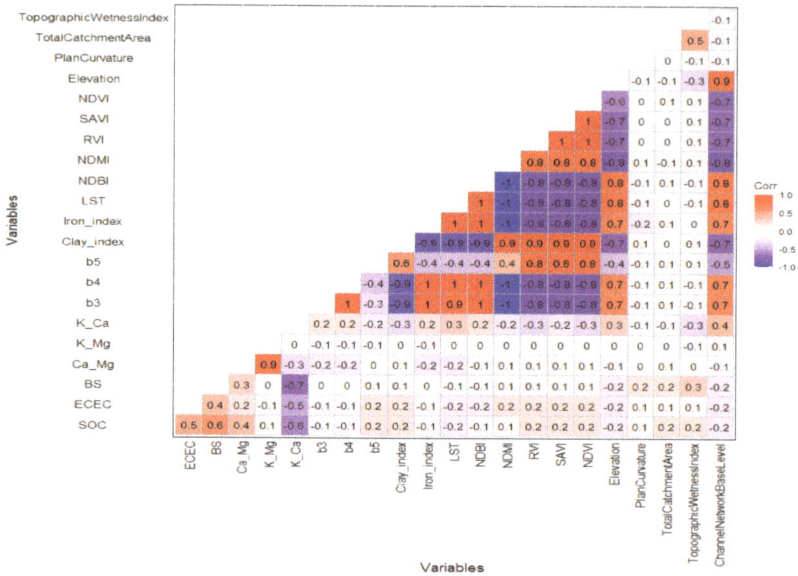

Figure 2. Correlation matrix between SOC (%) and environmental variables and soil nutrient indicators.

3.3. Modeling Approach and Variables of Importance in the Individual Models

The optimum selection strategy of covariates is that the correlation between the covariates and the response variable is significant or high, and the covariates are obtained effortlessly [16]. Among 22 explanatory variables, only 14 of the explanatory variables that showed a significant correlation with SOC were selected ($p < 0.01$). These variables were b5, clay_Index, LST, RVI, SAVI, NDVI, elevation, total catchment area, topographic wetness index, channel network base level, ECEC, BS, K_Ca and Ca_Mg.

For RF prediction model, as shown in Figure 3, Ca_Mg, BS, ECEC, K_Ca, topographic wetness index best predictors to explain the variability of SOC in a flat terrain system. In addition, the result reveals that the soil nutrient indicators contribute much more compared to environmental variables in estimating SOC in a flat topographic system.

Similarly, the environmental variables show their inability to contribute to SOC prediction in low relief conditions. This result is supported by Mosleh et al. [10]. They conducted a study in Iran and stated that environmental variables are not essential relative variables in low relief conditions. Furthermore, Solly et al.'s [60] report supported this current study through the study done in Switzerland on the preservation of SOC using cation exchange capacity plus mean annual temperature, mean annual precipitation, and leaf area index. Their study concludes that soil physical and chemical properties serve as better predictors in a homogenous terrain. Similar conclusions were reported by Song et al. [61], who noted that local environmental attributes play a less significant role than other predictors on a flat terrain system. Li et al. [62] inferred that environmental attributes could capture large-scale influences of soil transport but not those occurring at a flat topographic condition. Thus, the over-employment of environmental factors in small-scale flat terrain areas reduces the prediction accuracy and increases the calculation complexity.

Presented in Figure 4 is the cubist model prediction using the calibration set. The plot showed a similar output with RF. Ca_Mg was the best predictor with BS, b5, ECEC and topographic wetness index following. Similarly, in the artificial neural network model (Figure 5), the best predictor is ECEC, closely followed by BS, Ca_Mg, K_Ca and b5. Landsat near-infrared band (b5) gave a 50% contribution to SOC prediction, and this is reverse to a region with strong undulating topography as reported by Emadi et al. [15] for complex terrain. This output is also similar to the previous models and still powerfully reveals the dominance of soil nutrient indicators in the estimation of SOC.

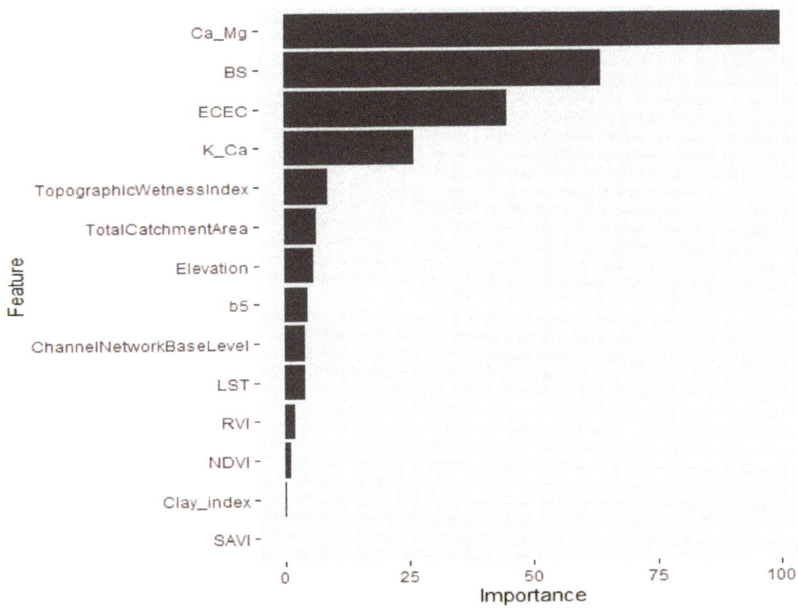

Figure 3. Relative importance variable for SOC using random forest (RF) model.

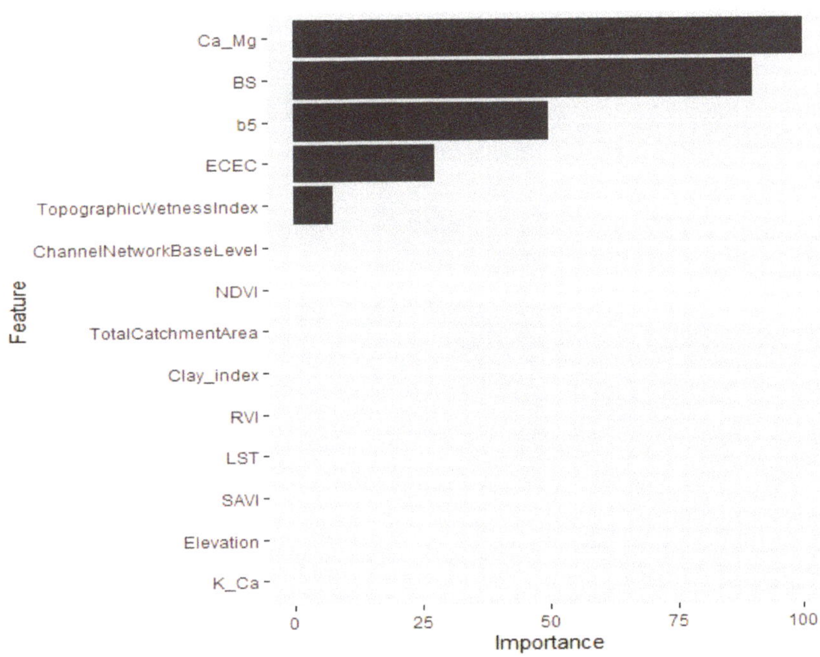

Figure 4. Relative importance variable for SOC using cubist model.

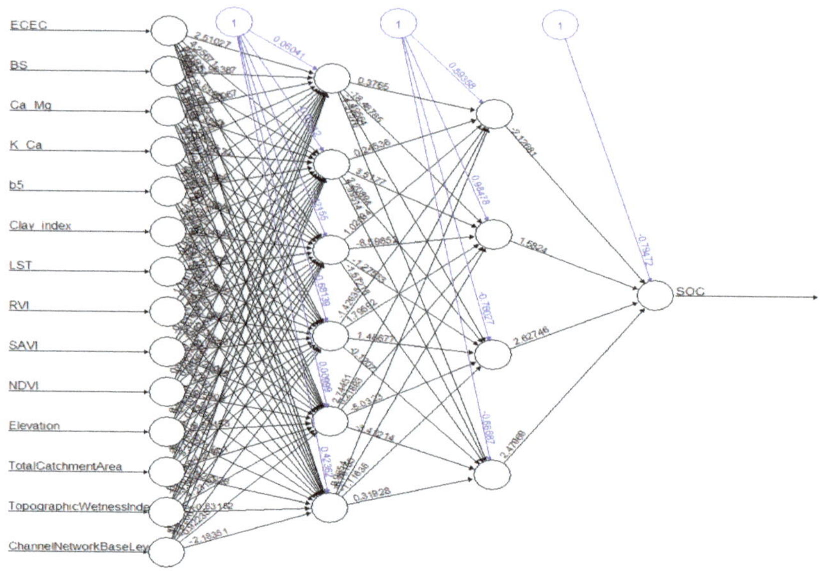

Figure 5. Relative importance variable for SOC using an artificial neural network model.

According to Figure 6, MLR presented high relative importance (>50%) of the explanatory variables. ECEC was the best predicting variable and then followed by BS, Ca_Mg, clay_index and LST. In Figure 7, the support vector machine model followed a similar pattern as compared to other models (i.e., RF and cubist). That is, soil nutrient indicators do a better job in estimating SOC to environmental variables in flat terrain condition under small-scale.

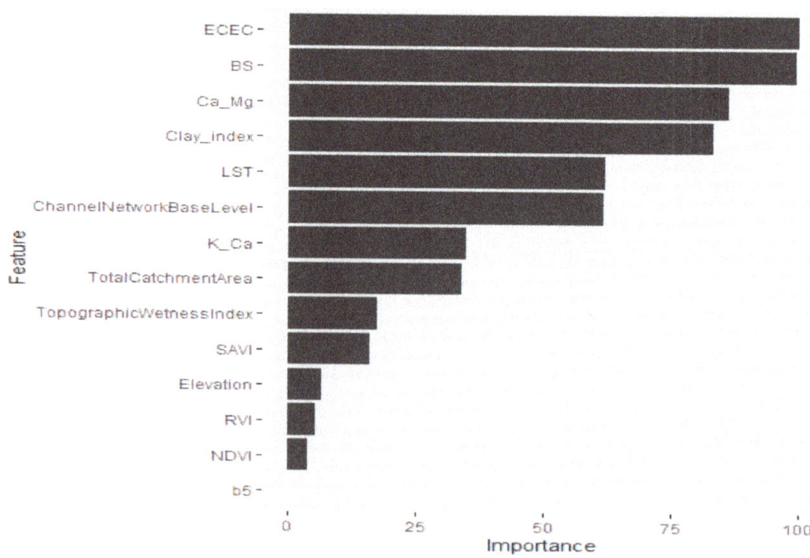

Figure 6. Relative importance variable for SOC using multiple linear regression (MLR).

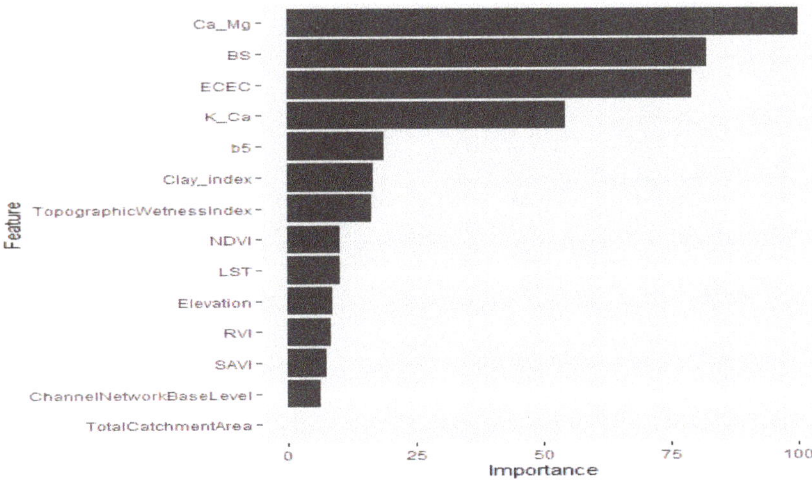

Figure 7. Relative importance variable for SOC using support vector machine (SVM).

Support vector machine model yielded Ca_Mg as the best predictor and then followed by BS, ECEC, K_Ca, b5, clay_index and topographic wetness index. The percentage contribution by topographic wetness index to SOC prediction is above the value reported by Emadi et al. [15]. However, they follow a similar pattern in that they contribute a little amount in SOC variability in low relief conditions. In all the five MLAs, NDVI made little or no contribution to SOC estimation, and this is contrary to what is experienced in more complex terrain.

3.4. SOC Estimation Using Different MLAs

Prediction model accuracy was assessed using standard validation indices such as MAE, RMSE and R^2 by 10-fold cross-validation and repeated five times. The results for both the calibration and the validation datasets are listed in Table 3. The model output was good using the calibration dataset (n = 48) except for MLR that gave an unacceptable prediction with calibration datasets ($0 < R^2 < 0.50$). In the calibration, the best performing model was ANN followed by RF, cubist, SVM and MLR with R^2 values of 0.94, 0.64, 0.54, 0.52 and 0.42, respectively. Using the validation dataset, the proposed MLA models showed their capabilities to predict SOC contents at an unsampled location in the southeastern region of Nigeria. The best performing model was RF (R^2 = 0.68) followed by the cubist model (R^2 = 0.51), SVM (R^2 = 0.36), ANN (R^2 = 0.36) and MLR (R^2 = 0.17). According to Li et al.'s [47] proposed model accuracy classification, RF and cubist models gave acceptable prediction as they fell within $0.50 < R^2 < 0.75$, while ANN, MLR and SVM gave unacceptable prediction ($0 < R^2 < 0.50$) for SOC in flat terrain conditions. The R^2 value reported in the current study was higher than that of Wang et al. [20]. They achieved an R^2 mean value of 0.48 of the total spatial SOC variability using the RF algorithm in a flat terrain of semiarid pastures of eastern Australia. Using, MLR, ECEC was the most important variable with lower R^2 value when compared to Nath [63] who reported R^2 of 0.31 with curvature as the important variable.

The RF algorithm showed the lowest mean MAE value (0.17) of the five studied ML algorithms. The cubist algorithm had the highest error with mean RMSE values of 0.57 compared with other ML models; meanwhile, RF outperformed with the lowest mean RMSE value (0.20). Contrary to the report by Emadi et al. [15] who stated that ANN, RF and cubist models had a similar predictive ability to forecast SOC in the Mazandaran province of Iran, in this current study, only RF and cubist models showed similar predictive ability. In addition, the study also contradicts the report by Taghizadeh-Mehrjardi et al. [6] and Zhang et al. [45] that reported ANN as the best model. Concerning R^2, the low predictive ability of ANN has been reported by Mosleh et al. [10]. However, this model could be improved by the acquisition of large datasets and parameters in order to fit the model that yields good performance [64].

Table 3. SOC calibration and validation results of the five machine learning models by 10-fold cross-validation.

Model	Calibration (n = 48)			Validation (n = 12)		
	MAE	RMSE	R^2	MAE	RMSE	R^2
RF	0.15	0.17	0.64	0.17	0.20	0.68
Cubist	0.18	0.22	0.54	0.49	0.57	0.51
ANN	0.04	0.06	0.94	0.22	0.26	0.36
MLR	0.60	0.77	0.42	0.23	0.28	0.17
SVM	0.17	0.21	0.52	0.19	0.22	0.36

RF: random forest; ANN: artificial neural network; MLR: multiple linear regression; SVM: support vector machine.

Figure 8 shows the scattered plots of RF, cubist, ANN, MLR and SVM predicted versus the measured SOC, respectively. In the figures, the central lines (1:1 line in black color) represented (predicted = measured). In Figure 8A reveals that RF scattered plots were more closed to the measured line than others. The plot further substantiated the MAE, RMSE, R^2 values obtained here, indicating RF as the best model predicting SOC at point scale for both calibration and test datasets using both environmental and soil nutrient indicators as variables.

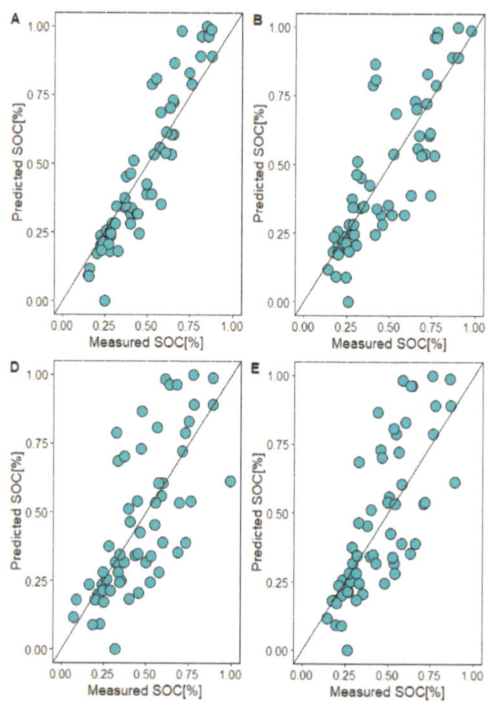

Figure 8. Measured vs. predicted values of soil organic carbon using five machine learning algorithms:(**A**) RF, (**B**) cubist, (**C**) ANN, (**D**) MLR and (**E**) SVM. (RF: random forest; cubist: regression tree; ANN: artificial neural networks; MLR: Multiple linear regression; SVM: support vector machine).

Generally, Bou Kheir et al. [65] reported that SOC variation in the floodplain of Denmark is explained by both environmental variables, remote sensing data, and soil-related data.

Wiesmeier et al. [66] reported that land use, soil types, and parent materials were the most critical variables controlling SOC distribution. Adhikari et al. [67] demonstrated the usefulness of environmental variables plus soil-related variables in explaining the SOC distribution down the soil depth in flat terrain. Besides the works mentioned above, this current study seems to contribute to the variables of choice in SOC prediction by including soil nutrient indicators (Ca_Mg, ECEC, BS, K_Ca) and these soil nutrient indicators are vital in crop growth and development. What happens in a flat terrain condition is that there is a slow rate of organic matter degradation and since soil organic matter has a large exchangeable site, basic cations (Ca^{2+}, Mg^{2+}, K^{+} and Na^{+}) are absorbed into the soil solution [68–70]. On the other hand, environmental variables that are supposed to facilitate the process of soil organic matter decomposition are impeded because these activities are carried on a homogenous terrain. Thus, they make very little or no contribution to SOC prediction as exposed in this current paper.

3.5. Digital Soil Mapping of SOC

The spatial result of digital SOC maps was produced with extracted cultivated land via the different models (RF, cubist, ANN, MLR and SVM) (Figure 9). RF and cubist models' predicted SOC maps (Figure 8A) were relatively similar to the measured SOC map (Figure 9B) and showed substantial spatial variability of SOC. High predicted SOC values occurred in the center, northeastern, eastern and northwestern and southern parts of the research area, where the land was mainly covered by groundnut, pumpkin, litter falls as a result of dense vegetation cover. In addition, long-term application of organic manure could explain the high SOC contents in these parts of the research area used for cultivation. Similarly, the dominant low values were observed in all the parts of the maps were possible because of the loss of soil nutrients in the area through active cultivation without proper management procedures.

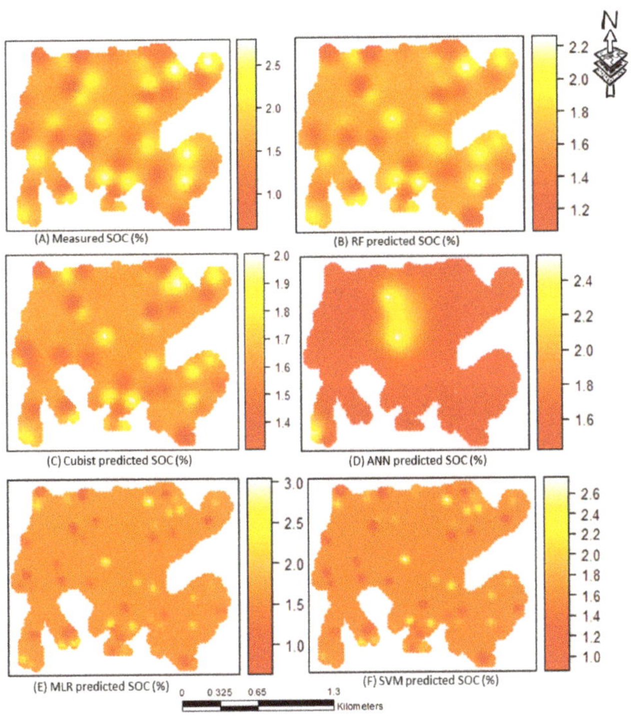

Figure 9. Prediction maps for soil organic carbon.

The maps generated by the MLR and SVM models are presented in Figure 9E,F, which highlight the high and low values in all the geographical positions of the maps. Compared with the RF, cubist, MLR and SVM models, the map of ANN more strongly manifested low SOC values in all the parts with high values at the center of the study area. Moreover, the map obtained by MLR resembled that of the SVM model (however, the map acquired by MLR ranged from 1.0 to 3.0% while SVM ranged from 1.0 to 2.6%).

4. Conclusions

In conclusion, among several predictors considered in this current study, environmental variables (b5, topographic wetness index and total catchment area), and soil nutrient indicators (Ca_Mg, ECEC, BS, K_Ca) had a significant influence on SOC distribution in the study area. They are valuable indicators in SOC prediction in flat homogenous topography. The RF model was the best model in the study. The resulting SOC map from RF prediction showed low SOC in the east and high SOC in the west direction of the site. The map suggests the gradual transportation and deposition of soil sediments. The study confirmed that SOC distribution could be digitally mapped through the five models as expected but more accurately with either RF or cubist models. Moreover, soil nutrient indicators, topographic wetness index and total catchment area were closely related to the SOC content in flat slope conditions.

From the study, soil nutrient program for SOC improvement could be implemented via RF and cubist models, incorporated into the digital soil mapping approach. However, RF showed to be a useful tool in prediction. The accuracy indicated that they act to reduce bias, and they can accommodate random inputs and random features to produce good results in classification—less so in regression. Cubist models generally give better results than those produced by simple techniques such as multivariate linear regression, while also being easier to understand than ANN.

Typically, low SOC levels require the application of organic manures, fallow cropping systems, organic fertilizer application and residual cropping to increase SOC levels. Through the application of MLAs in conjunction with digital soil mapping, the proper understanding of existing soil conditions may be gathered and thus allow precise soil management for sustainable crop production. This research sets a precedent for future digital soil mapping in other regions of Nigeria. Future studies should include other auxiliary predictors (e.g., soil physical and chemical properties, and lithological data) as well as cover a broader range of soil types to improve model performance.

Author Contributions: Conceptualization, I.A.I., and K.J.; methodology, K.J., I.A.I. and S.M.A.; software, K.J., P.C.A.; validation, K.J.; formal analysis, I.A.I., and K.J.; investigation, E.O.A and N.M.K., resources, K.J.; data curation, S.M.A.; writing—original draft preparation, I.A.I., and K.J.; writing—review and editing, I.A.I., E.O.A., N.M.K., P.C.A.; visualization, K.J.; supervision, S.M.A. All authors have read and agreed to the published version of the manuscript.

Funding: This research was funded by Internal PhD grant of the Faculty of Agrobiology, Food and Natural Resources of the Czech University of Life Sciences Prague (CZU), grant number no. 21130/3131" and "The APC was funded by Katedra pedologie a ochrany půd, Fakulta agrobiologie, potravinových a přírodních zdrojů Česká zemědělská univerzita v Praze."

Acknowledgments: This study was supported by an Internal PhD grant no. 21130/3131 of the Faculty of Agrobiology, Food and Natural Resources of the Czech University of Life Sciences Prague (CZU). The support from the Ministry of Education, Youth and Sports of the Czech Republic (project No.CZ.02.1.01/0.0/0.0/16_019/0000845) is also acknowledged. The Centre of Excellence (Centre of the investigation of synthesis and transformation of nutritional substances in the food chain in interaction with potentially risk substances of anthropogenic origin: comprehensive assessment of the soil contami-nation risks for the quality of agricultural products, NutRisk Centre), European project no. CZ.02.1.01/0.0/0.0/16_019/0000845 is highly acknowledged. Finally, we would like to thank the anonymous reviewers for their valued contributions and insights on the betterment of the original manuscript.

Conflicts of Interest: The authors declare no conflict of interest.

References

1. Amalu, U.C.; Isong, I.A. Status and spatial variability of soil properties in relation to fertilizer placement for intercrops in an oil palm plantation in Calabar, Nigeria. *Niger. J. Crop Sci.* **2018**, *5*, 58–72.
2. Akpan, J.F.; Aki, E.E.; Isong, I.A. Comparative assessment of wetland and coastal plain soils in Calabar, Cross River State. *Glob. J. Agric. Sci.* **2017**, *16*, 17–30. [CrossRef]
3. Jenny, H. *Factors of Soil Formation: A System of Quantitative Pedology*, 1st ed.; McGraw-Hill Inc.: New York, NY, USA, 1941.
4. Chikezie, I.A.; Eswaran, H.; Asawalam, D.O.; Ano, A.O. Characterization of two benchmark soils of contrasting parent materials in Abia State, Southeastern Nigeria. *Glob. J. Pure Appl. Sci.* **2010**, *16*, 23–29. [CrossRef]
5. Amalu, U.C.; Isong, I.A. Land capability and soil suitability of some acid sand soil supporting oil palm (*Elaeis guinensis Jacq*) trees in Calabar, Nigeria. *Niger. J. Soil Sci.* **2015**, *25*, 92–109.
6. Taghizadeh-Mehrjardi, R.; Nabiollahi, K.; Kerry, R. Digital mapping of soil organic carbon at multiple depths using different data mining techniques in Baneh region, Iran. *Geoderma* **2016**, *266*, 98–110. [CrossRef]
7. Bian, Z.; Guo, X.; Wang, S.; Zhuang, Q.; Jin, X.; Wang, Q.; Jia, S. Applying statistical methods to map soil organic carbon of agricultural lands in northeastern coastal areas of China. *Arch. Agron. Soil Sci.* **2020**, *66*, 532–544. [CrossRef]
8. Chen, L.; Ren, C.; Li, L.; Wang, Y.; Zhang, B.; Wang, Z.; Li, L. A Comparative Assessment of Geostatistical, Machine Learning, and Hybrid Approaches for Mapping Topsoil Organic Carbon Content. *ISPRS Int. J. Geo-Information* **2019**, *8*, 174. [CrossRef]
9. Kingsley, J.; Lawani, S.O.; Esther, A.O.; Ndiye, K.M.; Sunday, O.J.; Penížek, V. Predictive Mapping of Soil Properties for Precision Agriculture Using Geographic Information System (GIS) Based Geostatistics Models. *Mod. Appl. Sci.* **2019**, *13*, 60. [CrossRef]
10. Mosleh, Z.; Salehi, M.; Jafari, A.; Esfandiarpour, I.; Mehnatkesh, A. The effectiveness of digital soil mapping to predict soil properties over low-relief areas. *Environ. Monit. Assess.* **2016**, *188*, 195. [CrossRef]
11. Zeraatpisheh, M.; Jafari, A.; Bodaghabadi, M.B.; Ayoubi, S.; Taghizadeh-Mehrjardi, R.; Toomanian, N.; Kerry, R.; Xu, M. Conventional and digital soil mapping in Iran: Past, present, and future. *Catena* **2020**, *188*, 104424. [CrossRef]
12. Akpan-Idiok, A.U.; Ukwang, E.E. Characterization and classification of coastal plain soils in Calabar, Nigeria. *J. Agric. Biotechnol. Econ.* **2012**, *5*, 19–33.
13. Baldock, J.A.; Wheeler, I.; McKenzie, N.; McBrateny, A. Soils and climate change: Potential impacts on carbon stocks and greenhouse gas emissions, and future research for Australianagriculture. *Crop Pasture Sci.* **2012**, *63*, 269–283. [CrossRef]
14. Minasny, B.; Setiawan, B.I.; Arif, C.; Saptomo, S.K.; Chadirin, Y. Digital mapping for cost effective and accurate prediction of the depth and carbon stocks in Indonesian peatlands. *Geoderma* **2016**, *272*, 20–31.
15. Emadi, M.; Taghizadeh-Mehrjardi, R.; Cherati, A.; Danesh, M.; Mosavi, A.; Scholten, T. Predicting and Mapping of Soil Organic Carbon Using Machine Learning Algorithms in Northern Iran. *Remote Sens.* **2020**, *12*, 2234. [CrossRef]
16. Wang, B.; Waters, C.; Orgill, S.; Cowie, A.; Clark, A.; Li, L.D.; Simpson, M.; McGowen, I.; Sides, T. Estimating soil organic carbon stocks using different modelling techniques in the semi-arid rangelands of eastern Australia. *Ecol. Indic.* **2018**, *88*, 425–438. [CrossRef]
17. Chen, D.; Chang, N.; Xiao, J.; Zhou, Q.; Wu, W. Mapping dynamics of soil organic matter in croplands with MODIS data and machine learning algorithms. *Sci. Total Environ.* **2019**, *669*, 844–855. [CrossRef]
18. Liu, S.; An, N.; Yang, J.; Dong, S.; Wang, C.; Yin, Y. Prediction of soil organic matter variability associated with different landuse types in mountainous landscape in southwestern Yunnan province, China. *Catena* **2015**, *133*, 137–144. [CrossRef]
19. Forkuor, G.; Hounkpatin, O.K.; Welp, G.; Thiel, M. High resolution mapping of soil properties using remote sensing variables in southwestern Burkina Faso: A comparison of machine learning and multiple linear regression models. *PLoS ONE* **2017**, *12*, e0170478. [CrossRef]
20. Wang, X.; Zhang, Y.; Atkinson, P.M.; Yao, H. Predicting soil organic carbon content in Spain by combining Landsat TM and ALOS PALSAR images. *Int. J. Appl. Earth Obs. Geoinf.* **2020**, *92*, 102182. [CrossRef]

21. Gehl, R.J.; Rice, C.W. Emerging technologies for in situ measurement of soil carbon. *Clim. Chang.* **2007**, *80*, 43–54. [CrossRef]
22. Al-Abbas, A.H.; Swain, P.H.; Baumgardner, M.F. Relating organic matter and clay content to the multi-spectral radiance of soils. *Soil Sci.* **1972**, *114*, 477–485. [CrossRef]
23. McMorrow, J.M.; Cutler, M.E.J.; Evans, M.G.; Al-Roichdi, A. Hyperspectral indices for characterizing upland peat composition. *Int. J. Remote Sens.* **2004**, *25*, 313–325. [CrossRef]
24. Viscarra Rossel, R.A.; Behrens, T. Using data mining to model and interpret soil diffuse reflectance spectra. *Geoderma* **2010**, *158*, 46–54. [CrossRef]
25. Fathololoumi, S.; Vaezi, A.R.; Alavipanah, S.K.; Ghorbani, A.; Saurette, D.; Biswas, A. Improved digital soil mapping with multitemporal remotely sensed satellite data fusion: A case study in Iran. *Sci. Total. Environ.* **2020**, *721*, 137703. [CrossRef]
26. Amalu, U.C.; Isong, I.A. Long-term impact of climate variables on agricultural lands in Calabar, Nigeria, I. Trend analysis of rainfall, temperature and relative humidity. *Niger. J. Crop Sci.* **2017**, *4*, 79–94.
27. Afu, S.M.; Isong, I.A.; Awaogu, C.E. Agricultural potentials of floodplain soils with contrasting parent material in Cross River State, Nigeria. *Glob. J. Pure Appl. Sci.* **2019**, *25*, 13–22. [CrossRef]
28. USDA NRCS. *Soil Survey Staff. Keys to Soil Taxonomy*, 12th ed.; United States Department of Agriculture, Natural Resources Conservation Service: Washington, DC, USA, 2014.
29. Akpan-Idiok, A.U.; Ogbaji, P.O. Characterization and Classification of Onwu River Floodplain Soils in Cross River State, Nigeria. *Int. J. Agric. Res.* **2013**, *8*, 107–122. [CrossRef]
30. Udo, E.J.; Ibia, T.O.; Ogunwale, J.A.; Ano, A.O.; Esu, I.E. *Manual of Soil, Plant and Water Analyses*; Sibon Books Limited: Lagos, Nigeria, 2009.
31. Roy, D.P.; Wulder, M.A.; Loveland, T.R.C.E.W.; Allen, R.G.; Anderson, M.C.; Helder, D.; Irons, J.R.; Johnson, D.M.; Kennedy, R.; Scambos, T.A.; et al. Landsat-8: Science and product vision for terrestrial global change research. *Remote Sens. Environ.* **2014**, *145*, 154–172. [CrossRef]
32. Breiman, L. Random forests. *Mach. Learn.* **2001**, *45*, 5–32. [CrossRef]
33. Heung, B.; Bulmer, C.E.; Schmidt, M.G. Predictive soil parent material mapping at a regional scale: A Random Forest approach. *Geoderma* **2014**, *214*, 141–154. [CrossRef]
34. Gislason, P.O.; Benediktsson, J.A.; Sveinsson, J.R. Random forests for land cover classification. *Pattern Recogn. Lett.* **2006**, *27*, 294–300. [CrossRef]
35. Díaz-Uriarte, R.; De Andres, S.A. Gene selection and classification of microarray data using random forest. *BMC Bioinform.* **2006**, *7*, 1–13. [CrossRef]
36. Zhou, J.; Li, E.; Wei, H.; Li, C.; Qiao, Q.; Armaghani, D.J. Random Forests and Cubist Algorithms for Predicting Shear Strengths of Rockfill Materials. *Appl. Sci.* **2019**, *9*, 1621. [CrossRef]
37. Rodríguez, J.D.; Perez, A.; Lozano, J.A. Sensitivity Analysis of k-Fold Cross Validation in Prediction Error Estimation. *IEEE Trans. Pattern Anal. Mach. Intell.* **2010**, *32*, 569–575. [CrossRef]
38. Quinlan, J.R. Learning with continuous classes. In Proceedings of the 5th Australian Joint Conference on Artificial Intelligence, Tasmania, Australia, 16–18 November 1992; pp. 343–348.
39. Kuhn, M.; Johnson, K. *Applied Predictive Modeling*; Springer: Berlin, Germany, 2013; Volume 26.
40. Wang, Y.W.I. Inducing Model Trees for Continuous Classes. In *Proceedings of the 9th European Conference on Machine Learning*; Springer: Berlin/Heidelberg, Germany, 1997; pp. 128–137.
41. Behrens, T.; Forster, H.; Scholten, T.; Steinrücken, U.; Spies, E.D.; Goldschmitt, M. Digital soil mapping using artificialneural networks. *J. Plant Nutr. Soil Sci.* **2005**, *168*, 21–33. [CrossRef]
42. Vapnik, V. *The Nature of Statistical Learning Theory*; Springer: New York, NY, USA, 1995.
43. Cherkassky, V.; Mulier, F. *Learning from Data: Concept, Theory and Methods*; John Wiley and Sons: New York, NY, USA, 1998.
44. Siewert, M.B. High-resolution digital mapping of soil organic carbon in permafrost terrain using machine learning: A case study in a sub-Arctic peatland environment. *Biogeosciences* **2018**, *15*, 1663–1682. [CrossRef]
45. Zhang, Y.; Sui, B.; Shen, H.; Ouyang, L. Mapping stocks of soil total nitrogen using remote sensing data: A comparison of random forest models with different predictors. *Comput. Electron. Agric.* **2019**, *160*, 23–30. [CrossRef]
46. R Core Team. *R: A Language and Environment for Statistical Computing*; R Foundation for Statistical Computing: Vienna, Austria, 2019; Available online: https://www.r-project.org/ (accessed on 13 May 2020).

47. Li, L.; Lu, J.; Wang, S.; Ma, Y.; Wei, Q.; Li, X.; Cong, R.; Ren, T. Methods for estimating leaf nitrogen concentration of winter oilseed rape (*Brassica napus L.*) using in situ leaf spectroscopy. *Ind. Crop. Prod.* **2016**, *91*, 194–204. [CrossRef]
48. Wilding, L.P.; Drees, L.R. Spatial variability and pedology. In *Pedogenesis and Soil Taxonomy: Concepts and Interactions*; Wilding, L.P., Smeck, N.E., Hall, G.F., Eds.; Elsevier: New York, NY, USA, 1983; pp. 83–116.
49. Denton, O.; Modupe, A.; Ojo, V.O.A.; Adeoyolanu, A.O.; Are, O.D.; Adelana, K.S.; Oyedele, A.O.; Adetayo, A.O.; Oke, A.O. Assessment of spatial variability and mapping of soil properties for sustainable agricultural production using geographic information system techniques. *Cogent Food Agric.* **2017**, *3*, 1–12. [CrossRef]
50. Reza, S.K.; Nayak, D.C.; Chattopadhyay, T.; Mukhopadhyay, S.; Singh, S.K.; Srinivasan, R. Spatial distribution of soil physical properties of alluvial soils: A geostatistical approach. *Arch. Agron. Soil Sci.* **2016**, *62*, 972–981. [CrossRef]
51. Landon, J.R. *Booker Tropical Soil Manual: A Handbook for Soil Survey and Agricultural Land Evaluation in the Tropics and Subtropics*; Longman: New York, NY, USA, 1991.
52. Bednářa, M.; Šarapatkaa, B. Relationships between physical–geographical factors and soil degradation on agricultural land. *Environ. Res.* **2018**, *164*, 660–668. [CrossRef]
53. Gelaw, A.M.; Singh, B.R.; Lal, R. Organic carbon and nitrogen associated with soil aggregates and particle sizes under different land uses in Tigray, Northern Ethiopia. *Land Degrad. Dev.* **2013**, *26*, 690–700. [CrossRef]
54. Anikwe, M.A.N. Carbon storage in soils of southeastern Nigeria under different management practices. *Carbon Balance Manag.* **2010**, *5*, 5. [CrossRef]
55. Six, J.; Feller, C.; Denef, K.; Ogle, S.M.; Sa, J.C.M.; Albrecht, A. Soil organic matter, biota and aggregation in temperate and tropical soils- Effect of no-tillage. *Agronomie* **2002**, *22*, 755–775. [CrossRef]
56. Lal, R. Soil Carbon sequestration impacts on global climate change and food security. *Science* **2004**, *30*, 1623–1627. [CrossRef]
57. Purwanto, B.H.; Alam, S. Impact of intensive agricultural management on carbon and nitrogen dynamics in the humid tropics. *Soil Sci. Plant Nutr.* **2020**, *66*, 50–59. [CrossRef]
58. Florinsky, I.V.; McMahon, S.; Burton, D.L. Topographic control of soil microbial activity: A case study of denitrifiers. *Geoderma* **2004**, *119*, 33–53. [CrossRef]
59. Akpan-Idiok, A.U.; Ogbaji, P.O.; Antigha, N.R.B. Infiltration, degradation rate and vulnerability potential of Onwu River floodplain soils in Cross River State, Nigeria. *J. Agric. Biotechnol. Ecol.* **2012**, *5*, 62–74.
60. Solly, E.F.; Weber, V.; Zimmermann, S.; Walthert, L.; Hagedorn, F.; Schmidt, M.W.I. A Critical Evaluation of the Relationship between the Effective Cation Exchange Capacity and Soil Organic Carbon Content in Swiss Forest Soils. *Front. For. Glob. Chang.* **2020**, *3*, 98. [CrossRef]
61. Song, Y.Q.; Yang, L.A.; Li, B.; Hu, Y.M.; Wang, A.L.; Zhou, W.; Cui, X.S.; Liu, Y.L.; Song, Y.Q.; Yang, L.A.; et al. Spatial prediction of soil organic matter using a hybrid geostatistical model of an extreme learning machine and ordinary kriging. *Sustainability* **2017**, *9*, 754. [CrossRef]
62. Li, X.; McCarty, G.W.; Du, L.; Lee, S. Use of Topographic Models for Mapping Soil Properties and Processes. *Soil Syst.* **2020**, *4*, 32. [CrossRef]
63. Nath, D.A. Soil Landscape Modeling in the Northwest Iowa Plains Region of O'Brien County, Iowa. Master's Thesis, Iowa State University, Ames, IA, USA, 2006.
64. Padarian, J.; Minasny, B.; McBratney, A.B. Using deep learning for digital soil mapping. *Soil* **2019**, *5*, 79–89. [CrossRef]
65. Bou Kheir, R.; Greve, M.H.; Bøcher, P.K.; Greve, M.B.; Larsen, R.; McCloy, K. Predictive mapping of soil organic carbon in wet cultivated lands using classification-tree based models: The case study of Denmark. *J. Environ. Manag.* **2010**, *91*, 1150–1160. [CrossRef]
66. Wiesmeier, M.; Barthold, F.; Blank, B.; Kögel-Knabner, I. Digital mapping of soil organic matter stocks using Random Forest modeling in a semi-arid steppe ecosystem. *Plant. Soil* **2011**, *340*, 7–24. [CrossRef]
67. Adhikari, K.; Hartemink, A.E.; Minasny, B.; Kheir, R.B.; Greve, M.B.; Greve, M.H. Digital mapping of soil organic carbon contents and stocks in Denmark. *PLoS ONE* **2014**, *9*, e105519. [CrossRef]
68. Andersson, S.; Nilsson, I.; Valeur, I. Influence of dolomiticlime on DOC and DON leaching in a forest soil. *Biogeo-Chemistry* **1999**, *47*, 295–315. [CrossRef]

69. Chan, K.Y.; Heenan, D.P. Lime-induced loss of soil organiccarbon and effect on aggregate stability. *Soil Sci. Soc. Am. J.* **1999**, *63*, 1841–1844. [CrossRef]
70. Thirukkumaran, C.M.; Morrison, I.K. Impact of simulated acid rain on microbial respiration, biomass, and metabolic quotient in a mature sugar maple (*Acer saccharum*) forest floor. *Can. J. For. Res.* **1996**, *26*, 1446–1453. [CrossRef]

Publisher's Note: MDPI stays neutral with regard to jurisdictional claims in published maps and institutional affiliations.

land

MDPI

Article

Growth of Common Plants of Boreal Reclamation Sites in Oil Sands Tailings Cake Mixes and Process Water

Kwadwo Omari [1,*], Bradley D. Pinno [2], Nicholas Utting [3] and Edith H.Y. Li [1]

1 Canadian Forest Service, Northern Forestry Centre, Natural Resources Canada, Edmonton, AB T6H 3S5, Canada; edith.li@hotmail.com
2 Department of Renewable Resources, University of Alberta, Edmonton, AB T6G 2E3, Canada; bpinno@ualberta.ca
3 CanmetENERGY, Devon Research Centre, Natural Resources Canada, Devon, AB T9G 1A8, Canada; nicholas.utting@canada.ca
* Correspondence: kwadwo.omari@gmail.com

Abstract: Oil sands surface mining and processing in Alberta generate large volumes of fluid tailings and process water high in salts and metals, which must be reclaimed. We investigated growth of four common plants (two native and two non-native) found in boreal oil sands reclamation sites as influenced by substrate type (tailings cake, and mixtures of cake-sand, cake-peat, and cake-forest floor mineral mix) and water quality (0%, 50%, and 100% oil sands process water). Overall, cake-peat supported the highest aboveground biomass among substrates whereas cake and cake-sand performed poorly, possibly due to high sodium and chloride concentrations. Adding process water to substrates generally reduced growth or increased mortality. Grasses had greater growth than forbs, and for each functional group, non-native species performed better than native species. *Hordeum vulgare* had the highest overall growth with no mortality followed by *Agropyron trachycaulum* with negligible (0.5%) mortality. *Chamerion angustifolium* was most affected by the treatments with the lowest growth and highest mortality (56%). *Sonchus arvensis* had higher growth than *C. angustifolium* but its slow growth makes it less suitable for reclaiming tailings. Our results indicate that *H. vulgare* and *A. trachycaulum* could be good candidates for use in initial reclamation of oil sands tailings.

Keywords: boreal plants; forest land reclamation; oil sands; process water; tailings cake

Citation: Omari, K.; Pinno, B.D.; Utting, N.; Li, E.H. Growth of Common Plants of Boreal Reclamation Sites in Oil Sands Tailings Cake Mixes and Process Water. *Land* **2021**, *10*, 25. https://doi.org/10.3390/land10010025

Received: 23 November 2020
Accepted: 23 December 2020
Published: 30 December 2020

1. Introduction

The oil sands deposits in northern Alberta, Canada, represent the world's third largest oil deposit, with proven reserves of 165.4 billion barrels [1]. Oil sands surface mining results in severe forest disturbance. Following mine closure, disturbed lands are to be returned to an equivalent land capability, which can support land uses similar to the pre-disturbed land [2]. The Government of Alberta has also implemented a directive for progressive reclamation to ensure that all fluid tailings from a mining project are ready to reclaim ten years after the end of mine life [3].

The extraction process generates large volumes of fluid fine tailings comprised of connate and process water, sand, silt, clay, residual bitumen, inorganic salts, and organic compounds [4–6]. Process water is classified as free water, residing on top of the tailings material, or pore water, trapped within the fine spaces of tailings deposits. Oil sands fluid fine tailings is generally composed of 70–80% water, 20–30% solids, and 1–3% residual bitumen [5], and is alkaline and slightly brackish with high concentrations of organic acids [4]. The suspended solids in oil sands tailings are dominated by quartz and clays from the McMurray Formation, predominantly kaolinite, illite, and montmorillonite [7]. The total dissolved solids (TDS) in process water vary and change over time (for example, ranging from 600 to 2200 mg L^{-1}) as TDS mostly depends on the type of ore being mined.

The aqueous cations are dominated by sodium while the anions are a mix of chloride, sulphate, and bicarbonate [8].

A variety of chemical, physical, and mechanical methods are used or are being tested to speed up the tailings dewatering process, with the objective of more quickly producing a solid deposit with sufficient strength for reclamation. One of the first commercially implemented tailings management method was composite or consolidated tailings where fine and coarse tailings are mixed along with sufficient gypsum to create a non-segregating solid mixture, which can settle to approximately 70 wt.% solids after one to two years. Another commercially used method is centrifugation in combination with chemical amendments to rapidly dewater tailings to produce tailings cake, with a typical solids content of 55 to 60 wt.%. In this study, we focus on tailings cake produced by centrifugation. As tailings are being reclaimed, it is important to understand the effects of dewatered tailings (as substrates) and process water (as groundwater seepage) on plant development and growth.

Response of plant growth to tailings or process water have been shown to vary depending on species. Above- and belowground biomass of one-month-old jack pine seedlings treated with process water was reduced by 31% and 20%, respectively, compared to seedlings irrigated with deionized water [9]. A positive correlation was also found between needle necrosis and tissue sodium and chloride for seven-month-old seedlings treated with process water [9]. Pouliot et al. [10], however, observed no stress signs after two growing seasons for fen vascular plants irrigated with process water, but groundwater discharge of process water adversely affected mosses under dry conditions. For raspberry grown in soil amended with 15% (by volume) fluid fine tailings, shoot and root dry weights reduced by more than 50%, but in conifer seedlings, shoot and root dry weights were not significantly different from those in control soils with no fluid fine tailings after 3 months [11]. Although plant responses to tailings or process water have been documented it is unclear how the synergistic stress of tailings and process water may affect growth and development of common species in boreal reclamation sites. Knowledge of the combined effect of tailings and process water on plant growth will help identify species that would be suitable for consideration in reclaiming oil sands tailings.

A desirable goal for reclaimed land in the oil sands region in northern Alberta is to have a functioning forest ecosystem composed of native plant species. However, substantial changes in forest ecosystems due to mining activities, e.g., increased soil salinity [12], may hinder growth of native species and favor non-native ones. Mixing contaminated sediments with soil [13] or modifying the physiochemical environment of contaminated sites through the addition of organic matter and nutrients in addition to planting native species acclimated to contaminated soils [14,15] may reduce the concentration of contaminants such as excess salts and improve plant growth. There is limited literature on the effect of heavy metals from process water on boreal plant health. However, it is well known that some heavy metals can accumulate within plants at high concentrations without any indication of stress [16–18]. Increase in concentrations of these metals above plant threshold levels would modify plant physiological processes [19]. Consequences include visible changes in plant morphology, such as chlorosis and necrosis in leaves, stunted plant growth, and changes in root structure [19–22].

In the current study, we investigated the response of four native and non-native plant species commonly found in newly reclaimed areas in the boreal forest region of Canada: *Chamerion angustifolium* (L.) Holub (fireweed, native forb), *Sonchus arvensis* L. (perennial sow thistle, non-native forb), *Agropyron trachycaulum* (Link) Malte (slender wheatgrass, native grass), and *Hordeum vulgare L.* (barley, non-native grass). Our objectives were to determine the effect of oil sands tailings, mixtures of treated oil sands tailings and reclamation substrates, and oil sands process water on aboveground biomass and mortality of these four plants.

2. Materials and Methods

2.1. Experimental Set Up

This was a randomized, complete block design, greenhouse pot study with 4 species × 4 substrates × 3 water quality treatments × 6 blocks (gradient of sunlight and temperature within greenhouse as influenced by distance of pots to greenhouse window) for a total of 288 pots. Additionally, control pots with no plants were set up for each substrate × water combination, with 6 replicates for each combination for a total of 72 control pots. The greenhouse temperature was set at 22–24 °C during the day (1000–2000 h) and 18–22 °C at night (0100–0600 h), relative humidity was set to 30% and 40% during the day and night, respectively, and an artificial light source (LumiGrow Pro 650) was turned on automatically, within a 16-h period (0500–2100), when natural light intensity fell below 200 W m^{-2}.

2.2. Substrates and Process Water

We used four tailings substrates: (i) pure centrifuge tailings cake, and mixtures (1:1 by volume) of (ii) tailings cake and forest floor mineral mix (FFMM) (cake-FFMM), (iii) tailings cake and sphagnum peat moss (cake-peat), and (iv) tailings cake and sand (cake-sand).

The tailings cake was created by centrifuging a mixture of fluid fine tailings (obtained from an operational mine site in northern Alberta), gypsum (~900 ppm), and a high molecular weight anionic polymer, A3338 polymer (~1000 ppm). The resulting tailings cakes had 55.7 wt.% solids. The sand and sphagnum peat moss were commercially obtained, and FFMM was obtained from an operational mine site in northern Alberta and consisted of forest floor materials mixed with the underlying mineral soil.

Three types of irrigation water were used, which differed in quality: 0%, 50%, and 100% process water. The 0% process water consisted of reverse osmosis water whereas the 100% process water was the centrate water obtained from the centrifugation process used to produce the tailings cake. The 50% process water was made up of equal proportions of reverse osmosis water and 100% process water.

Chemical characterization of the substrates and process water used for the experiment (Table 1) was done by CanmetENERGY, Natural Resources Canada, Devon, AB, Canada. Tailings cake and tailings cake mixtures were slightly alkaline to alkaline (pH of 7.2–8.2). In general, concentrations of ions in the 100% process water were approximately double that of the 50% process water and were both substantially greater than the 0% process water. The 0% process water was slightly acidic (pH of 6.7) and the 50% and 100% process water were alkaline (pH of 8.2). To estimate soil nutrient supply rates during the period of the experiment, a pair of anion and cation plant root simulator (PRS; Western Ag Innovations, Saskatoon, SK, Canada) probes were installed to a depth of 9–10 cm in the control pots. PRS probes give estimates of soil nutrient supply rates by attracting and adsorbing ions on negatively and positively charged ion-exchange membranes [23,24]. The probes were removed after eight weeks, washed with reverse osmosis water and sent to Western Ag Innovations for extraction and laboratory analysis.

Table 1. Chemical characteristics of substrates and process water used for plant growth. EC, TDS, and SAR represent electrical conductivity, total dissolved solids and sodium adsorption ratio, respectively.

	Substrate				Water		
	Cake	**Sand + Cake**	**FFMM + Cake**	**Peat + Cake**	**Reverse Osmosis**	**50% Process Water**	**100% Process Water**
Percent solids (%)	56	78	73	53			
pH	7.90	8.21	7.87	7.17	6.71	8.24	8.26
EC (mS/cm)	2.79	4.95	4.73	2.39	0.02	1.16	2.18
TDS Calculated (g/L)	3.69	5.64	4.82	2.86	0.02	1.06	1.87
SAR Concentrations		mg/kg mineral solids				mg/L	
Na	867	910	754	559	4.1	285	495
Cl^-	215	212	202	190	1.11	133.5	238
$CO_3{}^{2-}$	5.5	5.5	<3.8	<3.8	<3.8	5.9	13
$HCO3^-$	1266	1567	1678	471	12.7	466	858

Table 1. *Cont.*

	Substrate				Water		
SO_4^{2-}	837	1766	1214	1017	2.30	94.9	151
Ca	85	368	294	86	0.33	11.4	21.4
K	42	56	78	44	<0.01	7.73	13.1
Mg	35	97	116	53	0.37	7.73	12.4
S	319	614	435	383	0.88	36.5	58.9
B	5.49	6.67	6.16	6.27	0.14	0.81	1.34
Ba	0.96	0.83	0.71	8.21	0.00	0.22	0.39
Al	0.07	0.04	0.63	1.49	0.00	0.03	0.02

2.3. Plants

The species selected for this experiment were *Chamerion angustifolium* (L.) Holub (native forb), *Sonchus arvensis* L. (non-native forb), *Agropyron trachycaulum* (Link) Malte (native grass), and *Hordeum vulgare* L. (non-native grass). These are common plant species found in newly reclaimed areas or used as cover species in reclamation and are fast growing herbaceous species suitable for short-term greenhouse experiments. They are also representatives of both native and non-native grasses and forbs.

Seeds of *S. arvensis* were obtained from Canadian Natural Resources Limited (Fort McKay, AB) and seeds of the remaining species were obtained from commercial sources across Canada and the United States of America. The seeds were germinated under greenhouse conditions in styroblocks (plug size of 2.5 cm in diameter and 11.3 cm in length) using a commercial garden soil and watered as needed. Five average-sized seedlings of each species were transplanted into 1.5-L pots filled with the experimental substrates and allowed to settle for one week before watering with the process water. During this period, plants were manually watered to field capacity with greenhouse irrigation water and those that died were replaced. At the end of this period, three healthy plants were selected for the experiment, and the remaining plants were uprooted from the pots. The plants were then watered manually each day with the process water on an as needed basis.

We applied a 20-20-20 nitrogen, phosphorus, and potassium fertilizer at rates of 75 mL per week (recommended rates by manufacturers) for the first four weeks and 15 mL biweekly (equivalent to the rate used in reclamation practices, i.e., 100 kg nitrogen ha yr^{-1}) for the last four weeks of the experiment. Fertilization is commonly used in oil sands reclamation in Alberta to ensure that planted or naturally regenerated plants have adequate nutrients for establishment and early growth [25].

Plants were grown under the experimental conditions for eight weeks and mortality was recorded in the final week of the experiment. At the end of the 8-week period, plants were clipped at the soil surface and dried to a constant weight at 40 °C to obtain aboveground biomass.

2.4. Data Analysis

Mixed model analysis of variance (ANOVA), with block as the random factor, was used to test for differences in aboveground biomass and mortality among substrates and watering treatments for each species at the end of the study. Tukey's procedure was used for pairwise comparisons. Cube root transformation was applied to *C. angustifolium* aboveground biomass to meet ANOVA assumption of homoscedasticity. Differences in nutrient concentrations were also tested among substrates with one-way ANOVA. Correlation analyses (spearman rank correlation) were performed between nutrient supply rates from the control pot and aboveground biomass for each species. Mixed model ANOVA and multiple comparison tests were performed with nlme [26] and emmeans [27] packages, respectively. Correlation coefficients and associated probability values were calculated with psych package [28]. ANOVA and correlation analysis were performed with *R* statistical software [29], and statistical significance was considered at $p < 0.05$.

3. Results

Across all treatments, the grasses, *H. vulgare* (10.73 g) and *A. trachycaulum* (7.85 g), exhibited 2–8 times greater aboveground biomass than the forbs, *S. arvensis* (5.01 g) and *C. angustifolium* (1.28 g), and the introduced species performed better than the native species within each functional group (Figure 1a–d.) Among substrates, cake-peat supported the overall highest aboveground biomass (6.83 g), followed by cake-FFMM (6.24 g), cake (6.20 g), and cake-sand (5.61), and the 0% process water, overall, supported a higher aboveground biomass (6.72 g) than the 50% (6.45 g) and 100% (5.49 g) process water. There were varying responses of aboveground biomass to substrate and water treatments among the four species.

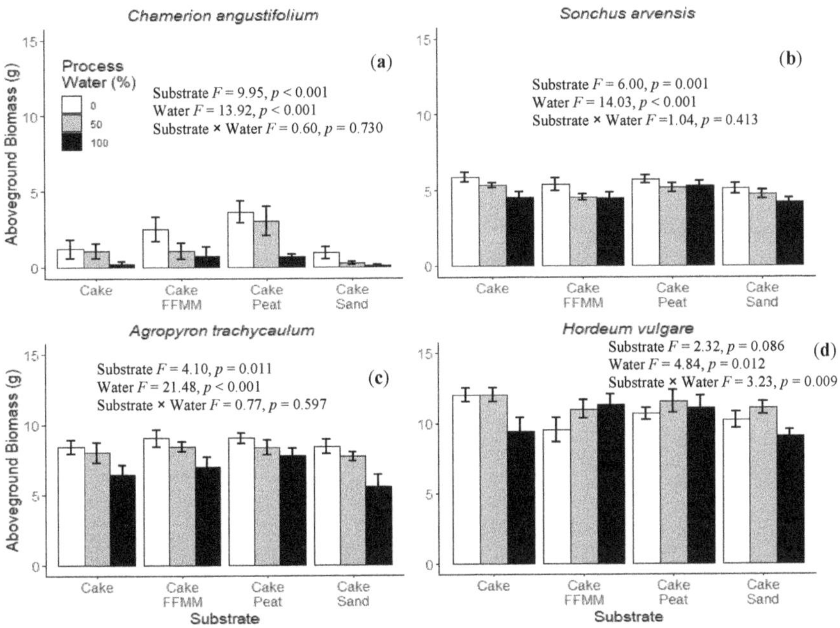

Figure 1. Total aboveground biomass (mean and standard error) of four understory species (**a**). *Chamerion angustifolium*; (**b**). *Sonchus arvensis*; (**c**). *Agroypyron trachycaulum*; (**d**) *Hordeum vulgare*, commonly found in newly reclaimed areas in the boreal forest region of Canada, under four soil amendments and three watering treatments.

The treatment effects were most pronounced in *C. angustifolium* (Figure 1a), with cake supporting 67% lower aboveground biomass than cake-peat ($p < 0.001$), and cake-sand supporting 70% or 82% lower aboveground biomass than cake mixed with FFMM ($p = 0.035$) or peat ($p < 0.001$), respectively. For the same species, watering with 100% process water reduced aboveground biomass by 79% ($p < 0.001$) and 68% ($p = 0.007$) compared to watering with 0% and 50% process water, respectively. For *S. arvensis*, cake-sand supported 13% and 11% lower biomass than cake-peat ($p = 0.004$) and cake ($p = 0.030$), respectively, and cake-FFMM supported 11% lower biomass than cake-peat ($p = 0.021$). Watering with 100% and 50% process water also reduced aboveground biomass by 16 % ($p < 0.001$) and 10% ($p = 0.003$), respectively, compared to watering *S. arvensis* with 0% process water (Figure 1b). Differences in aboveground biomass among substrates was only found between cake-peat (8.40 g) and cake-sand (7.24 g) ($p = 0.013$) for *A. trachycaulum*, and for the same species, watering with 100% process water reduced aboveground biomass by 23% ($p < 0.001$) and 18% ($p = 0.005$) compared to watering with 0% and 50% process water, respectively (Figure 1c). Substrate × process water interaction effect on variation in aboveground biomass was only observed for *H. vulgare*. Cake-sand watered with 100%

process water had 25% or 26% lower aboveground biomass than cake watered with 50% ($p = 0.023$) or 0% ($p = 0.020$) process water (Figure 1d).

In relation to supply rates of nutrients (Table 2), *C. angustifolium* aboveground biomass was positively correlated with nitrate ($p = 0.030$), phosphorus ($p = 0.040$), and magnesium ($p = 0.010$), and *S. arvensis* was positively correlated with magnesium ($p = 0.040$). No significant relationships were found between the other species and nutrients supply rates (Table 2).

Table 2. Spearman rank correlations between plant aboveground biomass and nutrient supply rates across substrates and watering treatments. Statistically significant correlations are marked with asterisk.

Species	NH_4^+	NO_3^-	P	K	Ca	S	Mg
Chamerion angustifolium	0.56	0.64 *	0.61 *	0.30	0.13	−0.10	0.75 *
Sonchus arvensis	0.55	0.20	0.49	0.34	−0.13	−0.13	0.61 *
Agropyron trachycaulum	0.29	0.53	0.57	0.00	0.38	0.13	0.52
Hordeum vulgare	0.34	0.15	0.44	0.42	−0.34	−0.31	0.34

Mortality was only observed among the native species. *C. angustifolium* had the highest mortality across treatments (56%), followed by a negligible amount for *A. trachycaulum* (0.5%) (Figure 2). For *C. angustifolium*, 100% and 50% process water had 36% ($p = 0.004$) and 29% ($p = 0.022$), respectively, higher mortality than 0% process water, but neither substrate nor substrate × water interaction effect was significant.

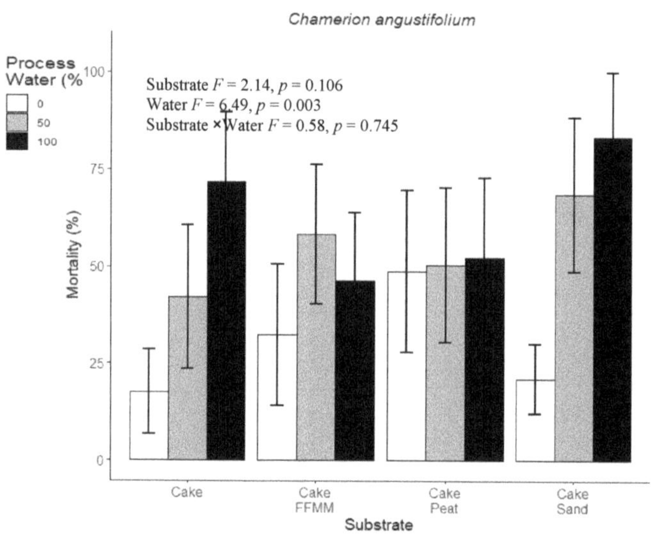

Figure 2. Mortality (mean and standard error) for *Chamerion angustifolium* under four soil amendments and three watering treatments.

Mixing cake with peat substantially reduced concentrations of sodium, chloride, and the carbonate, HCO_3^-, from 867 mg kg^{-1}, 215 mg kg^{-1}, and 1266 mg kg^{-1}, respectively, to 559 mg kg^{-1} ($p = 0.003$), 190 mg kg^{-1} ($p = 0.033$), and 471 mg kg^{-1} ($p < 0.001$), respectively (Table 1). On the other hand, electrical conductivities and concentrations of macronutrients (calcium, potassium, magnesium, and sulfur) tended to increase when cake was mixed with sand or FFMM but not with peat (Table 1). Supply rates of nitrogen, phosphorus, and potassium were generally lower for cake-sand compared to that of the other substrates (Table 3).

Table 3. Mean values (associated standard errors) of supply rates of nutrients over an 8-week period for tailings cake and cake-amendments under three watering treatments.

		Nutrient Supply Rate (µg 10 cm^{-2} 8 Weeks^{-1})						
Substrate	Process Water (%)	NH_4^+	NO_3^-	P	K	Ca	S	Mg
Cake	0	8.2	0.0	7.4	44.9	1700.6	270.3	505.2
		(2.6)	(0.0)	(2.5)	(3.2)	(51.9)	(54.3)	(25.1)
Cake	50	6.0	0.0	3.0	47.3	1592.4	238.2	490.5
		(1.3)	(0.0)	(0.9)	(3.2)	(84.3)	(39.6)	(24.3)
Cake	100	6.3	0.0	1.3	45.7	1444.2	266.8	469.3
		(1.8)	(0.0)	(0.3)	(3.2)	(43.90)	(66.1)	(21.8)
Cake + FFMM	0	4.6	2.8	5.5	30.4	2144.3	315.8	530.0
		(1.6)	(2.8)	(0.7)	(1.1)	(71.0)	(56.2)	(22.0)
Cake + FFMM	50	5.2	2.6	3.7	32.5	2025.5	302.7	508.6
		(2.0)	(1.8)	(0.7)	(1.7)	(84.1)	(24.0)	(19.2)
Cake + FFMM	100	3.09	5.7	3.9	32.0	1997.4	336.0	479.0
		(0.9)	(3.9)	(0.3)	(1.7)	(111.9)	(58.0)	(18.0)
Cake + Peat	0	19.8	0.2	2.1	49.6	1845.5	638.4	569.8
		(3.0)	(0.2)	(0.4)	(2.3)	(59.8)	(82.6)	(17.5)
Cake + Peat	50	16.3	0.07	2.0	51.8	1742.3	500.4	540.2
		(2.9)	(0.1)	(0.1)	(2.2)	(46.7)	(65.9)	(12.0)
Cake + Peat	100	12.9	0.0	2.0	53.0	1659.2	599.3	530.3
		(3.6)	(0.0)	(0.4)	(2.3)	(86.2)	(51.5)	(17.0)
Cake + Sand	0	2.4	0.0	2.2	27.1	2117.6	875.8	379.2
		(0.3)	(0.0)	(0.3)	(1.4)	(62.7)	(80.4)	(16.7)
Cake + Sand	50	2.0	0.00	1.3	29.9	2083.4	723.6	395.7
		(0.4)	(0.0)	(0.1)	(2.1)	(109.3)	(28.0)	(23.7)
Cake + Sand	100	2.5	0.0	1.0	30.8	1990.2	615.7	382.8
		(0.8)	(0.0)	(0.1)	(1.5)	(86.4)	(45.5)	(10.0)

4. Discussion

We examined the effect of oil sands tailings, mixtures of treated oil sands tailings and reclamation substrates, and oil sands process water on aboveground biomass and mortality of four plants (*C. angustifolium*, *S. arvensis*, *A. trachycaulum* and *H vulgare*) commonly found in boreal oil sands reclamation sites. Overall, cake-peat supported the highest aboveground biomass among substrates whereas cake and cake-sand performed poorly. Another study also reported that consolidated tailings amended with peat improved germination, survival, and growth compared to plants growing directly in consolidated tailings [30]. In the present study, mixing cake with peat reduced pH and substantially reduced the concentrations of sodium, chloride, and carbonates. The high pH of tailings could result in plant mineral deficiency by reducing available macronutrients (e.g., phosphorus and nitrogen) and trace elements [30]. Salts are also known to adversely affect plant water balances by targeting the osmotic gradient across cells [31,32]. In particular, chloride has been observed to accumulate in shoots while the buildup of sodium in plant tissue has the potential to interfere with enzymes participating in chlorophyll production, and the accumulation of both ions within plant tissue can reduce photosynthesis and growth [33–35]. The better growth performance of plants grown in cake-peat in our study may be due to the reduced pH and salt content of the cake-peat substrate. Organic contaminants in the process water were not measured; however, the higher organic carbon content of the peat is expected to also cause sorption of dissolved organics to the peat, reducing the toxicity of the water [36,37].

On the other hand, mixing cake with sand resulted in sodium, chloride, and carbonates concentrations comparable to levels found in cake. This is possibly due to the presence of soluble mineral material in the sand. Supply rates of nitrogen, phosphorus, and potassium were also generally low for cake-sand compared to the other substrates. Consequently, this caused poor growth of plants grown in the cake-sand substrate. Compensating for nutrient deficiencies, e.g., through fertilization, should be combined with processes that reduce salinity, e.g., addition of organic matter to reduce evapotranspiration, in tailings reclamation.

Addition of process water to the tailings and tailings mixes adversely affected plant performance by reducing plant growth or increasing mortality. This may be due to the presence of naphthenic acids and salts in process water [35]. The combined impact of naphthenic acids and excess salts could exceed the sum of the individual effects of each of them [38] and increase water stress, interfere with respiration, and be toxic for organisms [35,39]. Leaf tip necrosis was observed in common herbaceous and woody forest plants grown hydroponically and subjected to undiluted process water treatment [40], possibly due to buildup of toxic compounds in the process water or nutritional deficiencies resulting from excess salts [40,41]. In our study, similar growth levels were observed between plants watered with reverse-osmosis water and those watered with equal proportions of centrate water and reverse-osmosis water. This suggests some interaction with process water will not be overly detrimental to the growth of plants.

We also found that the grasses had better growth performance than did forbs. Similar findings have been reported by Naeth and Wilkinson [42]. *H. vulgare* also had the best growth performance among all species. The ability of *H. vulgare* to germinate and establish on tailings under controlled conditions suggest that it is a good candidate for early tailings reclamation efforts, such as erosion control and phytoremediation [34]. *A. trachycaulum* showed good health across all treatments, but its aboveground biomass accumulation was less than *H. vulgare* over the study period. The slow growth of *A. trachycaulum* restricts the quantity of potentially toxic ions it can remove from contaminated soils [30]. It can, however, be used in combination with *H. vulgare* in reclamation efforts to increase vegetation cover, and consequently long-term stabilization [43] of tailings.

The native forb, *C. angustifolium*, has been suggested as a suitable species for reclaiming disturbed forests because it can establish on reclaimed soils (especially, a forest floor mineral mix) and capture soil nutrients effectively [44]. However, its growth on tailings and tailing mixes was the poorest, exhibiting the greatest mortality, and surviving plants showed average-to-poor health. This may be due in part to nutritional deficiencies of the tailings and tailing mixtures since seedling establishment of the species may be confined to areas that are rich in nutrients [45]. *C. angustifolium* growth was positively correlated with macronutrients (nitrates, phosphorus, magnesium), which supports the observation that the factors that influence successful establishment of *C. angustifolium* may be site and soil specific [44]. Reclaiming tailings with *C. angustifolium* will be a challenge because of its poor performance on tailings and potential soil specificity. *S. arvensis* exhibited better growth performance than *C. angustifolium*. However, its very slow growth on tailings makes it potentially less suitable for phytoremediation in these substrates compared to *H. vulgare* or *A. trachycaulum*. It should be noted that while *S. arvensis* may occur in reclaimed areas, it would not be specifically planted as it is classified as a noxious weed in Alberta [46].

Within each functional group, non-native species had better growth than native species. Naeth and Wilkinson [42] also found that non-native species had higher emergence and establishment on consolidated tailings than native species. Because native species are suited to the pre-disturbed ecosystem, their decline in novel environments following mining can be expected. However, native boreal species have been shown to exhibit varying tolerance to salt [47]. Salt tolerant population of *A. trachycaulum* can be found in a dry area, in Southern Alberta, with an underlying marine shale formation [48]. Understanding salt tolerance levels of native species as well as differences among accessions is important to determine their suitability for land reclamation [30].

5. Conclusions

This study tested the suitability of common boreal plants to the combined effect of fluid fine tailings cake, mixtures of tailings cake and reclamation substrates, and process water. Our results showed that mixing peat with cake tailings can reduce salinity and improve plant growth. Additionally, *H. vulgare* and *A. trachycaulum* exhibited greater overall aboveground biomass and lower mortality and could therefore be suitable for initial reclamation of oil sands tailings. Because our study was performed under controlled greenhouse conditions, caution must be taken when extrapolating these studies to field sites where conditions such as extreme temperatures and competition exist.

Author Contributions: Conceptualization, B.D.P., E.H.Y.L. and N.U.; formal analysis, K.O.; investigation, E.H.Y.L.; data curation, E.H.Y.L.; writing—original draft preparation, K.O.; writing—review and editing, K.O., B.D.P., N.U., and E.H.Y.L., supervision, B.D.P. and N.U; project administration, B.D.P. and N.U.; funding acquisition, B.D.P. and N.U. All authors have read and agreed to the published version of the manuscript.

Funding: This study was funded by the Energy Innovation Program of Natural Resources Canada.

Institutional Review Board Statement: Not applicable.

Informed Consent Statement: Not applicable.

Data Availability Statement: Data may be available on request.

Acknowledgments: We thank Christopher Wohl, Sophia Yang, Stephanie Jean, Ruth Errington, Sanatan Das Gupta, Jim Weber, Craig McMullen, Michelle Morin, and Pamela Muñoz for assistance with the project.

Conflicts of Interest: The authors declare no conflict of interest.

References

1. Government of Alberta. Oil Sands Facts and Statistics. Available online: https://www.alberta.ca/oil-sands-facts-and-statistics.aspx (accessed on 6 August 2019).
2. Alberta Government. *Conservation and Reclamation Information Letter Guidelines for Reclamation to Forest Vegetation in the Athabasca Oil Sands Region*; C and R/IL/99-1; Government of Alberta: Edmonton, AB, Canada, 1999.
3. Alberta Energy Regulator. *Directive 085: Fluid Tailings Management for Oil Sands Mining Projects*; Alberta Energy Regulator: Alberta, AB, Canada, 2017.
4. MacKinnon, M.D.; Sethi, A. A comparison of the physical and chemical properties of the tailings ponds at the Syncrude and Suncor oil sands plants. In Proceedings of the Our Petroleum Future Conference, Edmonton, AB, Canada, 4–7 April 1993.
5. Allen, E.W. Process water treatment in Canada's oil sands industry: I. Target pollutants and treatment objectives. *J. Environ. Eng. Sci.* **2008**, *7*, 123–138. [CrossRef]
6. Bedair, O. Engineering Challenges in the Design of Alberta's Oil Sands Projects. *Pr. Period. Struct. Des. Constr.* **2013**, *18*, 247–260. [CrossRef]
7. Chalaturnyk, R.J.; Scott, J.D.; Ozum, B. Management of Oil Sands Tailings. *Pet. Sci. Technol.* **2002**, *20*, 1025–1046. [CrossRef]
8. Mahaffey, A.; Dube, M. Review of the composition and toxicity of oil sands process-affected water. *Environ. Rev.* **2016**, *25*, 97–114. [CrossRef]
9. Franklin, J.A.; Renault, S.; Croser, C.; Zwiazek, J.J.; MacKinnon, M. Jack pine growth and elemental composition are affected by saline tailings water. *J. Environ. Qual.* **2002**, *31*, 648–653. [CrossRef]
10. Pouliot, R.; Rochefort, L.; Graf, M.D. Impacts of oil sands process water on fen plants: Implications for plant selection in required reclamation projects. *Environ. Pollut.* **2012**, *167*, 132–137. [CrossRef]
11. Renault, S.; Zwiazek, J.J.; Fung, M.; Tuttle, S. Germination, growth and gas exchange of selected boreal forest seedlings in soil containing oil sands tailings. *Environ. Pollut.* **2000**, *107*, 357–365. [CrossRef]
12. Audet, P.; Pinno, B.D.; Thiffault, E. Reclamation of boreal forest after oil sands mining: Anticipating novel challenges in novel environments. *Can. J. For. Res.* **2015**, *45*, 364–371. [CrossRef]
13. Hanslin, H.M.; Eggen, T. Salinity tolerance during germination of seashore halophytes and salt-tolerant grass cultivars. *Seed Sci. Res.* **2005**, *15*, 43–50. [CrossRef]
14. Iglesia, R.; Castro, D.; Ginocchio, R.; Lelie, D.; González, B. Factors influencing the composition of bacterial communities found at abandoned copper-tailings dumps. *J. Appl. Microbiol.* **2006**, *100*, 537–544. [CrossRef]
15. Lam, E.J.; Keith, B.F.; Montofré, Í.L.; Gálvez, M.E. Copper uptake by *Adesmia atacamensis* in a Mine Tailing in an Arid Environment. *Air Soil Water Res.* **2018**, *11*, 1178622118812462. [CrossRef]
16. Guala, S.D.; Vega, F.A.; Covelo, E.F. The dynamics of heavy metals in plant–soil interactions. *Ecol. Model.* **2010**, *221*, 1148–1152. [CrossRef]
17. Chibuike, G.U.; Obiora, S.C.; Chibuike, G.U.; Obiora, S.C. Heavy Metal Polluted Soils: Effect on Plants and Bioremediation Methods. *Appl. Environ. Soil Sci.* **2014**, *2014*, 1–12. [CrossRef]
18. Gall, J.E.; Boyd, R.S.; Rajakaruna, N. Transfer of heavy metals through terrestrial food webs: A review. *Environ. Monit. Assess.* **2015**, *187*, 201. [CrossRef]

19. Goyal, D.; Yadav, A.; Prasad, M.; Singh, T.B.; Shrivastav, P.; Ali, A.; Dantu, P.K.; Mishra, S. Effect of Heavy Metals on Plant Growth: An Overview. In *Contaminants in Agriculture*; Springer: Cham, Switzerland, 2020; pp. 79–101.
20. Schutzendubel, A.; Polle, A. Plant responses to abiotic stresses: Heavy metal-induced oxidative stress and protection by mycorrhization. *J. Exp. Bot.* **2002**, *53*, 1351–1365. [CrossRef]
21. Prasad, M.N.V. *Heavy Metal Stress in Plants from Biomolecules to Ecosystems*; Springer: Berlin, Germany, 2010; 462p.
22. Mourato, M.; Moreira, I.N.; Leitão, I.; Pinto, F.; Sales, J.R.; Martins, L. Effect of Heavy Metals in Plants of the Genus Brassica. *Int. J. Mol. Sci.* **2015**, *16*, 17975–17998. [CrossRef]
23. Davies, K.G.; Nafus, A.M.; Sheley, R.L. Non-native competitive perennial grass impedes the spread of an invasive annual grass. *Biol. Invasions* **2010**, *12*, 3187–3194. [CrossRef]
24. Errington, R.C.; Pinno, B.D. Early successional plant community dynamics on a reclaimed oil sands mine in comparison with natural boreal forest communities. *Écoscience* **2015**, *22*, 133–144. [CrossRef]
25. Howell, D.M.; Das Gupta, S.; Pinno, B.D.; MacKenzie, M.D. Reclaimed soils, fertilizer, and bioavailable nutrients: Determining similarity with natural benchmarks over time. *Can. J. Soil Sci.* **2016**, *97*, 149–158. [CrossRef]
26. Pinheiro, J.; Bates, D.; DebRoy, S.; Sarkar, D.; R Core Team. *nlme: Linear and Nonlinear Mixed Effects Models*; R Package Version 3.1-140; R Core Team: Evanston, IL, USA, 2019.
27. Russell, L. *Emmeans: Estimated Marginal Means, aka Least-Squares Means*; R Package Version 1.4.1 2019; R Core Team: Evanston, IL, USA, 2019.
28. Revelle, W. *psych: Procedures for Personality and Psychological Research*; R Packaage Version 1.8.12; R Core Team: Evanston, IL, USA, 2018.
29. R Core Team. *R: A Language and Environment for Statistical Computing*; R Foundation for Statistical Computing: Vienna, Austria, 2018; Available online: https://www.R-project.org/ (accessed on 10 November 2020).
30. Renault, S.; Qualizza, C.; MacKinnon, M. Suitability of altai wildrye (*Elymus angustus*) and slender wheatgrass (*Agropyron trachycaulum*) for initial reclamation of saline composite tailings of oil sands. *Environ. Pollut.* **2004**, *128*, 339–349. [CrossRef]
31. Volkmar, K.M.; Hu, Y.; Steppuhn, H. Physiological responses of plants to salinity: A review. *Can. J. Plant Sci.* **1998**, *78*, 19–27. [CrossRef]
32. Gosselin, P.; Hrudey, S.E.; Naeth, A.; Plourde, A.; Therrien, R.; Van Der Kraak, G.; Xu, Z. *Environmental and Health Impacts of Canada's Oil Sands Industry*; The Royal Society of Canada: Ottawa, ON, Canada, 2010; 440p.
33. Popova, L.P.; Stoinova, Z.G.; Maslenkova, L.T. Involvement of abscisic acid in photosynthetic process in *Hordeum vulgare* L. during salinity stress. *J. Plant Growth Regul.* **1995**, *14*, 211–218. [CrossRef]
34. Renault, S.; MacKinnon, M.; Qualizza, C. Barley, a Potential Species for Initial Reclamation of Saline Composite Tailings of Oil Sands. *J. Environ. Qual.* **2003**, *32*, 2245–2253. [CrossRef] [PubMed]
35. Apostol, K.G.; Zwiazek, J.J.; MacKinnon, M.M. Naphthenic acids affect plant water conductance but do not alter shoot Na+ and Cl- concentrations in jackpine (*Pinus banksiana*) seedlings. *Plant Soil* **2004**, *263*, 183–190. [CrossRef]
36. Fetter, C.W. *Contaminant Hydrogeology*, 2nd ed.; Prentice Hall: Upper Saddle River, NJ, USA, 1999.
37. Jandada, A. A Laboratory Evaluation of the Sorption of Oil Sands Naphthenic Acids on Soils. Masters Thesis, University of Saskatchewan, Saskatoon, SK, Canada, 2007.
38. Walker, C.; Sibly, R.; Peakall, D. *Principles of Ecotoxicology*, 3rd ed.; Taylor and Francis Group: New York, NY, USA, 2016.
39. Armstrong, S.A.; Headley, J.V.; Peru, K.M.; Germida, J.J. Phytotoxicity of oil sands naphthenic acids and dissipation from systems planted with emergent aquatic macrophytes. *J. Environ. Sci. Health Part A* **2007**, *43*, 36–42. [CrossRef]
40. Renault, S.; Lait, C.; Zwiazek, J.; MacKinnon, M. Effect of high salinity tailings waters produced from gypsum treatment of oil sands tailings on plants of the boreal forest. *Environ. Pollut.* **1998**, *102*, 177–184. [CrossRef]
41. Maas, E.V. Salinity and citriculture. *Tree Physiol.* **1993**, *12*, 195–216. [CrossRef]
42. Naeth, M.A.; Wilkinson, S.R. Plant species suitability for reclamation of oil sands consolidated tailings. In Proceedings of the 26th Annual British Columbia Mine Reclamation Symposium, Dawson Creek, BC, Canada, 9–13 September 2002; pp. 226–229.
43. Ripley, E.A.; Redmann, E.R.; Crowder, A.A.; Ariano, T.C.; Corrigan, C.A.; Farmer, R.J.; Jackson, L.M.; Redmann, R. *Environmental Effects of Mining*; St. Lucie Press: Boca Raton, FL, USA, 2018.
44. Pinno, B.D.; Landhäusser, S.M.; Chow, P.S.; Quideau, S.; MacKenzie, M.D. Nutrient uptake and growth of fireweed (*Chamerion angustifolium*) on reclamation soils. *Can. J. For. Res.* **2014**, *44*, 1–7. [CrossRef]
45. Myerscough, P.J.; Whitehead, F.H. Comparative Biology of *Tussilago farfara* L., *Chamaenerion angustifolium* (L.) Scop., *Epilobium montanum* L., and *Fpilobium Adenocaulon* hausskn. II. Growth and Ecology. *New Phytol.* **2006**, *66*, 785–823. [CrossRef]
46. Government of Alberta. About Invasive Plants and Weeds. 2019. Available online: https://www.alberta.ca/about-invasive-plants-and-weeds.aspx (accessed on 23 August 2019).
47. Bois, G.; Bigras, F.J.; Bertrand, A.; Piché, Y.; Fung, M.Y.; Khasa, D.P. Ectomycorrhizal fungi affect the physiological responses of *Picea glauca* and *Pinus banksiana* seedlings exposed to an NaCl gradient. *Tree Physiol.* **2006**, *26*, 1185–1196. [CrossRef]
48. Acharya, S.N.; Darroch, B.A.; Hermesh, R.; Woosaree, J. Salt stress tolerance in native Alberta populations of slender wheatgrass and alpine bluegrass. *Can. J. Plant Sci.* **1992**, *72*, 785–792. [CrossRef]

land

MDPI

Article

Soil-Quality Assessment during the Dry Season in the Mun River Basin Thailand

Chunsheng Wu [1], Erfu Dai [1], Zhonghe Zhao [2], Youxiao Wang [2] and Gaohuan Liu [2,*]

[1] Lhasa Plateau Ecosystem Research Station, Key Laboratory of Ecosystem Network Observation and Modeling, Institute of Geographic Sciences and Natural Resources Research, CAS, Beijing 100101, China; wucs@igsnrr.ac.cn (C.W.); daief@igsnrr.ac.cn (E.D.)

[2] State Key Laboratory of Resources and Environment Information System, Institute of Geographic Sciences and Natural Resources Research, CAS, Beijing 100101, China; zhaozh.16b@igsnrr.ac.cn (Z.Z.); wangyx.19b@igsnrr.ac.cn (Y.W.)

* Correspondence: liugh@igsnrr.ac.cn

Abstract: The Mun River Basin is one of Thailand's major grain-producing areas, but the production is insufficient, and most of the cultivated lands are rain-fed and always unused in the dry season. All this makes it necessary to determine the status of soil nutrients and soil quality in the dry season to improve soil conditions, which will be useful for cultivation in the farming period. The aim of this study was to construct a soil-quality assessment based on soil samples, and in the process the minimum data set theory was introduced to screen the assessment indicators. The geographically weighted regression method was used to complete the spatial interpolation process of indicators, and the fuzzy logic model was constructed to evaluate the soil quality. The results showed that the spatial distributions of soil quality and indicators were similar. The soil quality was the best in the upstream while poor in the downstream, and the dry fields in the west and the forests in the east of the basin were better than other areas nearby. However; the soil qualities of paddy fields in the middle and east of the basin were poor due to the lack of soil nutrient supply when the fields were unused

Keywords: Mun River; soil quality; GWR; fuzzy logic model; dry season

Citation: Wu, C.; Dai, E.; Zhao, Z.; Wang, Y.; Liu, G. Soil-Quality Assessment during the Dry Season in the Mun River Basin Thailand. *Land* **2021**, *10*, 61. https://doi.org/10.3390/land10010061

Received: 9 December 2020
Accepted: 9 January 2021
Published: 12 January 2021

1. Introduction

Soil is an indispensable resource and the basis of most natural ecological and social environments [1]. Soil quality has a great influence on the vegetation that grows in it, especially for crops, which make it important to maintain soil attributions for food security and sustainable development [2]. There is no definition of soil quality that is universally accepted so far, but most scholars believe that soil productivity accounts for a great portion of soil quality [3,4].

There has been much research on soil quality, whose objects include forests [5], grasslands, farmland, and other types [6–8]. The research methods have also developed from qualitative expressions in the past to statistics and model construction based on quantitative data [9]; for example, commonly used methods include principal component analysis, analytic hierarchy process, regression analysis, fuzzy analysis, and artificial neural networks [10–14]. Comparing the processes, methods, and results of previous research, it is found that there are still some problems and defects: first, there is a lot of redundancy among the indicators selected in the evaluation process, and there is a lack of a screening mechanism [15]. Second, most of the research is based on the data of sampling points, and the research results on the point scale are used to replace the entire area; some of the methods used in the spatial expansion of the research are mostly geostatistical methods [16,17], which are very dependent on the number of sampling points, otherwise the accuracy of the result is difficult to guarantee. Additionally, the determination of index thresholds and the division of the quantitative classification range of research results in most evaluation processes being unreasonable. Most of the indicators are standardized and graded directly according to some rules [18], and these grades are directly used for evaluation [19,20]. This strict classification method is very questionable, and it is necessary to make some improvements.

The Mun River Basin is in the northeast Thailand and occupies a large part of the Nakhon Ratchasima Plateau. It is one of Thailand's major grain-producing areas, but the average yield is low. This area is divided into dry season and rainy season because of the tropical monsoon climate [21], and rice is grown on most farmland during the rainy season, but the farmland is unused in dry season [22]. The soil-quality research in tropical regions is significantly less than in other climate regions, let alone the area with obvious tropical monsoon climate such as the Mun River Basin, and there has been no research on soil quality in the dry season in the Mun River Basin until now. Thus, it is necessary to carry out relevant research in this area, which will not only provide scientific reference for identifying tropical soil characteristics, but also provide practical basis for regional land improvement and agricultural development.

Based on the above description of the evaluation method and process, this study aims to evaluate the soil quality in the dry season in the Mun River Basin, and introduces the minimum data theory [23–25], geographic weighted regression model [26] and fuzzy logic model [27] to process and analyze the process of the indicator selection, indicator spatialization, and comprehensive evaluation respectively. The results of the study can provide basic information for soil improvement in the rainy season and will hopefully be helpful in improving the soil in the study area, especially for the rainy season when the crops are growing.

2. Data and Materials

2.1. Study Area

The Mun River Basin is in northeast Thailand and includes 10 provinces. The Mun River is a tributary of the Mekong River, and the basin is approximately in the range of 14°07′–16°23′ N and 101°16′–105°38′ E with an area of 70,435.94 km^2. The terrain generally shows a trend of higher west and lower east with the elevation in the range of 17–1300 m, and the mountains are mainly distributed on the southern boundary of the basin (Figure 1). It has a tropical monsoon climate, and the dry season is from mid-October to the end of April of the next year with a lower precipitation than rainy season. The soil texture types of the basin mainly include light clay, loam, sandy clay loam, and sandy loam, the proportions are about 18.70%, 17.97%, 10.90% and 52.44% respectively, and sandy loam is the main soil. Approximately 78% of the study area is farmland, and 75% of which are paddy fields, and approximately 90% of the paddy fields are rain-fed, which makes many arable lands unused during the dry season.

2.2. Soil Sampling

The surface soil was used for the quality assessment, and the samples were collected from 19 February 2017, to 1 March 2017. Considering that there are few land-use types and soil types in the Mun River Basin, and the spatial distribution of each land-use type also has a certain pattern, most of which are farmland, and forests are mainly distributed in the southern region, moreover, combine the terrain conditions of the basin, road distribution, and other factors, the research laid out a 10 km × 10 km grid throughout the study area for sampling. However, some sample points were moved to adjacent positions because of the accessibility or operability limitations, and some locations are not even allowed to enter, which resulted in the spatial inhomogeneity of final samples. The soil layer of 0–20 cm under the surface was collected, and each soil sample was placed in a sealed bag. The surrounding characteristics of each sampling point were recorded, including latitude and longitude, and a total of 67 samples were collected. Some samples outside the study area were collected because of the accessibility limitations (Figure 1). All samples were dried naturally or in a dryer, ground, and sieved before analyses in the laboratory.

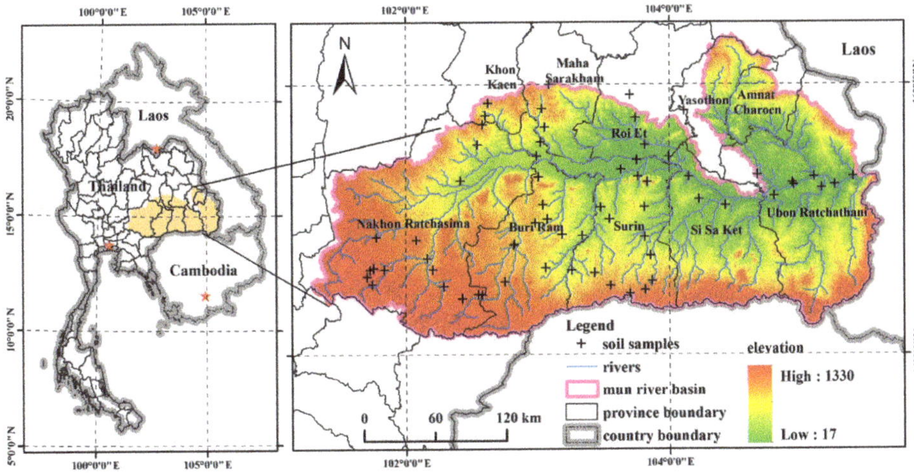

Figure 1. Scope of the study area and sampling locations.

According to household surveys, data inquiries and consultations with native experts about the soil properties, 8 indicators in advance were chosen for the soil-quality assessment, which included soil pH, total nitrogen (TN), available phosphorus (AP), soil particle composition (clay, silt and sand), soil organic matter (SOM) and soil electronic conductivity (EC). The soil particle composition was detected using a laser particle analyzer, and the SOM and TN were measured by Walkley-Black method and Kjeldahl digestion method respectively [28,29], while AP was obtained through extracting samples with a 0.5 mol/L sodium bicarbonate solution and detecting with a spectrophotometer [30]. Soil pH was measured using the electrometric method on a soil/water suspension, and EC was detected by a conductivity meter in the field.

2.3. Auxiliary Data

The auxiliary data is mainly used in the processes of evaluation indicators screening, indicator interpolation, and comprehensive evaluation, which mainly included elevation, topography, distance from river, land-use type, soil type, normalized differential vegetation index (NDVI), environmental vegetation index (EVI), modified soil adjusted vegetation index (MSAVI) and meteorological data. The land-use status of 2017 was generated by interpreting remote sensing images based on the land-use type of 2016, which was obtained from the Land Development Department of Thailand, and from which the distance from river was extracted through distance model of ArcGIS software. The spatial analyst tools were used to obtain the elevation and topography indexes based on the digital elevation model (DEM), which was downloaded from the Geospatial Data Cloud (http://www.gscloud.cn/). The NDVI, EVI, and MSAVI were generated from the remote sensing images, or could be downloaded from the United States Geological Survey (USGS)/Earth Resources Observation and Science (EROS) Center. The soil type, meteorological data, and other data were obtained from different government departments of Thailand. The projection systems of all spatial datasets were converted to the WGS84-based Transverse Mercator orthographic projection coordinate system, and the spatial resolution was set to 250 m $\times$ 250 m. Moreover, the questionnaire surveys about crop fertilization and yield were conducted aimed to analyze the soil conditions properly.

3. Methods

3.1. Geographically Weighted Regression

Geographically Weighted Regression (GWR) was selected for the spatial interpolation and it is similar to the traditional multiple linear regression, but the difference is that the sample locations are considered in the model [31].

$$y = \beta_0 + \sum_{j=1}^{m} \beta_j x_j$$

where y is the dependent variable, x_j represents independent variable values, β_0 is an intercept, β_j indicates regression coefficients of different independent variables. The coefficient is unique in in each location, which can be obtained by weighted least squares approach:

$$\hat{\beta} = \left[X^T W X\right]^{-1} X^T W Y$$

where Y is a $(n \times 1)$ dependent data vector, n is the number of observation data for the location to be calculated, X is a $[n \times (m \times 1)]$ independent variable matrix, one column of which is intercepts, while W is a spatially weighted diagonal matrix:

$$W_{ij} = e^{-0.5(d_{ij}/r)^2}$$

where W_{ij} is the weight of the observed data at location j for determining the dependent variable at location i and r is a bandwidth. The equation indicates that the weight of the observed data is a continuous distance attenuation function, and a modified Akaike information criterion (AICc) is introduced to obtain a reasonable r, which can reduce the complexity of the model and instances of under-smoothing [32]. All soil samples were divided for training (50 samples) and verification (17 samples), and the elevation, topography, distance from river, NDVI, EVI, MSAVI and meteorological data were used as auxiliary data in the interpolation process [26].

3.2. Fuzzy Logic Model

The fuzzy membership of an indicator refers to the possibility that the indicator belongs to a certain grade, and a fuzzy function is introduced to obtain the fuzzy membership of the indicator and then which specific grade the indicator belongs to is determined according to some principles [20,33]. The common fuzzy membership function is a bell-shaped function:

$$MF_{x_i} = \left[1/\left(1 + ((x_i - b)/d)^2\right)\right]$$

where $0 < MF_{x_i} \leq 1$, represents the fuzzy membership of indicator i; x_i is the specific value of i and d is the transition width of i, while the d is always set to be the difference of indicator values when the membership is 0.5 and 1 [14,33], and b is the indicator value when the membership is 1 (Figure 2).

According to the description above, it is an important process to set a suitable range for each indicator, which can be used to gain the membership value through functions while the indicator value belongs to the range, otherwise, it will be set to be 0 or 1. The suitable ranges of the indicators are summarized through consulting previous studies, documentations, standards, and specifications. The integrated weighting method is used to get the final evaluation:

$$MF = \sum_{i=1}^{n} MF_{x_i} w_i$$

where w_i is the weight of different indicator, which can be obtained by the analytic hierarchy process. Additionally, some indicators cannot be used for the soil-quality evaluation because of the redundancy among the primary indicators, and the indicator screening process is necessary. The minimum dataset (MDS) theory was selected in the study, in which a principal component analysis (PCA) is the main method used for the MDS establishing, and the indicators with high factor loadings in the components with eigenvalues ≥ 1 were selected to reflect the soil quality, and the land-use types and soil types were used as the auxiliary data in the screening process [12].

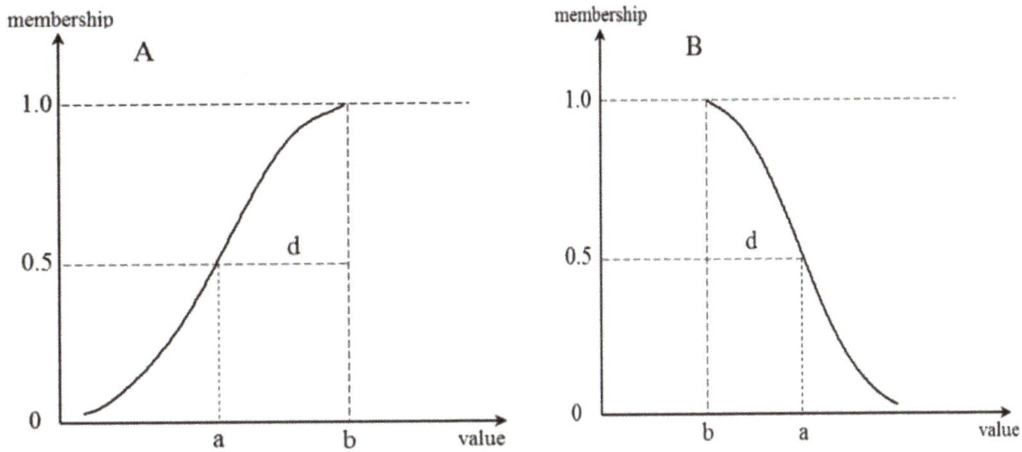

Figure 2. The fuzzy logic models. (**A**) and (**B**) represent the positive and negative indicators, respectively.

4. Results

4.1. Descriptive Statistics

First, outlier tests were conducted on the indicators, and the values that exceeded the threshold range $(u - 3s,\ u + 3s)$ (where u and s are the mean and standard deviation of the indicator value respectively) would be regarded as outliers, which would be set to the maximum or minimum of the remaining values. Table 1 was the descriptive statistics of the data after removing the outliers and Table 2 showed the correlation among different indicators.

Table 1. Basic statistics of the indicators.

	Minimum	Maximum	Mean	SD	Skewness	Kurtosis	K-S Test	CV
pH	4.60	8.00	6.02	0.71	0.81	0.06	0.04	11.84
EC (us/cm)	21.67	732.00	182.73	167.67	1.72	2.48	0.01	91.76
Clay (%)	2.88	46.46	14.49	9.86	1.11	0.46	0.01	68.07
Sand (%)	47.10	96.54	78.47	12.33	−0.60	−0.54	0.16	15.71
Silt (%)	0.00	15.95	7.04	4.71	0.26	−1.19	0.22	66.88
SOM (%)	0.10	3.56	1.26	0.81	1.21	1.18	0.09	64.02
AP (mg/kg)	24.93	284.70	64.93	64.82	2.65	6.35	0.00	99.84
TN (%)	0.01	0.15	0.06	0.03	1.30	1.55	0.13	54.69

SD: Standard deviation; CV: Coefficient of variation.

Table 1 indicated that all indicators showed moderate variation, as all the CV values were less than 100, but the AP was so high that it would be strong variation. Table 2 showed that the correlation coefficients between most indicators were significant at 0.01 and 0.05 levels. There was a high correlation between SOM and TN, and the physical properties of soil had a serious impact on SOM and TN as the correlation coefficients between them were highly significant at 0.01 level.

Table 2. Correlation coefficients between indicators.

	SOM	AP	TN	Clay	Sand	Silt	pH	EC
SOM	1							
AP	0.23	1						
TN	0.89 **	0.25 *	1					
Clay	0.62 **	−0.01	0.57 **	1				
Sand	−0.60 **	0.05	−0.55 **	−0.93 **	1			
Silt	0.25 *	−0.09	0.25 *	0.35 **	−0.66 **	1		
pH	0.12	0.21	0.14	0.16	−0.07	−0.16	1	
EC	0.36 **	0.21	0.44 **	0.22	−0.20	0.05	0.27 *	1

**, *: Correlation is significant at the 0.01 level and 0.05 level, respectively.

4.2. Interpolation of Soil Indicators

According to the MDS model process, this research had screened out four indicators for soil-quality assessment, including TN, AP, SOM, and soil pH, while the seven auxiliary indicators including elevation, terrain curvature, topographic index, distance to rivers, NDVI, EVI, and MSAVI that were preselected for GWR construction were not all used as the multicollinearity among other variables exceeded the tolerance of the model. Moreover, there were no auxiliary indicators selected for the interpolation of pH, and the kriging method was used to obtain the spatial distribution of pH.

Figure 3 showed that all indicators had certain spatial distribution characteristics and their prediction accuracies were reasonable, with mean error of each indicator was close to 0 and root mean square error of each indicator did not exceed 0.5. The SOM had an obvious ladder-like distribution in space, with its content gradually declined from upstream to downstream of the river, and the mountainous area in the south edge had a higher content than the internal flat area of the basin. The areas near the Mun River and its tributaries displayed different spatial characteristics of SOM content in upstream and downstream areas; it was lower along the rivers than the other regions upstream, while it was higher along the rivers than other regions downstream. The content of TN was very low throughout the basin, and its spatial distribution was similar to that of SOM, which declined from west to east gradually, and the highest content was concentrated in the mountains of the southwest, but it was lower in the south edge of the basin. Compared to SOM, the contents along the rivers were not very different from other areas near the rivers. The AP, of which the content was higher over the basin than the other indicators, displayed high values in upstream and downstream areas and low values in the middle of the stream; it was at the lowest level in the southwest especially. The content of AP along the rivers in downstream areas was higher than in the surrounding areas. The soil was mainly acidic over the basin according to the spatial distribution of pH, whose value declined from the periphery to the inside of the basin, and the lowest value was about 5.3, which was strongly acidic soil.

Furtherly, the land use (Figure 4) was used for analysis, overlaid with the spatial distribution of assessment indicators, and the mean of the indicators in different land-use type are shown in Table 3:

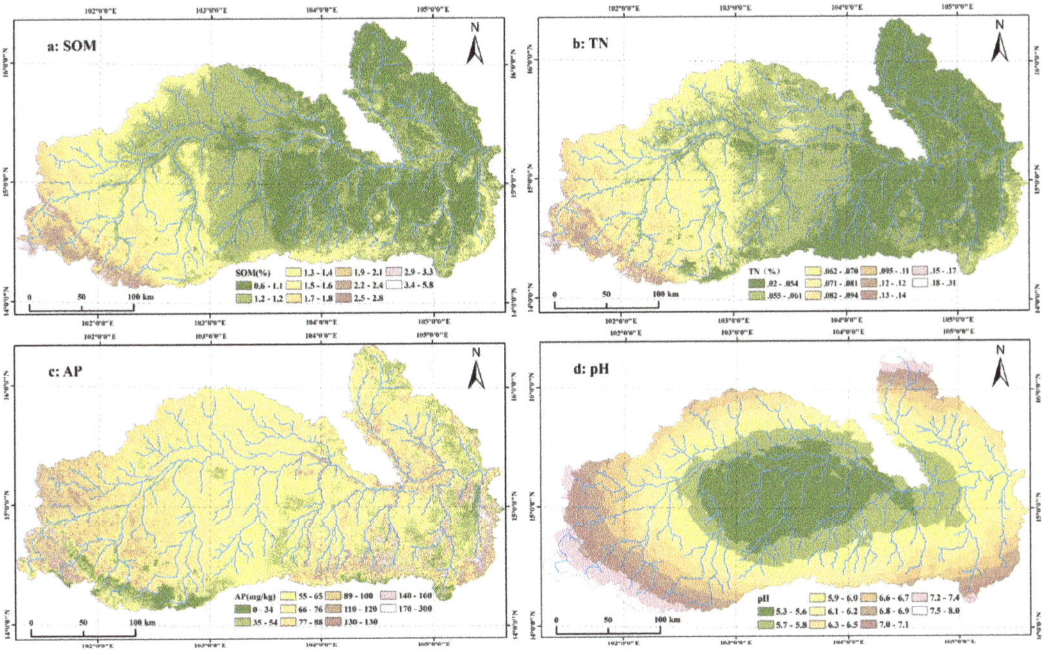

Figure 3. The spatial distributions of the four assessment indicators.

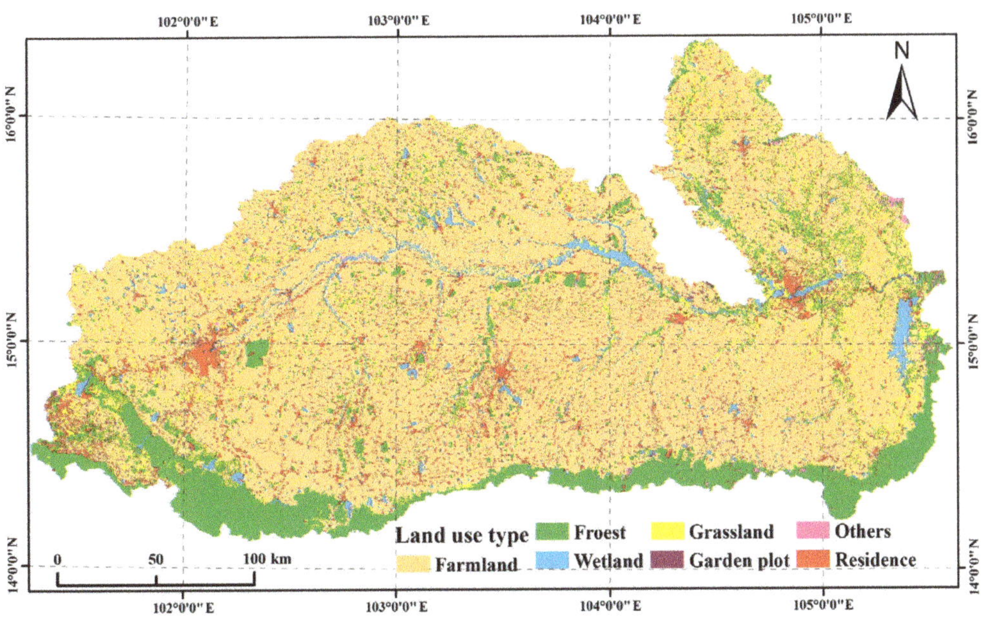

Figure 4. The spatial distribution of different land-use type.

Table 3. The statistics of the indicators in different land-use type.

Land-Use Type	Area Proportion (%)	SOM (%)	TN (%)	AP (mg/kg)	pH
Farmland	72.125	1.181	0.060	72.503	6.058
Forest	13.076	1.475	0.069	70.202	6.572
Grassland	3.422	1.236	0.062	77.452	6.236
Wetland	4.059	1.147	0.059	71.954	5.983
Garden plot	0.825	1.477	0.073	88.948	6.473
Others	0.430	1.115	0.057	66.917	6.274
Residence	6.062	1.204	0.061	72.478	6.074

Figure 4 and Table 3 show that the main land-use types are farmland, followed by the forest, and their total area proportion exceeded 85%. From the figure, we could also find that the forest was mainly distributed in the southern part of the basin, where there were many mountains and the terrain is too steep to be used as farmland. The contents of SOM and TN were plenty in the forest, and its soil did not show strong acidity or alkalinity most of which are neutral according to the measurement of soil pH. However, the content of AP in the forest was lowest than that in other land-use types. the soil condition of the farmland was not very good because the content of SOM and TN was very low, and the soil was strongly acidic. In addition, the soil condition value of all indicators in paddy and dry fields had large differences. The mean values of SOM, TN, AP, and pH in the dry fields were 1.433, 0.074, 83.491 and 6.424, and they were 1.115, 0.057, 68.296 and 5.957 in paddy field, respectively.

4.3. Result of Soil-Quality Assessment

The suitable ranges of all indicators selected for the assessment were determined through summarizing the research results, expert opinions, standards, and literature [34,35] The parameter b and d were obtained for the fuzzy logic function (Table 4), and then the membership of the four indicators were generated. The pH had a double trend, and it was positive when its value was less than 7, on the contrary it was negative.

Table 4. The parameters of the fuzzy logic function.

Indicator	Range	*b*	*d*	Tendency
TN	0.01–0.075	0.075	0.025	Positive
AP	20–120	120	50	Positive
pH	5.5–7	7	1	Positive
pH	7–8.5	7	1	Negative
SOM	0.6–1.5	1.5	0.5	Positive

The indicator weight was obtained based on the communality of each indicator generated in the MDS construction process, and the soil-quality assessment result was generated through integrated weighting method (Table 4). The result was divided into six grades (I–VI) according to the natural breakpoint method, where the ranges were $\leq$0.49, 0.49–0.56, 0.56–0.65, 0.65–0.76, 0.76–0.88 and $\geq$0.88, and grade VI represented the best soil quality. The result is shown in Figure 5.

Figure 5 showed that the best soil quality was distributed in the upstream area of the basin and the soil quality were bad in most of the downstream, but it showed a different situation in the southeast edge and some areas along the rivers, where it was mainly in grade II in the middle of the basin. The statistics showed that grades II, III, and IV were the most widely distributed, with areas of 19,571.13 km^2, 14,413.06 km^2 and 10,478.63 km^2, respectively, and grades I, and VI were the smallest, with areas of 8863.25 km^2 and 7974.19 km^2, which were distributed in the east and west of the basin, respectively. Grade V was mainly distributed in the upstream, and its area was about 9135.69 km^2, especially in the southeast mountains of the study area (Table 5). The soil quality of farmland and forest is better with their mean values of soil quality of 0.63 and 0.72, which belonged to grades III and IV, respectively.

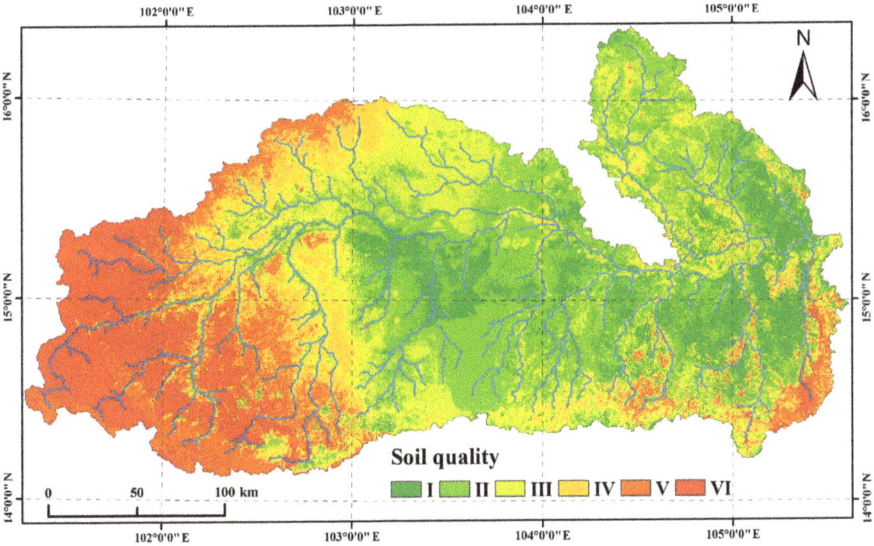

Figure 5. Assessment result of soil quality in the dry season.

Table 5. The area of different land-use type in different soil-quality grade.

Land-Use Type \ Grade	I	II	III	IV	V	VI	Total
Farmland	7019.88	15,631.63	10,095.18	7216.27	5361.92	5477.20	50,802.08
Forest	457.05	1186.60	1939.27	1680.03	2783.37	1164.23	9210.55
Grassland	269.63	598.27	591.99	359.79	204.06	386.92	2410.66
Wetland	574.97	675.62	691.34	439.14	222.46	255.75	2859.27
Garden plot	21.42	78.39	98.72	102.26	79.24	200.84	580.87
Others	66.31	71.41	70.95	51.87	24.20	18.27	303.01
Residence	453.99	1329.19	925.61	629.27	460.45	470.98	4269.49
Total	8863.25	19,571.13	14,413.06	10,478.63	9135.69	7974.19	70,435.94

Table 5 showed the areas of different land-use type in different grade, and we could find that most farmland was in grade II and III, which indicated that the soil quality of farmland was in bad condition and some optimization policies should be carried out to improve the land, fortunately, about 35% of farmland distributed in high grades, which were mainly distributed in the western areas of the basin according to Figure 5. The dry fields were in higher grades than paddy fields, and approximately 77% of dry fields were in grade IV, V, and VI, while approximately 76% of paddy fields were in grades I, II, and III. The forest had better soil quality than farmland, as most forests were in high grades, which were also mainly distributed in the upstream areas. Additionally, the elevation was divided into five levels (<170 m, 170–240 m, 240–370 m, 370–580 m and >580 m) according to the natural breakpoint method and used to calculate the mean value of soil-quality membership. The result showed that the higher the elevation, the better the soil quality.

5. Discussion

The verification of the study results was not constructed because of the lack of related studies, but we could get a general judgment according to the spatial similarity between the indicators and soil quality based on that the spatial accuracy of all indicators had been confirmed. The heavy weights of SOM and TN in the assessment process resulted in the spatial distribution of soil quality being more similar to that of SOM and TN. Additionally, the AP mainly influenced the southwest of the basin, but the low-grade soil quality in the middle of the basin was mainly caused by soil pH. From the above, we thought the assessment result was credible and reasonable.

The soil quality of the basin showed some obvious regularity in space and different land-use type. First, most forests were undisturbed, which made the soil nutrients accumulate through the decomposition of dead branches and leaves year by year, and which furtherly led to rich SOM and TN. The abundance of SOM limited the conversion of AP, which, coupled with the impact of rain, made AP less abundant in the forest. Secondly, the soil quality in farmland was poor because most farmland in the dry season was unused. The soil was dry, and there was not a soil nutrient supply, as the crop residues could not be decomposed. However, there were some dry fields in the upstream where the terrain was very undulating that were not suitable for paddy fields, so artificial fertilization activities would increase soil quality, and the assessment result also showed that the soil-quality value of the dry field was obviously higher than that of paddy fields. Thirdly, the western part of the basin had a complex topography and was the main forest distribution area, which led to the better soil quality than other areas, while the terrain was gentle in the central and eastern part of the basin, and the main land-use type was paddy field there, but it was unused in dry season, and all above made the soil quality poor. However, the areas near the rivers were still available for cultivation because of fertilization played an important role in improving the soil quality.

The assessment result was reasonable as the spatial regularity was consistent with questionnaire surveys, but there still were some insufficiencies. First, the number of sample points was insufficient, and the samples were not distributed evenly in the study area, which would make the assessment process imprecise and lead to the absence of spatial details, especially the number of samples were very small in the east of the basin. Secondly, the number of assessment indicators in this study were fewer than other studies on soil-quality assessment, mover over, there were no biological indicators because of the restrictions of experimental conditions. We will continue our studies in the Mun River Basin, and we will do our best to solve these problems in the near future.

6. Conclusions

The soil nutrient indicators of the Mun River Basin were regularly distributed in space. The contents of SOM and TN were very low in the basin, but they had similar spatial distributions rules, with higher values in the west of the basin than other areas, and their contents were high in mountainous forests and dry fields but low in the paddy fields of the flat terrain area. The content of AP was very high in the basin, but it was very different between forests and farmland, with the lowest values distributed in forest areas. The pH showed that the land was very acidic in the middle of the basin. The assessment results of soil quality also had a decreasing trend from the west to east area, and the dry fields in the west and the forests in the east of the basin were better than other surrounding areas; however, the soil quality of paddy fields in the middle and east of the basin was poor due to the lack of soil nutrient supply when the fields were unused, so the assessment result was useful for soil-quality improvement in the rainy season according to the spatial distributions of soil nutrient indicators. In addition, the limited soil samples and incomplete indicator system would cause the imprecise assessment result, and these shortcomings are what we should solve in future studies.

Author Contributions: C.W. was responsible for the conception and writing of the entire article, E.D. provided great help to the revision of the article, Z.Z. and Y.W. mainly processed and analyzed the data, and G.L. was the main instructor of this article. All authors have read and agreed to the published version of the manuscript.

Funding: This research was funded by the National Natural Science Foundation of China under grant number [41801354] and [41661144030].

Institutional Review Board Statement: Not applicable.

Informed Consent Statement: Not applicable.

Data Availability Statement: The data presented in this study are available on request from the corresponding author.

Conflicts of Interest: The authors declare no conflict of interest.

References

1. Gavrilenko, E.G.; Ananyeva, N.D.; Makarov, O.A. Assessment of soil quality in different ecosystems (with soils of Podolsk and Serpukhov districts of Moscow oblast as examples). *Eurasian Soil Sci.* **2013**, *46*, 1241–1252. [CrossRef]
2. Andrews, S.S.; Karlen, D.L.; Mitchell, J.P. A comparison of soil quality indexing methods for vegetable production systems in Northern California. *Agric. Ecosyst. Environ.* **2002**, *90*, 25–45. [CrossRef]
3. Doran, J.W.; Parkin, T.B. Defining and Assessing Soil Quality. In *Defining Soil Quality for a Sustainable Environment*; Special Publication; Soil Science Society of America: Madison, WI, USA, 1994; pp. 3–21.
4. Karlen, D.L.; Mausbach, M.J.; Doran, J.W.; Cline, R.G.; Harris, R.F.; Schuman, G.E. Soil quality: A concept, definition, and framework for evaluation. *Soil Sci. Soc. Am. J.* **1997**, *61*, 4–10. [CrossRef]
5. Volchko, Y.; Norrman, J.; Rosen, L.; Norberg, T. A minimum data set for evaluating the ecological soil functions in remediation projects. *J. Soils Sediments* **2014**, *14*, 1850–1860. [CrossRef]
6. Biswas, S.; Hazra, G.C.; Purakayastha, T.J.; Saha, N.; Mitran, T.; Roy, S.S.; Basak, N.; Mandal, B. Establishment of critical limits of indicators and indices of soil quality in rice-rice cropping systems under different soil orders. *Geoderma* **2017**, *292*, 34–48. [CrossRef]
7. Firdous, S.; Begum, S.; Yasmin, A. Assessment of soil quality parameters using multivariate analysis in the Rawal Lake watershed. *Environ. Monit. Assess.* **2016**, *188*. [CrossRef] [PubMed]
8. Cui, Q.; Xia, J.; Yang, H.; Liu, J.; Shao, P. Biochar and effective microorganisms promote Sesbania cannabina growth and soil quality in the coastal saline-alkali soil of the Yellow River Delta, China. *Sci. Total Environ.* **2021**, *756*, 143801. [CrossRef]
9. van Hall, R.L.; Cammeraat, L.H.; Keesstra, S.D.; Zorn, M. Impact of secondary vegetation succession on soil quality in a humid Mediterranean landscape. *Catena* **2017**, *149*, 836–843. [CrossRef]
10. Gholami, V.; Sahour, H.; Amri, M.A.H. Soil erosion modeling using erosion pins and artificial neural networks. *Catena* **2021**, *196*. [CrossRef]
11. Mosaffaei, Z.; Jahani, A.; Chahouki, M.A.Z.; Goshtasb, H.; Etemad, V.; Saffariha, M. Soil texture and plant degradation predictive model (STPDPM) in national parks using artificial neural network (ANN). *Modeling Earth Syst. Environ.* **2020**, *6*, 715–729. [CrossRef]
12. Wu, C.; Liu, G.; Huang, C.; Liu, Q. Soil quality assessment in Yellow River Delta: Establishing a minimum data set and fuzzy logic model. *Geoderma* **2019**, *334*, 82–89. [CrossRef]
13. Zuber, S.M.; Behnke, G.D.; Nafziger, E.D.; Villamil, M.B. Multivariate assessment of soil quality indicators for crop rotation and tillage in Illinois. *Soil Tillage Res.* **2017**, *174*, 147–155. [CrossRef]
14. Baja, S.; Chapman, D.M.; Dragovich, D. A conceptual model for defining and assessing land management units using a fuzzy modeling approach in GIS environment. *Environ. Manag.* **2002**, *29*, 647–661. [CrossRef] [PubMed]
15. Ashwood, F.; Butt, K.R.; Doick, K.J.; Vanguelova, E.I. Interactive effects of composted green waste and earthworm activity on tree growth and reclaimed soil quality: A mesocosm experiment. *Appl. Soil Ecol.* **2017**, *119*, 226–233. [CrossRef]
16. Emadi, M.; Baghernejad, M. Comparison of spatial interpolation techniques for mapping soil pH and salinity in agricultural coastal areas, northern Iran. *Arch. Agron. Soil Sci.* **2014**, *60*, 1315–1327. [CrossRef]
17. Li, H.Y.; Shi, Z.; Webster, R.; Triantafilis, J. Mapping the three-dimensional variation of soil salinity in a rice-paddy soil. *Geoderma* **2013**, *195*, 31–41. [CrossRef]
18. Rodríguez, E.; Peche, R.; Garbisu, C.; Gorostiza, I.; Epelde, L.; Artetxe, U.; Irizar, A.; Soto, M.; Becerril, J.M.; Etxebarria, J. Dynamic Quality Index for agricultural soils based on fuzzy logic. *Ecol. Indic.* **2016**, *60*, 678–692. [CrossRef]
19. Kaufmann, M.; Tobias, S.; Schulin, R. Quality evaluation of restored soils with a fuzzy logic expert system. *Geoderma* **2009**, *151*, 290–302. [CrossRef]
20. Liu, Y.L.; Jiao, L.M.; Liu, Y.F.; He, J.H. A self-adapting fuzzy inference system for the evaluation of agricultural land. *Environ. Model. Softw.* **2013**, *40*, 226–234. [CrossRef]
21. Akter, A.; Babel, M.S. Hydrological modeling of the Mun River basin in Thailand. *J. Hydrol.* **2012**, *452*, 232–246. [CrossRef]

22. Prabnakorn, S.; Maskey, S.; Suryadi, F.X.; de Fraiture, C. Rice yield in response to climate trends and drought index in the Mun River Basin, Thailand. *Sci. Total Environ.* **2018**, *621*, 108–119. [CrossRef] [PubMed]
23. Chen, Y.D.; Wang, H.Y.; Zhou, J.M.; Xing, L.; Zhu, B.S.; Zhao, Y.C.; Chen, X.Q. Minimum Data Set for Assessing Soil Quality in Farmland of Northeast China. *Pedosphere* **2013**, *23*, 564–576. [CrossRef]
24. de Lima, A.C.R.; Hoogmoed, W.; Brussaard, L. Soil quality assessment in rice production systems: Establishing a minimum data set. *J. Environ. Qual.* **2008**, *37*, 623–630. [CrossRef] [PubMed]
25. Rahmanipour, F.; Marzaioli, R.; Bahrami, H.A.; Fereidouni, Z.; Bandarabadi, S.R. Assessment of soil quality indices in agricultural lands of Qazvin Province, Iran. *Ecol. Indic.* **2014**, *40*, 19–26. [CrossRef]
26. Wu, C.S.; Liu, G.H.; Huang, C. Prediction of soil salinity in the Yellow River Delta using geographically weighted regression. *Arch. Agron. Soil Sci.* **2017**, *63*, 928–941. [CrossRef]
27. Wu, C.; Liu, G.; Huang, C.; Liu, Q.; Guan, X. Ecological Vulnerability Assessment Based on Fuzzy Analytical Method and Analytic Hierarchy Process in Yellow River Delta. *Int. J. Environ. Res. Public Health* **2018**, *15*, 855. [CrossRef]
28. Nie, X.; Zhao, T.; Su, Y. Fossil fuel carbon contamination impacts soil organic carbon estimation in cropland. *Catena* **2021**, *196*. [CrossRef]
29. Singh, P.; Singh, R.K.; Song, Q.-Q.; Li, H.-B.; Yang, L.-T.; Li, Y.-R. Methods for Estimation of Nitrogen Components in Plants and Microorganisms. *Methods Mol. Biol.* **2020**, *2057*, 103–112. [CrossRef]
30. Wen, J.; Jiang, T.; Arken, S. Selective leaching of vanadium from vanadium-chromium slag using sodium bicarbonate solution and subsequent in-situ preparation of flower-like VS2. *Hydrometallurgy* **2020**, *198*. [CrossRef]
31. Wang, K.; Zhang, C.R.; Li, W.D. Predictive mapping of soil total nitrogen at a regional scale: A comparison between geographically weighted regression and cokriging. *Appl. Geogr.* **2013**, *42*, 73–85. [CrossRef]
32. Hurvich, C.M.; Simonoff, J.S.; Tsai, C.L. Smoothing parameter selection in nonparametric regression using an improved Akaike information criterion. *J. R. Stat. Soc. Ser. B-Stat. Methodol.* **1998**, *60*, 271–293. [CrossRef]
33. Joss, B.N.; Hall, R.J.; Sidders, D.M.; Keddy, T.J. Fuzzy-logic modeling of land suitability for hybrid poplar across the Prairie Provinces of Canada. *Environ. Monit. Assess.* **2008**, *141*, 79–96. [CrossRef] [PubMed]
34. Obade, V.d.P.; Lal, R. Towards a standard technique for soil quality assessment. *Geoderma* **2016**, *265*, 96–102. [CrossRef]
35. Xu, J.; Zhang, G.; Xie, Z. *Soil Indices and Soil Assessment*; Science Press: Beijing, China, 2010.

land

MDPI

Article

Spatial Variations of Vegetation Index from Remote Sensing Linked to Soil Colloidal Status

Marco Bascietto *, Enrico Santangelo and Claudio Beni

Consiglio per la Ricerca in Agricoltura e L'analisi dell'Economia Agraria (CREA)-Centro di Ricerca Ingegneria e Trasformazioni Agroalimentari, CREA-IT, via della Pascolare 16, 00015 Monterotondo, Italy; enrico.santangelo@crea.gov.it (E.S.); claudio.beni@crea.gov.it (C.B.)
* Correspondence: marco.bascietto@crea.gov.it

Abstract: Recent decades have seen a progressive degradation of soils owing to an intensification of farming practices (weeding and high trafficking), increasing use of pesticides and fertilizers, mainly nitrogen, resulting in a steady decline in soil organic matter, a key component to maintain soil fertility. The work has coupled the normalized difference vegetation index (NDVI) of wheat cultivation in Central Italy to soil properties where the wheat was grown to identify the properties linked to within-field variability in productivity. NDVI was assessed through Copernicus Sentinel-2 (S-2) data during the wheat anthesis phase. The main outcome showed a significant correlation of NDVI variability to soil colloidal status and to the relative quantity in the exchange complex of the Ca^{2+} ions. No relationship emerged between NDVI and soil macronutrients (nitrogen, phosphorus, and potassium) concentration. The work suggested that such elements (nitrogen, especially) should not be provided solely considering the vegetation index spatial variations. Rational and sustainable management of soil fertility requires the integration of the NDVI data with the whole complex of soil physical/chemical status. In this way, the identification of the real key factors of fertility will avoid the negative impact of overfertilization. As an example, a fertilization plan was simulated for the sunflower–wheat sequence. The results showed that in the study area additional supplies of N and K would be unnecessary.

Keywords: remote sensing; soil degradation; vegetation index; colloid index; fertilization; nitrogen; phosphorus; potassium; calcium

check for updates

Citation: Bascietto, M.; Santangelo, E.; Beni, C. Spatial Variations of Vegetation Index from Remote Sensing Linked to Soil Colloidal Status. *Land* **2021**, *10*, 80. https://doi.org/10.3390/land10010080

Received: 6 December 2020
Accepted: 13 January 2021
Published: 17 January 2021

1. Introduction

Approximately 81% of the organic carbon resources that are actively involved in the global carbon cycle are stored in soils [1]. Soil organic matter (SOM) represents one of the largest reservoirs of carbon on the global scale; its quantity and quality are important in the management of soil fertility, nutrient supply, and carbon dynamics [2]. Preserving and/or increasing the SOM pool ensures favorable nutrient conditions for field crops and, in turn, contributes to securing food security [3,4]. It has been remarked that soil degradation should be recognized, alongside climate change, as one of the most pressing problems facing humanity particularly in arid and semi-arid regions where salt-induced soil degradation coupled to intensive farming is a major cause of soil organic carbon (SOC) loss [5], the main component of SOM.

Although poor irrigation tilling practices coupled to the extensive use of pesticides and fertilizers represent a major cause of soil organic matter loss, degradation of the fertile layer of the soil, and its desertification [6,7], a few farming practices can be employed to improve soil functional qualities and increase SOC, including optimal fertilization, crop–grassland rotation, and amendments [5].

In the Mediterranean area, a typical agriculture soil is characterized by a high content of clay and active limestone; declining SOM; and by a high level of degradation due to intensive use, frequent mineral fertilization, weeding, no addition of organic matter, and intense traffic of heavy-weight machinery [6].

Available multi-layered studies associate vegetation indexes from remote sensing to nitrogen fertilization [8], to the content [9] or to the supply of N, P, and K [10], to soil texture [11], or soil type [12]. Less frequent (or absent) is the association between remote data and specific soil chemical and physical features.

Remote sensing technologies have proven to be viable for assessing and mapping the availability of agro-forestry resources such as productivity, residual biomass, and crop yields [13] as well as to optimize the interaction crop–soil environment with the final objective of increasing the sustainability of yield by means of more judicious input management [14,15]. Focused mainly on site-specific fertilization [16,17], remote sensing application has progressively been extended to the analysis of all the sources of intra-field variation [18,19] such as yield, soil, crop, anomalous factor (i.e., weeds or pathogens), and management variability (from tillage to the application of fertilizers or pesticides).

Depending on the properties of the instrument and platform, remote sensing data are available from coarse (more than one kilometer) to fine (sub-meter) spatial resolution, and at variable temporal resolution, daily to monthly [20]. The main remote and proximal sensing technologies employed in precision agriculture include satellite platforms, drones, and sensors installed on tractors [15]. Earth observation satellites record and collect spatial information regularly, with wide coverage and low cost, and therefore represent an advantageous tool for the detection of natural and agricultural resources over the last decades [21]. Sentinels 1 and 2 or lower resolution satellite missions have been used to create dynamic cropland masks [22]; perform crop type mapping [23]; estimate soil moisture [24,25]; monitor rice production [26]; estimate plant parameters such as leaf area index [27,28]; or aid in within-field decisions in precision agriculture settings for crop yield mapping, fertilizer use, and minimizing nitrogen loss to water [8,29,30].

We hypothesized that the mineral status of the soil layer where the great part of wheat roots are distributed, due to its influence on plant nutritional status [31,32], should be picked up by remotely sensed normalized difference vegetation index (NDVI) layer.

An exploratory investigation was applied to infer a fundamentally general relationship among exploratory (soil mineral status) and response variables (NDVI) from remote sensing in a typical Mediterranean field. A wheat cultivation was set up in late 2018 to evaluate the contribution of NDVI, calibrated by soil chemical properties monitoring (Figure 1). The study attempted to answer two key questions: (1) which soil nutritional elements and physical features account for the within-field spatial variation of crop productivity, and (2) whether and how the NDVI can be used as a proxy of soil mineral status.

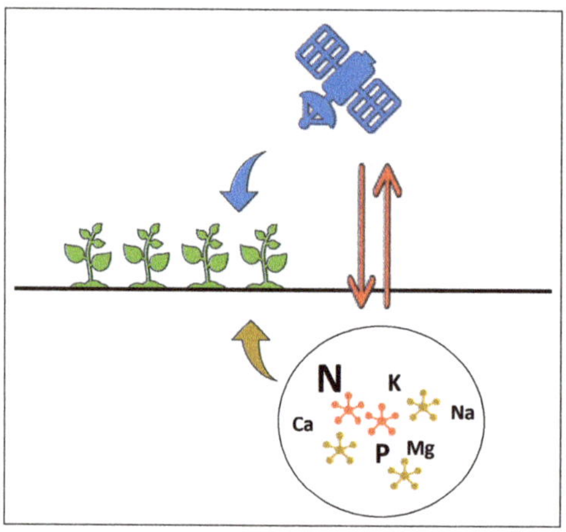

Figure 1. Graphical rendering of the multi-layered concept of the study: remote sensing sensor and soil mineral status to spatially characterize crop productivity.

2. Methods

2.1. Study Site

A 5.0 ha experimental field crop, spatially located in the northeastern outskirts of Rome, central Italy (Figure 2, Latitude 42.103° N, 12.628° E), was set up in 2018. The field is property of the Research Center for Engineering and Agro-Food Processing (CREA) to pursue academic investigations.

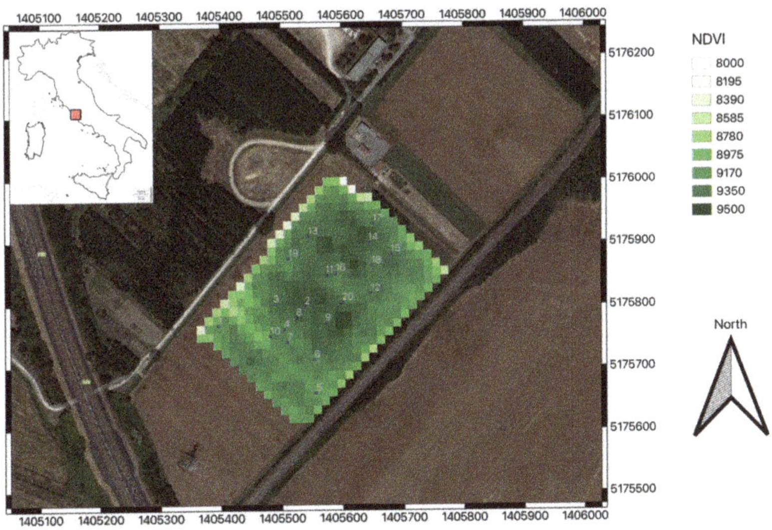

Figure 2. The study area is located in Italy (inset), region of Lazio, province of Rome, and it is part of the CREA experimental farm. Sentinel 2 Normalized Difference Vegetation Index (NDVI) map for the date 16 May 2019 is overlaid on the study area, the random sampling points are numbered 1 to 20; metric coordinates in WGS 84/UTM zone 33N coordinate reference system shown on frame.

Geologic formation outcrops throughout the area consist of volcanic tuff effusive types of lower and middle Pleistocene. These formations are connected to the intense volcanic activity of the northern Lazio region at the time, particularly associated with the calderas activity of the area. The predominant rock types are pyroclastic launch products, mainly composed of loose sand-lapilli levels and sometimes with the presence of cineritic levels more or less cemented. The pyroclastics are leucititic type, which is of significant importance in the process of alteration, show an intense activity of quaternary hydrothermal type, which led to the formation of Analcime (natural zeolite). This confers special properties to the soil in terms of water retention and nutrients release (less water and nutrients available in the real conditions with respect to the analytical data, as captured by the zeolite lattice structure). The southern part of the experimental farm of CREA also includes a portion of an alluvial valley of a secondary stream of the Tevere River. The alluvio-colluvial deposits here consist of fine sandy loam and fine sediment resulting from erosion and reworking of the deposits and soils of the slopes [33]. Soils are of volcanic origin and are classified as Typic Argixeroll [34], soil profile was described in [35].

From the climatic and pedoclimatic point of view, according to the long-term data (30 years) of the ISIS 1.0 database, the average annual air temperature is 13.7 °C, the average annual rainfall is 890 mm, equivalent to an aridity index calculated with the De Martonne equation of 37.5 class (35–40) moderately humid. The thermal regime of the soil is thermal (15–22) with an average soil temperature of 16.3 °C at 0.50 m depth. The water regime is xeric (80–115 days), with 88 cumulated dry days per year. Climatic data were collected by the monitoring station adjacent to the study area [33].

The soil being tested was characterized through a complete physical-chemical analysis, in order to evaluate the characteristics of the mineralogical components and the relationships between them; the reactions and the electrical conductivity that influence the bioavailability of many nutrients; the level of chemical fertility, through the determination of the nutrient content and the colloidal capacity of the soil, which indicates the state of aggregation of the particles; the drainage ratio; and the buffering power of the soil. The analysis involved the collection within the area of twenty random sampling points considered as replicates. The sampling points were randomly drawn in a downsized study area to account for a 10 m border effect (Figure 2).

A 2 kg soil sample from the 0–20 cm layer in each sampling point was collected. At wheat heading 50% of roots are localized in the 0–20 cm soil layer while a further 10% being found in the 20–40 cm layer [36]. The figures were confirmed by the authors of [37] by modeling root distributions of eleven temperate crops: at least half of the root biomass could be found in the upper 20 cm of soil, 61–68% of wheat roots are found in the 0–30 cm soil layer.

Soil characterization was carried out according to the Italian official method of analysis [38] by a UNI CEI EN ISO/IEC 17025:2005-certified laboratory. Particle size distribution was determined by gravimetric method; pH in H_2O with a potentiometer; total organic carbon through Walkley and Black's method; total nitrogen with Kjeldahl's method; available phosphorus by means of Olsen's procedure; cation exchange capacity and exchangeable cations measured in the extracted soil solution (ammonium acetate) by using the atomic absorption spectroscopy (AAS, model AA240FS, Varian, Crawley—UK); assimilable metals by extraction with diethylenetriaminepentaacetic acid (DTPA); and spectrometric analysis with a inductive coupled plasma spectrophotometer model iCAP Pro (ICP-OES, Thermo Fisher Scientific, Waltham—USA). Soil analyses included macro- and micronutrients as well as soil particle size distribution (Table 1).

According to the US classification standards, soil particles are divided into three grades: clay particles <2 μm, silt particles $\geq$ 2 μm < 50 μm, and sand $\geq$ 50 μm < 2000 μm.

Among the derived chemical properties, exchangeable potassium, magnesium, sodium and calcium are defined in % of the CEC; colloids index is defined according to [39]

$$CI = 10 \cdot SOM_{\%} + Clay_{\%} \quad (1)$$

Table 1. Soil analyses and associated units of measure carried out on the 20 sampling set points. Percentages are mass over mass (m/m). Abbreviations: Tot., Total; Act., Active; Av., Available; Ass., Assimilable; Sol., soluble; Exc., exchangeable; CEC, Cation Exchange Capacity; EC, Electrical Conductivity; CI, Colloids Index; SOM, Soil Organic Matter.

Physical Properties	Chemical Properties	Derived Chemical Properties
Sand [%]	Tot. and Act. limestone [%]	Magnesium/potassium ratio
Silt [%]	Tot. organic carbon [%]	Carbon/nitrogen ratio
Clay [%]	Tot. nitrogen [%]	SOM [%]
Particles size distribution [%]	Av. phosphorus [mg kg^{-1}]	CI [%]
pH	Ass. iron [mg kg^{-1}]	Exc. potassium [% of CEC]
EC [mS cm^{-1}]	Ass. manganese [mg kg^{-1}]	Exc. magnesium [% of CEC]
	Ass. copper [mg kg^{-1}]	Exc. sodium [% of CEC]
	Ass. zinc [mg kg^{-1}]	Exc. calcium [% of CEC]
	Sol. boron [mg kg^{-1}]	
	Sol. cobalt [mg kg^{-1}]	
	Exc. calcium [mg kg^{-1}]	
	Exc. magnesium [mg kg^{-1}]	
	Exc. potassium [mg kg^{-1}]	
	Exc. sodium [mg kg^{-1}]	
	CEC [meq 100 g^{-1}]	

2.2. Vegetation Model

At the end of 2018, the study area was sown with durum wheat. The cultivation was conducted following the common farming practices of the area. A linear model was set up to investigate the effect of chemical and physical properties of the soil and soil/vegetation moisture on the Normalized Difference Vegetation Index (NDVI) of the wheat crop growing during winter and spring of 2019, on the same 20 sampling points identified for soil analysis. NDVI is directly related to the photosynthetic capacity and therefore to the energy absorption of plant canopies [40,41], thus proving to be an excellent predictor of productivity and yield [30].

NDVI was estimated for the study area on all available passes of the satellites of the Copernicus Sentinel-2 (S-2) mission in 2019 as

$$NDVI_{t,x,y} = \frac{\rho_{842,t,x,y} - \rho_{490,t,x,y}}{\rho_{842,t,x,y} + \rho_{490,t,x,y}} \tag{2}$$

where $NDVI_{t,x,v}$ is NDVI at time t and spatial coordinates x, y; $\rho_{490,t,x,y}$ and $\rho_{842,t,x,y}$ are the spectral reflectances of the central wavelengths of the near-infrared and red bands of S-2 recorded at time t and at x, y coordinates. These spectral reflectances are themselves ratios of the reflected over the incoming radiation in each spectral band.

The S-2 satellites aim at providing multispectral data with a 5-day revisit frequency and 10 meters spatial resolution [42]. The medium-to-high spatial resolution granted the independence assumption of the sampling points locations (i.e., a one-to-one correspondence links the soil sampling point set and the S-2 pixel set). Cloud and cirrus formations were detected and removed through the quality assurance metadata provided and the resulting pixels masked from NDVI calculation.

The NDVI profile of the study area (Figure 3) helped in tracing the timing of phenology of the crop. At the latitudes of the study, wheat sowing takes place between the end of October and the beginning of November. Field observation [43], phenological model [44,45], or analysis of vegetation index [46] confirmed that in Mediterranean environments, the anthesis occurs between the end of spring frosts and the beginning of the summer drought, corresponding to the end of April–first half of May. The passage from anthesis to maturity is a crucial phase shift because the photosynthates accumulated in the photosynthetic organs (source) relocate towards the ear (sink) for grain filling [47,48]. NDVI was further calculated for the 20 sampling points set, on the peak season day image, this $NDVI_{x,y}$ variable was used as a model predictor. The estimation of the NDVI profiles was performed in Google Earth Engine [49].

Soil chemical and physical features on the 20 sampling points set along with vegetation and soil moisture were included in the model as potential explanatory variables. Moisture was proxied by extracting the C-band Synthetic Aperture Radar (SAR) Ground Range Detected (GRD) single bands on the sampling points for the Sentinel-1 image available on 15 May 2019. Vertical–Vertical (VV) and Vertical–Horizontal (VH) bands report the portion of the outgoing radar signal that the target redirects directly back towards the radar antenna; in VV mode, the microwaves of the electric field are oriented in the vertical plane for both signal transmission and reception whereas in VH mode the backscatter signal is received in the horizontal plane.

The image was preprocessed on Google Earth Engine (apply orbit file, GRD border noise removal, thermal noise removal, radiometric calibration, and terrain correction). The process to define the optimal minimal model of NDVI prediction from the starting comprehensive model formed by all explanatory variables included four steps. First, all variables were standardized (i.e., centered on their mean and scaled by their standard deviation) to ease interpretation of model coefficients and avoid issues due to multicollinearity of explanatory variables. In these standardized models, a unit increase in an explanatory variable is equal to its standard deviation, and it affects the predictive variable by a unit of its standard deviation. Intercept term was dropped from the standardized models of the successive steps.

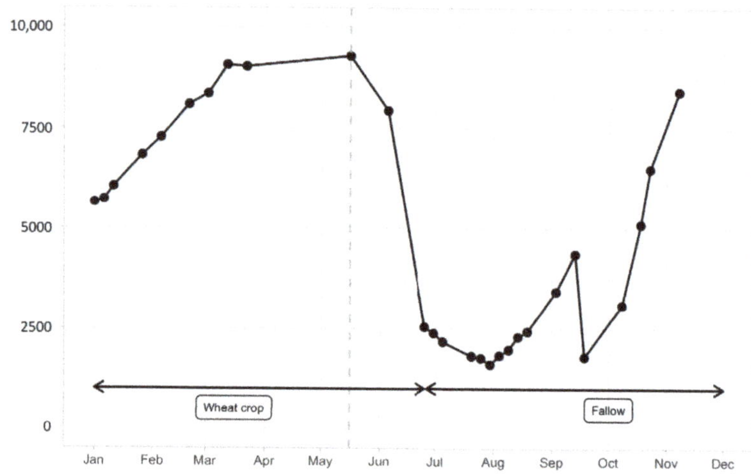

Figure 3. Average NDVI profile for 2019 in the study area. Dots are NDVI values averaged over the study area on available S-2 date (i.e., not masked due to low QA metadata). Crop peak phase was detected to be on 16 May, vertical dashed segment.

Second step concerned feature selection of explanatory variables. Firstly highly correlated variables were removed. If two variables had a high correlation (pairwise absolute correlation cutoff: 0.95), the variable with the largest mean absolute correlation was removed. Highly correlated removed variables included carbon/nitrogen ratio, organic carbon, SOM, and exchangeable calcium (mg kg^{-1}) (Figure 4). Second, important variables were selected by fitting an unsupervised random forest classification model over different tuning parameters and filtering out the least significant variables based on the importance measure, on a percentage scale (cutoff: 40%). Among the remaining explanatory variables, Colloids Index (CI), CEC, exchangeable Calcium (% of CEC), and total nitrogen were selected in order of decreasing importance. This intermediate linear model explained 67% of variance in NDVI (Adjusted R-squared: 0.6782); residual standard error was 0.55 on 16 degrees of freedom.

The third step involved stepwise model selection, based on Akaike's An Information Criterion (AIC) value, of a set of linear models fitted using generalized least squares. Each variable was considered for subtraction from the set of explanatory variables based on AIC value.

Heteroskedasticity and spatial correlation were accounted for in the fourth step. A slightly increasing linear relationship in residuals vs. NDVI values for the explanatory variables was accounted for by weighting observations by selecting variance functions that minimized AIC while being not significantly different from the optimal model. Variance functions chosen were

- an exponential function for CI (where denotes the variance function evaluated at CI and t is the variance function coefficient, t = −0.39);
- a power function for exchangeable Ca (t = −0.32).

Similarly, spatial autocorrelation was accounted for by evaluating the better performing correlation structure (longitude + latitude) in terms of AIC that resulted the spherical spatial correlation (where d is the range and n is the nudge, d = 91.2, n = 0.004). Model estimation was performed in R 3.6.3 [50], stepwise procedure by package MASS 7.3–51.5 [51], GLS modeling by package nlme 3.1–144 [52], variable importance by packages caret 6.0–85 [53], and randomForest 4.6–14 [54]; general data table management was performed by package data.table 1.12.8 [55].

Figure 4. Correlation matrix among exploratory variables. Correlation values are color-coded as squares whose size is proportional to their significance, insignificant ($p < 0.01$) correlation values are struck with a black cross. VV: backscatter of single co-polarization, Vertical transmit/Vertical receive, VH: backscatter of Vertical transmit/Horizontal receive polarisation. See Table 1 for soil abbreviations.

2.3. Fertilization Plans

For applying the procedure to a case study, nitrogen, phosphorus, and potassium fertilization plans were computed from soil chemical and physical analyses of the 20 sampling points set, following the indications formulated in the regulation drawn up by the Lazio Region [56]. Soil nutrient balances taking into account inputs and losses of N, P, and K were computed to satisfy the nutrient demands of sunflower and wheat. As an example, nitrogen balance (a dynamic element considered fundamental for productivity) included seven components: N crop demand for sunflower and for wheat, availability of N for the crop, N leakage caused by rainfall, N leakage due to immobilization processes, residual N supply from previous crop, residual N supply from previous organic fertilizations, and supply of N from natural and anthropic sources. Expected yields were estimated from 2019 average yields in the province of Rome (http://dati.istat.it/): 1330 kg/ha for sunflower and 3000 kg/ha for wheat. The fertilization plan was estimated by fertplan 0.1 [57], an R package specifically developed; spatialization of the fertilization plan from the sampling set was performed by ordinary kriging in R 3.6.3 by package gstat 2.0–4 [58].

3. Results

All the analyzed soil samples fall within the clayey loam or silty clayey loam USDA classification (Table 2; Figure 5) and have a sub-alkaline reaction. Workability is difficult with a tendency to retain too much water, often resulting in stagnant water after heavy rainfall. Water stagnation can be a serious problem for most crops. It often causes stunted growth and rot or other diseases. Therefore, this type of soil requires drainage of surface waters.

Table 2. Descriptive statistics for remote sensing and soil physical properties on sampling point set. Abbreviations: AVG: average, MIN: minimum value, MAX: maximum value, STD: standard deviation, CV: coefficient of variation (i.e., $STD/AVG \cdot 100$); NDVI: Normalized Difference Vegetation Index. See Table 1 for soil physical abbreviations and units of measures and Figure 4 for remote sensing abbreviations.

	Remote Sensing Properties			Soil Physical Properties					
	NDVI $\times 10^4$	VH Backscatter	VV Backscatter	Moisture %	Sand %	Silt %	Clay %	pH	EC mS cm^{-1}
AVG	9288	−21.4	−15.08	20.5	25	38	37	7.9	0.47
MIN	9124	−22.9	−16.72	18.2	16	21	34	7.4	0.32
MAX	9404	−19.6	−13.16	23.8	41	48	40	8.0	0.65
STD	76.27	0.98	0.89	1.29	7.0	7.6	2.2	0.10	0.06
CV	0.82%	−4.50%	−5.90%	6.28%	28%	20%	5.9%	1.7%	13%

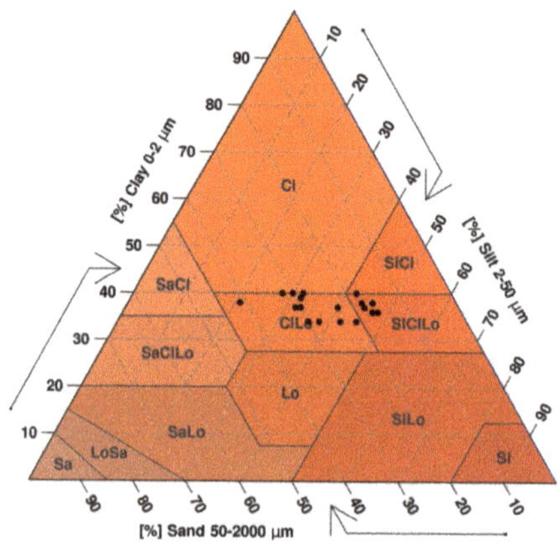

Figure 5. Ternary plot of soil texture (USDA) on the sampling point set. Soil texture classes are Cl (clay), SiCl (silty clay), SaCl (sandy clay), ClLo (clay loam), SiClLo (silty clay loam), SaClLo (sandy clay loam), Lo (loam), SiLo (silty loam), SaLo (sandy loam), Si (silt), LoSa (loamy sand), Sa (sand).

Total limestone is high, with a high content of active limestone (Table 3). The cation exchange capacity is high (>30 meq 100 g^{-1}), as the basic cations saturation rate and colloids index. Calcium is the most present exchangeable cation, and the activity of limestone causes this metal to completely saturate the exchange complex. The content of organic matter and total nitrogen can be classified as medium; given the sub-alkaline nature of the soil, potassium is also well represented. The available phosphorus is low, highlighting a high degree of immobilization of the element due to the excess of Ca and the presence of active limestone. Metallic microelements are represented in good quantities, in particular as regards Fe and Mn. Among the macronutrients, N and K concentrations are widely above the levels of sufficiency for an average demanding crop, so their supply is not required. Phosphorus, on the other hand, is affected by the high degree of immobilization of the soil, so it must necessarily be added with specific fertilizations. In addition to water, soil also retains nutrients, greatly increasing its chemical fertility. The presence of calcium carbonates derived from the degradation of the original or secondary minerals, associated with mineral or organic colloids, contribute to the formation of a stable structure. On the other side, the release of sodium from sodium salts, represents a highly destructuring factor.

Table 3. Descriptive statistics for soil chemical properties on sampling point set. See Tables 1 and 2 for abbreviations and units of measures.

		AVG	MIN	MAX	STD	CV
Tot. lime	[%]	9.69	0.6	17.4	4.01	41.40%
Act. lime	[%]	4.30	0.0	6.5	1.55	36.00%
SOM	[%]	2.63	1.97	3.56	0.352	13.40%
Org. C	[%]	1.52	1.14	2.06	0.204	13.40%
Tot. N	[%]	0.157	0.122	0.205	0.019	11.80%
Ass. P	[mg kg^{-1}]	14.1	11.0	21.0	2.57	18.20%
Ass. Fe	[mg kg^{-1}]	23.4	16.4	32.0	3.99	17.10%
Ass. Mn.	[mg kg^{-1}]	36.3	22.8	70.4	12	33.00%
Ass. Cu	[mg kg^{-1}]	4.91	4.0	6.40	0.676	13.80%
Ass. Zn	[mg kg^{-1}]	1.31	0.60	3.40	0.626	48.00%
Exc. Ca	[mg kg^{-1}]	5852	4940	6640	390	6.67%
Exc. Mg	[mg kg^{-1}]	149.5	124	218	20.2	13.50%
Exc. K	[mg kg^{-1}]	365.3	285	492	58.7	16.10%
Exc. Na	[mg kg^{-1}]	87.95	66	124	14.2	16.20%
Exc. Ca	[% of CEC]	91.71	89.17	92.93	0.909	0.99%
Exc. Mg	[% of CEC]	3.875	3.254	6.141	0.588	15.20%
Exc. K	[% of CEC]	2.953	2.174	4.211	0.549	18.60%
Exc. Na	[% of CEC]	1.201	0.894	1.637	0.173	14.40%
Sol. B	[mg kg^{-1}]	0.872	0.5	1.36	0.21	24.00%
Sol. Co	[mg kg^{-1}]	0.014	0.01	0.02	0.005	36.20%
CEC	[meq 100 g^{-1}]	31.8	27.3	35.8	1.95	6.11%
Mg/K		1.37	0.9	2	0.27	19.70%
C/N		9.71	9.34	10	0.153	1.58%
CI	[%]	63.2	57	71.6	3.7	5.85%

These soils, very common in the Mediterranean area, are generally used for the cultivation of cereals, oil, or industrial crops, resulting in a massive use of fertilizers and pesticides to prevent a decrease of their fertility. Their tendency to lose their structure can be contrasted with the addition of organic matter to counteract the excessive presence of clay and silt.

The NDVI from Sentinel-2 optical bands is very close to the theoretical maximum (1×10^4) and with fairly low variability among the sampling set (Coefficient of Variation 1%, Table 2), whereas microwave bands from remote sensing exhibit higher spatial diversity.

Colloids Index (CI) and exchangeable Calcium (% of CEC) were the explanatory variables selected in the optimal vegetation model after pruning of all the other explanatory variables:

$$NDVI = f(\beta_1 \cdot CI + \beta_2 \cdot Ca) \quad (3)$$

Standardized beta coefficients along with standard deviation of the explanatory variables are given in Table 4. Residual standard error of the optimal model decreased to 0.31 on 18 degrees of freedom while ANOVA confirmed it to be not significantly different from the intermediate model, built after variable selection. Neither of the three soil macronutrients (N, P, and K) nor carbon and any of the micronutrients were able to explain variation of NDVI in the field, during the key phenological phase of anthesis-start of grain filling of the wheat grown in 2019.

Table 4. Vegetation model results. β coefficients and their standards errors in standardized form together with their significance values and cross-correlation. β coefficients are converted to unstandardized form by multiplying their variable standard deviations. Colloid Index is defined in Equation (1). Exc. Ca. is defined as % of Cation Exchange Capacity (Table 1). Abbreviations: Expl., Exploratory; Var., Variable; std. err., standard error; std. dev., standard deviation.

Expl. Var.	Standardized $\beta\pm$ std. Err.	*t*-Value	*p*-Value	Correlation	Var. Std. Dev.	$\beta\pm$ Std. Err.
CI	0.66 ± 0.04	15.91	0		3.7	2.5 ± 0.15
Exc. Ca	0.29 ± 0.05	5.4	0	−0.08	0.91	0.27 ± 0.01

Nevertheless, among the sampling point set, although the NDVI range is very narrow (Figure 6), its variability is largely related to soil colloidal status (CI) and, to a lesser extent, to relative quantity in the exchange complex of the Ca^{2+} ions. Colloids index, in turn, heavily depends on SOM variations ($\times 10$) coupled to clay soil quantity [39] so that a limited increase in SOM can greatly improve soil colloid status leading to an increase in NDVI.

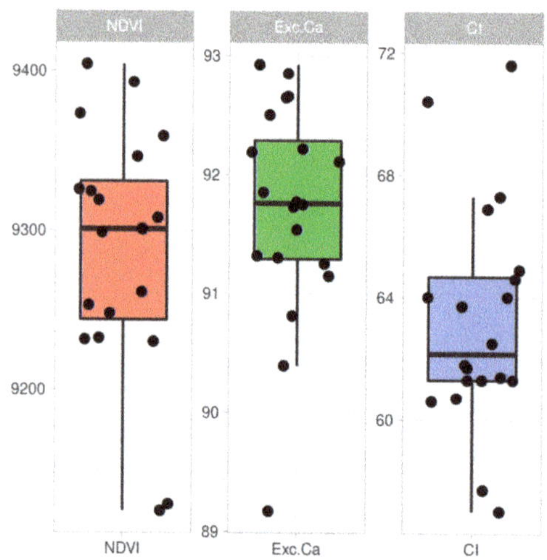

Figure 6. Data distribution (jittered dots) and boxplot of the predictive model variable (NDVI) and explanatory variables. See Table 1 for abbreviations.

The spatialized fertilization plans and the relative doses, calculated in accordance with the regional guidelines, showed a different pattern of within-field variability for N, P, and K. The maps confirmed as N and K concentrations were above the demand for both sunflower and wheat. Phosphorus, instead, must be supplied at concentrations ranging from 37 kg P_2O_5 ha^{-1} to 77 kg P_2O_5 ha^{-1} for sunflower, and with concentrations ranging from 56 to 98 kg P_2O_5 ha^{-1} for wheat (Figure 7).

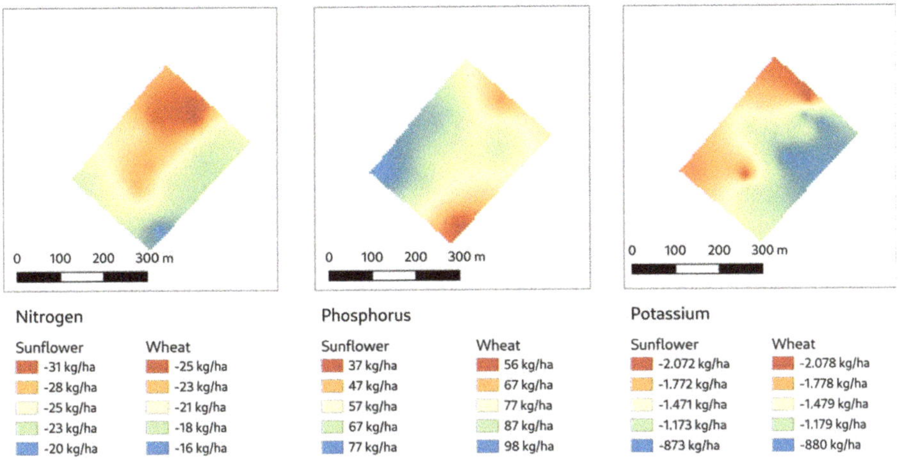

Figure 7. Spatialized fertilization plans for the three main macronutrients for sunflower and wheat crops: from left-most panel to right-most panel: nitrogen, phosphorus, and potassium. Excess nutrient concentration in soil is shown in negative figures (the lower concentration the more excess of nutrient), whereas a demand for nutrient is shown in positive figures (the higher concentration the more demand for nutrient).

4. Discussion

Normalized Difference Vegetation Index (NDVI) was estimated from Sentinel 2 satellites constellation data on a 5 ha wheat field crop during the phenological time-step of the start of grain filling period to test whether it can be a reliable proxy for the mineral status of the soil. NDVI from Sentinel 2 is commonly used for crop yield mapping, fertilizer use, and minimizing nitrogen loss to water [8], usually in precision agriculture settings [59] despite its medium-to-low spatial resolution. A limitation of this study concerns the saturation effect that may affect NDVI by losing its linear relationships with aboveground biomass at higher biomass values. The saturation effects may hinder its estimation performance and confound its relationship to soil chemical/physical properties. A multi-sensor approach might help overcoming this effect (see, e.g., in [60]).

Soil mineral status was sampled in the layer where the great part of wheat roots are distributed [37], particularly in clayey, unstructured soils, with high bulk density and worked at shallow depths in the Mediterranean climate, characterized by low rainfall and low soil moisture content [61].

On the one hand, despite the fact that the availability of the three macronutrients (nitrogen, phosphorus, and potassium) is commonly associated to soil fertility and crop growth, they did not account for the spatial variation of NDVI over the study field nor did any of the micronutrients sampled or any other soil physical features. On the other hand, NDVI variability was associated with the soil colloidal status and, in particular, with the components most active in determining the flocculation of the clays and the aggregation of the soil particles, i.e., the organic matter and the Ca^{2+} ions adsorbed on the exchange complex [62].

The role of SOM is highlighted by the positive correlation between NDVI and the colloids index: even small variations in organic matter can significantly influence the structure of the soil, inducing improvements in soil physical-chemical fertility and in plant nutrition. The presence of higher quantities of Ca^{2+} ions induces the formation of a larger number of bonds between the mineralogical component and the organic matter, which causes the formation of a higher number of stable soil aggregates [63].

An important ecosystem property that contributes to soil organic carbon (SOC) stabilization and soil structure stability is the interaction between SOC and cations or minerals.

SOC can be stabilized by organo-cation or organo-mineral interactions [64]. When the polyvalent cations concentration is high, it becomes sufficient to flocculate and precipitate soluble organic matter. In particular, research in Ca-rich field environments has highlighted a positive correlation between exchangeable Ca^{2+} and SOC concentration. Ca is a plant macronutrient, and it has a localized positive effect on net primary productivity and soil organic matter inputs both for aboveground and belowground biomass. Exchangeable Ca concentration is correlated with a reduction of SOC leaching, photo-oxidation and respiration [65]. Furthermore, the role of clays can also have different effects on the stabilization of soil aggregates, depending on clay mineralogy, particularly at large clay contents. Wuddivira and Camps-Roach [66], by treating a clayey-kaolinitic soil and a sandy-kaolinitic soil with Ca^{2+} and organic matter, improved aggregation within a short time, while the same treatment on a clayey-smectic soil gave the opposite effect, suggesting the need for adequate time for aggregate improvement through Ca^{2+} bridging.

The lack of relation between N, P, and K content in the soil and NDVI has been validated by elaborating fertilization plans both for a successive sunflower crop and for a successive wheat crop elaborated following the current regulations enacted by the competent regional administration. An excess of nitrogen and potassium (K_2O) and a slight demand for phosphorus (P_2O_5) were highlighted by both fertilization plans, although with spatial variations within the field. However, it must be stressed that most of the phosphorus added to the soil is likely to be immobilized, due to the high active limestone content of the soil.

Often, spatial variability of NDVI is commonly associated with nitrogen demand by the crop so that N fertilization plans are deployed by thresholding NDVI into spatially explicit classes and assigning them different N concentrations [17,67]. Although limited to the specific condition of the study, a lower demand for macronutrients may be a less frequent condition than one might think and hence to be worthy of investigation. Should our results be confirmed on wider soil contexts, in Mediterranean intensively used soils the traditional macronutrient fertilization practices could be limited. As the vegetation model has suggested that NDVI variability is to be associated with the variability in SOM, an organic fertilization could be better suited to increase soil matter and crop yield than classic N or NP, or NPK fertilization. Higher SOM tends to mean a larger soil microbial population and therefore potentially higher N supply through mineralization [68]. Second, it should be emphasized that careful evaluation of soil chemical and physical properties should be instrumental to the deployment of properly conceived fertilizations treatments even in precision agriculture frameworks.

5. Conclusions

The commonly used Normalized Difference Vegetation Index from the Sentinel-2 satellite constellation has demonstrated to be very sensible even to the narrow crop productivity variations in the field. For the specific conditions of the study, NDVI variability was influenced by the colloidal status of the soil more than its nutrient availability. Further research is needed to confirm whether the relationship between NDVI and colloid index reported still holds in other clayey soils and in other soil contexts.

Fertilization plans that do not take into consideration soil chemical and physical features may wrongly supply one or more macronutrients under the simplifying assumption that NDVI is solely correlated to nitrogen availability. Variations in crop productivity can be associated with different functional qualities of the soil. Nitrogen, due to its dynamism and its mobility, is the main factor of crop production variability in soils with low fertility. In clay soils, functional qualities can be connected to other factors, first of all the content of organic matter, exchangeable cations, and the quantity and composition of the clay minerals.

Soil and fertility degradation caused by the extensive use of pesticides and fertilizers can be tackled by applying a reasoned analysis of soil properties viewed as a whole system. An integrated agro-ecological assessment coupled to remote sensing can provide useful insights into a more sustainable and targeted fertilization approach. This is particularly true in intensively used soils, such as those the study was based on, where soil organic matter is steadily declining. This approach (i.e., use the NDVI/soil characteristics association) can play a remarkable role to better target the nutrient inputs and to avoid unjustified use of fertilizer.

Author Contributions: M.B.: Data curation, Software, Visualization, Writing—review and editing. E.S.: Validation, Writing—review and editing. C.B.: Conceptualization, Methodology, Writing—review and editing. All authors have read and agreed to the published version of the manuscript.

Funding: This research was funded by the Italian Ministry of Agriculture, Food and Forestry Policies (MiPAAF), grant DM 36503.7305.2018, 20/12/2018 sub-project "Tecnologie digitali integrate per il rafforzamento sostenibile di produzioni e trasformazioni agroalimentari (AgroFiliere)" (AgriDigit programme).

Institutional Review Board Statement: Not applicable.

Informed Consent Statement: Not applicable.

Data Availability Statement: The data presented in this study are openly available in Zenodo/GitHub at doi 10.5281/zenodo.4442166.

Conflicts of Interest: The authors declare no conflict of interest.

References

1. Schlesinger, W. An overview of the C cycle. In *Soils and Global Change*; Lal, R., Kimble, J., Levin, J., Stewart, B.A., Eds.; CRC: Boca Raton, FL, USA, 1995; pp. 9–26.
2. Paul, E.A. The nature and dynamics of soil organic matter: Plant inputs, microbial transformations, and organic matter stabilization. *Soil Biol. Biochem.* **2016**, *98*, 109–126. [CrossRef]
3. Gomiero, T. Soil Degradation, Land Scarcity and Food Security: Reviewing a Complex Challenge. *Sustainability* **2016**, *8*, 281. [CrossRef]
4. Branca, G.; Lipper, L.; McCarthy, N.; Jolejole, M.C. Food security, climate change, and sustainable land management. A review. *Agron. Sustain. Dev.* **2013**, *33*, 635–650. [CrossRef]
5. Ritsema, C.J.; Lynden, G.W.J.V.; Jetten, V.G.; Jong, S.M.D. DEGRADATION. In *Encyclopedia of Soils in the Environment*; Hillel, D., Ed.; Elsevier: Oxford, UK, 2005; pp. 370–377. [CrossRef]
6. Pereira, P.; Brevik, E.C.; Oliva, M.; Estebaranz, F.; Depellegrin, D.; Novara, A.; Cerdà, A.; Menshov, O. Chapter 3—Goal Oriented Soil Mapping: Applying Modern Methods Supported by Local Knowledge. In *Soil Mapping and Process Modeling for Sustainable Land Use Management*; Pereira, P., Brevik, E.C., Muñoz-Rojas, M., Miller, B.A., Eds.; Elsevier: Oxford, UK, 2017; pp. 61–83. [CrossRef]
7. Purwanto, B.H.; Alam, S. Impact of intensive agricultural management on carbon and nitrogen dynamics in the humid tropics. *Soil Sci. Plant Nutr.* **2020**, *66*, 50–59. [CrossRef]
8. Vizzari, M.; Santaga, F.; Benincasa, P. Sentinel 2-Based Nitrogen VRT Fertilization in Wheat: Comparison between Traditional and Simple Precision Practices. *Agronomy* **2019**, *9*, 278. [CrossRef]
9. Song, Y.Q.; Zhao, X.; Su, H.Y.; Li, B.; Hu, Y.M.; Cui, X.S. Predicting Spatial Variations in Soil Nutrients with Hyperspectral Remote Sensing at Regional Scale. *Sensors* **2018**, *18*, 3086. [CrossRef]
10. Khitrov, N.B.; Rukhovich, D.I.; Koroleva, P.V.; Kalinina, N.V.; Trubnikov, A.V.; Petukhov, D.A.; Kulyanitsa, A.L. A study of the responsiveness of crops to fertilizers by zones of stable intra-field heterogeneity based on big satellite data analysis. *Arch. Agron. Soil Sci.* **2020**, *66*, 1963–1975. [CrossRef]
11. Tewes, A.; Hoffmann, H.; Nolte, M.; Krauss, G.; Schäfer, F.; Kerkhoff, C.; Gaiser, T. How Do Methods Assimilating Sentinel-2-Derived LAI Combined with Two Different Sources of Soil Input Data Affect the Crop Model-Based Estimation of Wheat Biomass at Sub-Field Level? *Remote Sens.* **2020**, *12*, 925. [CrossRef]
12. Hongo, C.; Sigit, G.; Shikata, R.; Niwa, K.; Tamura, E. The Use of Remotely Sensed Data for Estimating of Rice Yield Considering Soil Characteristics. *J. Agric. Sci.* **2014**, *6*, 13.
13. Bascietto, M.; Sperandio, G.; Bajocco, S. Efficient Estimation of Biomass from Residual Agroforestry. *ISPRS Int. J. Geo-Inf.* **2020**, *9*, 21. [CrossRef]
14. Mulla, D.J. Twenty five years of remote sensing in precision agriculture: Key advances and remaining knowledge gaps. *Biosyst. Eng.* **2013**, *114*, 358–371. [CrossRef]
15. Weiss, M.; Jacob, F.; Duveiller, G. Remote sensing for agricultural applications: A meta-review. *Remote Sens. Environ.* **2020**, *236*, 111402. [CrossRef]
16. Auernhammer, H. Precision farming—The environmental challenge. *Comput. Electron. Agric.* **2001**, *30*, 31–43. [CrossRef]

17. Basso, B.; Fiorentino, C.; Cammarano, D.; Schulthess, U. Variable rate nitrogen fertilizer response in wheat using remote sensing. *Precis. Agric.* **2016**, *17*, 168–182. [CrossRef]
18. Pallottino, F.; Antonucci, F.; Costa, C.; Bisaglia, C.; Figorilli, S.; Menesatti, P. Optoelectronic proximal sensing vehicle-mounted technologies in precision agriculture: A review. *Comput. Electron. Agric.* **2019**, *162*, 859–873. [CrossRef]
19. Zhang, N.; Wang, M.; Wang, N. Precision agriculture—A worldwide overview. *Comput. Electron. Agric.* **2002**, *36*, 113–132. [CrossRef]
20. Ginaldi, F.; Bajocco, S.; Bregaglio, S.; Cappelli, G. Spatializing Crop Models for Sustainable Agriculture. In *Innovations in Sustainable Agriculture*; Farooq, M., Pisante, M., Eds.; Springer International Publishing: Cham, Switzerland, 2019; pp. 599–619. [CrossRef]
21. Kasampalis, D.; Alexandridis, T.; Deva, C.; Challinor, A.; Moshou, D.; Zalidis, G. Contribution of Remote Sensing on Crop Models: A Review. *J. Imaging* **2018**, *4*, 52. [CrossRef]
22. Valero, S.; Morin, D.; Inglada, J.; Sepulcre, G.; Arias, M.; Hagolle, O.; Dedieu, G.; Bontemps, S.; Defourny, P.; Koetz, B. Production of a Dynamic Cropland Mask by Processing Remote Sensing Image Series at High Temporal and Spatial Resolutions. *Remote Sens.* **2016**, *8*, 55. [CrossRef]
23. Inglada, J.; Arias, M.; Tardy, B.; Hagolle, O.; Valero, S.; Morin, D.; Dedieu, G.; Sepulcre, G.; Bontemps, S.; Defourny, P.; et al. Assessment of an Operational System for Crop Type Map Production Using High Temporal and Spatial Resolution Satellite Optical Imagery. *Remote Sens.* **2015**, *7*, 12356–12379. [CrossRef]
24. Boke-Olén, N.; Ardö, J.; Eklundh, L.; Holst, T.; Lehsten, V. Remotely sensed soil moisture to estimate savannah NDVI. *PLoS ONE* **2018**, *13*, e0200328. [CrossRef]
25. Taktikou, E.; Bourazanis, G.; Papaioannou, G.; Kerkides, P. Prediction of Soil Moisture from Remote Sensing Data. *Procedia Eng.* **2016**, *162*, 309–316. [CrossRef]
26. Torbick, N.; Chowdhury, D.; Salas, W.; Qi, J. Monitoring Rice Agriculture across Myanmar Using Time Series Sentinel-1 Assisted by Landsat-8 and PALSAR-2. *Remote Sens.* **2017**, *9*, 119. [CrossRef]
27. Campos-Taberner, M.; García-Haro, F.; Camps-Valls, G.; Grau-Muedra, G.; Nutini, F.; Busetto, L.; Katsantonis, D.; Stavrakoudis, D.; Minakou, C.; Gatti, L.; et al. Exploitation of SAR and Optical Sentinel Data to Detect Rice Crop and Estimate Seasonal Dynamics of Leaf Area Index. *Remote Sens.* **2017**, *9*, 248. [CrossRef]
28. Clevers, J.; Kooistra, L.; van den Brande, M. Using Sentinel-2 Data for Retrieving LAI and Leaf and Canopy Chlorophyll Content of a Potato Crop. *Remote Sens.* **2017**, *9*, 405. [CrossRef]
29. Hunt, M.L.; Blackburn, G.A.; Carrasco, L.; Redhead, J.W.; Rowland, C.S. High resolution wheat yield mapping using Sentinel-2. *Remote Sens. Environ.* **2019**, *233*, 111410. [CrossRef]
30. Kayad, A.; Sozzi, M.; Gatto, S.; Marinello, F.; Pirotti, F. Monitoring Within-Field Variability of Corn Yield using Sentinel-2 and Machine Learning Techniques. *Remote Sens.* **2019**, *11*, 2873. [CrossRef]
31. Jonard, M.; Fürst, A.; Verstraeten, A.; Thimonier, A.; Timmermann, V.; Potočić, N.; Waldner, P.; Benham, S.; Hansen, K.; Merilä, P.; et al. Tree mineral nutrition is deteriorating in Europe. *Glob. Chang. Biol.* **2015**, *21*, 418–430. [CrossRef]
32. Fageria, N.K.; Baligar, V.C.; Jones, C.J. *Growth and Mineral Nutrition of Field Crops*, 3rd ed.; CRC Press: Boca Raton, FL, USA, 2010.
33. Bazzoffi, P.; Francaviglia, R.; Neri, U.; Napoli, R.; Marchetti, A.; Falcucci, M.; Pennelli, B.; Simonetti, G.; Barchetti, A.; Migliore, M.; et al. Environmental effectiveness of GAEC cross-compliance Standard 1.1a (temporary ditches) and 1.2g (permanent grass cover of set-aside) in reducing soil erosion and economic evaluation of the competitiveness gap for farmers. *Ital. J. Agron.* **2016**, *10*. [CrossRef]
34. Soil Survey Staff. *Keys to Soil Taxonomy*, 12th ed.; USDA-Natural Resources Conservation Service: Washington, DC, USA, 2014.
35. Mecella, G; Scandella, P.; Di Blasi, N.; Pierandrei, F.; Biondi, F.A. Land classification and climatic aspects of upper Tiber Valley territory [Latium]-Land classification ed aspetti climatici del territorio dell'Alta Valle del Tevere [Lazio]. *Annali dell'Ist. Sper. Nutr. Piante* **1985**, *13*, 1–172.
36. Motzo, R.; Attene, G.; Deidda, M. Genotypic variation in durum wheat root systems at different stages of development in a Mediterranean environment. *Euphytica* **1992**, *66*, 197–206. [CrossRef]
37. Fan, J.; McConkey, B.; Wang, H.; Janzen, H. Root distribution by depth for temperate agricultural crops. *Field Crop. Res.* **2016**, *189*, 68–74. [CrossRef]
38. MiPAAF. Official Methods of Chemical Analysis of Soil. Italian Ministry of Food, Agriculture and Forestry Policies Decree 13 september 1999. Available online: https://www.gazzettaufficiale.it/eli/gu/1999/10/21/248/so/185/sg/pdf (accessed on 13 January 2021).
39. Beni, C.; Servadio, P.; Marconi, S.; Neri, U.; Aromolo, R.; Diana, G. Anaerobic Digestate Administration: Effect on Soil Physical and Mechanical Behavior. *Commun. Soil Sci. Plant Anal.* **2012**, *43*, 821–834. [CrossRef]
40. Myneni, R.B.; Hall, F.G.; Sellers, P.J.; Marshak, A.L. The Interpretation of Spectral Vegetation Indexes. *Trans. Geosci. Remote Sens.* **1995**, *33*, 481–486. [CrossRef]
41. Sellers, P.J. Canopy reflectance, photosynthesis and transpiration. *Int. J. Remote Sens.* **1985**, *6*, 1335–1372. [CrossRef]
42. Drusch, M.; Del Bello, U.; Carlier, S.; Colin, O.; Fernandez, V.; Gascon, F.; Hoersch, B.; Isola, C.; Laberinti, P.; Martimort, P.; et al. Sentinel-2: ESA's Optical High-Resolution Mission for GMES Operational Services. *Remote Sens. Environ.* **2012**, *120*, 25–36. [CrossRef]

43. Motzo, R.; Giunta, F. The effect of breeding on the phenology of Italian durum wheats: From landraces to modern cultivars. *Eur. J. Agron.* **2007**, *26*, 462–470. [CrossRef]
44. Ceglar, A.; van der Wijngaart, R.; de Wit, A.; Lecerf, R.; Boogaard, H.; Seguini, L.; van den Berg, M.; Toreti, A.; Zampieri, M.; Fumagalli, D.; et al. Improving WOFOST model to simulate winter wheat phenology in Europe: Evaluation and effects on yield. *Agric. Syst.* **2019**, *168*, 168–180. [CrossRef]
45. Di Paola, A.; Ventura, F.; Vignudelli, M.; Bombelli, A.; Severini, M. A generalized phenological model for durum wheat: application to the Italian peninsula. *J. Sci. Food Agric.* **2020**, *100*, 4093–4100. [CrossRef]
46. Magney, T.S.; Eitel, J.U.; Huggins, D.R.; Vierling, L.A. Proximal NDVI derived phenology improves in-season predictions of wheat quantity and quality. *Agric. For. Meteorol.* **2016**, *217*, 46–60. [CrossRef]
47. Ali, M.A.; Hussain, M.; Khan, M.I.; Ali, Z.; Zulkiffal, M.; Anwar, J.; Sabir, W.; Zeeshan, M. Source-Sink Relationship between Photosynthetic Organs and Grain Yield Attributes during Grain Filling Stage in Spring Wheat (Triticum aestivum). *Int. J. Agric. Biol.* **2010**, *12*, 8.
48. Borghi, B.; Corbellini, M.; Cattaneo, M.; Fornasari, M.E.; Zucchelli, L. Modification of the Sink/Source Relationships in Bread Wheat and its Influence on Grain Yield and Grain Protein Content*. *J. Agron. Crop Sci.* **1986**, *157*, 245–254. [CrossRef]
49. Gorelick, N.; Hancher, M.; Dixon, M.; Ilyushchenko, S.; Thau, D.; Moore, R. Google Earth Engine: Planetary-scale geospatial analysis for everyone. *Remote Sens. Environ.* **2017**, *202*, 18–27. [CrossRef]
50. R Core Team. *R: A Language and Environment for Statistical Computing*; R Foundation for Statistical Computing: Vienna, Austria, 2018.
51. Venables, W.N.; Ripley, B.D. *Modern Applied Statistics with S*, 4th ed.; Springer: New York, NY, USA, 2002.
52. Pinheiro, J.; Bates, D.; DebRoy, S.; Sarkar, D.; R Core Team. *nlme: Linear and Nonlinear Mixed Effects Models*. 2020. Available online: https://CRAN.R-project.org/package=nlme (accessed on 13 January 2021).
53. Kuhn, M. *Caret: Classification and Regression Training*. 2020. Available online: https://CRAN.R-project.org/package=caret (accessed on 13 January 2021).
54. Liaw, A.; Wiener, M. Classification and Regression by randomForest. *R News* **2002**, *2*, 18–22.
55. Dowle, M.; Srinivasan, A. *Data.Table: Extension of 'Data.frame'*. 2019. Available online: https://CRAN.R-project.org/package=data.table (accessed on 13 January 2021).
56. Assessorato Agricoltura, Promozione della Filiera e della Cultura del Cibo, Ambiente e Risorse Naturali. *Parte Agronomica, Norme Generali; Disciplinare di Produzione Integrata della Regione Lazio—SQNPI*; Regione Lazio: Roma, Italy, 2020; Volume 1.
57. Bascietto, M. Fertplan: Compute NPK Fertilization Plans. 2020. Available online: https://github.com/mbask/fertplan (accessed on 13 January 2021).
58. Gräler, B.; Pebesma, E.; Heuvelink, G. Spatio-Temporal Interpolation using gstat. *R J.* **2016**, *8*, 204–218. [CrossRef]
59. Bongiovanni, R.; Lowenberg-Deboer, J. Precision Agriculture and Sustainability. *Precis. Agric.* **2004**, *5*, 359–387. [CrossRef]
60. Cao, Q.; Miao, Y.; Feng, G.; Gao, X.; Li, F.; Liu, B.; Yue, S.; Cheng, S.; Ustin, S.L.; Khosla, R. Active canopy sensing of winter wheat nitrogen status: An evaluation of two sensor systems. *Precis. Agric.* **2015**, *112*, 54–67. [CrossRef]
61. Martino, D.L.; Shaykewich, C.F. Root penetration profiles of wheat and barley as affected by soil penetration resistance in field conditions. *Can. J. Soil Sci.* **1994**, *74*, 193–200. [CrossRef]
62. Rengasamy, P.; Marchuk, A. Cation ratio of soil structural stability (CROSS). *Soil Res.* **2011**, *49*, 280–285. [CrossRef]
63. Oades, J.M. Soil organic matter and structural stability: Mechanisms and implications for management. *Plant Soil* **1984**, *76*, 319–337.
64. Dexter, A.R. Advances in characterization of soil structure. *Soil Tillage Res.* **1988**, *11*, 199–238. [CrossRef]
65. Rowley, M.C.; Grand, S.; Verrecchia, E.P. Calcium-mediated stabilisation of soil organic carbon. *Biogeochemistry* **2018**, *137*, 27–49. [CrossRef]
66. Wuddivira, M.N.; Camps-Roach, G. Effects of organic matter and calcium on soil structural stability. *Eur. J. Soil Sci.* **2007**, *58*, 722–727. [CrossRef]
67. Robertson, M.; Isbister, B.; Maling, I.; Oliver, Y.; Wong, M.; Adams, M.; Bowden, B.; Tozer, P. Opportunities and constraints for managing within-field spatial variability in Western Australian grain production. *Field Crop. Res.* **2007**, *104*, 60–67. [CrossRef]
68. Colaço, A.; Bramley, R. Do crop sensors promote improved nitrogen management in grain crops? *Field Crop. Res.* **2018**, *218*, 126–140. [CrossRef]

land

MDPI

Article

Characterization of Soil Carbon Stocks in the City of Johannesburg

Kelebohile Rose Seboko [1,*], Elmarie Kotze [1], Johan van Tol [1] and George van Zijl [2]

[1] Department of Soil, Crop and Climate Sciences, Faculty of Natural and Agricultural Sciences, University of the Free State, Bloemfontein 9301, South Africa; KotzeE@ufs.ac.za (E.K.); vantolJJ@ufs.ac.za (J.v.T.)

[2] Unit for Environmental Sciences and Management, North-West University, Potchefstroom 2520, South Africa; george.vanzijl@nwu.ac.za

* Correspondence: 2014079897@ufs4life.ac.za; Tel.: +27-67-282-3975

Abstract: Soil organic carbon (SOC) is a crucial indicator of soil health and soil productivity. The long-term implications of rapid urbanization on sustainability have, in recent years, raised concern. This study aimed to characterize the SOC stocks in the Johannesburg Granite Dome, a highly urbanized and contaminated area. Six soil hydropedological groups; (recharge (deep), recharge (shallow), responsive (shallow), responsive (saturated), interflow (A/B), and interflow (soil/bedrock)) were identified to determine the vertical distribution of the SOC stocks and assess the variation among the soil groups. The carbon (C) content, bulk density, and soil depth were determined for all soil groups, and thereafter the SOC stocks were calculated. Organic C stocks in the A horizon ranged, on average, from 33.55 ± 21.73 t C ha^{-1} for recharge (deep) soils to 17.11 ± 7.62 t C ha^{-1} for responsive (shallow) soils. Higher C contents in some soils did not necessarily indicate higher SOC stocks due to the combined influence of soil depth and bulk density. Additionally, the total SOC stocks ranged from 92.82 ± 39.2 t C ha^{-1} for recharge (deep) soils to 22.81 ± 16.84 t C ha^{-1} for responsive (shallow) soils. Future studies should determine the SOC stocks in urban areas, taking diverse land-uses and the presence of iron (Fe) oxides into consideration. This is crucial for understanding urban ecosystem functions.

Keywords: soil quality; soil organic carbon stocks; and urban areas

Citation: Seboko, K.R.; Kotze, E.; van Tol, J.; van Zijl, G. Characterization of Soil Carbon Stocks in the City of Johannesburg. *Land* **2021**, *10*, 83. https://doi.org/10.3390/land10010083

Received: 10 December 2020
Accepted: 13 January 2021
Published: 18 January 2021

1. Introduction

An increasing urban population has stimulated interest in the status and sustainable use of soil resources in urban areas. Anthropogenic activities have largely contributed to the variation of urban soil properties [1]. Urban soils are generally characterized by increased bulk density, pH levels, and carbon (C) content due to organic pollutants at industrial sites, increased residential waste, traffic, and infrastructure [2–5]. Conversely, a study by [6] found that soil pH and bulk density were not significantly different in urban areas. However, soils in the city of Johannesburg, the economic hub of Africa, are diverse in both physical and biochemical aspects. Gold mine tailings across the city are often laden with lead and arsenic [7,8], reducing agricultural productivity and overall soil quality.

Soils provide essential ecological services, such as nutrient cycling, biomass production, a habitat for soil organisms, storage and filtration of water, and C storage [9]. Soil organic carbon (SOC), a measurable component of soil organic matter (SOM), is an important soil quality indicator, as it influences climate change mitigation [10,11], soil fertility, porosity, aggregation, and water-holding capacity [12,13]. Soils act as either a source or sink of atmospheric carbon dioxide (CO_2) and thus play a crucial role in the storage of carbon (C). Extensive cultivation and deforestation are part of the largest anthropogenic sources of CO_2 [14]. Carbon dioxide emissions in areas affected by deforestation were 0.4 Pg C yr^{-1}, approximately 0.3 Pg C yr^{-1} less than the average from 1997 to 2008. Soil organic matter is primarily derived from plant residues [15] and is a source of essential plant nutrients, such as nitrogen, phosphorus, potassium, calcium, and magnesium. Furthermore, SOM binds soil particles and forms aggregates, improving the water-holding capacity of soils [16]. Previous research found urban soils often are water deficient [17], but Mao et al. [6] discovered that soil moisture content and SOC increased notably from the urban periphery to the core.

In recent decades, environmental policy-making has integrated the protection of soil resources to promote sustainability. There is a need for detailed soil quality data in urban areas due to unsustainable development. Although urban areas occupy about 0.5% of the global land surface [18], they are responsible for approximately 70–75% of global anthropogenic CO_2 emissions [19]. Rapid urbanization leads to environmental challenges, such as the modification of local and regional climate [20], the loss of biodiversity caused by the destruction of natural habitats [21], as well as the degradation of water resources [22]. The characterization of urban SOC stocks is, therefore, crucial for understanding urban ecosystem functions.

Carbon storage in South African cities has hardly been researched. Generally, 58% of soils in South Africa have less than 0.5% SOC and only 4% of soils have more than 2% SOC [23]. Recently, there has been increased interest in quantifying C storage in urban areas [24–26]. Some research indicated that SOC stocks are not significantly different across urban areas [6,24]. Moreover, in the United Kingdom, the total SOC storage was estimated to be about 17.6 kg m^{-2} for urban areas [25]. One study [26], which aimed to describe the impacts of urbanization on the SOC stocks in north-eastern China, found a decrease from 2.77 ± 1.09 kg m^{-2} to 2.16 ± 0.93 kg m^{-2} over two decades. The loss in carbon was attributed to rapid urbanization.

Here, we hypothesize that an increase in the retention of water and a higher bulk density result in greater SOC stocks. This study aims to determine the vertical distribution of SOC stocks in six hydropedological groups identified in the Johannesburg Granite Dome area and assesses the variation among the groups.

2. Materials and Methods

2.1. Study Area

The study was conducted in the Johannesburg Granite Dome area (Figure 1), partly located in the Upper Crocodile catchment. The catchment area is part of the Witwatersrand Supergroup, underlain by a combination of granitic, gneissic, and granodiorite rocks, which have been weathered and modified by tectonic processes [27]. The research site has soils that have been severely altered by sand mining [28]. Due to the soil-forming process of ferrolysis (reduction of free iron (III) oxides to Fe^{2+} followed by the oxidation of Fe^{2+}), soils in this area have limited clay accumulation. Johannesburg has a semi-arid environment. It is a summer rainfall area, receiving most of its rainfall between October and March. The region has a mean annual precipitation of 682 mm and mean annual evaporation of approximately 1700 mm [27].

Johannesburg is the primary economic hub of sub-Saharan Africa, with a population of 5.7 million people [29]. The Johannesburg Granite Dome area is drained by the Jukskei and Crocodile Rivers, recognized as a stressed catchment in South Africa [27]. The study area, with a total surface area of 768 km^2, was identified by the South African Department of Environmental, Forestry, and Fisheries as one of the catchments that required urgent attention in terms of environmental quality. The main drivers of a decline in environmental quality are agricultural, domestic, industrial, and mining processes [10]. Catchments in this area have been studied previously [30–32]; however, quite a few focus on soil quality.

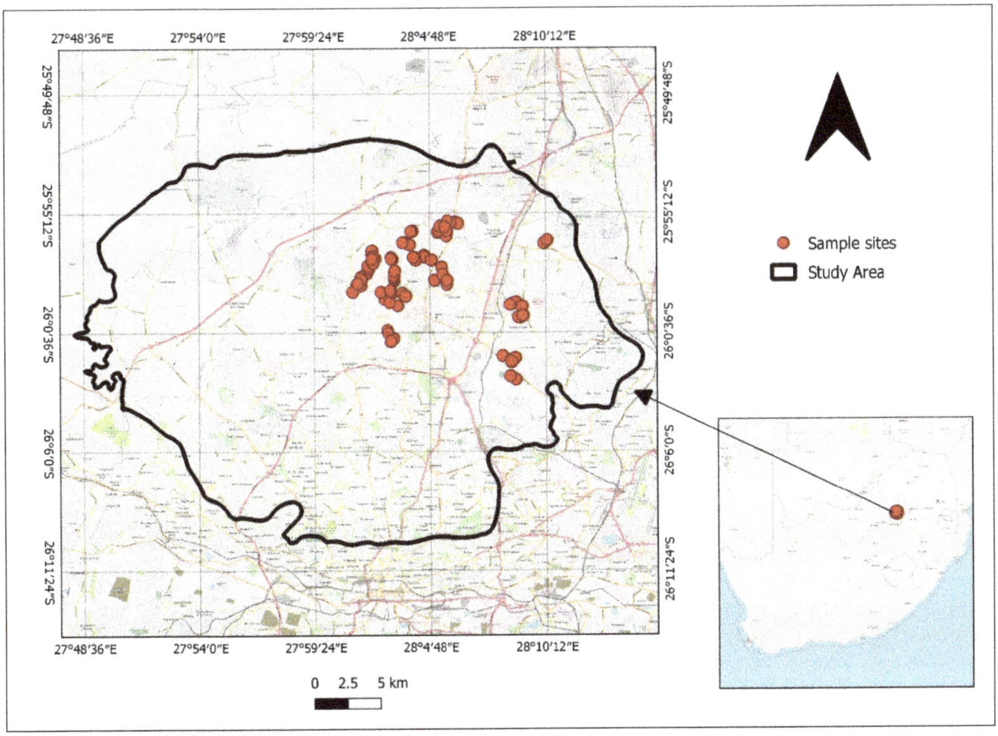

Figure 1. Location of the sampling sites within Johannesburg Granite Dome.

2.2. Soil Data and Analysis

A total of 93 sampling sites were randomly identified within the study area as part of a previous research project to determine the hydropedological behaviour of the catchment [33]. Soil samples were collected throughout the soil profiles using an auger and then classified according to the South African Soil Classification System [34] and the Food and Agricultural Organization (FAO) World Reference Base (WRB) [35]. At each sampling site, all the diagnostic soil horizons (A, B, and C) were sampled, irrespective of soil depth. Using the South African Soil Classification System, a total of 17 soil forms were identified (Table 1). For simplification, these soil forms were regrouped into six hydropedological groups, as described by [36], namely: recharge (deep), recharge (shallow), responsive (saturated), responsive (shallow), interflow (A/B), and interflow (soil/bedrock) soils. These hydropedological groupings were selected for this study because they capture some of the key factors affecting soil C content [37], such as soil moisture, drainage, texture, aeration, and position in the landscape. A brief description of the dominant properties of these hydropedological groups is presented in Table 1. Soil samples were dried at room temperature, sieved through a 2 mm screen, and then analysed for SOC content.

Table 1. Description of the six hydropedological groups and the respective South African and FAO WRB soil classification systems.

Hydropedological Group	Soil Forms [34]	WRB [35]	Description
Recharge (deep)	Glencoe, Clovelly, Hutton, Constantia, Griffin	Plinthosols, Acrisol, Arenosols	These soils show limited signs of saturation and water flows vertically through and out of the soil profile.
Recharge (shallow)	Glenrosa, Mispah	Leptosols	These soils also show limited signs of wetness, where water flows vertically into the underlying bedrock.
Responsive (saturated)	Westleigh, Kroonstad, Katspruit	Plinthosols, Gleysols	These soils show prolonged periods of saturation and typically result in overland flow due to limited water storage capacity.
Responsive (shallow)	Dresden, Mispah	Plinthosols, Leptosols	Shallow soils with a permeable underlying rock. These soils have limited water storage capacity, promoting overland flow.
Interflow (A/B)	Longlands, Wasbank	Plinthosols, Planosols, Acrisols, Luvisols	These are typically duplex soils, with clayey topsoil horizons, facilitating build-up of water.
Interflow (soil/bedrock)	Avalon, Pinedene, Fernwood, Tukulu	Plinthosols, Lixisols, Arenosols, Stagnosols, Acrisols,	In these soils, a freely drained soil horizon overlies a relatively impermeable bedrock, leading to periodic saturation.

2.2.1. Soil Bulk Density

A total of 14 undisturbed core samples (730 cm^3) were collected from the diagnostic horizons of the representative profiles to determine bulk density. Bulk density was calculated by dividing the oven-dried soil weight by total core volume, as described in [38]. Due to time and cost constraints, a limited number of samples were collected, which resulted in some standard deviation calculations being equal to zero (Table 2).

Table 2. Summary of the mean soil bulk density, C content, and SOC stocks for each hydropedological group at the respective soil horizons (mean ± standard deviation).

Hydropedological Group	Average Soil Depth (cm)	Horizon	Bulk Density (g cm^{-3})	Soil C (%)	SOC Stocks (t C ha^{-1})	n
Recharge (deep)	20	A	1.51 ± 0	1.33 ± 0.85	35.05 ± 22.41	29
	50	B	1.56 ± 0.72	0.53 ± 0.13	41.42 ± 10.87	28
	180	C	1.41 ± 0.11	0.24 ± 0.06	16.35 ± 5.92	20
Total	250				92.82 ± 39.2	77
Recharge (shallow)	15	A	1.46 ± 0	1.14 ± 0.44	25.50 ± 9.77	9
	80	B	1.26 ± 0	0.34 ± 0.37	27.84 ± 29.89	9
	15	C	1.40 ± 0.16	0.11 ± 0.14	2.83 ± 3.97	9
Total	110				56.17 ± 43.63	27
Responsive (saturated)	15	A	1.21 ± 0	1.58 ± 0.76	30.16 ± 14.55	13
	30	B	1.55 ± 0	0.58 ± 0.18	27.61 ± 8.51	13
	110	C	1.50 ± 0.07	0.32 ± 0.20	29.93 ± 21.55	13
Total	155				87.7 ± 44.61	39

Table 2. *Cont.*

Hydropedological Group	Average Soil Depth (cm)	Horizon	Bulk Density ($g\ cm^{-3}$)	Soil C (%)	SOC Stocks ($t\ C\ ha^{-1}$)	n
Responsive (shallow)	15	A	1.41 ± 0	1.00 ± 0.44	17.11 ± 7.62	5
	10	B	1.55 ± 0	0.28 ± 0.39	4.27 ± 6.02	5
	30	C	1.45 ± 0	0.10 ± 0.22	1.43 ± 3.20	5
Total	55				22.81 ± 16.84	15
Interflow (A/B)	15	A	1.39 ± 0	1.17 ± 0.59	19.54 ± 9.83	15
	30	B	1.45 ± 0.01	0.45 ± 0.20	19.44 ± 8.49	15
	15	C	1.47 ± 0.07	0.19 ± 0.16	5.51 ± 4.66	15
Total	60				44.49 ± 22.98	45
Interflow (soil/bedrock)	15	A	1.43 ± 0	1.01 ± 0.35	23.21 ± 8.11	22
	35	B	1.57 ± 0.06	0.47 ± 0.26	26.23 ± 13.25	22
	60	C	1.46 ± 0.10	0.25 ± 0.08	12.30 ± 5.59	18
Total	110				61.74 ± 26.95	62

2.2.2. Soil C Content and Soil C Stocks

A total of 265 soil samples were collected from the diagnostic horizons to analyse SOC content. Soil organic carbon content was analysed by dry combustion adapted from [39], with a TruSpec Leco CN analyser.

To calculate the soil C stocks in terms of the mass of C in tons per hectare ($t\ C\ ha^{-1}$), the soil C at each sampling location was multiplied by the corresponding soil bulk density and soil horizon thickness, as presented in Equation (1).

$$\text{SOC stocks} = \text{Soil C} \times \text{BD} \times \text{t}, \quad (1)$$

where:

SOC stocks = soil carbon stocks ($t\ C\ ha^{-1}$);
Soil C = soil C carbon (%);
BD = soil bulk density ($g\ cm^{-3}$);
t = soil depth (cm).

2.2.3. Spatial Representation of Carbon Stocks

The calculated SOC stocks data were applied to the map units of an existing Johannesburg Granite Dome hydropedological soil map by [40]. The map was created using multinomial logistic regression, with a point accuracy of 80% and a Kappa statistic value of 0.71. A Kappa coefficient value closer to one is preferred, as it indicates that the map provides a good representation of reality. Two choropleth SOC stocks maps were created; one representing the SOC stocks in the A horizon and the other for the total SOC stocks. Although soil observations were only made in selected hillslopes, the C stocks could be mapped for the entire area.

2.2.4. Statistical Analysis

All data were tested for normality and homogeneity using the Shapiro–Wilk and Levene's test, respectively. Thereafter, statistical analyses were then performed with the IBM SPSS Statistics Version 26 software package (SPSS Inc. IBM Corp. Armonk, New York, USA). All measured and calculated parameters were subjected to a two-way analysis of variance (ANOVA) to determine the significant differences between SOC and SOC stocks among the soil hydropedological groupings. Since the data were unbalanced (sample sizes are not all equal), Welch's test was used to determine the equivalence of the standardized means. Furthermore, means were compared with Tukey's honestly significant difference (HSD) post-hoc tests at the 95% confidence level.

3. Results

Soil C Content and Soil C Stocks

A summary of the measured soil C, bulk density, and the calculated C stocks for all soil hydropedological groups at the different soil depths, is indicated in Table 2. There was no significant difference between the soil group for the C content; however, there were significant differences among the diagnostic soil horizons ($F = (2, 247) = 105.35$, $p = 0.000$). In addition, there was a statistical correlation between the interaction of the soil groups and diagnostic soil horizons and SOC stocks ($F = (10, 247) = 2.58$, $p = 0.005$).

(a) Recharge (deep) soils

The results showed that the soil C content ranged, on average, between 0.24 ± 0.06% for Horizon C to 1.33 ± 0.85% for Horizon A. The steepest decline (0.8%) was observed from Horizon A to Horizon B (Figure 2a). Additionally, the B and A horizons contained the majority of the SOC stocks, with 41.42 ± 10.87 and 35.05 ± 22.41 t C ha^{-1}, respectively. However, the mean soil C content measured in Horizon A (1.33 ± 0.85%) was significantly higher than that in Horizon B (0.53 ± 0.13%). These soils recorded the highest mean C stocks in the A and B horizons as well as total C stocks (92.82 ± 39.2 t C ha^{-1}) of all the hydropedological groups.

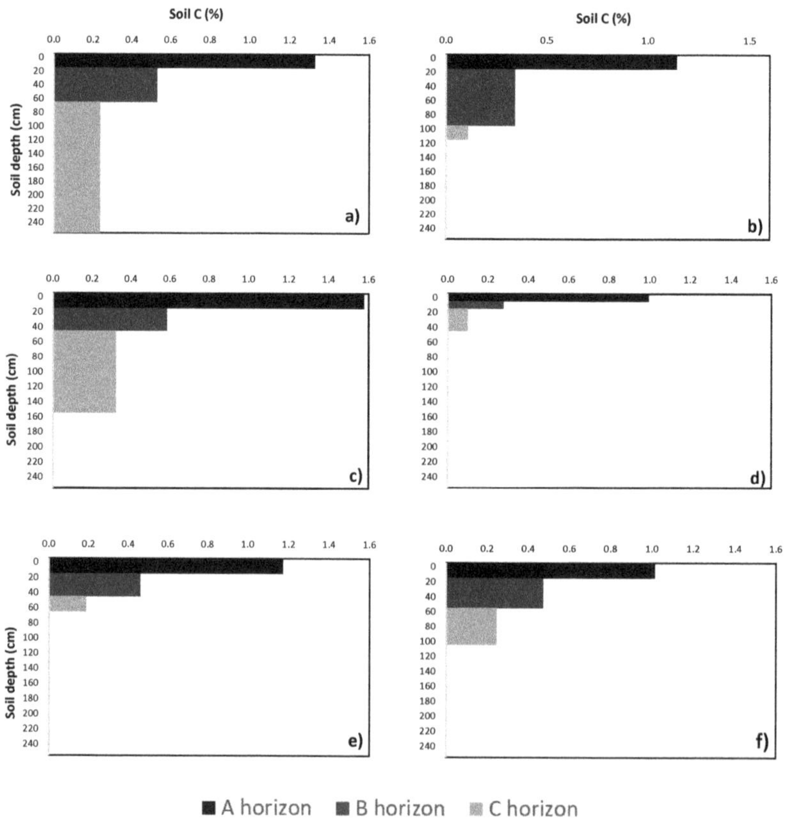

Figure 2. The vertical distribution of soil carbon within the (**a**) recharge (deep); (**b**) recharge (shallow); (**c**) responsive (saturated); (**d**) responsive (shallow); (**e**) interflow (A/B); and (**f**) interflow (soil/bedrock) soil profiles for the A, B, and C horizons.

(b) Recharge (shallow) soils

The A horizon contained the most soil C (1.14 ± 0.44%); however, this was not reflected in the SOC stocks (25.50 ± 9.77 t C ha^{-1}). The B horizon showed the highest SOC stocks (27.84 ± 29.89 t C ha^{-1}). Similar to recharge (deep) soils, the steepest decline (0.8%) was observed from the A to B Horizon (Figure 2b). These soil profiles did not follow the trend of a decrease in SOC stocks with an increase in soil depth. This highlights the contrast between C (%) and C stocks (t C ha^{-1}). The SOC stocks ranged between 2.83 ± 3.97 (C horizon) and 27.84 ± 29.89 t C ha^{-1} (B horizon). The total C stocks recorded for these soils were 56.17 ± 43.63 t C ha^{-1}.

(c) Responsive (saturated) soils

These soils displayed the highest mean soil C content (1.58 ± 0.76%), in comparison to all other hydropedological groups, in the A Horizon. The steepest decline (1%) was observed from Horizon A to Horizon B, which was the steepest decline from any horizon in all the soil groups (Figure 2c).

These soils were the only group to record mean soil C stocks over 25 t C ha^{-1} in all soil horizons. Although these soils also recorded the highest mean C contents in all horizons, this did not translate into the highest mean SOC stocks available. The soil C content ranged, on average, between 0.32 ± 0.20% for the C horizon and 1.58 ± 0.76% for the A horizon. The average SOC stocks in these soils ranged between 27.61 ± 8.51 t C ha^{-1} for the B horizon and 30.16 ± 14.55 t C ha^{-1} for the A horizon. These soils further recorded the second highest total C stocks, 87.7 ± 44.61 t C ha^{-1}.

(d) Responsive (shallow) soils

These soils recorded the lowest mean soil C content in all three horizons (1.00 ± 0.44, 0.28 ± 0.39, and 0.10 ± 0.22%, respectively) as well as the lowest total C content (0.46 ± 0.52%). Soil C content in these soils gradually decreased by 0.72% from the A horizon to the B horizon and by 0.18% from the B to C horizon (Figure 2d).

The average SOC stocks in these soils were mainly located in the A horizon, with 17.11 ± 7.62 t C ha^{-1}. The B and C horizons both recorded values below 5 t C ha^{-1}, 4.27 ± 6.02 and 1.43 ± 3.20 t C ha^{-1}, respectively. Out of all the hydropedological groups, these were also the lowest SOC stocks recorded. The mean SOC stocks in all horizons failed to exceed 20 t C ha^{-1}, with total SOC stocks of 22.81 ± 16.84 t C ha^{-1}. These soils followed the trend of a decrease in SOC stocks with an increase in soil depth.

(e) Interflow (A/B) soils

The results showed that the soil C content ranged, on average, between 0.19 ± 0.16% for Horizon C to 1.17 ± 0.59% for Horizon A. The steepest decline (0.72%) was observed from Horizon A to Horizon B, with a 0.26% decline from Horizon B to Horizon C (Figure 2e).

Furthermore, the A and B horizons showed the highest SOC stocks, with 19.54 ± 9.83 and 19.44 ± 8.49 t C ha^{-1}, respectively. These soils followed the trend of a decrease in SOC stocks with an increase in soil depth, with the C horizon recording 5.51 ± 4.66 t C ha^{-1}. Similar to responsive (shallow) soils, the mean soil C stock values in all horizons failed to exceed 20 t C ha^{-1}, with total C stocks of 44.49 ± 22.98 t C ha^{-1}.

(f) Interflow (soil/bedrock) soils

Similar to all other hydropedological groups, the soil C content was mainly concentrated in the A horizon. It ranged, on average, between 0.25 ± 0.08% (horizon C) and 1.01 ± 0.35%. The steepest decline (0.54%) was observed from Horizon A to Horizon B (Figure 2f). Like the recharge (deep) and recharge (shallow) soils, these soils also recorded the highest SOC stocks in the B horizon (26.23 ± 13.25 t C ha^{-1}). The mean SOC stocks ranged, on average, between 12.30 ± 5.59 and 26.23 ± 13.25 t C ha^{-1}. The total SOC stocks recorded for these soils were 61.74 ± 26.95 t C ha^{-1}.

Soil organic carbon stocks, both in the topsoil and entire soil profile, are graphically presented in Figure 3.

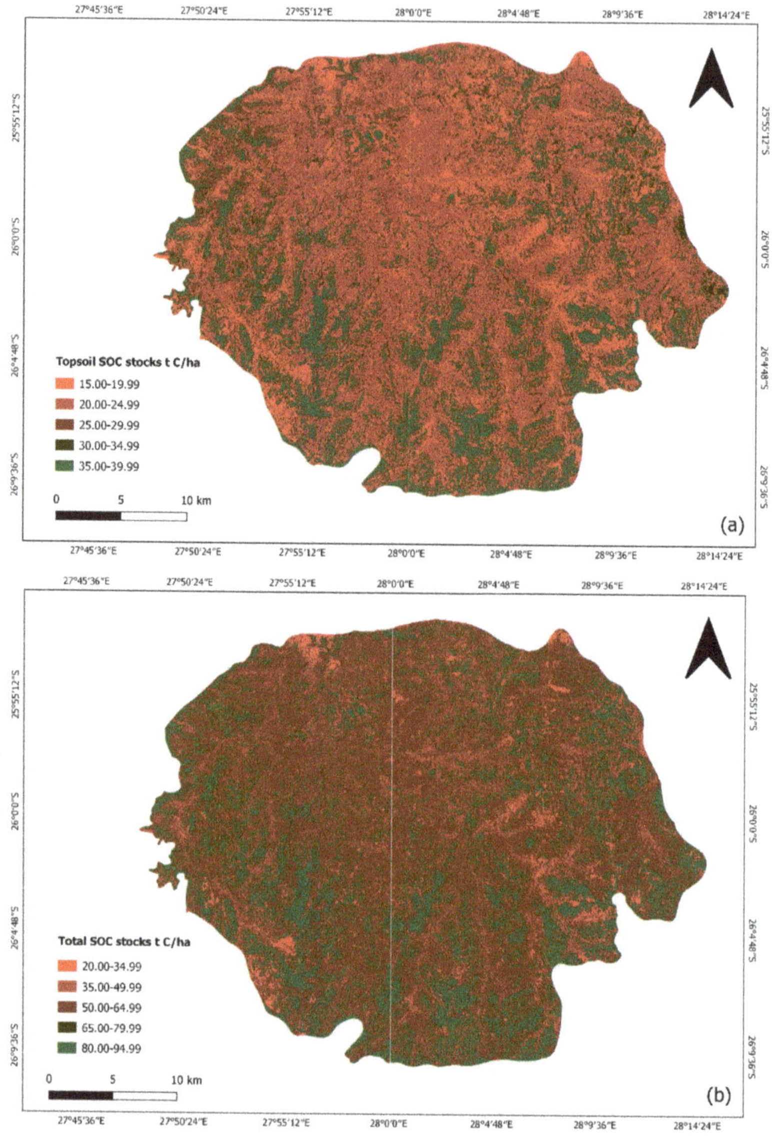

Figure 3. The spatial distribution of soil carbon stocks within (**a**) the A horizon; and (**b**) the entire soil profile.

4. Discussion

In this study, the results showed that soil groups had no significant impact on the soil C content, but the diagnostic soil horizons did have a significant impact on soil C. Furthermore, the interaction between soil groups and diagnostic soil horizons had a significant impact on the SOC stocks. The amount of soil C is dependent on soil moisture, drainage, texture, aeration, vegetation, and position in the landscape [41]. In urban areas, population changes and population density affect CO_2 emissions. Research in the United States found that large cities with a greater population size resulted in increased emissions [42]. Additionally, Rybski et al. [43] presented evidence to support that urban emissions are largely dependent on the degree of the city's development. In this study, however, we have no data to discuss the effect of population changes and density, soil texture, land management, and vegetation on urban emissions.

4.1. Soil Carbon Content

Soil organic matter is a key component of any terrestrial ecosystem, and any variation in its abundance and composition has important effects on many of the processes that occur within soil systems [44]. Soil organic matter decomposition is influenced by soil moisture content [44,45]. Soil organic matter decomposition is inadequate in dry conditions, resulting in relatively low C contents. In this study, soil groups had no significant impact on soil C contents and, therefore, differences in SOC stocks could not be due to soil moisture. This was observed in recharge (deep) and recharge (shallow) soils with limited signs of saturation, which had important soil C pools within the first 70 and 95 cm, respectively. Similarly, Parajuli & Duffy [46] found that the C amount was not influenced by soil moisture. Conversely, Liu et al. and Hobley et al. [47,48] described the accumulation of SOC at increasing levels of soil moisture. It is worth noting that the shallow soils in the catchment, namely, recharge (shallow), responsive (shallow), and interflow (A/B), had lower amounts of C than the deep soils at relatively the same depths. This could be due to fewer disturbances in the soils. Future studies in the city of Johannesburg could research the effect of land-use and land management on C capture and storage. Understanding the effect of land management practices on SOC sequestration is crucial for adopting effective management strategies.

4.2. Soil Carbon Stocks

Organic carbon stocks can differ significantly due to the influence of land management [24,49]. Traffic in urban areas has an impact on soil bulk density. Previous studies reported a strong correlation between bulk density and SOC stocks [50–52], which is consistent with the results of the present study in which the SOC stocks increased with increasing bulk density (Table 2). Soil C contents were significantly different in all diagnostic soil horizons, influencing the amount of SOC stocks in the individual horizons. Soil depth also played an important role in the calculation of SOC stocks. The deep soil groups (recharge (deep), responsive (saturated), and interflow (soil/bedrock)) had the three highest SOC stocks because deeper diagnostic horizons have a high capacity to store significant amounts of C.

Soils in the study area were formed during the process of ferrolysis. During the dry season, Fe^{2+} is oxidized to produce Fe^{3+} oxides. The presence of Fe^{3+} oxides affects the C dynamics in soils [53,54]. A study [55] found that soils with a higher clay content have higher SOC stocks because clay protects SOM from decomposition, whereas [56] determined the opposite to be true. They found that clay-textured soils accelerated SOM decomposition. To determine the role of Fe^{3+} oxides on SOC stocks in this area, analysis should be done on samples collected during the wet season. In wet conditions, Fe^{3+} is reduced to produce Fe^{2+}. Ferrolysis results in the seasonal destruction and translocation of clay [57].

5. Conclusions

The findings in this study demonstrated that the majority of SOC stocks were contained near the surface layer of the soil groups studied. It was also shown that there is a combined effect of bulk density, soil depth, and soil C in diagnostic horizons on the accumulation of SOC stocks within the soils. The presence of Fe^{3+} oxides may have contributed to the observed SOC stocks; however, this needs to be researched further. Soil C was significantly higher in the deeper soils at relatively similar depths, possibly due to fewer soil disturbances. It was also discovered that a relatively high C content did not necessarily translate to high SOC stocks. Therefore, it can be concluded that SOC stocks should be used to determine the C storing capacity of soils, rather than C content, because bulk density and soil depth are also taken into consideration. The role of land management on the SOC stocks in Johannesburg should be monitored as it may be necessary to continually revise management strategies for modified urban environments.

Author Contributions: Conceptualization, K.R.S., E.K. and J.v.T.; methodology, K.R.S., E.K., J.v.T. and G.v.Z.; software, K.R.S., J.v.T. and G.v.Z.; validation, K.R.S., E.K., J.v.T. and G.v.Z.; formal analysis, K.R.S.; investigation, E.K. and J.v.T.; resources, E.K. and J.v.T.; writing—original draft preparation, K.R.S.; writing—review and editing, E.K., J.v.T. and G.v.Z.; visualization, K.R.S.; supervision, E.K., J.v.T. and G.v.Z.; project administration, E.K.; funding acquisition, K.R.S. and J.v.T. All authors have read and agreed to the published version of the manuscript.

Funding: This research was funded by The Department of Environment, Forestry, and Fisheries in South Africa as part of the Natural Resource Management Programme.

Institutional Review Board Statement: Not applicable.

Informed Consent Statement: Not applicable.

Data Availability Statement: The data presented in this study are available on request from the corresponding author. The data are not publicly available as it is still being analyzed and interpreted for several other reports and publications.

Acknowledgments: The authors would like to acknowledge the South African National Research Foundation for their financial support (Unique Grant No.: 117030).

Conflicts of Interest: The authors declare no conflict of interest. The funders had no role in the design of the study; in the collection, analyses, or interpretation of data; in the writing of the manuscript, or in the decision to publish the results.

References

1. Davies, R.; Hall, S.J. Direct and indirect effects of urbanization on soil and plant nutrients in desert ecosystems of the Phoenix metropolitan area, Arizona (USA). *Urban Ecosyst.* **2010**, *13*, 295–317. [CrossRef]
2. Asabere, S.B.; Zeppenfeld, T.; Nketia, K.A.; Sauer, D. Urbanization leads to increases in pH, carbonate, and soil organic matter stocks of arable soils of Kumasi, Ghana (West Africa). *Front. Environ. Sci.* **2018**, *6*, 1–17. [CrossRef]
3. Lovett, G.M.; Traynor, M.N.; Pouyat, R.V.; Carreiro, M.M.; Zhu, W.; Baxter, J.W. Atmospheric deposition to oak forests along an urbanrural gradient. *Environ. Sci. Technol.* **2000**, *34*, 4294–4300. [CrossRef]
4. Vodyanitskii, Y.N. Organic matter of urban soils: A review. *Eurasian Soil Sci.* **2015**, *8*, 802–811. [CrossRef]
5. Lorenz, K.; Lal, R. Biogeochemical C and N cycles in urban soils. *Environ. Int.* **2009**, *35*, 1–8. [CrossRef]
6. Mao, Q.; Huang, G.; Buyantuev, A.; Wu, J.; Luo, S.; Ma, K. Spatial heterogeneity of urban soils: The case of the Beijing metropolitan region, China. *Ecol. Process.* **2014**, *3*, 1–11. [CrossRef]
7. Ebenebe, P.C.; Shale, K.; Sedibe, M.; Tikilili, P.; Achilonu, M.C. South African mine effluents: Heavy metal pollution and impact on the ecosystem. *Int. J. Chem. Sci.* **2017**, *15*, 1–12.
8. Fashola, M.O.; Ngole-Jeme, V.M.; Babalola, O.O. Heavy metal pollution from gold mines: Environmental effects and bacterial strategies for resistance. *Int. J. Environ. Res. Public Health* **2016**, *13*, 1047. [CrossRef]
9. Ghaley, B.B.; Porter, J.R.; Sandhu, H.S. Soil-Based ecosystem services: A synthesis of nutrient cycling and carbon sequestration assessment methods. *Int. J. Biodivers. Sci. Ecosyst. Serv. Manag.* **2014**, *10*, 177–186. [CrossRef]
10. Schwilch, L.; Bernet, L.; Fleskens, E.; Giannakis, J.; Leventon, T.; Marañón, J.; Mills, C.; Short, J.; Stolte, H.; van Delden, S. Operationalizing ecosystem services for the mitigation of soil threats: A proposed framework. *Ecol. Indic.* **2016**, *67*, 586–597. [CrossRef]
11. Oertel, C.; Matschullat, J.; Zurba, K.; Zimmermann, F.; Erasmi, S. Greenhouse gas emissions from Soils—A review. *Geochemistry* **2016**, *76*, 327–352. [CrossRef]
12. Jobbágy, E.; Jackson, R. The vertical distribution of soil organic carbon and its relation to climate and vegetation. *Ecol. Appl.* **2000**, *10*, 423–436. [CrossRef]
13. Batjes, N.H. Total carbon and nitrogen in the soils of the world. *Eur. J. Soil Sci.* **1996**, *47*, 151–163. [CrossRef]
14. Le Quéré, C.; Raupach, M.R.; Canadell, J.G.; Marland, G.; Bopp, L.; Ciais, P.; Conway, T.J.; Doney, S.C.; Feely, R.A.; Foster, P.; et al. Trends in the sources and sinks of carbon dioxide. *Nat. Geosci.* **2009**, *2*, 831–836. [CrossRef]
15. Tommaso, S.; Mouratiadou, I.; Gaiser, T.; Berg-Mohnicke, M.; Wallor, E.; Ewert, F.; Nendel, C. Estimating the contribution of crop residues to soil organic carbon conservation. *Environ. Res. Lett.* **2019**, *14*, 094008. [CrossRef]
16. Obalum, S.; Chibuike, G.; Peth, S.; Ouyang, Y. Soil organic matter as sole indicator of soil degradation. *Environ. Monit. Assess.* **2017**, *189*, 176. [CrossRef]
17. Jim, C.Y. Soil characteristics and management in an urban park in Hong Kong. *Environ. Manag.* **1998**, *22*, 683–695. [CrossRef]
18. Schneider, A.; Friedl, M.A.; Potere, D. A new map of global urban extent from MODIS satellite data. *Environ. Res. Lett.* **2009**, *4*, 044003. [CrossRef]

19. Intergovernmental Panel on Climate Change. Human settlements, infrastructure, and spatial planning. In *Climate Change 2014: Mitigation of Climate Change: Working Group III Contribution to the IPCC Fifth Assessment Report*; Cambridge University Press: Cambridge, UK; New York, NY, USA, 2015; pp. 923–1000.
20. Yang, X.; Yue, W.; Xu, H.; Wu, J.; He, Y. Environmental consequences of rapid urbanization in Zhejiang province, East China. *Int. J. Environ. Res. Public Health* **2014**, *11*, 7045–7059. [CrossRef]
21. Alberti, M. The effects of urban patterns on ecosystem function. *Int. Reg. Sci. Rev.* **2005**, *28*, 168–192. [CrossRef]
22. Cullis, J.D.S.; Horn, A.; Rossouw, N.; Fisher-Jeffes, L.; Kunneke, M.M.; Hoffman, W. Urbanisation, climate change and its impact on water quality and economic risks in a water scarce and rapidly urbanising catchment: Case study of the Berg River Catchment. *H2Open J.* **2019**, *2*, 146–167. [CrossRef]
23. Du Preez, C.C.; van Huyssteen, C.W.; Mnkeni, P.N.S. Land use and soil organic matter in South Africa 1: A review on spatial variability and the influence of rangeland stock production. *S. Afr. J. Sci.* **2011**, *107*, 27–34. [CrossRef]
24. Toru, T.; Kibret, K. Carbon stock under major land use/land cover types of Hades sub-watershed, eastern Ethiopia. *Carbon Balance Manag.* **2019**, *14*, 1–14. [CrossRef]
25. Lindén, L.; Riikonen, A.; Setälä, H.; Yli-Pelkonen, V. Quantifying carbon stocks in urban parks under cold climate conditions. *Urban For. Urban Green.* **2020**, *49*, 126633. [CrossRef]
26. Wang, S.; Adhikari, K.; Zhuang, Q.; Gu, H.; Jin, X. Impacts of urbanization on soil organic carbon stocks in the northeast coastal agricultural areas of China. *Sci. Total Environ.* **2020**, *721*, 137814. [CrossRef]
27. Zondi, S. Recharge Rates and Processes in the Upper Crocodile Catchment. Master's Thesis, University of the Witwatersrand, Johannesburg, South Africa, 2017.
28. Tinnefeld, M. *Conceptual Hydropedological Response Verification and Polygon Refinement on Portion 31 & 32 Blue Hills 397 JR, Midrand, Gauteng Province*; Hydro Pedo (Pty) Ltd: Johannesburg, South Africa, 2017.
29. The City of Johannesburg. *Drafted Integrated Development Plan 2020/21*; The City of Johannesburg: Johannesburg, South Africa, 2020.
30. Terblanche, A.P.S.; Opperman, L.; Nel, C.M.E.; Reinach, S.G.; Tosen, G.; Cadman, A. Preliminary results ofexposure measurements and health effects of the Vaal Triangle air pollution health study. *S. Afr. Med. J.* **1992**, *81*, 550–556.
31. Annegarn, H.; Scorgie, Y. Air quality management strategy for the Vaal Triangle: Part III. *Clean Air J.* **1997**, *9*, 11–19.
32. Scorgie, Y.; Kneen, M.A.; Annegarn, H.J.; Burger, L.W. Air pollution in the Vaal Triangle: Quantifying source contributions and identifying cost effective solutions. *Off. J. Natl. Assoc. Clean Air* **2003**, *13*, 2. [CrossRef]
33. Tinnefeld, M.; Le Roux, P.A.L.; Job, N.; Van Zijl, G.M.; Van Tol, J.J.; Lorentz, S.A. *Research Report on the Hydropedology of Part of the Halfway House Granite Dome*; Department of Infrastructure and Environment, City of Johannesburg Metropolitan Council: Johannesburg, South Africa; University of the Free State, Institute of Groundwater Studies: Bloemfontein, South Africa, 2017.
34. Soil Classification Working Group. *Soil Classification: A Taxonomic System for South Africa*; Department of Agricultural Development: Pretoria, South Africa, 1991.
35. IUSS Working Group WRB. *World reference base for soil resources 2014. Update 2015 International Soil Classification System for Naming Soils and Creating Legends for Soil Maps*; World Soil Resources Reports No. 106; FAO: Rome, Italy, 2015.
36. Van Tol, J.J.; Le Roux, P.A.L. Hydropedological grouping of South African soil forms. *S. Afr. J. Plant Soil* **2019**, 1–3. [CrossRef]
37. Xu, S.; Sheng, C.; Tian, C. Changing soil carbon: Influencing factors, sequestration strategy and research direction. *Carbon Balance Manag.* **2020**, *15*, 1–9. [CrossRef]
38. Blake, G.R.; Hartge, K.H. Bulk density. In *Methods of Soil Analysis, Part 1*; Klute, A., Ed.; Agronomy 9 ASA: Madison, WI, USA, 1986; pp. 363–375.
39. Nelson, D.W.; Sommers, L.E. Total carbon, organic carbon, and organic matter. In *Methods of Soil Analysis, Part 3: Chemical Methods*; Sparks, D.L., Page, A.L., Helmke, P.A., Loeppert, R.H., Soltanpour, P.N., Tabatabai, M.A., Johnston, C.T., Sumner, M.E., Eds.; Book Series No.5; Soil Science Society of America: Madison, WI, USA, 1996; pp. 961–1010.
40. Van Zijl, G.M.; Van Tol, J.J.; Bouwer, D.; Lorentz, S.; Le Roux, P.A.L. Combining historical remote sensing, digital soil mapping and hydrological modelling to produce solutions for infrastructure damage in Cosmo City, South Africa. *Remote Sens.* **2020**, *12*, 433. [CrossRef]
41. Noppol Arunrat, N.; Pumijumnong, N.; Sereenonchai, S.; Chareonwong, U. Factors controlling soil organic carbon sequestration of highland agricultural areas in the Mae Chaem Basin, Northern Thailand. *J. Agron.* **2020**, *10*, 305. [CrossRef]
42. Ribeiro, H.V.; Rybski, D.; Kropp, J.P. Effects of changing population or density on urban carbon dioxide emissions. *Nat. Commun.* **2019**, *10*, 3204. [CrossRef] [PubMed]
43. Rybski, D.; Sterzel, T.; Reusser, D.; Fichtner, C.; Kropp, J. Cities as nuclei of sustainability? *Environ. Plan. B* **2016**, *44*, 425–440. [CrossRef]
44. Ontl, T.A.; Schulte, L.A. Soil carbon storage. *Nat. Educ. Knowl.* **2012**, *3*, 35.
45. Henry, M.; Valentini, R.; Bernoux, M. Soil carbon stocks in ecoregions of Africa. *Biogeosci. Discuss.* **2009**, *6*, 797–823. [CrossRef]
46. Parajuli, P.B.; Duffy, S. Evaluation of soil organic carbon and soil moisture content from agricultural fields in Mississippi. *Open J. Soil Sci.* **2013**, *3*, 81–90. [CrossRef]
47. Liu, Z.; Shao, M.; Wang, Y. Effect of environmental factors on regional soil organic carbon stocks across the Loess Plateau region, China. *Agric. Ecosyst. Environ.* **2011**, *142*, 184–194. [CrossRef]
48. Hobley, E.; Wilson, B.; Wilkie, A.; Gray, J.; Koen, T. Drivers of soil organic carbon storage and vertical distribution in Eastern Australia. *Plant Soil* **2015**, *390*, 111–127. [CrossRef]

49. Davies, Z.G.; Edmondson, J.L.; Heinemeyer, A.; Leake, J.R.; Gaston, K.J. Mapping an urban ecosystem service: Quantifying above-ground carbon storage at a city-wide scale. *J. Appl. Ecol.* **2011**, *48*, 1125–1134. [CrossRef]
50. Wilson, T.M.; Warren, J.G. Bulk Density and carbon concentration variance influence on soil carbon stock measurements. *Commun Soil Sci. Plant Anal.* **2015**, *46*, 2342–2356. [CrossRef]
51. Chaudhari, P.R.; Ahire, D.V.; Ahire, V.D.; Chkravarty, M.; Maity, S. Soil bulk density as related to soil texture, organic matter content and available total nutrients of coimbatore soil. *Int. J. Sci. Res.* **2013**, *3*, 1–8.
52. Ghimire, P.; Bhatta, B.; Pokhrel, B.; Kafle, G.; Paudel, P. Soil organic carbon stocks under different land uses in chure region of makawanpur district, Nepal. *SAARC J. Agric.* **2018**, *16*, 13–23. [CrossRef]
53. Wen, Y.; Xiao, J.; Goodman, B.A.; He, X. Effects of organic amendments on the transformation of Fe (Oxyhydr) oxides and soil organic carbon storage. *Front. Earth Sci.* **2019**, *7*, 1–15. [CrossRef]
54. Weber, K.A.; Urrutia, M.M.; Churchill, P.F.; Kukkadapu, R.K.; Roden, E.E. Anaerobic redox cycling of iron by freshwater sediment microorganisms. *Environ. Microbiol.* **2006**, *8*, 100–113. [CrossRef]
55. Plante, A.F.; Conant, R.T.; Stewart, C.E.; Paustian, K.; Six, J. Impact of soil texture on the distribution of soil organic matter in physical and chemical fractions. *Soil Sci. Soc. Am. J.* **2006**, *70*, 287–296. [CrossRef]
56. Azlan, A.; Aweng, E.R.; Ibrahim, C.O.; Noorhaidah, A. Correlation between soil organic matter, total organic matter and water content with climate and depths of soil at different land use in Kelantan, Malaysia. *J. Appl. Sci. Environ. Manag.* **2012**, *16*, 353–358
57. Wei, H.; Guenet, B.; Vicca, S.; Nunan, N.; Asard, H.; AbdElgawad, H.; Shen, W.; Janssens, I.A. High clay content accelerates the decomposition of fresh organic matter in artificial soils. *Soil Biol. Biochem.* **2014**, *77*, 100–108. [CrossRef]

land

MDPI

Review

Soil Protection in Floodplains—A Review

Mariam El Hourani * and Gabriele Broll

Department of Soil Science and Agroecology, Institute of Geography, University of Osnabrueck, 49074 Osnabrueck, Germany; Gabriele.Broll@uni-osnabrueck.de
* Correspondence: mariam.elhourani@uni-osnabrueck.de

Abstract: Soils in floodplains and riparian zones provide important ecosystem functions and services. These ecosystems belong to the most threatened ecosystems worldwide. Therefore, the management of floodplains has changed from river control to the restoration of rivers and floodplains. However, restoration activities can also negatively impact soils in these areas. Thus, a detailed knowledge of the soils is needed to prevent detrimental soil changes. The aim of this review is therefore to assess the kind and extent of soil information used in research on floodplains and riparian zones. This article is based on a quantitative literature search. Soil information of 100 research articles was collected. Soil properties were divided into physical, chemical, biological, and detailed soil classification. Some kind of soil information like classification is used in 97 articles, but often there is no complete description of the soils and only single parameters are described. Physical soil properties are mentioned in 76 articles, chemical soil properties in 56 articles, biological soil properties in 21 articles, and a detailed soil classification is provided in 32 articles. It is recommended to integrate at least a minimum data set on soil information in all research conducted in floodplains and riparian zones. This minimum data set comprises soil types, coarse fragments, texture and structure of the soil, bulk density, pH, soil organic matter, water content, rooting depth, and calcium carbonate content. Additionally, the nutrient and/or pollution status might be a useful parameter.

Keywords: soil protection; restoration; floodplain; soil bioengineering

Citation: El Hourani, M.; Broll, G. Soil Protection in Floodplains—A Review. *Land* **2021**, *10*, 149. https://doi.org/10.3390/land10020149

Academic Editors: Chiara Piccini and Manuel López-Vicente

Received: 17 December 2020
Accepted: 29 January 2021
Published: 3 February 2021

1. Introduction

Floodplains and their soils are an important part of the river system and fulfil important ecological, economic, and social functions like natural flood protection, sustaining high biological diversity or filtering and storing water [1,2]. Floodplains can be regarded as hotspots for biogeochemical processes such as denitrification [3,4] or eutrophication [1]. Floodplains are regularly flooded by the adjacent river [5]. Thus, the lateral connection to the river is essential for the functioning of a floodplain [6]. The riparian zone is characterized as the zone between the low-water and the high-water mark [7,8]. Both represent ecotones at the transition between aquatic and terrestrial environments [6]. Riparian zones hence are the last point in the landscape where nutrients can be intercepted before they enter the rivers [9]. Often, the terms floodplain and riparian zone are treated as synonyms in the literature or are not clearly differentiated from each other. Floodplains do not only provide a wide range of ecosystem services, but also are one of the most threatened ecosystems in the world [2,10]. Today, many floodplains worldwide are degraded because of high hydromorphological and diffuse pollution pressures, dam building, diversion, or abstraction of water or clearing of land and cannot deliver the ecosystem services in the same extent as a natural floodplain [1,11,12]. Approximately 70–90% of Europe's floodplains are degraded [12]. The dynamic flow regime of the river is essential not only to the river functioning, but also to the ability of the floodplain to provide ecosystem services [11].

Soils in the floodplains and the riparian zone are strongly influenced by the adjacent river. These soils are often called alluvial soils as their physical, morphological, chemical, and mineralogical properties are influenced by the alluvial parent material derived from the river. The development of alluvial soils strongly depends on the flow regime [13]. Sediment transport and deposition are characteristic processes for the development of alluvial soils [14]. Recent alluvial soils are often classified into the reference soil group of Fluvisols in the world reference base for soil resources or into the order of Entisols (suborder Fluvents) in the US soil taxonomy [13,15,16]. Older alluvial soils can be transformed into multiple different soil types [13]. Fluvisols are characterized by fluvic material and can occur on any continent and in any climate zone. They occupy less than 350 million ha worldwide [15]. Naturally Fluvisols are fertile soils having been used by humans since the prehistoric times. Soils in the floodplain or riparian zone influenced by groundwater and showing classic gleyic properties can also be classified as Gleysols. These are soils that typically occupy low positions in the landscape with high groundwater tables and can also occur on any continents and in any climate zones. The parent material on which Gleysols develop can be a wide range of unconsolidated deposits, but often they also develop on fluvial, marine, or lacustrine deposits like Fluvisols [15]. Through their special characteristics these alluvial soils are able to provide information on past and present fluvial dynamics and ecosystem structure through their morphology [17,18].

In the past decades, floodplain management has changed from river control to the restoration of floodplains and rivers which can reduce the pressures and restore related functions and services [1,2,10,19–21]. In Europe, several directives like the Water Framework Directive (Directive 2000/60/EC), the Habitat and Birds Directives (Council Directive 92/43/EEC and Directive 2009/147/EC) or the Floods Directive (Directive 2007/60/EC) foster the restoration of river and floodplain ecosystems [22]. The decade of 2021–2030 is also assigned as the United Nations decade on ecosystem restoration. It emphasizes that nowadays there is still an urgent need to restore degraded ecosystems (https://www.decadeonrestoration.org/).

Restoration activities in floodplains and riparian zones, however, can also affect soils in these areas through the use of heavy machinery, resulting in soil compaction, or the disturbance and mixing of the soil [23–25]. These negative effects and disturbances can persist, at least for a decade [23,25]. Soil development is, compared to the changes in vegetation or hydrology, a slow process [26,27] which explains why soils would not recover within a relatively shorter period after the restoration impact [25]. The assessment of the positive or negative impacts of restoration on riparian and floodplain soils, is of major importance [28] as crucial ecosystem services and functions are associated with soils in this zone [29].

The aim of this review is therefore to assess if and how riparian soils and soil properties are addressed in the research on floodplain and river restoration and in the research on floodplains and riparian zones with direct implications to future restoration projects.

The objectives of this review are:

1. To give an overview on research in floodplains and riparian zones of the world with implication to restoration projects in the last 20 years;
2. To assess in which kind and to what extent soils are addressed in the research;
3. To recommend further research needs on soil protection in floodplains.

2. Materials and Methods

This literature review is based on the principles of Pickering and Byrne [30] and the Preferred Reporting Items for Systematic Reviews and Meta-Analysis (PRISMA) guidelines [31]. In July 2020 a literature research was performed in Scopus and Web of Science. As the search for the terms "soil protection" in combination with "floodplain restoration" or "river restoration" resulted in only 12 or 10 articles, respectively, a broader understanding of soil protection had to be applied. In a first search article titles, keywords, and abstracts were searched for the terms soil, protection, river or floodplain, restoration, or construction and additionally water framework directive or WFD. A second search in the same databases in article titles, keywords, and abstracts with the terms soil, restoration, and riparian zone was performed. The review should cover all aspects of soil protection in floodplains and riparian zones and hence the search terms have not been further specified. The search was limited to literature published between the years 2000 and 2020 to focus on activities since the implementation of the Water Framework Directive in 2000. The results of the search are shown in a PRISMA flow diagram (Figure 1).

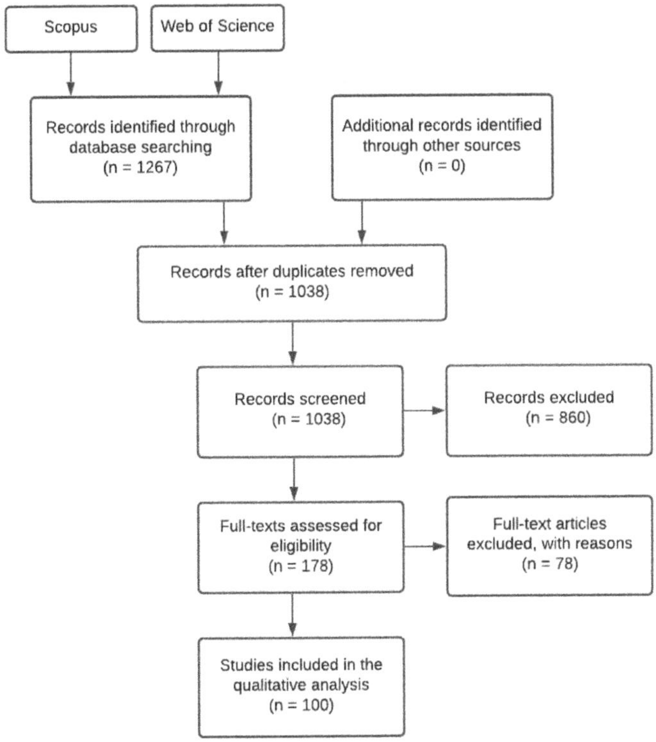

Figure 1. Flow diagram of the quantitative literature research performed in July 2020 (diagram adapted from Moher et al. [31]).

After duplicates were removed the search returned 1038 records. These articles were screened by abstract and 860 were excluded. Only journal articles were included. Books and conference proceedings were excluded from the beginning. Articles were excluded if the study area was different from rivers, streams, floodplains, or riverine/riparian wetlands. Water reservoirs, wetlands with no further specification (e.g., as riparian wetland) and artificial wetlands (e.g., treatment wetlands), coastal areas (like mangroves), and lakes were not considered for this review. Articles only concerning other topics like vegetation or forest growth, seedbanks, fish productivity, the functioning of a special geomaterial or geosynthetic, a construction work in a place different than a floodplain or river, landfills, etc., and no direct link to soil and soil protection were also excluded. The spatial scale was set to the floodplain or riparian zone. No restrictions were made to the geographic or climatic region. Articles at the spatial scale of river basins or watersheds and no direct reference to the soils in the riparian zone were also excluded. Only research articles fully written in English were considered for this review. This resulted in 178 full-text articles which were assessed for eligibility. Another 78 articles did not meet the criteria mentioned above. Finally, 100 full-text articles were included in the qualitative analysis.

The 100 articles were scanned for study region, year, and available soil information in the research. The soil information was grouped into categories including soil properties (physical, chemical, and biological), detailed soil classification, other type of classification like alluvial soils, and other soil information like the use of soil maps (Appendix A Table A1).

3. Results

3.1. Overview on Research in Floodplains and Riparian Zones of the World

Research on soil protection was conducted on every continent or geographic region respectively, with the exception of Antarctica (Table 1). Most research (44 published articles) focusses on soil protection in floodplains and riparian zones in North America. In second place, 25 articles have been published about study sites in Europe. In one article research was conducted in Europe and North America. Then, 12 articles focused on research in Asia 12 in Oceania, four in South America, and one in Africa. In three articles the geographic region was not specified, for example when research focused on models or frameworks without the need of a special study area.

Table 1. Number of articles on soil protection in floodplains or riparian zones per geographic region.

Africa	Asia	Europe	North America	South America	Oceania [2]	Not Specified	Total
1	12	25 [1]	44 [1]	4	12	3	100

[1] One article covered study sites in Europe and North America. [2] Oceania here only comprises Australia and New Zealand. For a detailed classification of the continents c.f. the United Nations definitions on geographic regions (https://unstats.un.org/unsd/methodology/m49/).

In total, research was conducted in over 24 different countries; half of them are in Europe. In most countries less than four studies have been realized. Most studies were carried out in the USA, followed by Australia with 11 studies and China with eight. Five studies were realized in Switzerland (Table 2). One article did not restrict the research to a specific country but focused on the whole Alpine area [32]. Studies in the USA were conducted in 22 different states.

Table 2. Number of study sites per country. Only countries with more than four studies are considered in this table.

USA	Australia	China	Switzerland
41	11	8	5

Regarding the climate zones after Schultz [33] approximately 50% of the articles covered study sites in the midlatitudes. Over 40% were carried out in the subtropics and dry tropics. In the boreal zone 2% of the studies were realized. In the humid tropics 3% of the studies were realized. In 2% of the studies no climate region could be assigned.

The number of articles published per year between 2000 and 2020 shows that only about one-third (33 articles) of the 100 articles has been published in the first decade between 2000 and 2010 (Figure 2). More than two-thirds of the considered papers have been published in the second decade between 2010 and July 2020 (67 articles), indicating an increasing interest in this topic. Most papers were published in 2017 and 2019 with 10 and nine papers each year.

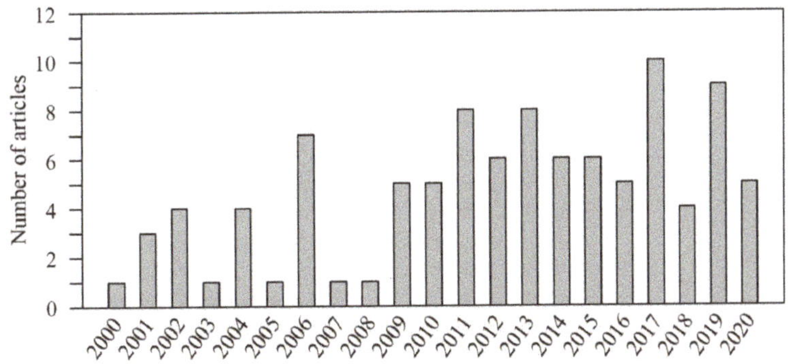

Figure 2. Number of articles on research on floodplains and riparian zones published each year between 2000 and 2020 (in 2020 until July).

Methods used over the period considered did not change significantly over time. Most research was done by field work (approx. 74%), e.g., soil surveys, field mapping, field experiments, and sampling. Laboratory experiments were carried out in about 10% of the studies. About 16% used models for the research, e.g., GIS-based models. Most studies included statistical analysis. Some studies used combined methods, e.g., field work and modeling.

3.2. Soil Information in the Articles on Soil Properties

Soil information in the articles was divided into physical, chemical, and biological soil properties, soil classification, and other soil information (Table 3). Soil information is vastly used in the examined research articles. Only three articles did not mention any soil information. In the remaining articles soil information is used to a different extent. A detailed table with the parameters of each soil information category is provided in the Appendix A (Table A1).

Table 3. Number of articles per soil information category (chemical, physical, biological properties, soil classification, other soil information, and no soil information).

Physical Properties	Chemical Properties	Biological Properties	Detailed Classification	Other Soil Classification	Other Soil Information	No Soil Information
76	56	21	32	6	9	3

In 76 articles some kind of physical soil parameters were used either to describe the study region or were investigated during the study. Physical soil parameters described by the different authors mainly contained classical soil physical parameters like texture and other descriptions of particle sizes and particle contents (e.g., fine material or coarse elements), electrical conductivity, porosity, soil temperature, or (dry) bulk density. In many cases soil parameters concerning the water household of soils like soil moisture content, (saturated) hydraulic conductivity, water holding capacity, infiltration, permeability, or field capacity are used, too. Some authors described more general parameters like the drainage situation or hydric conditions of the sites, but did not go into more detail. Other physical parameters mentioned were the pore-water pressure, the Atterberg limits, the specific gravity of the soil, (effective) cohesion, soil erodibility or an erosion coefficient, shear strength or shear stress, the (internal) friction angle, the van Genuchten parameters, and the rooting zone.

Chemical soil parameters were mentioned in 56 articles. Soil chemical parameters can be divided into several categories. In many articles nutrients were assessed, with focus on inorganic nitrogen (N) forms (NO_3^-, NO_2^-, NH_4^+, N_2O, total N), different phosphorus (P) speciations (e.g., plant available P, soluble reactive P, total P) and potassium (K) (e.g., total K, plant available K). Despite being nutrients, especially nitrogen and phosphorus are seen as non-point source pollutants, too. Other contaminants investigated are (heavy) metals like Cd, Pb, Hg, Zn, Cr, Cu, and others. In one paper organo-chlorine pesticides were examined. Another important soil chemistry category is soil organic matter (SOM). Here, different forms and types of SOM were addressed, like total carbon, inorganic and organic carbon, recalcitrant organic carbon (ROC), refractory index for carbon (RIC), or coarse particular organic matter (CPOM). Other parameters assessed were pH, salinity, $CaCO_3$, C/N, and isotopic ratios of C and N. One article mentioned the fertility of the soils investigated, but did not go further into detail.

Soil biological parameters were considered in 21 articles, containing data on soil organisms and processes driven by these inhabitants. In the research, soil invertebrates, soil microbial community structure (e.g., denitrifier and ammonium oxidizer density), and microbial number, species traits, operational taxonomic units and phylogenetic diversity, soil enzyme activity, denitrification enzyme activity (DEA), and actual denitrification were addressed. Other parameters were net potential nitrification, net potential N mineralization, potential mineralizable N, potential denitrification (rate), potential C mineralization, and microbial biomass C. Besides soil invertebrates and microorganisms, also root parameters, like root density, total belowground plant biomass, and root exudates, were examined. One article mentioned general biological activity features, but did not provide more details.

Some kind of soil classification/taxonomy is mentioned in 38 articles, whereas it has to be differentiated between a detailed classification from a common classification system or another soil description. Detailed soil description is provided in roughly one-third of the considered articles for this review (32 articles) and comprises descriptions on soil series, soil associations, soil types, soil map units, or soil orders based on the US Soil Taxonomy, the WRB, the Australian classification system, the French classification system, and others. In most articles these parameters are mentioned in detail (Which soil types? Which soil series?), but in few articles it is only mentioned that soil map units for example are used, but not which ones. In the remaining six articles soils are described more in general, for example as alluvial or hydric soils, but do not classify the soils in a common pedological classification system.

In the 32 articles that provide a detailed soil classification it is interesting in which combination and to which extent soil classification is combined with soil physical, chemical, and biological parameters (Table 4).

Table 4. Combination of physical, chemical, and biological soil properties in the 32 articles that provide a detailed soil classification [number of articles]. Articles that provide other soil information were not considered.

Physical + Chemical + Biological Properties + Classification	Physical + Chemical + Properties + Classification	Physical Properties + Classification	Chemical Properties + Classification	Biological Properties + Classification	Classification Only
6	12	5	3	0	6

Only six articles consider physical, chemical, and biological soil properties in combination with a detailed soil classification. Approximately one-third (12 articles) additionally mention soil physical and chemical parameters in their research. Five articles provide physical soil properties and three articles chemical soil properties in a combination with a detailed soil classification. Additional soil biological properties without chemical or physical properties were not covered in the research. Six articles provided a detailed soil classification only.

Good examples of the provision and use of soil information are mostly those articles that explicitly address soil properties in their research. For example, to describe the morphology of riparian soils in a restored floodplain in Switzerland as a restoration monitoring measure, Fournier et al. [34] provide not only detailed soil taxonomy, but also basic soil physical (texture, coarse soil), soil chemical (organic matter content and type, hydromorphological features), and soil biological parameters (root density and general biological activity features). In a comparison of the effects of different stream restoration practices (designed channel restoration vs. ecological buffer restoration) on riparian soils, beside USDA soil map units, the soil organic matter content, bulk density, soil moisture, texture, and root biomass were used and compared [25]. Other examples are the studies of Kauffman et al. [35], Clement et al. [36], Smith et al. [37], and Sutton-Grier et al. [38] which all provide soil information from all categories in their research.

In the 68 articles that do not provide a detailed soil description from a common soil classification, 11 articles, however, provide information on soil physical, soil chemical, and soil biological properties (Table 5).

Table 5. Combination of physical, chemical, and biological soil properties in the 68 articles that do not provide a detailed soil classification [number of articles].

Physical + Chemical + Biological Properties	Physical + Chemical + Properties	Physical + Biological Properties	Chemical + Biological Properties	Biological Properties Only	Chemical Properties Only	Physical Properties Only
11	16	3	1	0	7	24 [1]

[1] 15 out of the 24 covered engineering topics.

In 16 articles a combination of soil physical and soil chemical parameters is used. Soil physical parameters in combination with soil biological parameters were covered in three articles. Soil chemical parameters and soil biological parameters have been combined in one article only. If only one soil property was investigated or mentioned, most articles (24) provided information on soil physical parameters only, seven on soil chemical parameters only. Only soil biological parameters were used in none of the reviewed articles. Fifteen out of the 24 articles which provide soil physical parameters covered engineering topics only.

Soil information that could not be classified into the before mentioned categories is used in nine articles. These data comprise information on the use of soil maps or soil databases for example, the number and lower boundary of the soil layers or information on soil morphology (soil typicality, dynamism, and diversity). In some cases, soil properties that are taken from the maps or databases are further specified, but in other articles there is no further information on the kind of soil properties (chemical, physical, biological) or soil taxonomy.

3.3. Information on Soils in Articles in Connection with Engineering and Land Management

In total, 18 articles covered engineering topics, like soil bioengineering, river bank stability, or erosion control which can also be understood as some kind of soil protection. In these articles physical soil properties are considered only, e.g., shear strength, cohesion, texture or hydraulic conductivity. In the engineering articles neither soil chemical properties nor soil biological properties were used. None of the articles provided a detailed soil classification. One article considers additional soil biological properties (root system and root biomass) [39].

Another 32 articles deal with land management and land use, restoration planning, and the evaluation of restoration efficiency. In this category no clear pattern of the use of soil information is observable. Chemical and physical soil properties are described in the same extent in the articles as detailed soil classification (16, 24 and 16 articles, respectively). Soil biological properties play a minor role and are mentioned in six articles only. The provision of soil data differs between the 32 articles as few articles provide chemical, physical, biological soil properties in combination with a detailed soil classification (three articles), most do mention only parts of the different soil data types in a variable proportion.

4. Research Needs on Soil Protection in Floodplains

The results in Section 3.1 show that research on floodplains and riparian zones is not evenly distributed worldwide. Most research in the regarded period was conducted in North America and Europe, providing a broad base of knowledge on restoration of floodplains and the riparian zones in these areas. Other regions like Oceania, South America, Asia, and Africa are underrepresented in the research which leads to a lack of knowledge not only on restoration in riparian zones and floodplains, but also on soil information in these regions. More research in these regions of the world is highly recommended. When regarding the countries in which research on the individual continents is conducted it becomes clear that research mostly concentrates on single countries like the USA, Australia, Brazil, and China. The number of articles published on floodplain and riparian zone research was not distributed evenly over the two decades considered in this review. With two-thirds of the articles published in the second half of the reviewed period this shows the increasing concern and importance of research in the floodplains and riparian zones.

To protect soils and to interpret results of the research in the soil context it is important to know detailed properties of the regarded soils. Soil properties are described in most reviewed articles, but the extent of the provision and description of the soil properties varies considerably. Soil properties are important indicators when evaluating the soil quality and assessing soil functions [40]. Basically, soil quality is the capacity of a soil to function [41]. Soil quality depends on soil inherent and dynamic properties. Inherent properties are mostly influenced by the soil-forming factors (e.g., parent material, topography, time). Dynamic properties are influenced by human management and natural disturbances (e.g., land use or the construction of buildings or roads). Typical inherent soil properties are the soil texture or the drainage class. Management-dependent soil properties comprise among others the organic matter content, infiltration, biological activity, or soil fertility. The different soil properties can interact and limit other soil properties. Finally, the dynamic soil properties provide information about the ability of a soil to provide ecological functions and services [40]. Indicators for soil quality are traditionally divided into soil physical, soil chemical, and soil biological parameters [40,42]. In the reviewed articles over 75% provide information on soil physical parameters and hence information on the soil hydrologic status, on the availability of nutrients, on aeration, limitations on root growth, or the ability to withstand physical disturbances [40,42]. This information on soil physical parameters is very important for soil protection. Although not every article contains the same physical parameters, basic information on texture or particle sizes and soil moisture are given in most articles. Chemical parameters, mentioned in over 50% of the reviewed articles, are important to evaluate nutrient availability, water quality, buffer capacity, or the mobility of contaminants. Soil biological parameters, like abundance and biomass of soil organisms and their byproducts can also serve as an indicator for a functioning soil [42]. Biological soil parameters are assessed only in about 20% of the articles. It can be summarized that in current research in floodplain and riparian zones soil physical properties, chemical properties, and biological properties are used. There is a lack of information, especially on soil chemical and soil biological parameters. Both parameters can provide important insights in soil functioning and the reaction of the soils to certain conditions.

A detailed description from a common soil classification system like the WRB, the US soil taxonomy or a national classification system can be very informative not only for soil scientists. Soil classification systems are based on soil properties that are defined in diagnostic horizons, properties, and materials [15]. Therefore, when providing a detailed soil description from a common soil classification system, a lot of information on soil physical, soil chemical, and soil biological properties can be derived from using this classification. This information is missing, however, in about two-thirds of the reviewed literature. In these articles that do not provide a detailed soil description from a common soil classification system the majority of the authors though provide additional information on physical, chemical, and biological soil properties or combinations of these properties. The group of the articles with only physical soil data described mostly comprises articles dealing with engineering topics. In this group, except for one article that mentions some soil biological characteristics [39], soil is characterized by the physical characteristics only while other parameters like chemical or biological parameters are not considered. In this field, soil seems to be a granular medium only, serving as a building material, not as an important ecosystem compartment. But even if the physical and geotechnical properties of soils are most important for engineering purposes, a pedologicalview of soils, integrating some basic information on soil classification, on chemical and biological properties, might be valuable for engineers, too. As engineering measures usually comprise the use of (heavy) machinery, these measures can also be considered as a kind of construction work. This usually implies that the floodplain and riparian soils, adjacent to the riverbank or engineering site, are affected by these measures, too. Therefore, at least a minimum dataset on the soils of the whole site should be considered in projects, working in floodplains and riparian zones.

Other, more general, soil descriptions like the term "alluvial soils" for example, can give only general information on the soil development and on-site characteristics, but do not provide detailed information on the soil properties. As the physical, morphological, chemical, and mineralogical properties of these soils are strongly influenced by the alluvial parent material coming from the river, the soil characteristics, e.g., the soil texture and the related properties, can vary considerably [13]. In contrast, when a soil is classified within a common classification system, for example as a Gleysol (WRB), it is obvious that this soil must be saturated with groundwater long enough to develop these gleyic properties [15]. In the WRB, additional information on the soils and their properties can be deduced from the principal and supplementary qualifiers, such as the presence of an organic surface layer (qualifier: histic) or non-cemented secondary carbonates accumulated (qualifier: calcaric). Information on organic horizons or layers or waterlogging conditions due to high groundwater tables in floodplains and riparian zones are very valuable as especially these soils are highly susceptible to compaction for example [43]. So even if there is no additional information on physical, chemical, or biological soil properties, from a detailed soil description many soil characteristics can be deduced.

If a detailed investigation and description of the soils and their characteristics of the study sites is not possible there are other opportunities that should be considered to assess at least basic soil information of the site. For most regions of the world free soil information is available online from different organizations. A compendium of available data worldwide and for specific regions has been provided by ISRIC, the International Soil Reference and Information Centre for example [44]. They also maintain other useful sites and services like the World Soil Information Service (WoSIS) [45] and the SoilGrids platform [46] which can be helpful to consider.

As the results show, soil information is available in the large majority of the research papers, but it becomes also clear that in most cases soil information is incomplete or very specific only. To protect soils in floodplains and riparian zones, especially in the context of restoration works, a more pedological view of soils is necessary. This would not only be important for restoration projects directly, but also for all research in floodplains and riparian zones with the objective to contribute to restoration projects, for example in the prioritization of restoration areas.

Restoration projects impact soils in floodplains and riparian zones [25] and can therefore often be regarded as construction works. In recent years, soil protection on construction sites has become more and more important, for example in Switzerland or Germany. Known as "Bodenkundliche Baubegleitung" in the German-speaking area, it aims to protect soils from physical disturbance and contamination prior to and during construction. This means that after finishing the construction, the soil should be able to fulfil its natural functions again [47,48]. Detrimental soil changes that can occur on construction sites comprise soil compaction, erosion and discharge of substances, contamination, mixing of different soil substrates, and mixing of natural soil substrate with technogenic materials [48]. The soil protection on construction sites concept has not been developed for restoration projects, but as many restoration projects are comparable to construction sites, this concept is also applicable to restoration projects.

Soil protection on construction sites is not only applied during the construction works, but also prior to the construction in the planning process and is also involved post-construction in the monitoring and documentation of the project [47,48]. The lack of sound knowledge about soils has been identified as one of the factors hampering effective ecological restoration [49]. In the soil protection on construction sites concept various soil information is assessed for planning the construction work and appropriate soil protection measures during construction. This soil information comprises information on the soil types and their special characteristics (e.g., susceptibility to compaction or organic soils), coarse fragments, texture and structure of the soil, bulk density, pH, soil organic matter content, water content, rooting depth, and calcium carbonate content [47,48]. This soil information could be applied as a minimum dataset on soils in all research in floodplains and riparian zones and in restoration projects. Additionally, the nutrient and/or pollution status of the soil might be a useful parameter to be considered. The parameters proposed for the minimum soil data set contain stable and dynamic parameters. For dynamic parameters a continuous monitoring program might be useful. If not, many dynamic parameters like the physiological rooting depth for example can be deduced from easy to assess parameters like soil depth and soil texture. Also in the USDA stream restoration handbook [50] it is recommended to obtain background information on the sites, i.e., about soils. In general, to avoid detrimental soil changes many parts of the soil protection on construction sites concept could be easily integrated in the protocols for river, floodplain or riparian buffer restoration projects, as well as in soil bioengineering practices. In soil bioengineering practices there is great potential to integrate this minimum soil data set and soil protection measures during construction. Rey et al. [51] highlight the importance of the incorporation of current findings of the research in geosciences, for example soil science, in soil bioengineering practices. Further, scientist and practitioners should cooperate and exchange current issues and knowledge.

5. Conclusions

1. Research on floodplains and riparian zones of the world is not distributed evenly over the different continents, with the majority of research in this area conducted in North America, especially in the USA. The research on floodplains and riparian zones is also not distributed evenly over the time covered in this review with two-thirds of the research published in the second decade between 2010 and 2020.
2. Soils are somehow addressed in most articles, but the kind and extent of provided soil information varies significantly between the articles. Mostly physical soil information is provided, followed by chemical soil information. Only one-fifth provides soil biological information. One-third provides a detailed soil description from a common classification system. Soil information in the field of engineering is limited to physical data only.
3. Soils are addressed in the majority of the research, but soil information is often incomplete from a soil scientists' view. It is recommended to integrate at least a minimum data set on soil information in all research conducted in floodplains and riparian zones. This minimum data set comprises soil data used in the soil protection on construction sites concept: soil types and associated special characteristics (e.g., susceptibility to compaction), coarse fragments, texture and structure of the soil, bulk density, pH, soil organic matter content, water content, rooting depth, and calcium carbonate content. Additionally, the nutrient and/or pollution status might be a useful parameter. Further, at least the use of regional soil databases can give important information on the soils in the study area, if field work is not possible.

Author Contributions: Conceptualization, M.E.H. and G.B.; methodology, M.E.H., formal analysis, M.E.H.; writing—original draft preparation, M.E.H., writing—review and editing, M.E.H., visualization, M.E.H., supervision, G.B.; All authors have read and agreed to the published version of the manuscript.

Funding: This research received no external funding.

Institutional Review Board Statement: Not applicable.

Informed Consent Statement: Not applicable.

Data Availability Statement: The data presented in this study is available in Appendix A Table A1.

Conflicts of Interest: The authors declare no conflict of interest.

Abbreviations

As	Arsenic
ASC	Australian Soil Classification System
C	Carbon
$CaCO_3$	Calcium carbonate
Cd	Cadmium
C/N	Carbon/nitrogen ratio
CPOM	Coarse particular organic matter
Cr	Chromium
Cu	Copper
DEA	Denitrification enzyme activity
DOC	Dissolved organic carbon
DOM	Dissolved organic matter
DON	Dissolved organic nitrogen
EC	Electrical conductivity
Fe	Iron
Hg	Mercury
IC	Inorganic carbon
K	Potassium
N	Nitrogen
Ni	Nickel
NO_3^-	Nitrate
NO_3^--N	Nitrate nitrogen
NO_2^-	Nitrite
NH_4^+	Ammonium
NH_4^+-N	Ammonia nitrogen
N_2O	Nitrous oxide
NO	Nitric oxide
NZG	New Zealand Soil Classification
OC	Organic carbon
OM	Organic matter
P	Phosphorus
PB	Lead
PO_4^{3-}	Phosphate
RIC	Refractory index for carbon
ROC	Recalcitrant organic carbon
RO	Réferentiel Pédologique (=French Soil Classification)

S	Sulfur
Sb	Antimony
SiBCS	Sistema Brasileiro de Clasifição de Solos (=Brazilian Soil Classification System)
Sn	Tin
SOC	Soil organic carbon
SOM	Soil organic matter
SRP	Soluble reactive P
TC	Total carbon
TDC	Total dissolved carbon
TDN	Total dissolved nitrogen
TBGB	Total belowground biomass
TK	Total potassium
TN	Total nitrogen
TOC	Total organic carbon
TP	Total phosphorus
V	Vanadium
WRB	World Reference Base for Soil Resources
Zn	Zinc

Appendix A

Table A1. Articles selected for this review, continent, country, and soil information categories.

Source	#Article	Continent	Country	Category	Chemical Properties	Physical Properties	Biological Properties	Detailed Classification	Other Classification	Other Soil Data
Agouridis et al. 2005	[52]	North America	USA, Kentucky	Management	-	-	-	Hagerstown (Fine, mixed, mesic Typic Hapludalf); McAfee (Fine, mixed, mesic Mollic Hapludalf); Woolper (Fine, mixed, mesic Typic Argiudoll)	-	-
Amezketa & del Valle de Lersundi 2008	[53]	Europe	Spain	Management	OM, $CaCO_3$, salinity	Texture, moisture, temperature, EC	-	Loamy-skeletal, mixed, mesic, Aridic Ustorthent; Coarse-loamy, mixed, mesic Aridic Ustifluvent; Fine-salty, mixed, mesic, Aridic Ustifluvents; Entisols	-	-
Andrews et al. 2011	[54]	North America	USA, Kentucky	Other	Fertility	Permeability, water holding capacity, rooting zone	-	Fine-loamy, mixed, mesic Dystric Fluventic Eutrochrepts (USDA 1996)	-	-
Anstead et al. 2012	[55]	Europe	UK	Engineering	-	Cohesion, texture	-	-	-	-
Asghari & Cavagnaro 2011	[56]	Oceania	Australia	Other	pH, plant available P, TC, TN	Texture	-	-	-	-
Atkinson & Lake 2020	[57]	North America	USA, Texas	Management	-	Erodibility	-	-	-	-
Bariteau et al. 2013	[58]	North America	Canada	Engineering	-	Texture	-	-	-	-
Beauchamp et al. 2015	[59]	North America	USA, Maryland	Management	OM, pH, C/N, plant available macronutrients and micronutrients	Texture	-	-	-	-
Bedison et al. 2013	[60]	North America	USA, New Jersey	Other	Mottling	Texture, drainage	-	Mesic Entisols; Histosols; Inceptisols	-	-
Bissels et al. 2004	[61]	Europe	Germany	Management	Plant available P and K, TN, TC, $CaCO_3$, OM, C/N	Texture	-	-	Alluvial soils	-
Botero-Acosta et al. 2017	[62]	North America	USA, Oklahoma	Engineering	-	Water content, field capacity, wilting point, saturated hydraulic conductivity	-	-	-	STATSGO soil map (soil types); Soil Characterization Database (physical soil properties)
Brovelli et al. 2012	[63]	n.a.	n.a.	Other	Various (not further specified)	Various (not further specified)	Various (not further specified)	-	-	-
Buchanan et al. 2012	[64]	North America	USA, New York	Management	-	Erodibility, texture	-	-	-	-
Burger et al. 2010	[9]	Oceania	Australia	Management	NO_3^-, NO_2^-, NH_4^+, plant available P, EC, pH, TC, TN	-	-	Grey, Yellow, and Brown Sodosols and Chromosols (ASC 1996)	-	-
Buzhdygan et al. 2016	[65]	Asia	Ukraine	Other	SOC, pH, TN	Bulk density	-	-	-	-
Cabezas & Comín 2010	[66]	Europe	Spain	Other	TOC, TN, C/N, RIC, ROC	Bulk density	-	-	-	-
Clement et al. 2003	[36]	Europe	France	Other	Hydromorphological features, OM, pH	Texture, bulk density	Denitrification activity, roots	Fine silty-clay loam, mixed, mesic Typic Haplaquoll (USDA 1990)	-	-

Table A1. *Cont.*

Source	#Article	Continent	Country	Category	Chemical Properties	Physical Properties	Biological Properties	Detailed Classification	Other Classification	Other Soil Data
Das 2016	[67]	Asia	India	Engineering	-	Shear strength, dry density, Atterberg limits, specific gravity, texture	-	-	-	-
Davis et al. 2006	[68]	North America	USA, Nebraska	Other	OM, TN, TK, TP, pH	Temperature, moisture, texture	Soil invertebrates	-	-	Groundwater level
De Mello et al. 2017	[69]	South America	Brazil	Management	-	Bulk density, available water capacity, saturated hydraulic conductivity	-	SiBCS (2018) (WRB 2015): Gleissolos (Gleysols); Latossolos Vermelho (Ferralsols); Latossolos Vermelho-Amarelo (Ferralsols); Neosolos Regolíticos (Regosols); Neossolos Flúvicos (Fluvisols); Cambissolos (Cambisols)	-	Number of layers, lower boundary of layers
Del Tánago & de Jalón 2006	[70]	n.a.	n.a.	Management	-	Permeability	-	-	-	-
Dhondt et al. 2006	[71]	Europe	Belgium	Other	OC, TN, IC, pH, N_2O fluxes	Texture	DEA	-	-	-
Dietrich et al. 2014	[28]	Europe	Sweden	Management	OM, mass fraction of C and N, isotopic ratios ($\Delta^{13}C$; $\Delta^{15}N$), TC = TOC	Texture, water holding capacity	-	-	-	-
Duong et al. 2014	[72]	Asia	Vietnam	Engineering	-	Water content, bulk density, saturated shear strength, saturated hydraulic conductivity, dry density, main grain size, effective cohesion, texture	-	-	-	-
Duró et al. 2020	[73]	Europe	Netherlands	Engineering	-	Internal friction angle, cohesion, texture, shear stress	-	-	-	-
Dybala et al. 2019	[74]	North America	USA, California	Other	TC, carbon stock	Bulk density	-	Cosumnes (Fine, mixed, active, nonacid, thermic Aquic Xerofluvents)	-	-
Fernandes et al. 2020	[75]	Europe	Portugal	Engineering	-	Cohesion	-	-	-	-
Fournier et al. 2015	[76]	Europe	Switzerland	Other	-	Hydric conditions	Species traits	-	-	-
Fournier et al. 2013	[33]	Europe	Switzerland	Other	OM, OM-type, hydromorphological features	Texture, coarse elements	Root density, biological activity features	RP (2009) (WRB 2006): REDOXISOLS fluviques carbonatés (Gleyic Fluvisols (Calcaric)); FLUVIOSOLS brut carbonatés (Regosols (Calcaric)); FLUVIOSOLS typiques carbonatés (Fluvisols (Calcaric)); FLUVIOSOLS typiques redoxiques carbonatés (Fluvisols (Calcaric) with redoximorphic features); REDUCTISOLS fluviques carbonatés (Gleysols (Calcaric))	-	Soil morphology: soil diversity, soil dynamism, soil typicality
Franklin et al. 2020	[77]	Oceania	Australia	Other	TN, TC, $NH_4{}^+$-N, $NO_3{}^-$-N, pH, OC; (DOM, DOC, DON, C/N, TDC, TDN, inorganic N in leachate)	Texture, moisture	-	Hard pedal mottled-yellow-grey duplex soil (Atlas of Australian Soils 1960–1968); USDA (2014): Paleustalf	-	-
Gageler et al. 2014	[78]	Oceania	Australia	Management	TN, SOC, $NO_3{}^-$, $NH_4{}^+$	Texture, infiltration, bulk density	-	Red Ferrosols; Clay loamy (ASC 1996); WRB (2014): Nitisols	-	-
Garvin et al. 2017	[79]	North America	USA, Oklahoma	Other	Cd, Pb, Zn	-	-	-	-	-

Table A1. *Cont.*

Source	#Article	Continent	Country	Category	Chemical Properties	Physical Properties	Biological Properties	Detailed Classification	Other Classification	Other Soil Data
Giese et al. 2000	[80]	North America	USA, South Carolina	Other	SOC	-	-	Typic Endoaquepts; Typic Fluvaquents; Thapto-Histic Fluvaquents; Grossarenic Hapludults; Arenic Endoaquults	-	-
Gift et al. 2010	[3]	North America	USA, Maryland	Other	OM, N_2O	Moisture	DEA, root biomass	-	-	-
Gold et al. 2001	[81]	North America	USA, various	Other	hydromorphological features	Soil wetness	-	-	Hydric soils	-
Gumiero & Boz 2017	[82]	Europe	Italy	Management	Moderately calcareous	Water content, texture, drainage	-	-	-	-
Guo et al. 2018	[83]	Asia	China	Other	Organo-chlorine pesticides	Texture	Soil microbial community structure	-	Brown soil	-
Hale et al. 2018	[84]	Oceania	Australia	Management	TC, TN, C/N, plant available P, CPOM	-	-	-	-	-
Hale et al. 2014	[85]	Oceania	Australia	Management	EC, pH, inorganic N (NO_3^-, N_2O, NH_4^+), TC, TN, plant available P	Water content, bulk density, texture	-	Various soil types (ASC 1996)	-	-
Harrison et al. 2011	[86]	North America	USA, Maryland	Other	N_2O, N_2	-	-	-	-	-
Hasselquist et al. 2017	[87]	Europe	Sweden	Other	$\Delta^{15}N$, bulk C and N, C/N	Texture	-	-	-	-
Higgisson et al. 2019	[88]	Oceania	Australia	Management	-	Particle size	-	-	-	-
Jansen & Robertson 2001	[89]	Oceania	Australia	Management	-	Bank stability, soil structure	-	-	-	-
Janssen et al. 2019	[90]	Europe	France, Switzerland	Engineering	-	-	-	-	-	-
Juracek & Drake 2016	[91]	North America	USA, Kansas	Other	Pb, Zn	Particle size	-	-	-	-
Kauffman et al. 2004	[35]	North America	USA, Oregon	Management	SOM, mineral N (NO_3^--N, NH_4^+-N)	Texture, bulk density, porosity, infiltration rates, moisture	TBGB, net potential nitrification, net potential N mineralization	Cryofluvents	-	-
Korol et al. 2019	[92]	North America	USA, various	Management	pH, OM, NO_3^-, NH_4^+, TC, TN, SRP	Bulk density, moisture	Denitrification potential, DEA, potential C mineralization	-	-	-
Langendoen et al. 2009	[93]	North America	USA, Mississippi	Engineering	-	Shear strength, pore-water pressure, cohesion, friction angle, bulk density, texture, saturated hydraulic conductivity	-	-	-	-
Larsen & Greco 2002	[94]	North America	USA, California	Engineering	-	Bank cohesion, texture	-	-	-	-
Laub et al. 2013	[25]	North America	USA, Maryland	Management	SOM	Bulk density, moisture, texture	Root biomass	Zekiah (Coarse-loamy, siliceous, active, acid, mesic Typic Fluvaquents); Issue (Coarse-loamy, mixed, active, mesic Fluvaquentic Dystrudepts); Hatboro (Fine-loamy, mixed, active, nonacid, mesic Fluvaquentic Endoaquepts); Fallsington (Fine-loamy, mixed, active, mesic Typic Endoaquults), Widewater (Fine-loamy, mixed, active, acid, mesic Fluvaquentic Endoaquepts); Codorus (Fine-loamy, mixed, active, mesic Fluvaquentic Dystrudepts); Lindside (Fine-silty, mixed, active, mesic Fluvaquentic Eutrudepts)	-	-

Table A1. *Cont.*

Source	#Article	Continent	Country	Category	Chemical Properties	Physical Properties	Biological Properties	Detailed Classification	Other Classification	Other Soil Data
Lee et al. 2011	[95]	Asia	South Korea	Other	-	-	-	-	-	Soil information (= soil properties; not further specified) from soil maps is used in model
Li et al. 2006	[96]	Asia	China	Engineering	-	Moisture, shear stress	-	-	-	-
Lindow et al. 2009	[97]	n.a.	n.a.	Engineering	-	Texture, hydraulic conductivity, van Genuchten parameters, effective cohesion, internal friction angle, residual and saturated water content	-	-	-	-
Maffra & Sutili 2020	[98]	South America	Brazil	Engineering	-	-	-	-	-	-
Maroto et al. 2017	[99]	Europe	Spain	Engineering	-	Texture	-	-	-	Poorly developed soil
Marquez et al. 2017	[100]	North America	USA, Iowa	Other	-	-	-	Coland (Fine-loamy, mixed, superactive, mesic Cumulic Endoaquoll)	-	-
Matheson et al. 2002	[101]	Oceania	New Zealand	Other	NO_3^-, NH_4^+	Bulk density, moisture content	-	NZG (1948): Waingaro steepland soil (northern yellow-brown earth); USDA (1975): Umbric Dystrochrept	-	-
Meals & Hopkins 2002	[102]	North America	USA, Vermont	Management	-	-	-	-	Alluvial and lacustrine soils	-
Meynendonckx et al. 2006	[103]	Europe	Belgium	Other	-	Drainage, texture	-	-	-	-
Neilen et al. 2017	[104]	Oceania	Australia	Other	NO_3^--N, NH_4^+-N, DON, DOC, SRP in leachate	-	-	Haplic, Mesotrophic, Red Ferrosols (ASC 2016)	-	-
Orr et al. 2007	[105]	North America	USA, Wisconsin	Other	OM, NO_3^--N	Moisture, texture	Actual denitrification potential, DEA	-	-	-
Peter et al. 2012	[106]	Europe	Switzerland	Other	-	Texture	-	-	-	-
Petrone & Preti 2010	[107]	South America	Nicaragua	Engineering	-	Texture	-	-	-	-
Pinto et al. 2016	[108]	Europe	Portugal	Engineering	-	"physical riverbank conditions" not further specified	-	-	-	-
Rahe et al. 2015	[109]	North America	USA, Illinois	Management	TC, TN, C/N, plant available P, CPOM	Infiltration, bulk density, moisture, texture, drainage	-	Swanwick (Fine-silty, spolic, mixed, active, nonacid, mesic Anthroportic Udorthents); Lenzburg (Fine-loamy, spolic, mixed, active, calcareous, mesic Anthroportic Udorthents)	-	-
Rassam & Pagendam 2009	[110]	Oceania	Australia	Management	-	Hydraulic conductivity (subsoil)	Denitrification rates	-	-	-
Recking et al. 2019	[32]	Europe	"alpine context"	Engineering	-	Cohesion, texture	-	-	-	-
Reisinger et al. 2013	[111]	North America	USA, Kansas	Other	-	-	-	Ivan (Fine-silty, mixed, superactive, mesic Cumulic Hapludolls)	-	-
Remo et al. 2017	[112]	North America	USA, Illinois	Management	-	Texture, drainage class, water retention capacity	-	Soil order (not further specified)	-	Data obtained from SSURGO
Rheinhardt et al. 2012	[113]	North America	USA, North Carolina	Other	SOM, SOC content	Bulk density	-	-	-	-
Rimondi et al. 2019	[114]	Europe	Italy	Other	Hg, As, Cd, Pb, Sb, Cr, Zn, Cu, Sn, V	-	-	-	-	-

Table A1. *Cont.*

Source	#Article	Continent	Country	Category	Chemical Properties	Physical Properties	Biological Properties	Detailed Classification	Other Classification	Other Soil Data
Rosenblatt et al. 2001	[115]	North America	USA, Rhode Island	Management	-	-	-	Inceptisols; Histosols; Entisols	-	-
Rosenfeld et al. 2011	[116]	Europe/North America	Sweden, Finland, Canada	Management	-	-	-	-	-	-
Saad et al. 2018	[117]	South America	Brazil	Management	-	Erodibility of soil classes, texture	-	SiBCS (2018): Argissolo Vermelho-Amarelo; Cambissolo Húmico; Neossolo Litólico; Neossolo Flúvico; Cambissolo Háplico USDA (2014): Ultisol; Inceptisol; Udorthent; Fluvent USDA (1996): Ochrept	-	-
Samaritani et al. 2011	[118]	Europe	Switzerland	Other	pH, TN, TOC, TIC, available P, C pools and fluxes	Texture, temperature	-	-	-	-
Sgouridis et al. 2011	[119]	Europe	UK	Other	-	Texture (topsoil)	-	Pelo-stagnogley soils; Stagnogley soils; Brown rendzinas; Gleyic brown calcareous earths; Grey rendzinas	-	-
Shah et al. 2010	[120]	North America	USA, New Mexico	Other	-	-	-	Typic Ustifluvents (Gila-Vinton-Brazito association)	-	-
Silk et al. 2006	[121]	North America	USA, California	Other	Bioavailable Cu, oxide-bound Cu, pH	-	-	-	-	-
Smith et al. 2012	[37]	Oceania	Australia	Other	NO_3^-, NO_2^-, NH_4^+, TC, TN, chemical nature of soil C	Texture, bulk density, gravimetric moisture	Potential mineralizable N, net nitrification	Red Chromosol (ASC 1996)	-	-
Sutton-Grier et al. 2009	[38]	North America	USA, North Carolina	Other	SOM, NO_3^--N, NH_4^+-N, inorganic P, C/N	Bulk density	Microbial biomass C, DEA	Monacan (Fine-loamy, mixed, active, thermic Fluvaquentic Eutrudepts)	-	-
Tang et al. 2016	[122]	Europe	Netherlands	Other	OM, plant available P, amorphous Fe, Fe-bound P, aluminum-bound P	Bulk density, texture	-	-	-	-
Tererai et al. 2015	[123]	Africa	South Africa	Other	-	-	-	-	Deep greyish alluvial soils	-
Theriot et al. 2013	[124]	North America	USA, Arkansas	Other	TC, TN, TP	Bulk density, moisture	Microbial biomass N, potential mineralizable N, potential denitrification	-	-	-
Tian et al. 2004	[125]	North America	USA, North Carolina	Management	pH, TN, TC, NO_3^-	-	Microbial biomass, denitrifier density, ammonium oxidizer density	-	-	-
Tomer et al. 2015	[126]	North America	USA, Iowa, Illinois	Management	-	-	-	Tama (Typic Argiudolls); Saude (Typic Hapludolls); Webster (Typic Endoaquolls); Osco (Mollic Hapludalfs)	Hydric soils	-
Unghire et al. 2011	[4]	North America	USA, North Carolina	Management	SOM, inorganic nutrients (NO_2^-, NO_3^-, inorganic P)	Moisture, bulk density, clay content	-	Cartecay (Coarse-loamy, mixed, semiactive, nonacid, thermic Aquic Udifluvents); Chewacla (Fine-loamy, mixed, active, thermic Fluvaquentic Dystrudepts)	-	-

Table A1. *Cont.*

Source	#Article	Continent	Country	Category	Chemical Properties	Physical Properties	Biological Properties	Detailed Classification	Other Classification	Other Soil Data
Vandecasteele et al. 2004	[127]	Europe	Belgium	Other	Cd, Cr, Zn, Cu, Ni, Pb, P, S, TN, $CaCO_3$, OC, pH	EC, texture	-	-	-	-
Walker et al. 2002	[128]	North America	USA, Georgia	Other	NO_3^-, NH_4^+, NH_3, NO, N_2O	Water content	-	Saunook (Fine-loamy, mixed, superactive, mesic Humic Hapludults)	-	-
Walker et al. 2009	[129]	North America	USA, North Carolina	Other	NO_3^-, NH_4^+, NO_2^-, TN, TC	Moisture	-	Rosman (Coarse-loamy, mixed, superactive, mesic Fluventic Humudepts)	-	-
Wang et al. 2019	[130]	Asia	China	Other	pH, SOM, TN, TP, TK, available N/P/K	Texture, water content	Soil microbial number (bacteria, actinomycete, fungi), soil enzyme activity, operational taxonomic units, phylogenetic diversity			
Wang et al. 2014	[131]	Asia	China	Other	NH_4^+-N, NO_3^--N, NO_2^--N, TN, PO_4^{3-} in water	Texture	Diversity and distribution of microbial community	-	-	-
Weller & Baker 2014	[132]	North America	USA, various	Other	NO_3^-	-	-	-	-	-
Welsh et al. 2017	[133]	North America	USA, North Carolina	Other	pH, OM, NO_3^-, NH_4^+, TC, TN, SRP	Moisture, texture	DEA	-	-	-
Welsh et al. 2019	[134]	North America	USA, North Carolina	Other	-	Texture	-	-	-	-
Xiong et al. 2015	[135]	Asia	China	Other	pH, OM, TN	Texture, moisture, bulk density	-	-	-	-
Ye et al. 2019	[136]	Asia	China	Other	Hg, As, Cr, Cd, Pb, Cu, Fe, Mn, Zn, SOM, TP, pH	Moisture, texture	-	-	-	-
Young et al. 2013	[137]	North America	USA, Vermont	Other	TP, pH, OM, different P speciations	-	-	-	-	-
Zaimes et al. 2006	[138]	North America	USA, Iowa	Management	-	Texture, bulk density, permeability	-	Spillville (Fine loamy, mixed, superactive, mesic Cumulic Hapludolls); Coland (Fine-loamy, mixed, mesic, superactive Cumulic Endoaquolls)	-	-
Zhang et al. 2018	[39]	Asia	China	Engineering	-	Texture, shear strength	Root system, root biomass	-	-	-
Zhao et al. 2013	[139]	Asia	China	Management	-	Erodibility	-	-	-	Soil map (1: 1,000,000); China soil scientific database (soil properties not further specified)

References

1. Christiansen, T.; Azlak, M.; Ivits-Wasser, E. *Floodplains: A Natural System to Preserve and Restore*; EEA Report 24/2019; European Environment Agency: Luxembourg, 2020.
2. Malmqvist, B.; Rundle, S. Threats to the running water ecosystems of the world. *Environ. Conserv.* **2002**, *29*, 134–153. [CrossRef]
3. Gift, D.M.; Groffman, P.M.; Kaushal, S.S.; Mayer, P.M. Denitrification Potential, Root Biomass, and Organic Matter in Degraded and Restored Urban Riparian Zones. *Restor. Ecol.* **2010**, *18*, 113–120. [CrossRef]
4. Unghire, J.M.; Sutton-Grier, A.E.; Flanagan, N.E.; Richardson, C.J. Spatial Impacts of Stream and Wetland Restoration on Riparian Soil Properties in the North Carolina Piedmont. *Restor. Ecol.* **2010**, *19*, 738–746. [CrossRef]
5. Junk, W.J.; Welcomme, R. Floodplains. In *Wetlands and Shallow Continental Water Bodies*; Patten, B.C., Ed.; SPB Academic Publishers: The Hague, The Netherlands, 1990; Volume 1, pp. 491–524. ISBN 905-103-146-0.
6. Thoms, M. Floodplain–river ecosystems: Lateral connections and the implications of human interference. *Geomorphology* **2003**, *56*, 335–349. [CrossRef]
7. Nilsson, C.; Berggren, K. Alterations of Riparian Ecosystems Caused by River Regulation: Dam operations have caused global-scale ecological changes in riparian ecosystems. How to protect river environments and human needs of rivers remains one of the most important questions of our time. *Bioscience* **2000**, *50*, 783–792. [CrossRef]
8. Naiman, R.J.; Bilby, R.E.; Bisson, P.A. Riparian Ecology and Management in the Pacific Coastal Rain Forest. *Bioscience* **2000**, *50*, 996–1011. [CrossRef]
9. Burger, B.; Reich, P.; Cavagnaro, T.R. Trajectories of change: Riparian vegetation and soil conditions following livestock removal and replanting. *Austral. Ecol.* **2010**, *35*, 980–987. [CrossRef]
10. Tockner, K.; Stanford, J.A. Riverine flood plains: Present state and future trends. *Environ. Conserv.* **2002**, *29*, 308–330. [CrossRef]
11. Palmer, M.A.; Ruhi, A. Linkages between flow regime, biota, and ecosystem processes: Implications for river restoration. *Science* **2019**, *365*, eaaw2087. [CrossRef]
12. Vanneuville, W.; Wolters, H.; Scholz, M.; Werner, B.; Uhel, R. *Flood Risks and Environmental Vulnerability—Exploring the Synergies between Floodplain Restoration, Water Policies and Thematic Policies*; EEA Report 1/2016; European Environment Agency: Luxembourg, 2016.
13. Boettinger, J.L. Alluvium and alluvial soils. In *Encyclopedia of Soils in the Environment*; Hillel, D., Ed.; Elsevier: Oxford, UK, 2005; pp. 45–49. [CrossRef]
14. Gerrard, J. *Alluvial Soils*; Van Nostrand Reinhold Co.: New York, NY, USA, 1987; ISBN 044-222-742-6.
15. IUSS Working Group WRB. *World Reference Base for Soil Resources 2014, Update 2015. International Soil Classification System for Naming Soils and Creating Legends for Soil Maps*; World Soil Resources Reports 106; FAO: Rome, Italy, 2015.
16. Soil Survey Staff. *Keys to Soil Taxonomy*, 12th ed.; United States Department of Agriculture—Natural Resources Conservation Service: Washington, DC, USA, 2014.
17. Daniels, J. Floodplain aggradation and pedogenesis in a semiarid environment. *Geomorphology* **2003**, *56*, 225–242. [CrossRef]
18. Bullinger-Weber, G.; Gobat, J.-M. Identification of facies models in alluvial soil formation: The case of a Swiss alpine floodplain. *Geomorphology* **2006**, *74*, 181–195. [CrossRef]
19. Palmer, M.; Bernhardt, E.; Allan, J.D.; Lake, P.; Alexander, G.; Brooks, S.; Carr, J.; Clayton, S.; Dahm, C.N.; Follstad Shah, J.; et al. Standards for ecologically successful river restoration. *J. Appl. Ecol.* **2005**, *42*, 208–217. [CrossRef]
20. Roni, P.; Hanson, K.; Beechie, T. Global Review of the Physical and Biological Effectiveness of Stream Habitat Rehabilitation Techniques. *N. Am. J. Fish. Manag.* **2008**, *28*, 856–890. [CrossRef]
21. Hornung, L.K.; Podschun, S.A.; Pusch, M. Linking ecosystem services and measures in river and floodplain management. *Ecosyst. People* **2019**, *15*, 214–231. [CrossRef]
22. European Centre for River Restoration. Regional and National Policies. Available online: https://www.ecrr.org/River-Restoration/Regional-and-national-policies (accessed on 4 December 2020).
23. Bruland, G.L.; Richardson, C.J. Spatial Variability of Soil Properties in Created, Restored, and Paired Natural Wetlands. *Soil Sci. Soc. Am. J.* **2005**, *69*, 273–284. [CrossRef]
24. Jähnig, S.C.; Brabec, K.; Buffagni, A.; Erba, S.; Lorenz, A.W.; Ofenböck, T.; Verdonschot, P.F.M.; Hering, D. A comparative analysis of restoration measures and their effects on hydromorphology and benthic invertebrates in 26 central and southern European rivers. *J. Appl. Ecol.* **2010**, *47*, 671–680. [CrossRef]
25. Laub, B.G.; McDonough, O.T.; Needelman, B.A.; Palmer, M.A. Comparison of Designed Channel Restoration and Riparian Buffer Restoration Effects on Riparian Soils. *Restor. Ecol.* **2013**, *21*, 695–703. [CrossRef]
26. Ballantine, K.; Schneider, R. Fifty-five years of soil development in restored freshwater depressional wetlands. *Ecol. Appl.* **2009**, *19*, 1467–1480. [CrossRef]
27. Cole, C.A.; Kentula, M.E. Monitoring and Assessment—What to Measure ... and Why. In *Wetlands: Integrating Multidisciplinary Concepts*; LePage, B.A., Ed.; Springer: Dordrecht, The Netherlands, 2011; pp. 137–152. ISBN 978-94-007-0551-7.
28. Dietrich, A.L.; Lind, L.; Nilsson, C.; Jansson, R. The Use of Phytometers for Evaluating Restoration Effects on Riparian Soil Fertility. *J. Environ. Qual.* **2014**, *43*, 1916–1925. [CrossRef]

29. Haines-Young, R.; Potschin, M.B. *Common International Classification of Ecosystem Services (CICES) V5.1 and Guidance on the Application of the Revised Structure*; CICES: Sale, UK, 2018.
30. Pickering, C.; Byrne, J. The benefits of publishing systematic quantitative literature reviews for PhD candidates and other early-career researchers. *High. Educ. Res. Dev.* **2014**, *33*, 534–548. [CrossRef]
31. Moher, D.; Liberati, A.; Tetzlaff, J.; Altman, D.G.; The PRISMA Group. Preferred reporting items for systematic reviews and meta-analyses: The PRISMA statement. *PLoS Med.* **2009**, *6*, e1000097. [CrossRef] [PubMed]
32. Recking, A.; Piton, G.; Montabonnet, L.; Posi, S.; Evette, A. Design of fascines for riverbank protection in alpine rivers: Insight from flume experiments. *Ecol. Eng.* **2019**, *138*, 323–333. [CrossRef]
33. Schultz, J. *The Ecozones of the World*, 2nd ed.; Springer: Berlin/Heidelberg, Germany, 2005; ISBN 978-2-540-28527-4.
34. Fournier, B.; Guenat, C.; Bullingerweber, G.; Mitchell, E.A.D. Spatio-temporal heterogeneity of riparian soil morphology in a restored floodplain. *Hydrol. Earth Syst. Sci.* **2013**, *17*, 4031–4042. [CrossRef]
35. Kauffman, J.B.; Thorpe, A.S.; Brookshire, E.N.J. Livestock exclusion and belowground ecosystem responses in riparian meadows of Eastern Oregon. *Ecol. Appl.* **2004**, *14*, 1671–1679. [CrossRef]
36. Clement, J.C.; Holmes, R.M.; Peterson, B.J.; Pinay, G. Isotopic investigation of denitrification in a riparian ecosystem in western France. *J. Appl. Ecol.* **2003**, *40*, 1035–1048. [CrossRef]
37. Smith, M.; Conte, P.; Berns, A.E.; Thomson, J.R.; Cavagnaro, T.R. Spatial patterns of, and environmental controls on, soil properties at a riparian–paddock interface. *Soil Biol. Biochem.* **2012**, *49*, 38–45. [CrossRef]
38. Sutton-Grier, A.E.; Ho, M.; Richardson, C.J. Organic amendments improve soil conditions and denitrification in a restored riparian wetland. *Wetlands* **2009**, *29*, 343–352. [CrossRef]
39. Zhang, D.; Cheng, J.; Liu, Y.; Zhang, H.; Ma, L.; Mei, X.; Sun, Y. Spatio-Temporal Dynamic Architecture of Living Brush Mattress: Root System and Soil Shear Strength in Riverbanks. *Forests* **2018**, *9*, 493. [CrossRef]
40. Kuykendall, H. *Soil Quality Physical Indicators: Selecting Dynamic Soil Properties to Assess Soil Function*; Soil Quality Technical Note No. 10; United States Department of Agriculture—Natural Resources Conservation Service: Washington, DC, USA, 2008.
41. Karlen, D.L.; Mausbach, M.J.; Doran, J.W.; Cline, R.G.; Harris, R.F.; Schuman, G.E. Soil Quality: A Concept, Definition, and Framework for Evaluation (A Guest Editorial). *Soil Sci. Soc. Am. J.* **1997**, *61*, 4–10. [CrossRef]
42. National Soil Survey Center. *Soil Quality Information Sheet. Indicators for Soils Quality Evaluation*; United States Department of Agriculture—Natural Resources Conservation Service: Washington, DC, USA, 1996.
43. Häusler, S.; Salm, C. *Bodenschutz beim Bauen (Soil Protection and Construction)*; Leitfaden Umwelt Nummer 10; BUWAL Bundesamt für Umwelt, Wald und Landschaft: Bern, Switzerland, 2001.
44. Soil Geographic Databases. Available online: https://www.isric.org/index.php/explore/soil-geographic-databases (accessed on 4 December 2020).
45. WoSIS Soil Profile Database. Available online: https://www.isric.org/index.php/explore/wosis (accessed on 4 December 2020).
46. New Edition of Soil Property Estimates for the World with Associated Web Platform Released (SoilGrids250m). Available online: https://www.isric.org/news/new-edition-soil-property-estimates-world-associated-web-platform-released-soilgrids250m (accessed on 4 December 2020).
47. Bellini, E. *Boden und Bauen. Stand der Technik und Praktiken (Soil and Construction. State of the Knowledge)*; Umwelt-Wissen Nr. 1508; BAFU Bundesamt für Umwelt: Bern, Switzerland, 2015.
48. Bundesverband Boden. *Bodenkundliche Baubegleitung BBB (Soil Protection on Construction Sites). Leitfaden für die Praxis. BVB-Merkblatt. Band 2*; Erich Schmidt Verlag: Berlin, Germany, 2013.
49. Fisher, J.; Cortina-Segarra, J.; Grace, M.; Moreno-Mateos, D.; Rodíguez Gonzáles, P.; Baker, S.; Frouz, J.; Klimkowska, A.; Andres, P.; Kyriazopoulos, A.; et al. *What Is Hampering Current Restoration Effectiveness? An EKLIPSE Expert Working Group Report*; UK Centre for Ecology & Hydrology: Wallingford, UK, 2019.
50. United States Department of Agriculture; Natural Resources Conservation Service. Site Assessment and Investigation. In *Stream Restoration Design National Engineering Handbook (Part 654)*; USDA, NRCS, Eds.; United States Department of Agriculture—Natural Resources Conservation Service: Washington, DC, USA, 2007.
51. Rey, F.; Bifulco, C.; Bischetti, G.B.; Bourrier, F.; De Cesare, G.; Florineth, F.; Graf, F.; Marden, M.; Mickovski, S.B.; Phillips, C.J.; et al. Soil and water bioengineering: Practice and research needs for reconciling natural hazard control and ecological restoration. *Sci. Total Environ.* **2019**, *648*, 1210–1218. [CrossRef]
52. Agouridis, C.T.; Edwards, D.R.; Workman, S.R.; Bicudo, J.R.; Koostra, B.K.; VanZant, E.S.; Taraba, J.L. Streambank erosion associated with grazing practices in the humid region. *Trans. ASAE* **2005**, *48*, 181–190. [CrossRef]
53. Amézketa, E.; del Valle de Lersundi, J. Soil classification and salinity mapping for determining restoration potential of cropped riparian areas. *Land Degrad. Dev.* **2008**, *19*, 153–164. [CrossRef]
54. Andrews, D.M.; Barton, C.D.; Kolka, R.K.; Rhoades, C.C.; Dattilo, A.J. Soil and Water Characteristics in Restored Canebrake and Forest Riparian Zones1. *JAWRA J. Am. Water Resour. Assoc.* **2011**, *47*, 772–784. [CrossRef]
55. Anstead, L.; Boar, R.R.; Tovey, N.K. The effectiveness of a soil bioengineering solution for river bank stabilisation during flood and drought conditions: Two case studies from East Anglia. *Area* **2012**, *44*, 479–488. [CrossRef]
56. Asghari, H.R.; Cavagnaro, T.R. Arbuscular mycorrhizas enhance plant interception of leached nutrients. *Funct. Plant Biol.* **2011**, *38*, 219–226. [CrossRef] [PubMed]

57. Atkinson, S.F.; Lake, M.C. Prioritizing riparian corridors for ecosystem restoration in urbanizing watersheds. *PeerJ* **2020**, *8*, e8174. [CrossRef]
58. Bariteau, L.; Bouchard, D.; Gagnon, G.; Levasseur, M.; Lapointe, S.; Bérubé, M. A riverbank erosion control method with environmental value. *Ecol. Eng.* **2013**, *58*, 384–392. [CrossRef]
59. Beauchamp, V.B.; Swan, C.M.; Szlavecz, K.; Hu, J. Riparian community structure and soil properties of restored urban streams. *Ecohydrology* **2015**, *8*, 880–895. [CrossRef]
60. Bedison, J.E.; Scatena, F.N.; Mead, J.V. Influences on the spatial pattern of soil carbon and nitrogen in forested and non-forested riparian zones in the Atlantic Coastal Plain of the Delaware River Basin. *For. Ecol. Manag.* **2013**, *302*, 200–209. [CrossRef]
61. Bissels, S.; Hölzel, N.; Donath, T.W.; Otte, A. Evaluation of restoration success in alluvial grasslands under contrasting flooding regimes. *Biol. Conserv.* **2004**, *118*, 641–650. [CrossRef]
62. Botero-Acosta, A.; Chu, M.L.; Guzman, J.A.; Starks, P.J.; Moriasi, D.N. Riparian erosion vulnerability model based on environmental features. *J. Environ. Manag.* **2017**, *203*, 592–602. [CrossRef] [PubMed]
63. Brovelli, A.; Batlle-Aguilar, J.; Barry, D.A. Analysis of carbon and nitrogen dynamics in riparian soils: Model development. *Sci. Total. Environ.* **2012**, *429*, 231–245. [CrossRef] [PubMed]
64. Buchanan, B.P.; Walter, M.T.; Nagle, G.N.; Schneider, R.L. Monitoring and assessment of a river restoration project in central New York. *River Res. Appl.* **2010**, *28*, 216–233. [CrossRef]
65. Buzhdygan, O.Y.; Rudenko, S.S.; Kazanci, C.; Patten, B.C. Effect of invasive black locust (Robinia pseudoacacia L.) on nitrogen cycle in floodplain ecosystem. *Ecol. Model.* **2016**, *319*, 170–177. [CrossRef]
66. Cabezas, Á.; Comín, F.A. Carbon and nitrogen accretion in the topsoil of the Middle Ebro River Floodplains (NE Spain): Implications for their ecological restoration. *Ecol. Eng.* **2010**, *36*, 640–652. [CrossRef]
67. Das, U.K. A Case Study on Performance of Jia Bharali River Bank Protection Measure Using Geotextile Bags. *Int. J. Geosynth. Ground Eng.* **2016**, *2*, 2. [CrossRef]
68. Davis, C.A.; Austin, J.E.; Buhl, D.A. Factors influencing soil invertebrate communities in riparian grasslands of the central platte river floodplain. *Wetlands* **2006**, *26*, 438–454. [CrossRef]
69. De Mello, K.; Randhir, T.O.; Valente, R.A.; Vettorazzi, C.A. Riparian restoration for protecting water quality in tropical agricultural watersheds. *Ecol. Eng.* **2017**, *108*, 514–524. [CrossRef]
70. Del Tánago, M.G.; De Jalón, D.G. Attributes for assessing the environmental quality of riparian zones. *Limnetica* **2006**, *25*, 389–402.
71. Dhondt, K.; Boeckx, P.; Verhoest, N.E.C.; Hofman, G.; Van Cleemput, O. Assessment of Temporal and Spatial Variation of Nitrate Removal in Riparian Zones. *Environ. Monit. Assess.* **2006**, *116*, 197–215. [CrossRef]
72. Duong, T.T.; Komine, H.; Do, M.D.; Murakami, S. Riverbank stability assessment under flooding conditions in the Red River of Hanoi, Vietnam. *Comput. Geotech.* **2014**, *61*, 178–189. [CrossRef]
73. Duró, G.; Crosato, A.; Kleinhans, M.G.; Winkels, T.G.; Woolderink, H.A.G.; Uijttewaal, W.S.J. Distinct patterns of bank erosion in a navigable regulated river. *Earth Surf. Process. Landf.* **2019**, *45*, 361–374. [CrossRef]
74. Dybala, K.E.; Steger, K.; Walsh, R.G.; Smart, D.R.; Gardali, T.; Seavy, N.E. Optimizing carbon storage and biodiversity co-benefits in reforested riparian zones. *J. Appl. Ecol.* **2019**, *56*, 343–353. [CrossRef]
75. Fernandes, L.F.S.; Pinto, A.A.; Terêncio, D.P.; Pacheco, F.A.L.; Cortes, R.M. Combination of Ecological Engineering Procedures Applied to Morphological Stabilization of Estuarine Banks after Dredging. *Water* **2020**, *12*, 391. [CrossRef]
76. Fournier, B.; Gillet, F.; Le Bayon, R.-C.; Mitchell, E.A.D.; Moretti, M. Functional responses of multitaxa communities to disturbance and stress gradients in a restored floodplain. *J. Appl. Ecol.* **2015**, *52*, 1364–1373. [CrossRef]
77. Franklin, H.M.; Carroll, A.R.; Chen, C.; Maxwell, P.; Burford, M.A. Plant source and soil interact to determine characteristics of dissolved organic matter leached into waterways from riparian leaf litter. *Sci. Total Environ.* **2020**, *703*, 134530. [CrossRef]
78. Gageler, R.; Bonner, M.; Kirchhof, G.; Amos, M.; Robinson, N.; Schmidt, S.; Shoo, L.P. Early Response of Soil Properties and Function to Riparian Rainforest Restoration. *PLoS ONE* **2014**, *9*, e104198. [CrossRef]
79. Garvin, E.M.; Bridge, C.F.; Garvin, M.S. Screening Level Assessment of Metal Concentrations in Streambed Sediments and Floodplain Soils within the Grand Lake Watershed in Northeastern Oklahoma, USA. *Arch. Environ. Contam. Toxicol.* **2017**, *72*, 349–363. [CrossRef]
80. Giese, L.A.; Aust, W.M.; Trettin, C.C.; Kolka, R.K. Spatial and temporal patterns of carbon storage and species richness in three South Carolina coastal plain riparian forests. *Ecol. Eng.* **2000**, *15*, S157–S170. [CrossRef]
81. Gold, A.J.; Groffman, P.M.; Addy, K.; Kellogg, D.Q.; Stolt, M.; Rosenblatt, A.E. Landscape attributes as controls on groithd water nitrate removal capacity of riparian zones. *JAWRA J. Am. Water Resour. Assoc.* **2001**, *37*, 1457–1464. [CrossRef]
82. Gumiero, B.; Boz, B. How to stop nitrogen leaking from a Cross compliant buffer strip? *Ecol. Eng.* **2017**, *103*, 446–454. [CrossRef]
83. Guo, F.; Sun, L.; Hu, X.; Luo, Q. Correlation analysis of OCPs (organo-chlorine pesticides) and microbial community diversity in the riparian zone of Liaohe River Conservation Area. *Fresenius Environ. Bull.* **2018**, *27*, 6844–6852.
84. Hale, R.; Reich, P.; Daniel, T.; Lake, P.S.; Cavagnaro, T.R. Assessing changes in structural vegetation and soil properties following riparian restoration. *Agric. Ecosyst. Environ.* **2018**, *252*, 22–29. [CrossRef]
85. Hale, R.; Reich, P.; Daniel, T.; Lake, P.S.; Cavagnaro, T.R. Scales that matter: Guiding effective monitoring of soil properties in restored riparian zones. *Geoderma* **2014**, *228*, 173–181. [CrossRef]
86. Harrison, M.D.; Groffman, P.M.; Mayer, P.M.; Kaushal, S.S.; Newcomer, T.A. Denitrification in Alluvial Wetlands in an Urban Landscape. *J. Environ. Qual.* **2011**, *40*, 634–646. [CrossRef]

87. Hasselquist, E.M.; Hasselquist, N.J.; Sparks, J.P.; Nilsson, C. Recovery of nitrogen cycling in riparian zones after stream restoration using δ 15N along a 25-year chronosequence in northern Sweden. *Plant Soil* **2017**, *410*, 423–436. [CrossRef]
88. Higgisson, W.P.; Downey, P.O.; Dyer, F.J. Changes in Vegetation and Geomorphological Condition 10 Years after Riparian Restoration. *Water* **2019**, *11*, 1252. [CrossRef]
89. Jansen, A.; Robertson, A.I. Relationships between livestock management and the ecological condition of riparian habitats along an Australian floodplain river. *J. Appl. Ecol.* **2001**, *38*, 63–75.
90. Janssen, P.; Cavaillé, P.; Bray, F.; Evette, A. Soil bioengineering techniques enhance riparian habitat quality and multi-taxonomic diversity in the foothills of the Alps and Jura Mountains. *Ecol. Eng.* **2019**, *133*, 1–9. [CrossRef]
91. Juracek, K.E.; Drake, K.D. Mining-Related Sediment and Soil Contamination in a Large Superfund Site: Characterization, Habitat Implications, and Remediation. *Environ. Manag.* **2016**, *58*, 721–740. [CrossRef]
92. Korol, A.R.; Noe, G.B.; Ahn, C. Controls of the spatial variability of denitrification potential in nontidal floodplains of the Chesapeake Bay watershed, USA. *Geoderma* **2019**, *338*, 14–29. [CrossRef]
93. Langendoen, E.J.; Lowrance, R.R.; Simon, A. Assessing the impact of riparian processes on streambank stability. *Ecohydrology* **2009**, *2*, 360–369. [CrossRef]
94. Larsen, E.W.; Greco, S.E. Modeling Channel Management Impacts on River Migration: A Case Study of Woodson Bridge State Recreation Area, Sacramento River, California, USA. *Environ. Manag.* **2002**, *30*, 209–224. [CrossRef] [PubMed]
95. Lee, E.-J.; Choi, K.-S.; Kim, T.-G. Estimation of the Pollutant Removal Efficiency in a Buffer Strip Using a SWAT Model. *Environ. Eng. Res.* **2011**, *16*, 61–67. [CrossRef]
96. Li, X.; Zhang, L.; Zhang, Z. Soil bioengineering and the ecological restoration of riverbanks at the Airport Town, Shanghai, China. *Ecol. Eng.* **2006**, *26*, 304–314. [CrossRef]
97. Lindow, N.; Fox, G.A.; Evans, R.O. Seepage erosion in layered stream bank material. *Earth Surf. Process. Landf.* **2009**, *34*, 1693–1701. [CrossRef]
98. Maffra, C.; Sutili, F. The use of soil bioengineering to overcome erosion problems in a pipeline river crossing in South America. *Innov. Infrastruct. Solut.* **2020**, *5*, 1–8. [CrossRef]
99. Maroto, R.; Robredo, J.C.; García, J.L.; Giménez, M.; Tardío, G. Eresma river slope: Stabilization and restoration project (Coca, Segovia, Spain). *Procedia Environ. Sci. Eng. Manag.* **2017**, *4*, 245–254.
100. Márquez, C.O.; García, V.J.; Schultz, R.C.; Isenhart, T.M. Assessment of Soil Aggradation through Soil Aggregation and Particulate Organic Matter by Riparian Switchgrass Buffers. *Agronomy* **2017**, *7*, 76. [CrossRef]
101. Matheson, F.E.; Nguyen, M.L.; Cooper, A.B.; Burt, T.P.; Bull, D.C. Fate of 15N-nitrate in unplanted, planted and harvested riparian wetland soil microcosms. *Ecol. Eng.* **2002**, *19*, 249–264. [CrossRef]
102. Meals, D.; Hopkins, R. Phosphorus reductions following riparian restoration in two agricultural watersheds in Vermont, USA. *Water Sci. Technol.* **2002**, *45*, 51–60. [CrossRef] [PubMed]
103. Meynendonckx, J.; Heuvelmans, G.; Muys, B.; Feyen, J. Effects of watershed and riparian zone characteristics on nutrient concentrations in the River Scheldt Basin. *Hydrol. Earth Syst. Sci.* **2006**, *10*, 913–922. [CrossRef]
104. Neilen, A.D.; Chen, C.R.; Parker, B.M.; Faggotter, S.J.; Burford, M.A. Differences in nitrate and phosphorus export between wooded and grassed riparian zones from farmland to receiving waterways under varying rainfall conditions. *Sci. Total. Environ.* **2017**, *598*, 188–197. [CrossRef] [PubMed]
105. Orr, C.H.; Stanley, E.H.; Wilson, K.A.; Finlay, J.C. Effects of restoration and reflooding on soil denitrification in a leveed midwestern floodplain. *Ecol. Appl.* **2007**, *17*, 2365–2376. [CrossRef] [PubMed]
106. Peter, S.; Rechsteiner, R.; Lehmann, M.F.; Brankatschk, R.; Vogt, T.; Diem, S.; Wehrli, B.; Tockner, K.; Durisch-Kaiser, E. Nitrate removal in a restored riparian groundwater system: Functioning and importance of individual riparian zones. *Biogeosciences* **2012**, *9*, 4295–4307. [CrossRef]
107. Petrone, A.; Preti, F. Soil bioengineering for risk mitigation and environmental restoration in a humid tropical area. *Hydrol. Earth Syst. Sci.* **2010**, *14*, 239–250. [CrossRef]
108. Pinto, A.; Fernandes, L.F.S.; Maia, R. Monitoring Methodology of Interventions for Riverbanks Stabilization: Assessment of Technical Solutions Performance. *Water Resour. Manag.* **2016**, *30*, 5281–5298. [CrossRef]
109. Rahe, N.H.; Williard, K.W.; Schoonover, J.E. Restoration of Riparian Buffer Function in Reclaimed Surface Mine Soils. *JAWRA J. Am. Water Resour. Assoc.* **2015**, *51*, 898–909. [CrossRef]
110. Rassam, D.W.; Pagendam, D. Development and application of the Riparian Mapping Tool to identify priority rehabilitation areas for nitrogen removal in the Tully-Murray basin, Queensland, Australia. *Mar. Freshw. Res.* **2009**, *60*, 1165–1175. [CrossRef]
111. Reisinger, A.J.; Blair, J.M.; Rice, C.W.; Dodds, W.K. Woody Vegetation Removal Stimulates Riparian and Benthic Denitrification in Tallgrass Prairie. *Ecosystems* **2013**, *16*, 547–560. [CrossRef]
112. Remo, J.W.F.; Guida, R.J.; Secchi, S. Screening the Suitability of Levee Protected Areas for Strategic Floodplain Reconnection Along the LaGrange Segment of the Illinois River, USA. *River Res. Appl.* **2016**, *33*, 863–878. [CrossRef]
113. Rheinhardt, R.D.; Brinson, M.; Meyer, G.; Miller, K. Integrating forest biomass and distance from channel to develop an indicator of riparian condition. *Ecol. Indic.* **2012**, *23*, 46–55. [CrossRef]
114. Rimondi, V.; Costagliola, P.; Lattanzi, P.; Morelli, G.; Cara, G.; Cencetti, C.; Fagotti, C.; Fredduzzi, A.; Marchetti, G.; Sconocchia, A.; et al. A 200 km-long mercury contamination of the Paglia and Tiber floodplain: Monitoring results and implications for environmental management. *Environ. Pollut.* **2019**, *255*, 113191. [CrossRef]

115. Rosenblatt, A.E.; Gold, A.J.; Stolt, M.H.; Groffman, P.M.; Kellogg, D.Q. Identifying Riparian Sinks for Watershed Nitrate using Soil Surveys. *J. Environ. Qual.* **2001**, *30*, 1596–1604. [CrossRef] [PubMed]
116. Rosenfeld, J.S.; Hogan, D.; Palm, D.; Lundquist, H.; Nilsson, C.; Beechie, T.J. Contrasting Landscape Influences on Sediment Supply and Stream Restoration Priorities in Northern Fennoscandia (Sweden and Finland) and Coastal British Columbia. *Environ. Manag.* **2010**, *47*, 28–39. [CrossRef]
117. Saad, S.I.; Da Silva, J.M.; Silva, M.L.N.; Guimarães, J.L.B.; Júnior, W.C.S.; Figueiredo, R.D.O.; Da Rocha, H.R. Analyzing ecological restoration strategies for water and soil conservation. *PLoS ONE* **2018**, *13*, e0192325. [CrossRef]
118. Samaritani, E.; Shrestha, J.N.B.; Fournier, B.; Frossard, E.; Gillet, F.; Guenat, C.; Niklaus, P.A.; Pasquale, N.; Tockner, K.; Mitchell, E.A.D.; et al. Heterogeneity of soil carbon pools and fluxes in a channelized and a restored floodplain section (Thur River, Switzerland). *Hydrol. Earth Syst. Sci.* **2011**, *15*, 1757–1769. [CrossRef]
119. Sgouridis, F.; Heppell, C.M.; Wharton, G.; Lansdown, K.; Trimmer, M. Denitrification and dissimilatory nitrate reduction to ammonium (DNRA) in a temperate re-connected floodplain. *Water Res.* **2011**, *45*, 4909–4922. [CrossRef]
120. Shah, J.J.F.; Harner, M.J.; Tibbets, T.M. Elaeagnus angustifolia Elevates Soil Inorganic Nitrogen Pools in Riparian Ecosystems. *Ecosystems* **2010**, *13*, 46–61. [CrossRef]
121. Silk, W.K.; Bambic, D.G.; O'Dell, R.E.; Green, P.G. Seasonal and spatial patterns of metals at a restored copper mine site II. Copper in riparian soils and Bromus carinatus shoots. *Environ. Pollut.* **2006**, *144*, 783–789. [CrossRef]
122. Tang, Y.; Van Kempen, M.M.; Van Der Heide, T.; Manschot, J.J.; Roelofs, J.G.; Lamers, L.P.M.; Smolders, A.J.P. A tool for easily predicting short-term phosphorus mobilization from flooded soils. *Ecol. Eng.* **2016**, *94*, 1–6. [CrossRef]
123. Tererai, F.; Gaertner, M.; Jacobs, S.M.; Richardson, D.M. Eucalyptus Camaldulensis Invasion in Riparian Zones Reveals Few Significant Effects on Soil Physico-Chemical Properties. *River Res. Appl.* **2015**, *31*, 590–601. [CrossRef]
124. Theriot, J.M.; Conkle, J.L.; Pezeshki, S.R.; Delaune, R.D.; White, J.R. Will hydrologic restoration of Mississippi River riparian wetlands improve their critical biogeochemical functions? *Ecol. Eng.* **2013**, *60*, 192–198. [CrossRef]
125. Tian, G.L.; Vose, J.M.; Coleman, D.C.; Geron, C.D.; Walker, J.T. Evaluation of the effectiveness of riparian zone restoration in the southern Appalachians by assessing soil microbial populations. *Appl. Soil Ecol.* **2004**, *26*, 63–68. [CrossRef]
126. Tomer, M.D.; Boomer, K.M.B.; Porter, S.A.; Gelder, B.K.; James, D.E.; McLellan, E. Agricultural Conservation Planning Framework: 2. Classification of Riparian Buffer Design Types with Application to Assess and Map Stream Corridors. *J. Environ. Qual.* **2015**, *44*, 768–779. [CrossRef] [PubMed]
127. Vandecasteele, B.; Quataert, P.; De Vos, B.; Tack, F.M.G. Assessment of the pollution status of alluvial plains: A case study for the dredged sediment-derived soils along the Leie River. *Arch. Environ. Contam. Toxicol.* **2004**, *47*, 14–22. [CrossRef]
128. Walker, J.; Geron, C.D.; Vose, J.M.; Swank, W.T. Nitrogen trace gas emissions from a riparian ecosystem in southern Appalachia. *Chemosphere* **2002**, *49*, 1389–1398. [CrossRef]
129. Walker, J.; Vose, J.M.; Knoepp, J.; Geron, C.D. Recovery of Nitrogen Pools and Processes in Degraded Riparian Zones in the Southern Appalachians. *J. Environ. Qual.* **2009**, *38*, 1391–1399. [CrossRef] [PubMed]
130. Wang, J.; Wang, D.M.; Wang, B. Effects of hydrological environmental gradient on soil and microbial properties in Lijiang riparian zones of China. *Fresenius Environ. Bull.* **2019**, *28*, 1297–1307.
131. Wang, Z.; Wang, Z.; Pei, Y. Nitrogen removal and microbial communities in a three-stage system simulating a riparian environment. *Bioprocess Biosyst. Eng.* **2013**, *37*, 1105–1114. [CrossRef]
132. Weller, D.E.; Baker, M.E. Cropland Riparian Buffers throughout Chesapeake Bay Watershed: Spatial Patterns and Effects on Nitrate Loads Delivered to Streams. *JAWRA J. Am. Water Resour. Assoc.* **2014**, *50*, 696–712. [CrossRef]
133. Welsh, M.K.; McMillan, S.K.; Vidon, P.G. Denitrification along the Stream-Riparian Continuum in Restored and Unrestored Agricultural Streams. *J. Environ. Manag.* **2017**, *46*, 1010–1019. [CrossRef] [PubMed]
134. Welsh, M.K.; Vidon, P.G.; McMillan, S.K. Changes in riparian hydrology and biogeochemistry following storm events at a restored agricultural stream. *Environ. Sci. Process. Impacts* **2019**, *21*, 677–691. [CrossRef] [PubMed]
135. Xiong, Z.; Li, S.; Yao, L.; Liu, G.; Zhang, Q.; Liu, W. Topography and land use effects on spatial variability of soil denitrification and related soil properties in riparian wetlands. *Ecol. Eng.* **2015**, *83*, 437–443. [CrossRef]
136. Ye, C.; Butler, O.M.; Du, M.; Liu, W.Z.; Zhang, Q.F. Spatio-temporal dynamics, drivers and potential sources of heavy metal pollution in riparian soils along a 600 kilometre stream gradient in Central China. *Sci. Total. Environ.* **2019**, *651*, 1935–1945. [CrossRef]
137. Young, E.O.; Ross, D.S.; Cade-Menun, B.J.; Liu, C.W. Phosphorus Speciation in Riparian Soils: A Phosphorus-31 Nuclear Magnetic Resonance Spectroscopy and Enzyme Hydrolysis Study. *Soil Sci. Soc. Am. J.* **2013**, *77*, 1636–1647. [CrossRef]
138. Zaimes, G.N.; Schultz, R.C.; Isenhart, T.M. Riparian land uses and precipitation influences on stream bank erosion in central Iowa. *JAWRA J. Am. Water Resour. Assoc.* **2006**, *42*, 83–97. [CrossRef]
139. Zhao, P.; Xia, B.; Hu, Y.; Yang, Y. A spatial multi-criteria planning scheme for evaluating riparian buffer restoration priorities. *Ecol. Eng.* **2013**, *54*, 155–164. [CrossRef]

land

MDPI

Article

Soil Diversity (Pedodiversity) and Ecosystem Services

Elena A. Mikhailova [1,*], Hamdi A. Zurqani [1,2], Christopher J. Post [1], Mark A. Schlautman [3] and Gregory C. Post [4]

1 Department of Forestry and Environmental Conservation, Clemson University, Clemson, SC 29634, USA; hzurqan@clemson.edu (H.A.Z.); cpost@clemson.edu (C.J.P.)
2 Department of Soil and Water Sciences, University of Tripoli, Tripoli 13538, Libya
3 Department of Environmental Engineering and Earth Sciences, Clemson University, Anderson, SC 29625, USA; mschlau@clemson.edu
4 Economics Department, Reed College, Portland, OR 97202, USA; grpost@reed.edu
* Correspondence: eleanam@clemson.edu

Abstract: Soil ecosystem services (ES) (e.g., provisioning, regulation/maintenance, and cultural) and ecosystem disservices (ED) are dependent on soil diversity/pedodiversity (variability of soils), which needs to be accounted for in the economic analysis and business decision-making. The concept of pedodiversity (biotic + abiotic) is highly complex and can be broadly interpreted because it is formed from the interaction of atmospheric diversity (abiotic + biotic), biodiversity (biotic), hydrodiversity (abiotic + biotic), and lithodiversity (abiotic) within ecosphere and anthroposphere. Pedodiversity is influenced by intrinsic (within the soil) and extrinsic (outside soil) factors, which are also relevant to ES/ED. Pedodiversity concepts and measures may need to be adapted to the ES framework and business applications. Currently, there are four main approaches to analyze pedodiversity: taxonomic (diversity of soil classes), genetic (diversity of genetic horizons), parametric (diversity of soil properties), and functional (soil behavior under different uses). The objective of this article is to illustrate the application of pedodiversity concepts and measures to value ES/ED with examples based on the contiguous United States (U.S.), its administrative units, and the systems of soil classification (e.g., U.S. Department of Agriculture (USDA) Soil Taxonomy, Soil Survey Geographic (SSURGO) Database). This study is based on a combination of original research and literature review examples. Taxonomic pedodiversity in the contiguous U.S. exhibits high soil diversity, with 11 soil orders, 65 suborders, 317 great groups, 2026 subgroups, and 19,602 series. The ranking of "soil order abundance" (area of each soil order within the U.S.) expressed as the proportion of the total area is: (1) Mollisols (27%), (2) Alfisols (17%), (3) Entisols (14%), (4) Inceptisols and Aridisols (11% each), (5) Spodosols (3%), (6) Vertisols (2%), and (7) Histosols and Andisols (1% each). Taxonomic, genetic, parametric, and functional pedodiversity are an essential context for analyzing, interpreting, and reporting ES/ED within the ES framework. Although each approach can be used separately, three of these approaches (genetic, parametric, and functional) fall within the "umbrella" of taxonomic pedodiversity, which separates soils based on properties important to potential use. Extrinsic factors play a major role in pedodiversity and should be accounted for in ES/ED valuation based on various databases (e.g., National Atmospheric Deposition Program (NADP) databases). Pedodiversity is crucial in identifying soil capacity (pedocapacity) and "hotspots" of ES/ED as part of business decision making to provide more sustainable use of soil resources. Pedodiversity is not a static construct but is highly dynamic, and various human activities (e.g., agriculture, urbanization) can lead to soil degradation and even soil extinction.

Keywords: climate change; extinction; food; land use; market; pedocapacity; security; soil capacity

check for updates

Citation: Mikhailova, E.A.; Zurqani, H.A.; Post, C.J.; Schlautman, M.A.; Post, G.C. Soil Diversity (Pedodiversity) and Ecosystem Services. *Land* **2021**, *10*, 288. https://doi.org/10.3390/land10030288

Academic Editors: Chiara Piccini and Rosa Francaviglia

Received: 25 February 2021
Accepted: 9 March 2021
Published: 11 March 2021

1. Introduction

Soils are complex, dynamic bodies that form from interactions among the Earth's various spheres (atmosphere, biosphere, lithosphere, hydrosphere) within the ecosphere, which is modified by the anthroposphere (the sphere of human influence) (Figure 1a).

The uniqueness of soils is that they are not discrete entities; but instead, soils form a continuum (pedosphere), which varies both with depth and horizontal distance [1,2]. The concept and measures of soil diversity/pedodiversity (variability of soils) are highly complicated because pedodiversity (biotic + abiotic) results from atmospheric diversity (abiotic + biotic), biodiversity (biotic), hydrodiversity (abiotic + biotic), and lithodiversity (abiotic) within the ecosphere, which is modified by the anthroposphere (Figure 1b). According to Mattson, 1938 [3], soils can be a product of two- or three-sphere combinations; therefore, pedodiversity can be based on two- or three-sphere combinations as well (Figure 1b). A definition of pedodiversity from Odeh (1998) [4] is "variability of soil in a specific area or region, as determined by its constitution, types, attributes and the conditions under which the various types were formed." The concept of pedodiversity (biotic + abiotic) can be widely interpreted based on a range of definitions, depending on the type of pedodiversity (e.g., taxonomic, genetic, parametric, and functional) [5] (Table 1). Pedodiversity is influenced by intrinsic (within pedodiversity itself) and extrinsic factors (environmental factors from atmosphere, biosphere, lithosphere, the hydrosphere, ecosphere, and anthroposphere that control and influence pedogenesis) (Figure 1) [6].

Previous studies have examined the concept and measures of pedodiversity from a pedological point of view and concluded its importance for the sustainable use of soil resources [7–11]. Ibáñez et al. (1995) [11] examined pedodiversity from an ecological point of view and concluded that "patterns of biodiversity, geomorphological diversity and pedodiversity have great similarities, suggesting that there are universal regularities common to the organization of biotic and abiotic ecological structures."

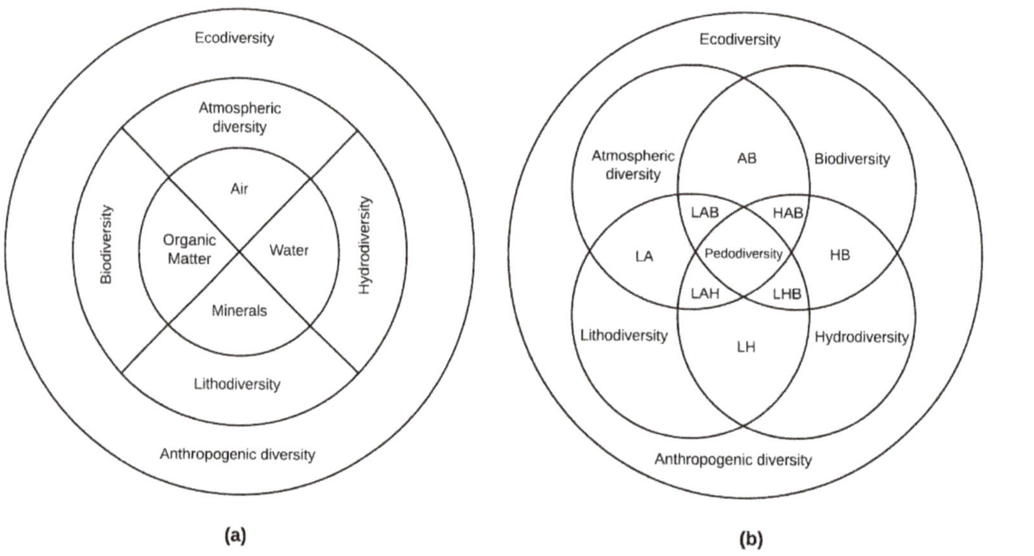

Figure 1. The scope of soil diversity (pedodiversity) (biotic + abiotic): (**a**) soil and relationship between soil components from the Earth's various diverse spheres; (**b**) formation of two-sphere, three-sphere, and four-sphere (e.g., pedodiversity) systems in nature (**A** = atmospheric diversity (abiotic + biotic); **B** = biodiversity (biotic); **H** = hydrodiversity (abiotic + biotic); **L** = lithodiversity (abiotic) (adapted from Mattson, 1938 [3]; Mikhailova et al., 2020 [12]; Ibáñez et al., 1998 [13]). Pedodiversity is influenced by intrinsic (within soil itself) and extrinsic (environmental) factors (e.g., atmospheric deposition).

Guo et al. (2003) [14] reported the taxonomic structure, spatial distribution, and relative abundance of soils in the contiguous United States (U.S.) by order (11), suborder (52), great groups (232), subgroups (1175), family (6226), and series (13,129) using the STATSGO (1997 version) database. Amundson et al. (2003) [2] examined "natural" soil diversity and land use in the U.S. based on U.S. Department of Agriculture (USDA) Soil Taxonomy [15], the State Soil Geographic (STATSGO) dataset (1997), and two numerical parameters: series density and series abundance. The same study quantified rare, unique, and endangered soils in the U.S. and listed directions for future research, including establishing the societal value of undisturbed soils, monitoring changes in pedodiversity, and focusing conservation efforts on soil diversity "hotspots" [2].

Table 1. Examples of soil diversity (pedodiversity) types.

Types of Soil Diversity (Pedodiversity)	Examples
Taxonomic (diversity of soil classes)	USDA Soil Taxonomy (e.g., soil order, series)
Genetic (diversity of genetic horizons)	A, B, etc.
Parametric (diversity of soil properties)	Soil organic matter (SOM), calcium carbonate ($CaCO_3$), etc.
Functional (soil behavior under different use)	Interpretive models to predict soil behavior

Soil and its diversity (pedodiversity) play significant roles in underlying ecosystem goods and services for humans [16–18], who have developed a human-centered ecosystem services framework [19] as an approach for valuing these goods and services in both economic and non-economic ways [20]. In fact, pedodiversity can be considered an ecosystem good and service in its own right [2]. According to Bartkowski (2017) [21], economic valuations of diversity are rare and often focus primarily on biodiversity [22,23]. Previous research on biodiversity and its significance lists the following benefits [24]: (1) biodiversity supports healthy ecosystems by increasing ecosystem stability, while the loss of biodiversity can reduce their function and efficiency; (2) the relationship between the loss of biodiversity and ecosystem function is not linear, with greater impact as the loss of biodiversity increases; (3) both variety of species and key individual species are critical for ecosystem functioning, with diversity across trophic levels potentially having a more important function compared to species within trophic levels. From a business point of view, Stephenson (2012) [25] describes the utility of biodiversity to: (1) identify the stock, its physical state, and spatial patterns of biodiversity in relation to the key ecosystem services (e.g., water and carbon sequestration), which are at risk and have a high value (e.g., social, economic); (2) assess biodiversity trends, high-risk biodiversity loss with its key drivers as well as a reference point against which progress can be measured; (3) develop a long-term coordinated vision assessing trade-offs and potential synergies including cost-benefit analyses; (4) identify and implement a cost-effective policy option; and (5) monitor progress towards objectives and reviewing and revising policies over time based on the progress.

Ecosystem services (ES) are goods and services provided by functioning ecological systems that directly and/or indirectly benefit human populations (e.g., food and climate regulation) [16–18]. At the same time, however, functioning ecological systems also can present detrimental effects for humans or so-called ecosystem disservices (ED) (e.g., social cost of carbon dioxide) [12]. Adhikari and Hartemink (2016) [16] examined the link between soil properties and ES without including the concept of pedodiversity and its measures in their literature review. Chandler et al. (2018) [26] proposed integrating soil analyses within frameworks for ES and the organizational hierarchy of soil systems. Mikhailova et al. (2020) [12] pointed out that applications of ES to soils are narrowly defined (e.g., soil-based, pedosphere-based), treating soil as a closed system instead of an open system, which requires a soil systems-based approach to ES. Mikhailova et al. (2020) [12] suggested including the contributions of the Earth's spheres (atmosphere, biosphere, hydrosphere, lithosphere, ecosphere, and anthroposphere) in the economic analysis of soil ES. Because most soils have been modified by humans, Mikhailova et al. (2020) [12] examined the business side of ecosystem services of soil systems and proposed to use the term "soil systems goods and services" (SSGS) instead of "soil ecosystem goods and services." Applications of biodiversity concepts and their measures to pedodiversity can be problematic, because they have not been designed explicitly for pedodiversity and its associated ES/ED valuations. Pedodiversity concepts and measures in their current forms have not been considered in ES/ED valuations and business decision making. Most likely the types of pedodiversity and measurement approaches listed in Table 1 cannot be used solely on their own in ES/ED but must be applied in combination with each other or even all together concerning specific ES/ED within a particular administrative extent.

The objective of this study is to illustrate the application of pedodiversity concepts and measures to value ES/ED, with examples provided primarily from the contiguous U.S., its administrative units, and the USDA Soil Taxonomy system of soil classification. Although the focus of the examples is on the U.S., the applications and measures described should be readily applicable to other geographic areas and market economies.

2. Materials and Methods

2.1. Data Compilation and Analyses

Soil survey information (including soil orders, suborders, great groups, subgroups, families and series) was obtained from Soil Survey Geographic (SSURGO) Database (2020) [27]. The information for each state in the contiguous U.S. was extracted using Zonal Statistics (Tables) spatial analyst tool in ArcGIS® Pro 2.6 (ESRI, Redlands, CA, USA), while the information for the regions and the Land Resource Regions (LRR) was computed by developing a Structured Query Language (SQL) code that was utilized in SSURGO webpage (https://sdmdataaccess.nrcs.usda.gov/, accessed on 10 October 2020). All this information was then used to create a Microsoft Excel file with the soil survey information for each boundary. Examples of soil ES/ED and their monetary valuations were obtained from various literature sources using the Web of Science [28]. These examples encompass the three major groups of ES commonly used in the literature: provisioning, regulation/maintenance, and cultural [29].

2.2. The Accounting Framework

Table 2 provides a conceptual overview of the accounting framework for market and non-market valuation of benefits/damages from three groups of ES (provisioning, regulation/maintenance, and cultural) based on biophysical and administrative accounts with examples primarily from the U.S. and its soils, as well as the related market-based information obtained from U.S. sources.

Table 2. A conceptual overview of the accounting framework for a systems-based approach in the ecosystem services (ES) valuation of various soil ecosystem goods and services based on soil diversity (pedodiversity) (adapted from Groshans et al., 2018 [30]).

STOCKS		FLOWS		VALUE
Biophysical Accounts (Science-Based)	**Administrative Accounts (Boundary-Based)**	**Monetary Accounts**	**Benefits/Damages**	**Total Value**
Soil extent:	Administrative extent:	Ecosystem good(s) and service(s):	Sector:	Types of value:
Examples of valuations based on soil diversity (pedodiversity)				
Examples of valuations based on the interaction of soil diversity (pedodiversity) and the Earth's spheres				
Soil diversity (pedodiversity), organizational hierarchy of soil systems	Administrative, organizational hierarchy	Provisioning, regulation/ maintenance and cultural	Environment, agriculture, industry, etc.	Market and non-market valuations

2.3. *The Total Economic Value (TEV) Framework with Insurance Value*

Table 3 provides a conceptual overview of the total economic value (TEV) framework with insurance value adapted from various sources to provide a general explanation of valuation methods used in the examples primarily from the U.S., which may apply to other market economies. It should be noted that the relevance and applications of economic valuation to soil systems are not always clearly defined and can be subject to interpretation.

Table 3. The total economic value (TEV) framework with insurance value (adapted from Nimmo-Bell (2011) [31], NZIER, 2018 [32], Baveye et al., 2016 [18], and Bartkowski et al., 2020 [20]).

<table>
<tr><td colspan="6">Total Economic Value (TEV)</td><td></td></tr>
<tr><td colspan="6">Instrumental Value
(Benefits to Humans)</td><td>Intrinsic Value
(Benefits to Nature)</td></tr>
<tr><td colspan="5">Use Values</td><td rowspan="2">Insurance Value</td><td rowspan="5">Unknown</td></tr>
<tr><td colspan="2">Actual Use Values</td><td colspan="3">Passive Use Values</td></tr>
<tr><td>Direct Use Value
(extractive and non-extractive uses)</td><td>Indirect Use Value
(functional benefits)</td><td>Altruistic Value
(for others)</td><td>Bequest Value
(for others)</td><td>Existence Value
(for life)</td><td>The amount available to replace lost value</td></tr>
<tr><td>Consumptive and non-consumptive (e.g., agriculture)</td><td>e.g., ecosystem services</td><td rowspan="2">e.g., preserving resource so others can use it now</td><td rowspan="2">e.g., preserving resource so others can use it in the future</td><td rowspan="2">e.g., resource preservation</td><td rowspan="2">e.g., buffering capacity</td></tr>
<tr><td colspan="2">Option Value</td></tr>
<tr><td colspan="7">—————— Decreasing Tangibility of Value to User ——————></td></tr>
</table>

3. Results

3.1. *Intrinsic Factors: Examples of Valuations Based on Pedodiversity and Ecosystem Services*

3.1.1. Examples of Taxonomic Pedodiversity and Ecosystem Services in the Contiguous U.S.

Pedodiversity is influenced by intrinsic (within the soil) factors, including taxonomic, genetic, parametric, and functional pedodiversity, which provide an important context for analyzing, interpreting, and reporting ES/ED within the ES framework. Although each approach can be used separately, three of these approaches (genetic, parametric, and functional) fall within the "umbrella" of taxonomic pedodiversity, which separates soils based on properties important for potential use. Pedodiversity in the U.S. can be quantified and valued within the framework of the USDA Soil Taxonomy (Soil Survey Staff, 1999 [15]), an international system of soil classification, with the purpose of organizing soils into groups with similar properties. Soil individual (a three-dimensional body) is the object of soil classification in Soil Taxonomy, which is based on a nested and hierarchical system with six taxonomic categories: order, suborder, great group, subgroup, family, and series (e.g., Cecil (Fine, kaolinitic, thermic Typic Kanhapludults) (Table 4). Table 4 demonstrates an example of criteria and sequence of taxonomic categories used to classify mineral soils (this sequence is different for organic soils and soils with permafrost). Soil Taxonomy is the basic system of soil classification for making and interpreting soil surveys, which can be used with the ES framework (Figure 2, Table 5).

Table 4. Example of criteria and sequence of taxonomic categories used to classify soils.

Taxonomic Category	Explanation	Example	Increase in Specificity
Order	Highest category, diagnostic horizons	Ultisols	↓
Suborder	The difference in moisture regimes	Udults	
Great Group	Presence of key horizons	Hapludults	
Subgroup	Proximity to "central concept"	Typic Kanhapludults	
Family	Particle-size classes and their substitutes	fine	
	Human-altered and human-transported material classes		
	Mineralogy classes	kaolinitic	
	Cation-exchange activity classes (CEC/% clay)		
	Calcareous and reaction classes		
	Soil temperature classes	thermic	
	Soil depth classes		
	Rupture-resistance classes		
	Classes of coatings on sands		
	Classes of permanent cracks		
Series	Smallest unit	Cecil	

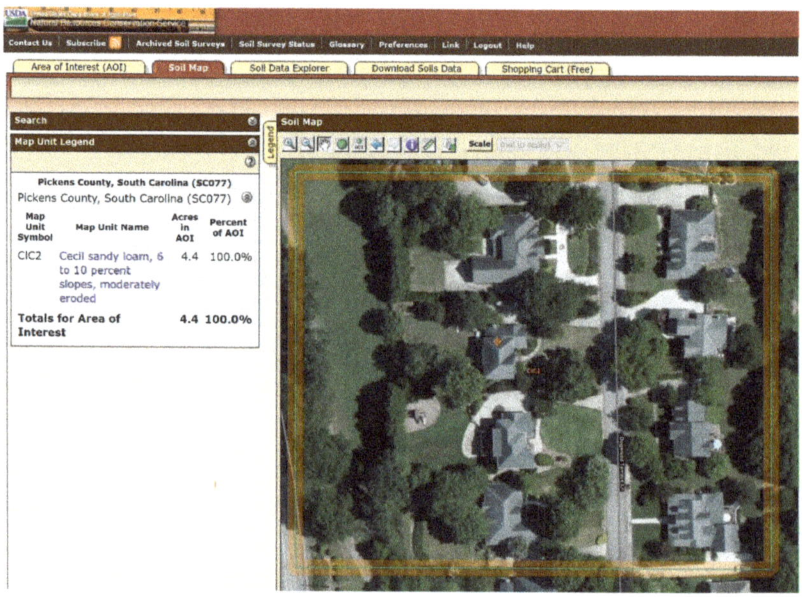

Figure 2. Example of a soil map generated with Web Soil Survey (WSS) [33] showing soil cover and land use.

Table 5. Example of different soil properties (e.g., physical soil properties) in the Web Soil Survey (WSS) [33].

Map Symbol and Soil Name	Depth	Sand	Silt	Clay	Organic Matter
	In	Pct	Pct	Pct	Pct
ClC2-Cecil sandy loam, 6 to 10 percent slopes, moderately eroded	0-5 5-54 54-80	59-70-75 15-30-40 35-40-50	10-20-35 6-16-26 17-27-40	5-10-15 35-54-59 20-33-34	0.5-0.5-1.0 0.0-0.1-0.5 0.0-0.1-0.5

Note: Three values are provided to identify the expected Low (L), Representative (R), and High (H).

Pedodiversity data in soil surveys and databases (e.g., maps, depth of soil horizons, and soil properties) are useful in business applications because they provide information about the pedodiversity of soil capital and the necessary data to calculate stocks of soil biotic (e.g., organic carbon) and abiotic (e.g., sand, silt, clay, and calcium carbonate) resources within different extents (e.g., science-based, administrative based, or in combination; by soil depth, by soil horizon, etc.). This information is essential for various ES/ED applications (e.g., provisioning and regulating) and even cultural ecosystem services. The names of some soil series used in the U.S. reflect cultural and historical heritage. For example, the name of New Mexico State Soil "Penistaja" is derived from the Navajo name meaning "forced to sit" [34].

The most general category of soil orders in Soil Taxonomy provides a useful framework and description of soil, which can be applied to describe the soil stock and its composition, its potential for delivering key ES, and constraints (ED) at several soil system scales, for example, world, continent, region, country, and watershed (Table 6). General characteristics and constraints of these soil orders provide both qualitative and quantitative measures regarding the ability of these soils to supply ES/ED within a geographic area.

Table 6. Soil diversity (pedodiversity) is expressed as taxonomic diversity at the level of soil order and ecosystem services types.

	Stocks	Ecosystem Services		
Soil Order	**General Characteristics and Constraints**	**Provisioning**	**Regulation/Maintenance**	**Cultural**
	Slight Weathering			
Entisols	Embryonic soils with ochric epipedon	x	x	x
Inceptisols	Young soils with ochric or umbric epipedon	x	x	x
Histosols	Organic soils with ≥20% of organic carbon	x	x	x
Gelisols	Frozen soils with permafrost	x	x	x
Andisols	Volcanic soils	x	x	x
	Intermediate Weathering			
Aridisols	Dry soils. Common in desert areas	x	x	x
Vertisols	Soils with swelling clays	x	x	x
Alfisols	Clay-enriched B horizon with B.S. ≥ 35%	x	x	x
Mollisols	Carbon-enriched soils with B.S. ≥ 50%	x	x	x
	Strong Weathering			
Spodosols	Coarse-textured soils with albic and spodic horizons	x	x	x
Ultisols	Highly leached soils with B.S. < 35%	x	x	x
Oxisols	Highly weathered soils rich in Fe and Al oxides	x	x	x

Note: B.S. = base saturation.

Taxonomic pedodiversity in the contiguous U.S. exhibits a wide range of soil diversity, with 11 soil orders, 65 suborders, 317 great groups, 2026 subgroups, and 19,602 series (Table 7). Table 7 shows the "soil order abundance"—total area of each soil order within the contiguous U.S. based on Soil Survey Geographic (SSURGO) Database (2020) with the following distribution: (1) Mollisols (27%), (2) Alfisols (17%), (3) Entisols (14%), (4) Inceptisols and Aridisols (11% each), (5) Spodosols (3%), (6) Vertisols (2%), and (7) Histosols and Andisols (1% each). In terms of the degree of weathering: slightly-weathered soils are 27%, intermediately-weathered soils are 58%, and strongly-weathered soils are 15% of the total area.

Information about taxonomic pedodiversity can be linked to various ES/ED. In terms of provisioning ES, 58% of the contiguous U.S. is occupied by soils with high and moderate fertility status (without taking into account the past and present land use). It also can be used to analyze the patterns of value for regulating ES. Mikhailova et al. (2019) [35] provided a valuation of soil organic carbon (SOC) stocks in the contiguous U.S. based on taxonomic pedodiversity and the avoided social cost of carbon (SC-CO_2) emissions, which varied by the degree of soil weathering as indicated by soil order information. This study found the following distribution of SC-CO_2 contribution within the contiguous U.S.: slightly-weathered soils (38%), intermediately-weathered soil (51%), and strongly-weathered soils (11%). In another example, according to Mikhailova et al. (2019) [36] Mollisols have the highest total soil carbon (TSC, soil organic + soil inorganic carbon) storage midpoint value ($7.78T) based on the social cost of carbon (SC-CO_2) and avoided emissions provided by carbon sequestration, which is about 30% of the total midpoint value for the contiguous U.S. These types of analyses are useful in identifying soil "hotspots" with regards to various ES/ED applications at different scales which has the potential to be managed with precision agriculture [2,37]. It can be concluded that taxonomic pedodiversity provides an important context for analyzing, summarizing, and presenting soil data for ES/ED applications.

Soil series is also a useful taxonomic category to describe pedodiversity regarding ES/ED at more detailed scales (e.g., farm and field), and this category is closely allied to interpretive uses (e.g., suitabilities and limitations for crop production and construction) (Table 7). Soil series consist of pedons that are grouped together based on similarity in pedogenesis, soil chemistry, and physical properties [38]. The number of soil series within the soil extent can describe its diversity (Table 7). According to Table 7, Mollisols have the highest number of soil series (5569), followed by Entisols (3700). Amundson et al. (2003) [2] proposed to apply a commonly used biodiversity parameter ("species density") to soil diversity, which they called a "series density" parameter (number of series divided by 100,000 ha) (Table 7).

Table 7. Soil diversity (pedodiversity) is expressed as the number of soil classes (taxonomic pedodiversity) within soil orders in the contiguous United States (U.S.) based on Soil Survey Geographic (SSURGO) Database (2020) [27].

Soil Order	Suborders	Great Groups	Subgroups	Series	Series Density
		Slight Weathering			
Entisols	25	56	246	3700	3.5
Inceptisols	26	67	386	3610	4.6
Histosols	7	25	73	334	3.1
Gelisols *	2	2	2	2	-
Andisols	13	26	90	642	9.3
		Intermediate Weathering			
Aridisols	17	44	283	2374	2.9
Vertisols	7	31	101	394	3.0
Alfisols	14	49	331	3242	2.5
Mollisols	23	55	422	5569	2.8
		Strong Weathering			
Spodosols	9	26	92	591	2.4
Ultisols	9	27	107	1091	1.3
Oxisols	-	-	-	-	-
Totals	**65**	**317**	**2026**	**19,602**	**2.7**

Note: * Soil order of Gelisols was reported for the state of Washington with an area of 11 m^2. Series density equals the number of series divided by 100,000 ha.

Taxonomic pedodiversity can also be used within administrative boundaries (e.g., Land Resource Regions, LRRs) (Table 8). Land Resource Regions (LRRs) are defined by the USDA using major land resource area (MLRA) and agricultural markets, which are denoted using capital letters (A, B, C, etc.; see Table 8 notes). The contiguous U.S. comprises 20 of the 28 LRRs. The LRRs with the highest number of soil orders are: (1) A—Northwestern Forest, Forage and Specialty Crop Region (11), and (2) E—Rocky Mountain Range and Forest Region (10). The LRRs with the highest number of series are: (1) D—Western Range and Irrigated Region (5739), (2) E—Rocky Mountain Range and Forest Region (3611), and (3) A—Northwestern Forest, Forage, and Specialty Crop Region (2065). The LRRs with the highest series density are: (1) A—Northwestern Forest, Forage and Specialty Crop Region (11.4), and (2) C—California Subtropical Fruit, Truck and Specialty Crop

Region (8.6), and (3) S—Northern Atlantic Slope Diversified Farming Region (7.2). According to Table 7, the average series density for the contiguous U.S. is 2.7 series/100,000 ha, and slightly-weathered soils have the highest series densities with Andisols in the lead (9.3 series/100,000 ha). Variation in soil series density can relate to ES/ED, but it depends on the properties of the soil series within an area and the interpretive uses. Soil ES related to agriculture can be reduced in some areas by high soil variability, which can impact the soil productivity at the farm scale. For example, in areas with soils derived from glacial materials, the high variability of soil properties can occur at the field scale limiting agricultural use and productivity.

Table 8. Soil diversity (pedodiversity) is expressed as the number of soil classes (taxonomic pedodiversity) within Land Resource Regions (LRRs) for the contiguous United States (U.S.) Soil Survey Geographic (SSURGO) Database (2020) [27].

LRRs	Orders	Suborders	Great Groups	Subgroups	Series	Series Density
A	11	53	159	567	2065	11.4
B	8	41	108	377	1482	5.7
C	9	38	107	294	1264	8.6
D	9	51	185	977	5739	4.5
E	10	52	165	783	3611	6.9
F	7	25	69	243	865	2.5
G	7	33	94	369	1957	3.8
H	8	27	69	270	1080	1.8
I	6	22	57	184	538	3.2
J	6	22	58	214	606	4.3
K	7	24	61	267	1265	4.2
L	6	19	53	185	819	6.8
M	8	31	89	352	1834	2.6
N	8	31	88	300	1700	2.8
O	7	17	43	128	346	3.7
P	8	27	88	316	1468	2.2
R	7	27	82	242	1321	4.4
S	8	23	66	192	712	7.2
T	9	29	93	295	854	3.7
U	8	22	50	127	279	3.3
Totals	**11**	**65**	**317**	**2026**	**19,602**	**2.7**

Note: A = Northwestern Forest, Forage and Specialty Crop Region; B = Northwestern Wheat and Range Region; C = California Subtropical Fruit, Truck and Specialty Crop Region; D = Western Range and Irrigated Region; E = Rocky Mountain Range and Forest Region; F = Northern Great Plains Spring Wheat Region; G = Western Great Plains Range and Irrigated Region; H = Central Great Plains Winter Wheat and Range Region; I = Southwest Plateaus and Plains Range and Cotton Region; J = Southwestern Prairies Cotton and Forage Region; K = Northern Lake States Forest and Forage Region; L = Lake States Fruit, Truck and Dairy Region; M = Central Feed Grains and Livestock Region; N = East and Central Farming and Forest Region; O = Mississippi Delta Cotton and Feed Grains Region; P = South Atlantic and Gulf Slope Cash Crops, Forest and Livestock Region; R = Northeastern Forage and Forest Region; S = Northern Atlantic Slope Diversified Farming Region; T = Atlantic and Gulf Cost Lowland Forest and Crop Region; U = Florida Subtropical Fruit, Truck Crop and Range Region. Series density equals the number of series divided by 100,000 ha.

Taxonomic pedodiversity within administrative boundaries (e.g., LRRs) can be broken down by soil orders with their corresponding areas for qualitative and quantitative assessments of soil stocks for ES/ED assessments (Table 9). For example, for the LRR A—Northwestern Forest, Forage and Specialty Crop Region, the area of soil orders were distributed as follows: Entisols (3%), Inceptisols (37%), Histosols (0%), Andisols (20%), Vertisols (1%), Alfisols (13%), Mollisols (11%), Aridisols (0%), Spodosols (5%), and Ultisols (2%). This type of analysis is useful in identifying hotspots (e.g., soils with high fertility status; soils with the high social cost of carbon ($SC\text{-}CO_2$) emissions) [37,39].

Table 9. Soil diversity (pedodiversity) by soil order (taxonomic pedodiversity) within Land Resource Regions (LRRs) for the contiguous United States (U.S.) Soil Survey Geographic (SSURGO) Database (2020) [27].

LRRs	**Slight <——— Degree of Weathering and Soil Development ———> Strong**									
	Slight Weathering				**Intermediate Weathering**				**Strong Weathering**	
	Entisols	**Inceptisols**	**Histosols**	**Andisols**	**Vertisols**	**Alfisols**	**Mollisols**	**Aridisols**	**Spodosols**	**Ultisols**
	Area (km^2)									
A	5517	58,562	756	31,792	869	20,490	17,235	21	7706	15,577
B	10,114	2118	75	735	536	1123	96,455	38,224	0	0
C	27,378	14,900	450	14	9701	32,638	35,314	5891	0	396
D	253,840	30,096	225	3286	10,548	44,608	173,838	439,983	0	4121
E	29,371	102,155	724	27,487	1825	58,240	171,044	13,110	124	0
F	39,138	12,568	916	0	14,337	12,880	277,240	3636	0	0
G	192,349	45,344	79	0	23,681	37,585	122,002	88,697	0	0
H	64,551	51,798	0.21	124	9249	83,914	332,943	12,012	0	0
I	3636	13,797	0	0	11,528	28,910	88,233	23,691	0	0
J	6432	9976	0	0	29,024	63,995	31,348	0	0	1058
K	35,700	22,652	47,791	0	64	86,599	21,525	0	56,948	0
L	13,173	14,287	5281	0	0	52,200	12,324	0	8804	0
M	43,269	32,020	4659	0	3295	256,429	365,036	0	14	2538
N	18,594	103,952	23	0	331	163,582	20,069	0	654	250,411
O	9761	19,274	703	0	28,771	29,086	2822	0	0	411
P	53,392	53,155	2556	0	10,452	115,700	2295	0	1251	385,496
R	14,067	130,799	10,428	0	0	25,480	638	0	97,131	2070
S	5883	31,348	617	0	1.50	17,149	315	0	325	36,382
T	27,168	12,076	18,587	0	14,965	35,179	8497	329	18,182	70,396
U	19,185	2085	9491	0	11	12,344	3476	0	23,952	4079
Totals	**872,518**	**762,962**	**103,361**	**63,438**	**169,189**	**1,178,131**	**1,782,649**	**625,594**	**215,091**	**772,935**

Note: A = Northwestern Forest, Forage and Specialty Crop Region; B = Northwestern Wheat and Range Region; C = California Subtropical Fruit, Truck and Specialty Crop Region; D = Western Range and Irrigated Region; E = Rocky Mountain Range and Forest Region; F = Northern Great Plains Spring Wheat Region; G = Western Great Plains Range and Irrigated Region; H = Central Great Plains Winter Wheat and Range Region; I = Southwest Plateaus and Plains Range and Cotton Region; J = Southwestern Prairies Cotton and Forage Region; K = Northern Lake States Forest and Forage Region; L = Lake States Fruit, Truck and Dairy Region; M = Central Feed Grains and Livestock Region; N = East and Central Farming and Forest Region; O = Mississippi Delta Cotton and Feed Grains Region; P = South Atlantic and Gulf Slope Cash Crops, Forest and Livestock Region; R = Northeastern Forage and Forest Region; S = Northern Atlantic Slope Diversified Farming Region; T = Atlantic and Gulf Cost Lowland Forest and Crop Region; U = Florida Subtropical Fruit, Truck Crop and Range Region.

For example, Mollisols have both high fertility status and a high potential for the social cost of carbon (SC-CO_2). Taxonomic pedodiversity within the boundaries of states and regions (Table 10) reveals that the states with the highest number of soil orders are: (1) Washington (11), (2) California and Oregon (10 each), and (3) Idaho and Montana (9 each). States with the highest number of soil series are: (1) California (2689), (2) Washington (1548), and (3) Idaho (1529). Table 11 provides a detailed distribution of various soil orders by state and region, with qualitative and quantitative assessments of soil stocks for ES/ED assessments. For example, for the state of Connecticut, the areas of soil orders are distributed as follows: Inceptisols (84%), Histosols (8%), Entisols (6%), and Mollisols (2%).

Table 10. Soil diversity (pedodiversity) expressed as the number of soil classes (taxonomic pedodiversity) within states (regions) in the contiguous United States (U.S.) based on Soil Survey Geographic (SSURGO) Database (2020) [27] with comparisons based on STATSGO (1997) [2].

State (Region)	Orders 2020 (1997)	Suborders	Great Groups	Subgroups	Series 2020 (1997)	Series Density 2020 (1997)
Connecticut	4	11	17	37	106 (86)	8.5 (6.7)
Delaware	6	13	23	46	70 (52)	13.9 (9.9)
Massachusetts	6 (5)	15	33	71	202 (129)	10.7 (6.2)
Maryland	7	19	47	105	287 (187)	11.4 (6.8)
Maine	5 (4)	16	36	79	231 (111)	2.9 (1.3)
New Hampshire	5 (4)	14	31	66	185 (127)	8.1 (5.3)
New Jersey	6 (7)	16	36	84	169 (148)	9.5 (7.5)
New York	7	25	64	168	823 (347)	6.9 (2.7)
Pennsylvania	7	17	46	109	391 (248)	3.4 (2.1)
Rhode Island	3	9	14	22	42 (45)	16.3 (15.9)
Vermont	6	18	39	102	231 (192)	9.7 (7.7)
West Virginia	7 (6)	17	40	88	250 (163)	4.1 (2.6)
(East)	**7**	**29**	**100**	**293**	**1763**	**3.5**
Iowa	6 (5)	35	17	118	486 (262)	3.4 (1.8)
Illinois	6	36	15	133	487 (358)	3.4 (2.5)
Indiana	7 (6)	35	16	126	451 (365)	4.8 (3.9)
Michigan	6	19	49	181	694 (371)	4.7 (2.5)
Minnesota	7 (6)	23	55	199	754 (620)	3.6 (2.8)
Missouri	6	18	41	142	403 (365)	2.3 (2.0)
Ohio	6	16	43	134	528 (339)	5.0 (3.2)
Wisconsin	7 (6)	19	48	176	663 (428)	4.7 (2.9)
(Midwest)	**8**	**30**	**93**	**403**	**3071**	**2.6**
Arkansas	6	18	47	129	397 (261)	2.9 (1.9)
Louisiana	7	15	39	115	253 (304)	2.3 (2.5)
Oklahoma	7	23	56	184	389 (463)	2.2 (2.6)
Texas	8	30	98	410	1512 (996)	2.3 (1.4)
(South Central)	**9**	**33**	**110**	**517**	**2249**	**2.1**
Alabama	8	19	48	127	365 (321)	2.8 (2.4)
Florida	8 (7)	22	49	150	340 (298)	2.5 (2.0)
Georgia	7	20	46	133	363 (250)	2.4 (1.6)
Kentucky	5 (6)	16	32	95	286 (211)	2.8 (2.0)
Mississippi	8 (7)	20	50	110	297 (220)	2.4 (1.8)
North Carolina	7 (6)	20	43	116	355 (228)	2.8 (1.8)
South Carolina	7	18	39	105	232 (214)	3.0 (2.7)
Tennessee	7 (6)	19	43	135	621 (344)	6.0 (3.2)
Virginia	8 (7)	20	54	129	529 (265)	5.2 (2.5)
(Southeast)	**8**	**27**	**87**	**371**	**2169**	**2.1**
Colorado	8	35	93	347	1292 (856)	5.1 (3.2)
Kansas	7	23	55	168	473 (370)	2.2 (1.7)
Montana	9	44	128	464	1465 (693)	4.2 (1.8)
North Dakota	6 (7)	20	43	128	282 (272)	1.6 (1.5)
Nebraska	7 (6)	21	44	131	428 (268)	2.2 (1.3)
South Dakota	7 (6)	28	72	225	751 (563)	3.9 (2.8)
Wyoming	8 (7)	40	108	403	1448 (794)	6.3 (3.1)
(Northern Plains)	**11**	**53**	**173**	**811**	**4619**	**2.9**
Arizona	7 (6)	26	68	222	915 (423)	3.4 (1.4)
California	10	52	161	597	2689 (1755)	7.6 (4.3)
Idaho	9	45	131	454	1529 (1083)	7.8 (5.0)
New Mexico	8 (7)	31	88	299	1174 (744)	4.1 (2.4)
Nevada	8	33	88	378	1361 (1354)	5.1 (4.7)
Oregon	10	47	139	462	1481 (1075)	6.2 (4.3)
Utah	8 (7)	35	95	369	1415 (1006)	7.6 (4.6)
Washington	11 (9)	45	132	438	1548 (912)	9.6 (5.1)
(West)	**11**	**63**	**260**	**1421**	**9375**	**4.8**
Totals	**11**	**65**	**317**	**2026**	**19,602**	**2.7**

Note: Series density equals the number of series divided by 100,000 ha.

Table 11. Soil diversity (pedodiversity) by soil order (taxonomic pedodiversity) in the contiguous United States (U.S.) based on Soil Survey Geographic (SSURGO) Database (2020) [27].

State (Region)	Slight <———— Degree of Weathering and Soil Development ————> Strong									
	Slight Weathering				Intermediate Weathering				Strong Weathering	
	Entisols	Inceptisols	Histosols	Andisols	Vertisols	Alfisols	Mollisols	Aridisols	Spodosols	Ultisols
	Area (km^2)									
Connecticut	784	10,374	1052	0	0	0	196	0	0	0
Delaware	1072	125	121	0	0	59	0	0	27	3639
Massachusetts	3832	11,552	1542	0	0	0	3	0	1977	13
Maryland	2162	3254	591	0	0	2602	34	0	47	16,576
Maine	1099	21,286	6286	0	0	0	18	0	51,895	0
New Hampshire	1206	8697	1617	0	0	0	3	0	11,277	0
New Jersey	3587	3180	724	0	0	1734	0	0	1484	7078
New York	7238	63,843	3518	0	0	20,233	856	0	22,167	576
Pennsylvania	4200	44,708	223	0	0	24,961	138	0	203	40,858
Rhode Island	489	2036	58	0	0	0	0	0	0	0
Vermont	905	9265	395	0	0	1010	137	0	12,053	0
West Virginia	4257	18,871	33	0	0	13,980	122	0	482	23,702
(East)	**29,768**	**197,828**	**13,844**	**0**	**0**	**63,022**	**1119**	**0**	**106,720**	**92,025**
Iowa	9611	12,070	152	0	295	34,439	87,234	0	0	0
Illinois	12,239	4947	380	0	0	61,155	65,121	0	0	107
Indiana	6276	9429	1301	0	0	51,962	21,045	0	3	3568
Michigan	18,137	12,051	13,295	0	0	44,231	12,865	0	46,952	0
Minnesota	16,942	20,714	28,759	0	4387	44,288	93,878	0	254	0
Missouri	8837	5657	0	0	2759	91,360	40,204	0	0	28,667
Ohio	5739	13,700	406	0	0	66,356	12,555	0	0	6685
Wisconsin	16,878	4976	14,587	0	0	63,450	15,799	0	24,849	2
(Midwest)	**93,424**	**78,531**	**60,744**	**0**	**6866**	**477,096**	**337,608**	**0**	**68,509**	**38,778**
Arkansas	7324	13,765	0	0	7097	35,779	3745	0	0	68,121
Louisiana	8525	12,317	7165	0	15,743	41,476	1168	0	0	22,879
Oklahoma	17,904	21,679	0	0	6501	45,022	71,197	266	0	14,078
Texas	41,454	64,235	0	0	61,723	170,569	218,194	79,732	15	24,727
(South Central)	**70,892**	**105,988**	**8092**	**0**	**95,568**	**297,126**	**296,443**	**75,817**	**10**	**132,467**
Alabama	21,800	20,410	1084	0	3168	7298	1296	0	19	75,872
Florida	35,568	5929	12,643	0	13	15,803	5477	0	33,349	27,708
Georgia	14,331	10,028	1582	0	0	3408	3	0	3286	116,647
Kentucky	3021	26,852	0	0	0	44,876	3233	0	0	23,865
Mississippi	21,348	18,906	761	0	8967	30,808	478	0	1	41,313
North Carolina	8450	25,796	4882	0	0	6675	363	0	2736	76,622
South Carolina	6663	8167	462	0	0	7287	232	0	1156	54,521
Tennessee	7234	21,321	0	0	100	28,366	4600	0	0	42,657
Virginia	5445	20,589	817	0	23	10,560	645	0	28	64,607
(Southeast)	**98,026**	**139,879**	**24,312**	**0**	**13,943**	**164,043**	**14,144**	**0**	**46,166**	**551,642**
Colorado	53,635	17,712	397	0	1824	41,700	91,424	47,089	107	0
Kansas	16,343	5552	0	0	9457	11,254	169,487	156	0	76
Montana	70,088	89,506	486	8600	11,800	41,922	115,914	12,518	5	0
North Dakota	13,271	7352	20	0	6962	832	150,151	0	0	0
Nebraska	92,172	5574	22	0	620	3165	96,746	119	0	0
South Dakota	30,742	9172	13	0	16,518	5851	126,070	3549	0	0
Wyoming	69,454	23,384	253	356	639	24,678	55,786	54,725	0	0
(Northern Plains)	**343,944**	**154,780**	**726**	**1551**	**42,374**	**121,581**	**836,422**	**113,683**	**112**	**75**
Arizona	88,659	2472	4	0	3483	11,706	33,412	127,130	0	0
California	83,218	64,545	734	2928	15,945	69,846	81,653	25,034	47	10,023
Idaho	9126	34,112	176	19,004	1111	12,205	77,220	44,184	0	17
New Mexico	63,846	8102	10	86	1898	32,215	61,969	11,6232	0	0
Nevada	74,116	5633	4	165	870	1306	62,434	12,4888	0	0
Oregon	5819	52,931	521	25,895	2265	14,876	95,277	29,023	692	12,577
Utah	68,382	5582	7	7	440	2700	47,003	60,909	0	0
Washington	9542	31,326	872	37,798	47	6443	53,974	9927	8847	3093
(West)	**369,521**	**174,653**	**2229**	**83,077**	**26,562**	**156,171**	**516,161**	**593,340**	**10,273**	**26,557**
Totals	**982,571**	**859,445**	**116,378**	**72,379**	**190,455**	**1326,421**	**2009,454**	**704,900**	**242,181**	**870,043**

Note: Soil order of Gelisols was reported for the state of Washington with an area of 11 m^2.

3.1.2. Examples of Genetic Pedodiversity and Ecosystem Services

Genetic pedodiversity refers to the diversity of genetic horizons (soil layers commonly parallel to the soil surface), which designation indicates a qualitative description of changes) (Figure 3, Table 12) [15,40]. Diagnostic horizons are quantitatively defined and not equivalent to the genetic horizons in Soil Taxonomy [15]. Soil horizons are integral components of the taxonomic pedodiversity and can be used to compute distinct or combined stocks within the soil (Figure 3, Table 12).

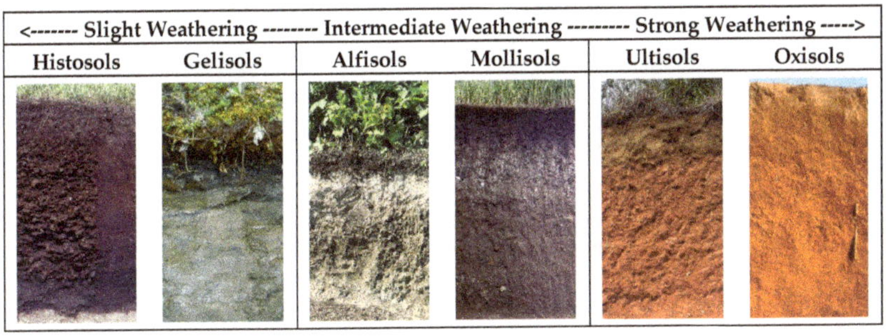

Figure 3. Examples of variations of soil horizons in different soil profiles.

Table 12. Examples of soil profile horizons (master horizons and corresponding lowercase letters commonly used with these horizons) and ecosystem services (adapted from Hartemink et al. (2020) [40].

Soil Profile Horizons		Ecosystem Services		
Master Horizons (Lowercase Letters)	**Description**	**Provisioning**	**Regulation and Maintenance**	**Cultural**
O	**Horizon with organic matter and plant litter**	x	x	ND
i	Slightly decomposed organic matter (fibric)	x	x	ND
e	Intermediately decomposed organic matter (hemic)	x	x	ND
a	Highly decomposed organic matter (sapric)	x	x	ND
A	**Zone of organic matter accumulation in the soil**	x	x	ND
p	Plowing or other disturbance	x	x	ND
E	**Zone of maximum eluviation**	x	x	ND
B	**Zone of maximum illuviation**	x	x	ND
c	Concretions or nodules	x	x	ND
b	Buried	x	x	ND
f	Frozen (permafrost)	x	x	ND
g	Strong gleying (mottling)	x	x	ND
h	Illuvial accumulation of organic matter (OM)	x	x	ND
j	Jarosite (yellow sulfur mineral)	x	x	ND
jj	Cryoturbation (frost churning)	x	x	ND
k	Accumulation of carbonate ($CaCO_3$)	x	x	ND
m	Cementation or induration	x	x	ND
n	Accumulation of sodium	x	x	ND
o	Accumulation of Fe and Al oxides	x	x	ND
q	Accumulation of silica	x	x	ND
s	Illuvial accumulation of OM and Fe and Al oxides	x	x	ND
ss	Slikenslides (shiny clay wedges)	x	x	ND
t	Accumulation of silicate clays	x	x	ND
v	Plinthite (high iron, red material)	x	x	ND
w	Distinctive color or structure	x	x	ND
x	Fragipan (high bulk density, brittle)	x	x	ND
y	Accumulation of gypsum ($CaSO_4 \cdot 2H_2O$)	x	x	ND
z	Accumulation of soluble salts	x	x	ND
C	**Weathered or soft rock**	x	x	ND
R	**Bedrock, consolidated rock**	x	x	ND

Note: ND = not determined. "True soil" = A, E, B. Regolith (weathered material) = O, A, E, B, C. Contributions of the different soil horizons to ES will vary with soil order, geographic location, environmental conditions, anthropologic setting, etc.

Numerous combinations of master horizons and their lowercase letters describe the incredible diversity of soil resources in the landscape. Soil horizons vary in thickness and exhibit within-horizon lateral and vertical variation [40]. Although all of the horizons in Table 12 present various types of value (e.g., market and non-market), some of these horizons can be over-exploited for their ES. For example, horizon A is commonly plowed for agricultural production and subject to nutrient depletion and erosion [41]. Ireland had to pass policies to protect peatlands, which often contain soil order of Histosols with horizon O in various stages of decomposition (lowercase letters: i, e, a, Table 12), which is used for multiple purposes (e.g., fuel and horticulture) [42]. Permafrost (indicated by the lowercase letter f in Table 12) is thawing rapidly, releasing large amounts of carbon dioxide and methane gases [43].

3.1.3. Examples of Parametric Pedodiversity and Ecosystem Services

Parametric pedodiversity refers to the diversity of soil properties, which are also often used in the context of taxonomic and genetic pedodiversities. Soil properties vary by soil type and exhibit within-horizon lateral and vertical variation [40]. Although there is no standardized list of soil properties, Adhikari and Hartemink (2016) [16] provided key soil properties related to ES: soil organic carbon; sand, silt, clay, and coarse fragments; pH; depth to bedrock; bulk density; available water capacity; cation exchange capacity; electrical conductivity; soil porosity and air permeability; hydraulic conductivity and infiltration; soil biota; soil structure and aggregation; soil temperature; clay mineralogy, and subsoil pans. This list seems to focus primarily on soil physical properties, but soil chemical properties and qualitative soil descriptions (e.g., official soil series descriptions) are also important in ES/ED valuations. Soil chemical properties (e.g., plant nutrients) are essential for agricultural production, and it is important to monitor the supply of these nutrients to meet the yearly recommended dietary allowances and intakes by population [44]. For example, Zurqani et al. (2019) [45] quantified the yearly human demand for major and trace elements in Libya by different administrative units with the future goal of linking it with the provisioning ES supply by the country's soils.

Very often, soil properties serve multiple ES/ED. For example, Groshans et al. (2018) [30] determined the provisioning value of soil inorganic carbon (SIC) based on liming replacement cost within the contiguous U.S. However, SIC can also be valued as a regulating ES/ED, and Groshans et al. (2019) [46] estimated the value of SIC stocks based on the avoided social cost of carbon emissions (SC-CO_2). There are numerous challenges in using appropriate soil data sources and valuation methods for soil properties. For example, Mikhailova et al. (2019) [47] compared field sampling and soil survey database for spatial heterogeneity in surface granulometry (sand, silt, and clay) for potential use in ES/ED valuations and concluded that field sampling provided more detailed information. The same study revealed that soil texture and coarse fragments are lithospheric-derived resources (Figure 1) and can be valued based on "soil" or "mineral" stock. Among soil properties, soil organic matter (SOM) and soil organic carbon (SOC) are the most researched soil properties because of their significance in provisioning (e.g., soil fertility) and regulating (e.g., carbon sequestration) ES. Guo et al. (2006) [39] reported spatial variability of soil carbon (SOC, SIC) in each of the soil orders within the contiguous U.S., and potential decline in SOC in the 0–20 cm depth (e.g., rooting depth of most crops) compared to 20–100 cm depth. Taxonomic pedodiversity and human demand for soil nutrients can adversely affect soil health and nutritional security [48].

3.1.4. Examples of Functional Pedodiversity and Ecosystem Services

Functional pedodiversity refers to soil behavior under different uses. Taxonomic and genetic pedodiversities influence soil behavior under current and potential uses. According to Amundson et al. (2003) [2], "in less than two centuries, the landscape of the U.S. has been transformed to the degree that would astound our 19th-century predecessors". According to Merrill and Leatherby (2018) [49], "the 48 contiguous states alone are

1.9 billion-acre jigsaw puzzle of cities, farms, forests and pastures that Americans use to feed themselves, power their economy and extract value for business and pleasure" with six major types of land: cropland (391.5M acres, where M = million = 10^6), forest (538.6M acres), pasture/range (654M acres), urban (69.4M acres), special use (168.6M acres), and miscellaneous (68.9M acres). Soils under different uses make a significant contribution to various ES/ED in the contiguous U.S., but monetary valuations of these ES/ED are rare at this scale. For example, Groshans et al. (2018) [46] assessed the value of SIC for provisioning ES for LRRs based on liming replacement cost (a 2014 U.S. average price of $10.42 per U.S. ton of agricultural limestone). The LRRs with the highest midpoint total replacement cost value of SIC storage were: (1) D—Western Range and Irrigated Region ($1.10T), (2) H—Central Great Plains Winter Wheat and Range Region ($926B), and (3) M—Central Feed Grains and Livestock Region ($635B) [46]. On an area basis, the highest replacement cost values were: (1) I—Southwest Plateaus and Plains Range and Cotton Region ($3.33 m^{-2}), (2) J—Southwestern Prairies Cotton and Forage Region ($2.83 m^{-2}), and (3) H—Central Great Plains Winter Wheat and Range Region ($1.59 m^{-2}) [46]. The LRRs with the highest mean replacement cost values per area over the depth interval 0–20 cm were: (1) I—Southwest Plateaus and Plains Range and Cotton Region ($0.43 m^{-2}), (2) J—Southwestern Prairies Cotton and Forage Region ($0.27 m^{-2}) and (3) D—Western Range and Irrigated Region ($0.11 m^{-2}) [46]. Over the depth interval 0–100 cm, the highest mean replacement cost values were: (1) I—Southwest Plateaus and Plains Range and Cotton Region ($1.86 m^{-2}), (2) J—Southwestern Prairies Cotton and Forage Region ($1.49 m^{-2}) and (3) F—Northern Great Plains Spring Wheat Region ($0.70 m^{-2}) [46].

Traditionally, provisioning ES have been seen as the primary value from the soil; however, in the face of potentially severe economic impacts from climate change, regulating ES should be recognized for their potential to provide ES through mitigation of net CO_2 release through different management regimes designed to maximize CO_2 uptake and minimize CO_2 release. Site-specific management of soil carbon "hotspots" through precision agriculture could serve to reduce CO_2 emission and provide regulating ES to humanity [37]. Although no economic system has been developed to coordinate long-term practices to sequester C [50], any contribution to CO_2 reduction would help mitigate emissions from fossil fuels. Different soil carbon (SOC, SIC, and TSC) should be accounted for in the mitigation efforts. A few studies attempted to put a monetary value on regulating ES/ED from SOC, SIC, and TSC within the contiguous U.S. using the social cost of carbon (SC-CO_2) of $42 per metric ton of CO_2, which is applicable for the year 2020 based on 2007 U.S. dollars and an average discount rate of 3% [35,36,46,51]. According to Mikhailova et al. (2019) [36], the LRRs with the highest TSC storage value (based on the avoided social cost of carbon emissions) were: (1) M—Central Feed Grains and Livestock Region ($2.82T), (2) D—Western Range and Irrigated Region ($2.64T), and (3) H—Central Great Plains Winter Wheat and Range Region ($2.48T). The value of TSC based on area density within LRR boundaries were ranked: (1) I—Southwest Plateaus and Plains Range and Cotton Region ($6.90 m^{-2}), (2) J—Southwestern Prairies Cotton and Forage Region ($6.38 m^{-2}), and (3) U—Florida Subtropical Fruit, Truck Crop and Range Region ($6.25 m^{-2}) [36].

3.2. *Extrinsic Factors: Examples of Monetary Valuations Based on Interaction of Soil Diversity (Pedodiversity) and the Earth's Spheres*

Pedodiversity is influenced by extrinsic (outside soil) factors from atmosphere, biosphere, hydrosphere, lithosphere, ecosphere, and anthroposphere (Figure 1), which can increase or decrease the value of ES/ED associated with pedodiversity. Valuations of both intrinsic and extrinsic factors can be made within biophysical accounts (e.g., within soil order boundaries) and then "translated" into the administrative accounts for decision making. For example, Figure 4 demonstrates the share between values of total soil carbon (intrinsic) and average annual total (extrinsic) monetary values of non-constrained potential soil inorganic carbon (SIC) sequestration from combined atmospheric Ca^{2+} and Mg^{2+} deposition (2000–2015) for different regions in the contiguous U.S. based on an avoided SC-CO_2 of $42 per metric ton of CO_2. In this case, both intrinsic (TSC) and extrinsic (potential for SIC sequestration from atmospheric Ca^{2+} and Mg^{2+} deposition) factors are spatially heterogeneous without considering physical and economic constraints for achieving the maximum potential for SIC sequestration from atmospheric sources [52]. Both intrinsic and extrinsic factors limit pedodiversity in its ability to supply ES/ED.

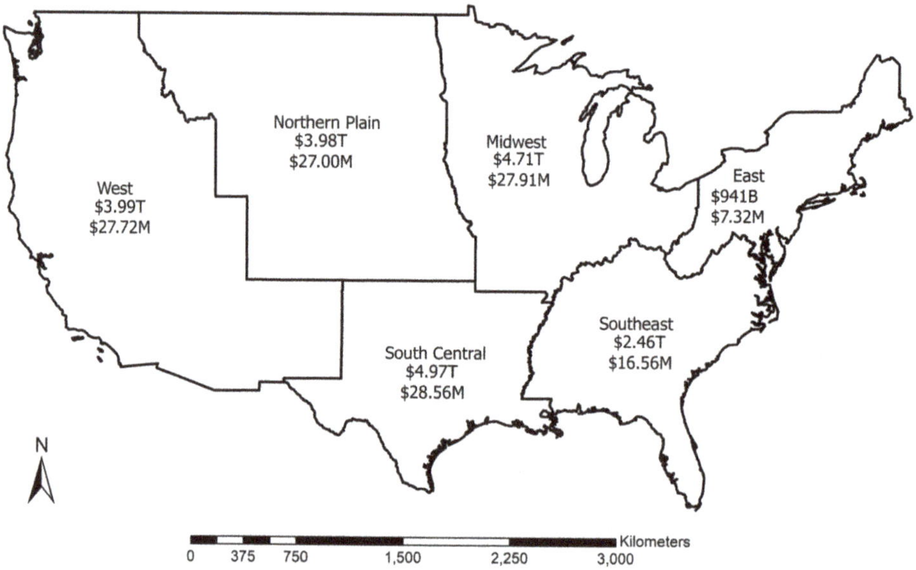

Figure 4. Total soil carbon **(intrinsic, top number)** and average annual total **(extrinsic, bottom number)** monetary values of non-constrained potential soil inorganic carbon (SIC) sequestration from combined atmospheric Ca^{2+} and Mg^{2+} deposition (2000–2015) for different regions in the contiguous United States (U.S.) based on an avoided SC–CO_2 of $42 per metric ton of CO_2 [51] (adapted from Mikhailova et al., 2019 [36]; Mikhailova et al., 2020 [52]). Note: M = million = 10^6, B = billion = 10^9, and T = trillion = 10^{12}.

Climate change is another set of extrinsic factors impacting the value of ES/ED derived from pedodiversity. Global warming threatens the existence of soil order of Gelisols because of thawing permafrost. Gelisols store large amounts of soil organic matter (SOM), and its decomposition can release large amounts of carbon dioxide and methane [53]. An increase in both ambient and soil temperature can intensify SOM decomposition and lead to self-ignition conditions in soil carbon-rich soils (e.g., Gelisols, Histosols), leading to wildfires [54]. Climate change is also predicted to increase global water erosion from 30 to 66% [55].

Urbanization, another example of extrinsic factors, alters soils in various ways (e.g., erosion and pollution) and creates ES/ED specific to urban soil diversity (urban pedodiversity), which requires adjustment of valuations to urban environments [56,57]. Urbanization alters the stocks and flows of ES/ED provided by soils in urban and non-urban environments since urban environments are not self-supporting, which creates a significant demand for ES from urban fringes and beyond [58].

3.3. Pedodiversity Threats and Losses in the Contiguous U.S. in Relation to Ecosystem Services

3.3.1. Land Cover Change (LCC) as a Threat to Pedodiversity

Pedodiversity in the contiguous U.S. is experiencing significant threats and losses, especially in agriculturally productive and important soils (e.g., Alfisols and Mollisols) and regions (e.g., Midwest, Northern Plains, South Central) that is driven by land cover change (e.g., urbanization, deforestation, and agricultural expansion) [2]. These soils and regions have high provisioning ES value, which can result in unsustainable use accompanied by the loss of regulating and provisioning ES services [59]. Although the economic value of ES from agriculturally productive soils is somewhat reflected in the total value of U.S. agricultural production, the social costs associated with this production and rates of soil diversity (pedodiversity) extinction are rarely reported (Table 13) [2]. The total social costs of agricultural production (present and past) are difficult to quantify because they are impacted by numerous factors, but on-going research on ES/ED provides useful insight into potential valuation methods. This type of analysis should account for ES and ED (actual or realized, and potential) provided by pedodiveristy using biophysical (e.g., soil orders) and administrative (e.g., states) accounts. For example, Groshans et al. (2018) [30,59] assessed the midpoint total provisioning value of SIC for ES for Mollisols in the contiguous U.S. based on liming replacement costs (average price of limestone in 2014, \$10.42 per U.S. ton) in 2 m soil depth at \$23.2B, USD (where B = billion = 10^9). A follow-up study estimated the regulating value of SIC for ES/ED for Mollisols in the contiguous U.S. based on a social cost of carbon (SC-CO_2) of \$42 U.S. dollars and reported a total midpoint value of \$3.57T, USD (where T = trillion = 10^{12}) [48].

Soil organic carbon is particularly important for agricultural production, and Mikhailova et al. (2019) [35] estimated the value of SOC stocks in the contiguous United States based on the avoided social costs of carbon emissions and reported that the total calculated monetary value for SOC storage in the contiguous U.S. was between \$4.64T and \$23.1T, with a midpoint value of \$12.7T [35]. Soil orders with the highest midpoint SOC storage values were (1): Mollisols (\$4.21T), (2) Histosols (\$2.31T), and (3) Alfisols (\$1.48T) [35]. The midpoint values of SOC normalized by area within soil order boundaries were ranked: (1) Histosols (\$21.58 m^{-2}), (2) Vertisols (\$2.26 m^{-2}), and (3) Mollisols (\$2.08 m^{-2}) [35]. States with the highest midpoint values of SOC storage were: (1) Texas (\$1.08T), (2) Minnesota (\$834B), and (3) Florida (\$742B) [35]. Midpoint values of SOC normalized by area within state boundaries were ranked: (1) Florida (\$5.44 m^{-2}), (2) Delaware (\$4.10 m^{-2}), and (3) Minnesota (\$3.99 m^{-2}) [35]. Regions with the highest midpoint values of SOC storage were: (1) Midwest (\$3.17T), (2) Southeast (\$2.44T), and (3) Northern Plains (\$2.35T) [35]. Midpoint values of SOC normalized by area within region boundaries were ranked: (1) Midwest (\$2.73 m^{-2}), (2) Southeast (\$2.31 m^{-2}), and (3) East (\$1.82 m^{-2}) [35].

Mapping can provide spatial context between population, urbanization, and endangered soil series (Table 13, Figures 5 and 6). Results show that there is a link between states and regions with a high number of endangered soil series and the value of provisioning and regulating ES (California; Midwest, Northern Plains, and South-Central regions). States and regions with zero endangered series represent areas with generally low productivity soils such as Aridisols (e.g., Arizona and New Mexico), Ultisols (e.g., Georgia, South Carolina, and North Carolina), and Entisols and Inceptisols (e.g., Maine and Vermont) (Table 13, Figure 5).

Some states and regions have low human populations, but some of the highest proportions of endangered rare soil series (Table 13, Figure 5). For example, the Northern Plains region has only 4.19% of the total U.S. population and some of the highest numbers of endangered soil series [60,61]. In the Midwest, Iowa has almost 81% of its rare soil series endangered, but it has only 0.95% of the total U.S. population (Table 13, Figure 5) [60,61]. Similarly, Kansas has nearly 43% of its rare soil series endangered, but it has only 0.88% of the U.S. population. These states have some of the highest values associated with provisioning ES from SOC, SIC, and TSC, as well as regulating ES/ED associated with these carbon stocks. A mismatch between "potential" and "realized" supply/demand of flow-dependent ES/ED [62] is not a new phenomenon for Kansas, which experienced soil

Table 13. Pedodiversity and selected population statistics in the contiguous United States (U.S.) (adapted from Amundson et al., 2003 [2], and Wikipedia [60]).

State (Region)	Orders	Series	Rare Series	Endangered Series	% of Rare Series Endangered	Extinct Soil Series	Percent of the Total U.S. Population (2019)	Percent Urban Population within State (2010)
Connecticut	4	86	8	4	50.0		1.07	88.0
Delaware	6	52	0	0	0.0		0.29	83.3
Massachusetts	5	129	5	0	0.0		2.09	92.0
Maryland	7	187	7	0	0.0		1.82	87.2
Maine	4	111	8	0	0.0		0.41	38.7
New Hampshire	4	127	10	0	0.0		0.41	60.3
New Jersey	7	148	22	2	9.1		2.68	94.7
New York	7	347	37	2	5.4		5.86	87.9
Pennsylvania	7	248	20	0	0.0		3.86	78.7
Rhode Island	3	45	2	0	0.0		0.32	90.7
Vermont	6	192	24	0	0.0		0.19	38.9
West Virginia	6	163	11	0	0.0		0.54	48.7
(East)	**n/a**	**n/a**	**n/a**	**n/a**	**n/a**	**n/a**	**19.54**	**n/a**
Iowa	5	262	26	21	80.8		0.95	64.0
Illinois	6	358	44	29	65.9	6	3.82	88.5
Indiana	6	365	44	36	81.8	2	2.03	72.4
Michigan	6	371	86	10	11.6		3.01	74.6
Minnesota	6	620	122	65	53.3	6	1.70	73.3
Missouri	6	365	27	12	44.4	4	1.85	70.4
Ohio	6	339	46	21	45.7	2	3.52	77.9
Wisconsin	6	428	51	8	15.7		1.75	70.2
(Midwest)	**n/a**	**n/a**	**n/a**	**n/a**	**n/a**	**n/a**	**18.63**	**n/a**
Arkansas	6	261	3	1	33.3		0.91	56.2
Louisiana	7	304	41	10	24.4	1	1.40	73.2
Oklahoma	7	463	46	3	6.5		1.19	66.2
Texas	8	996	176	6	3.4		8.74	84.7
(South Central)	**n/a**	**n/a**	**n/a**	**n/a**	**n/a**	**n/a**	**12.24**	**n/a**
Alabama	8	321	19	0	0.0		1.48	59.0
Florida	7	298	67	9	13.4	3	6.47	91.2
Georgia	7	250	4	0	0.0		3.20	75.1
Kentucky	6	211	14	0	0.0		1.35	58.4
Mississippi	7	220	17	2	11.8		0.90	49.3
North Carolina	6	228	18	0	0.0		3.16	66.1
South Carolina	7	214	13	0	0.0		1.55	66.3
Tennessee	6	344	44	3	6.8		2.06	66.4
Virginia	7	265	10	0	0.0		2.57	75.5
(Southeast)	**n/a**	**n/a**	**n/a**	**n/a**	**n/a**	**n/a**	**22.74**	**n/a**
Colorado	8	856	153	0	0.0		1.74	86.2
Kansas	7	370	14	6	42.9		0.88	74.2
Montana	9	693	188	21	11.2		0.32	55.9
North Dakota	7	272	26	10	38.5		0.23	59.9
Nebraska	6	268	23	14	60.9	2	0.58	73.1
South Dakota	6	563	61	18	29.5		0.27	56.7
Wyoming	7	794	121	0	0.0		0.17	64.8
(Northern Plains)	**n/a**	**n/a**	**n/a**	**n/a**	**n/a**	**n/a**	**4.19**	**n/a**
Arizona	6	423	27	0	0.0		2.19	89.9
California	10	1755	671	104	15.5	1	11.9	95.0
Idaho	9	1083	361	49	13.6		0.54	70.6
New Mexico	7	744	139	0	0.0		0.63	77.4
Nevada	8	1354	399	1	0.3		0.93	94.2
Oregon	10	1075	301	16	5.3		1.27	81.0
Utah	7	1006	279	5	1.8		0.97	90.6
Washington	9	912	462	25	5.4	3	2.29	84.0
(West)	**n/a**	**n/a**	**n/a**	**n/a**	**n/a**	**n/a**	**20.72**	**n/a**
Totals	**n/a**	**n/a**	**n/a**	**n/a**	**n/a**	**n/a**	**98.06**	**80.7**

Note: n/a = not available. For "series abundance," the following categories are defined [2]: (a) rare soils—less than 1000 ha total area in the US, and (b) endangered soils as those rare or rare-unique soil series that have lost more than 50% of their area to various land disturbances.

degradation and the Dust Bowl in the past due to a combination of prolonged drought and "suitcase farming," which was sponsored by "non-resident farmers" [63]. From the ES framework perspective, examining the potential for dry land farming in the 1920s, it would have indicated limitations due to intrinsic factors (high soil erodibility because of silty soil texture) and extrinsic factors (susceptibility of this area to regular droughts). According to Lee and Gill [64], the soil and drought conditions were compounded by an economic collapse that reduced crop value. Social costs associated with the Dust Bowl went far beyond the boundaries of the states where it originated, with almost half a million Dust Bowl refugees, massive quantities of topsoil being deposited in the Atlantic Ocean and impacting the air quality of Washington D.C. and other faraway states [64]. The ES framework, in combination with detailed spatial and temporal environmental data, can be used to inform sustainable decision-making to help avoid and mitigate similar and other disasters.

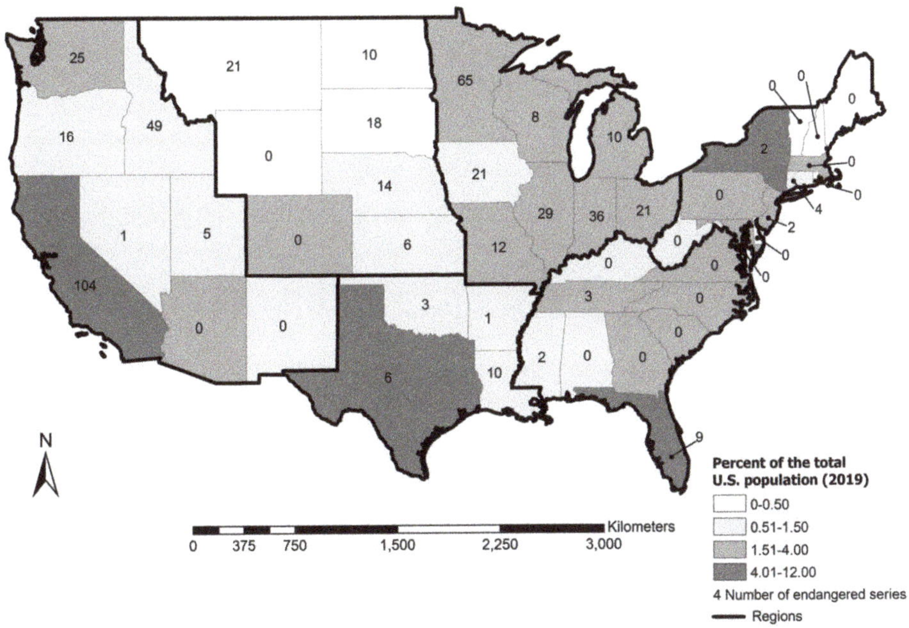

Figure 5. Number of endangered soil series [2] and percent of the total U.S. population (2019) [60] in the contiguous United States (U.S.).

For example, land cover change maps over time can provide insight into geographical patterns of ES stocks, flows, and values. Urbanization trends increase demand for ES, which are not always supplied by local soil resources and require soil ecosystem goods and services to be "imported" from soil stocks in other geographic areas (Figures 6 and 7). Figure 7 shows large urban area increases in the states of Texas, California, Florida, Arizona, Georgia, and North Carolina. Increases in urban areas in states can be accompanied by decrease in the agricultural areas (e.g., Florida, Arizona, and Georgia) and/or loss of forest areas (e.g., California, Arizona, Texas, and Georgia) (Figures 8 and 9).

In some cases, states increase the flow of ecosystem goods and services by increasing the agricultural area within the state based on available soil resources (e.g., Texas) (Figure 8). In other cases, an increase in the flow of ecosystem goods and services by increasing the agricultural area may be limited because of constraints associated with inherently low-fertility soils and other extrinsic factors (e.g., low precipitation).

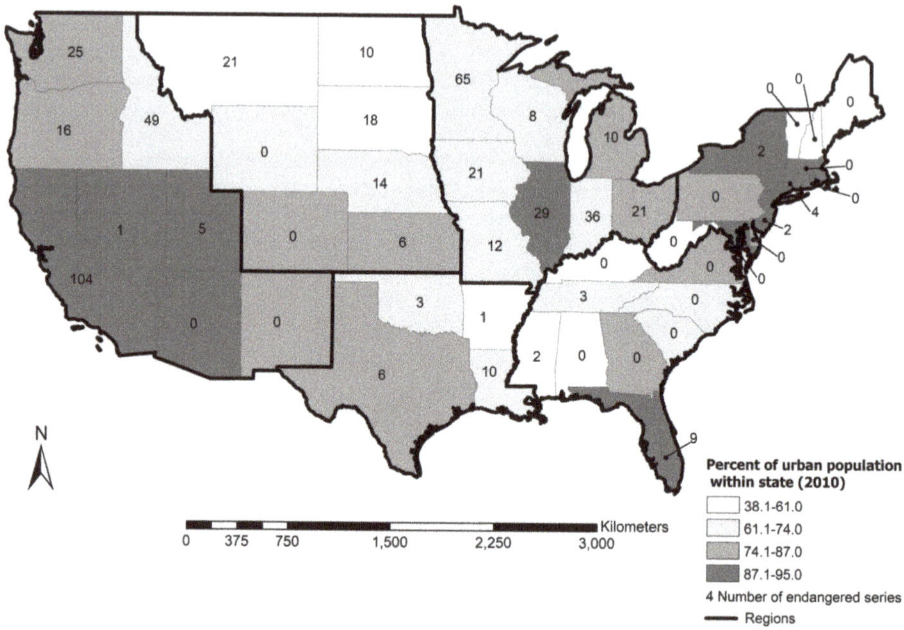

Figure 6. Number of endangered soil series [2] and percent of the urban population within each state (2010) [61] in the contiguous United States (U.S.).

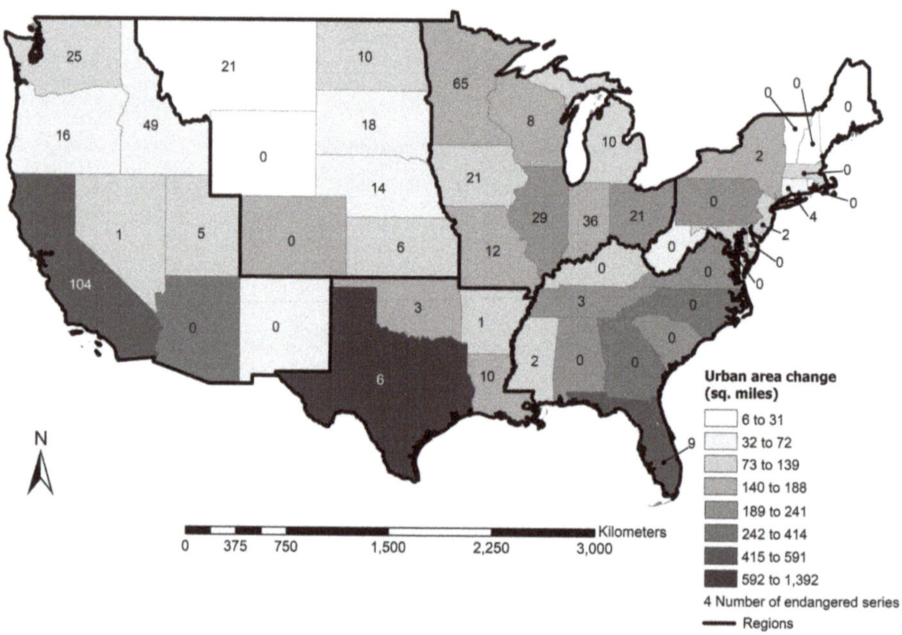

Figure 7. Urban land cover changes over time from 2001 to 2011 in the contiguous United States (U.S.) (adapted from Wentland et al. (2020) [65]).

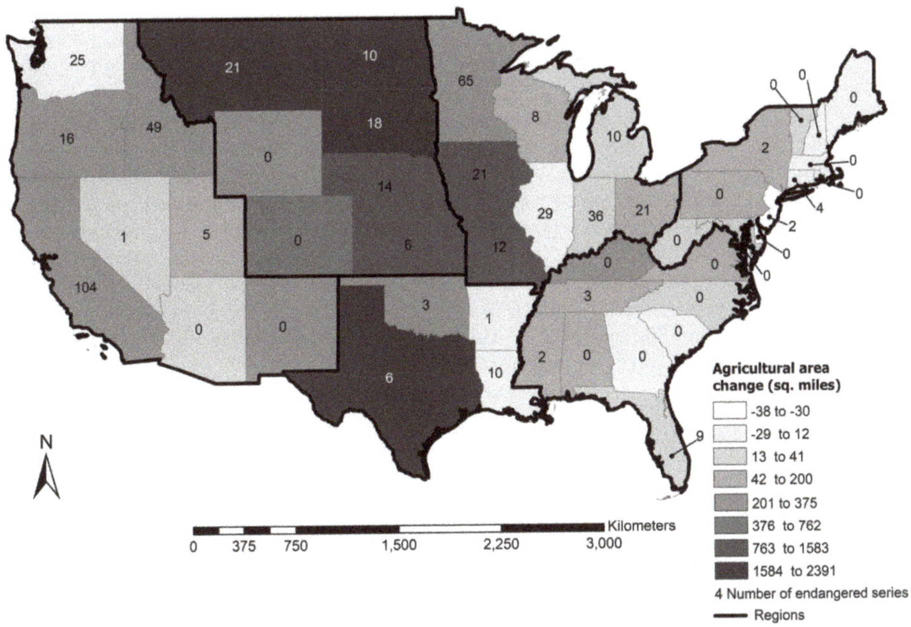

Figure 8. Agricultural land cover changes over time from 2001 to 2011 in the contiguous United States (U.S.) (adapted from Wentland et al. (2020) [65]).

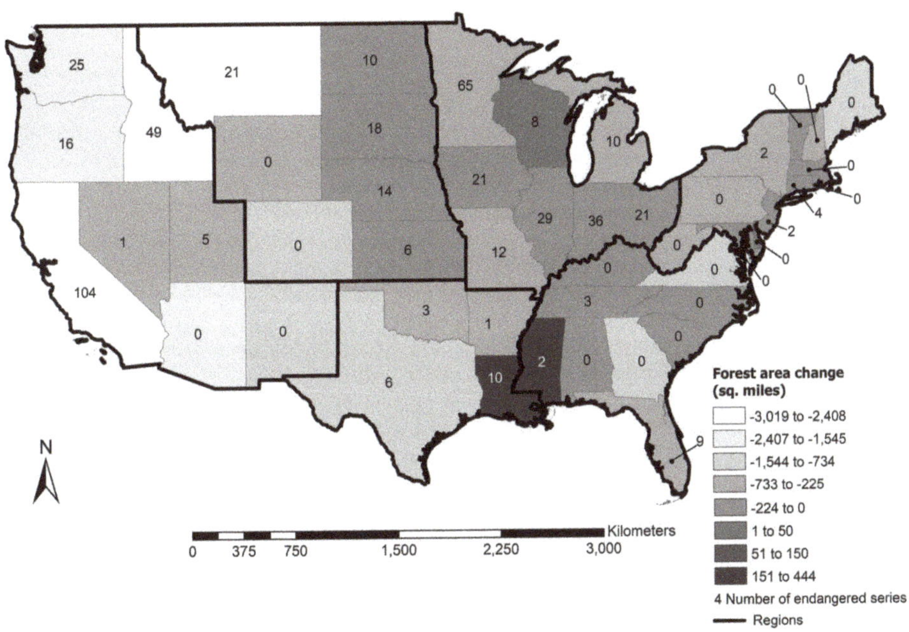

Figure 9. Forest land cover changes over time from 2001 to 2011 in the contiguous United States (U.S.) (adapted from Wentland et al. (2020) [65]).

According to Wentland et al. (2020) [65], overall U.S. land cover has seen declines in agricultural areas, forests, and pasture, and increases in developed areas as well as barren, and scrub/shrub land cover classes. These declines are mostly concentrated in the Southeastern U.S. [65].

Land cover and its change are important in ES valuations. Wetland et al. (2020) [65] estimated the total value of private land in the contiguous United States at $25.1T (T = trillion = 10^{12}) in 2016 using the Zillow Transaction and Assessment Dataset, which includes individual property attributes linked to market transaction price data. These values do not necessarily represent ES values (intrinsic and extrinsic) because there are no available tools for the public to appraise any facet of ES values. Land price values can vary based on times of economic growth or decline. For example, Wentland et al. (2020) [65] reported a 28% decline in land value during a financial crisis in the last decade without ES valuation. This is a clear evidence that land value and ES value are not connected, which can be an essential consideration for future research. There may be limited instances where the land value is tied to agricultural productivity (provisioning ES). Soil ecosystem goods and services are no longer used locally but are subject to a vast global distribution network, contributing to biogeochemical cycles' destabilization.

3.3.2. Climate Change as a Threat to Pedodiversity

Climate change poses a range of unique threats (e.g., changes in temperature, precipitation, and extreme conditions) to pedodiversity and its ES, which will be discussed following the concept of pedodiversity and its measures outlined in this study. Since pedodiversity (biotic + abiotic) forms from the interaction of various spheres (biosphere, lithosphere, hydrosphere, atmosphere, ecosphere, and anthroposphere), climate change threats will be multifaceted and complex. Both biotic and abiotic aspects of pedodiversity are sensitive to climate change and include the following examples relevant to ED:

- **Biotic** (e.g., increase in soil organic matter decomposition rates due to increase in temperature and precipitation [66] leading to increase in soil CO_2 emissions and associated social costs);
- **Abiotic** (e.g., increase in soil erosion due to an increase in precipitation and extreme rainfall events [67]).

Pedodiversity is influenced by intrinsic (within the soil) and extrinsic (outside soil) factors, where climate change is an extrinsic factor (e.g., changes in temperature, precipitation) with subsequent effect on alterations in intrinsic soil characteristics and properties (e.g., soil temperature and moisture regimes, and moisture content). In terms of taxonomic pedodiversity (diversity of soil classes), climate change poses an existential threat to the soil order of Gelisols. Climate change can lead to changes in soil classification, especially with regards to the use of soil temperature (e.g., pergelic, subgelic, cryic, and frigid) and moisture regimes (e.g., udic and ustic). An example of climate-induced changes in genetic diversity (diversity of soil horizons) includes the potential disappearance of permafrost, which is indicated by the lowercase letter "f" (frozen) in the soil profile. Climate change will impact parametric pedodiversity (diversity of soil properties) in various ways; for example, it can reduce soil organic matter content because of increased decomposition due to temperature. Soil pH can become more acidic because of increased precipitation and leaching, and in the case of agricultural soil, more liming material will need to be applied to compensate for the reduction in provisioning service provided by soil. Functional pedodiversity (soil behavior under different uses) will be affected in many parts of the world because of climate change. For example, global sea rise will influence soils under rice production, resulting in annual crop losses of up to $10.59 billion USD [68].

Projections of future U.S. climate change predict that the entire U.S. is likely to warm over the next 40 years, with an increase of 1–2 °C over much of the country and a 2–3 °C increase in the interior of the country [69]. Reilly et al. (2003) [70] examined the effects of climate change on provisioning ES from U.S. agriculture and reported a potential shift of crop production northwards and a positive overall increase in agricultural production with regional differences (e.g., possible declines in production in the Southern U.S.). These ES changes are likely to be accompanied by ED in the form of increased social costs associated with carbon dioxide emissions, soil erosion, depletion of soil nutrients, and others, which contribute to the issues of soil and human security worldwide [71]. Climate change in combination with population growth may increase demand for soil nutrients, which replacement from soil weathering is relatively slow in comparison with "anthropogenic use rate" [72]. Since pedodiversity is not evenly distributed within most geographic areas, soil nutrient depletion can be more acute in some places leading to prohibitively high replacement costs associated with fertilizer and liming applications. If the nutrients (e.g., base cations) are not replaced through liming and fertilization, it will alter the soil chemical composition, which can change its pedodiversity classification. Climate change will have a direct impact on the classification of soils, with some soil types disappearing and others changing in both extent and properties. Soil carbon changes associated with climate change and increased organic matter decomposition will also change how soils are classified as they are "decarbonized."

4. Discussion

Pedodiversity is a source of various ecosystem goods, services, and disservices, and its value is as complex as its concept. The total economic value (TEV) of pedodiversity is only a portion of the total system value (TSV) of pedodiversity because pedodiversity and ES form a multilayered relationship with the general trend of decreasing the tangibility of the value of soil to users (Table 3) from the monetary value (e.g., actual use values: consumptive food production) to pedodiversity value (e.g., intrinsic value) (Figure 10) [72,73].

Currently, there are four main approaches to analyze pedodiversity: taxonomic (diversity of soil classes), genetic (diversity of genetic horizons), parametric (diversity of soil properties), and functional (soil behavior under different use). The concept of pedodiversity and its classification varies by country; therefore, its applications to ES are country-specific [74]. According to Gerasimova (2010) [74], "the American Soil Taxonomy is the main and single classification in 45 countries, whereas, in 80 countries, it is used along with the national classifications". Despite differences in country-specific classifications, these soil classifications provide science-based soil information that can be integrated with administrative accounts (Table 2). Taxonomic, genetic, parametric, and functional pedodiversity provide an essential context for analyzing, interpreting, and reporting ES/ED within the ES framework for business applications. Although each approach can be used separately, three of these approaches (genetic, parametric, and functional) fall within the "umbrella" of taxonomic pedodiversity, which separates soils based on properties important to potential use.

Taxonomic pedodiversity provides a general description of the stock, its type, and spatial (both horizontal and vertical) distribution, which are particularly useful in agricultural business applications (e.g., soil productivity ratings in soil survey). For example, an area abundance of soil orders describes the spatial distribution within defined administrative boundaries (e.g., LRRs defined by the USDA based on MLRAs and agricultural markets) (Table 9). The phrases "portfolio effect" and "evenness effect" [75] are often applied to describe the theoretical links between biodiversity and ecosystem function. The "portfolio effect" is the analogy between the stock market and species diversity, where having more species allows a system to better respond to external stimuli. At the same time, the "evenness effect" finds that having similar numbers of species can help buffer against disturbances [76]. The concepts of "portfolio effect,", "evenness effect," and the newly proposed "distribution effect" can also be applied to pedodiversity with various degrees of interpretation (Figures 11 and 12).

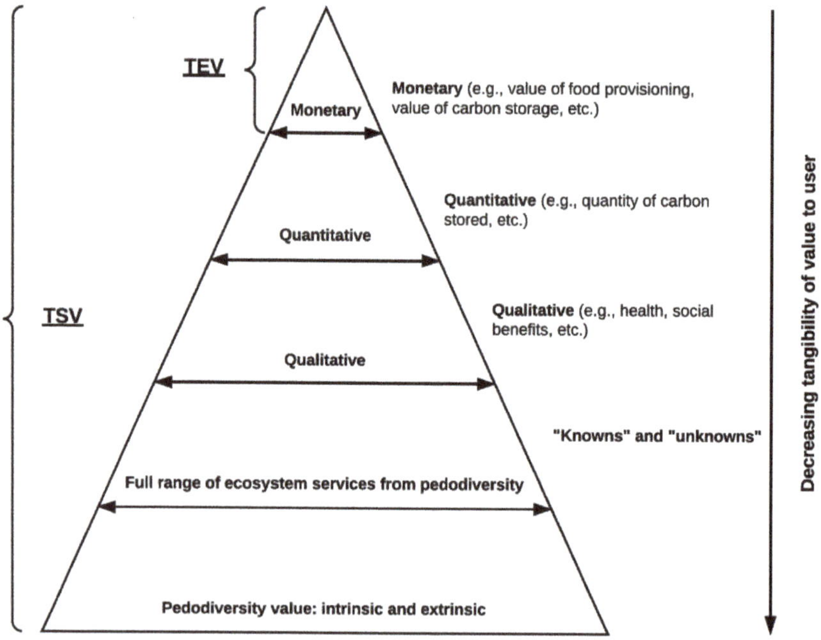

Figure 10. The newly expanded scope of pedodiversity valuation pyramid with a comparison of total economic value (TEV) and total system value (TSV) of ecosystem services (ES) and disservices (ED) (adapted from Gantioler et al., 2000 [76]).

Figure 11 illustrates these concepts using the contiguous United States as an example. In this context, the "portfolio effect" is defined as the number of different stocks (soil orders) within the country (Figure 11). The "distribution effect" shows the distribution of stocks (soil orders), its variation (e.g., slightly-weathered, intermediately-weathered, and strongly-weathered soils), and associated avoided or realized social costs of SOC, SIC, and TSC within the country.

Figure 12 illustrates these concepts using three states (Iowa, Rhode Island, and Georgia). In this context, the "portfolio effect" is defined as the number of different stocks (soil orders) within each state: Iowa (5), Rhode Island (3), and Georgia (7) (Figure 12). The "distribution effect" shows the distribution of stocks (soil orders) and its variation (e.g., slightly-weathered, intermediately-weathered, and strongly-weathered soils) within the state: Iowa (skewed towards intermediately-weathered soils), Rhode Island (skewed towards slightly-weathered soils), and Georgia (skewed towards strongly-weathered soils).

The "evenness effect" describes instances when similar soil types are evenly represented (an example is not shown) (e.g., Mollisols and Alfisols are both fertile soils). For each state, a paired graph shows the proportion of total area occupied by soil order and value of soil organic carbon (SOC) based on avoided social cost of CO_2, with Iowa having largest values, mainly from Alfisols and Mollisols, low total value in Rhode Island, and intermediate value in Georgia based on Ultisols and Histolsols.

Another pedodiversity measure, series density, provides important information about soil diversity, but its interpretation can differ from biological species density. While higher levels of species density are often seen as an advantage when describing biological systems [77], areas with higher soil series density (e.g., typical for soils derived from glacial parent material) may have less agricultural productivity compared with more homogenous and productive soils (e.g., typical for soils derived from loess parent material).

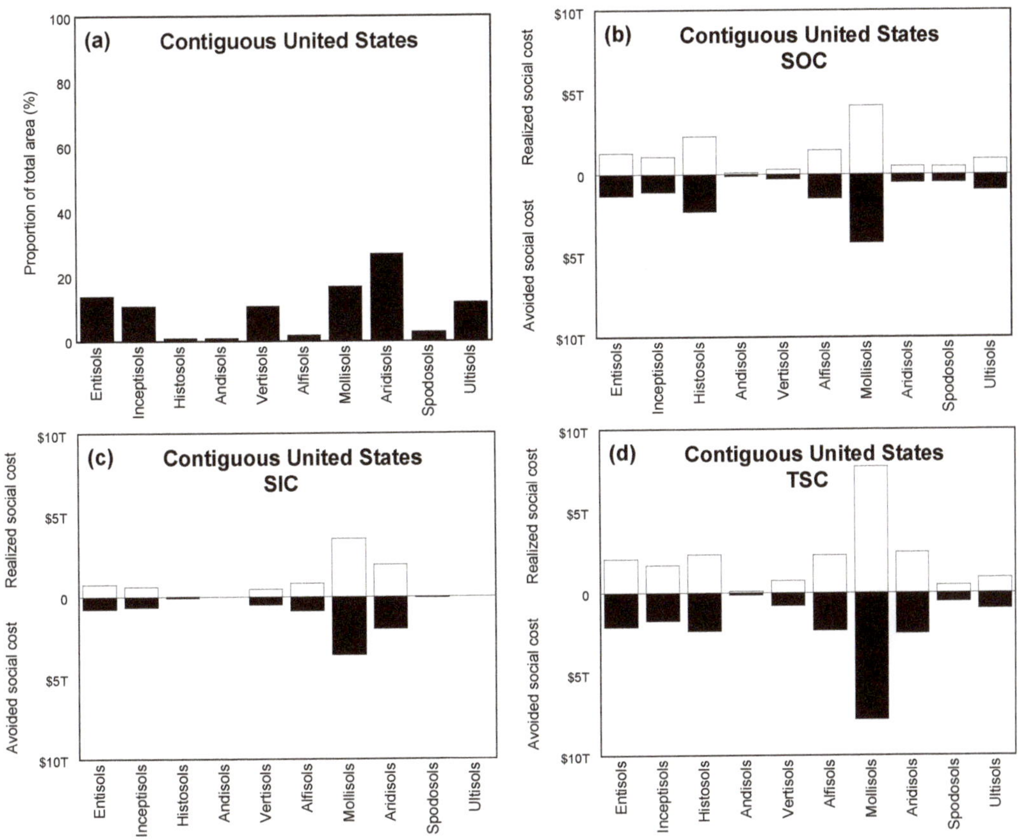

Figure 11. Diagram showing how "portfolio effect" and "distribution effect" of pedodiversity can vary within the country: (**a**) pedodiversity by soil order area; (**b**) value of soil organic carbon (SOC) storage, (**c**) value of soil inorganic carbon (SIC) storage, (**d**) value of total soil carbon (TSC) storage in the upper 2 m depth based on avoided or realized social cost of CO_2 (SC-CO_2) of $42 (USD) per metric ton of CO_2 [30,35,36,51] by soil order. Note: T = trillion = 10^{12}.

Some regions with homogenous soils are characterized by low ES and productivity (e.g., Aridisols). In terms of pedodiversity and ES, it is not just the density or numbers of different soils, but their properties (e.g., chemical and physical) as they relate to the effectiveness (level of performance) and reliability (consistency and predictability) to drive production including agriculture [78]. The soil-to-agricultural market value chain is heavily dependent on large homogeneous areas of soils with high agricultural productivity associated with "soil carbon hotspots" (e.g., Midwest, Northern Plains in the U.S.) [37]. These areas have the most significant pedodiversity loss (and even extinction) and some of the lowest proportion of the U.S. population (Table 12, Figure 5). There is a potential to manage these "hotspots" with precision agriculture technology. It should be noted that not all homogeneous soil areas necessarily have a high ES value.

Loss of pedodiversity may continue, considering projected world population growth [79,80], given that the Midwest and Northern Plains regions export large quantities of agricultural products to the world. Economic estimates focus on the profit from provisioning ES through the sales of agricultural products without considering regulating (e.g., the social cost of pollution including greenhouse emissions) and replacement cost associated with loss of soil nutrients (Figure 6) [80,81].

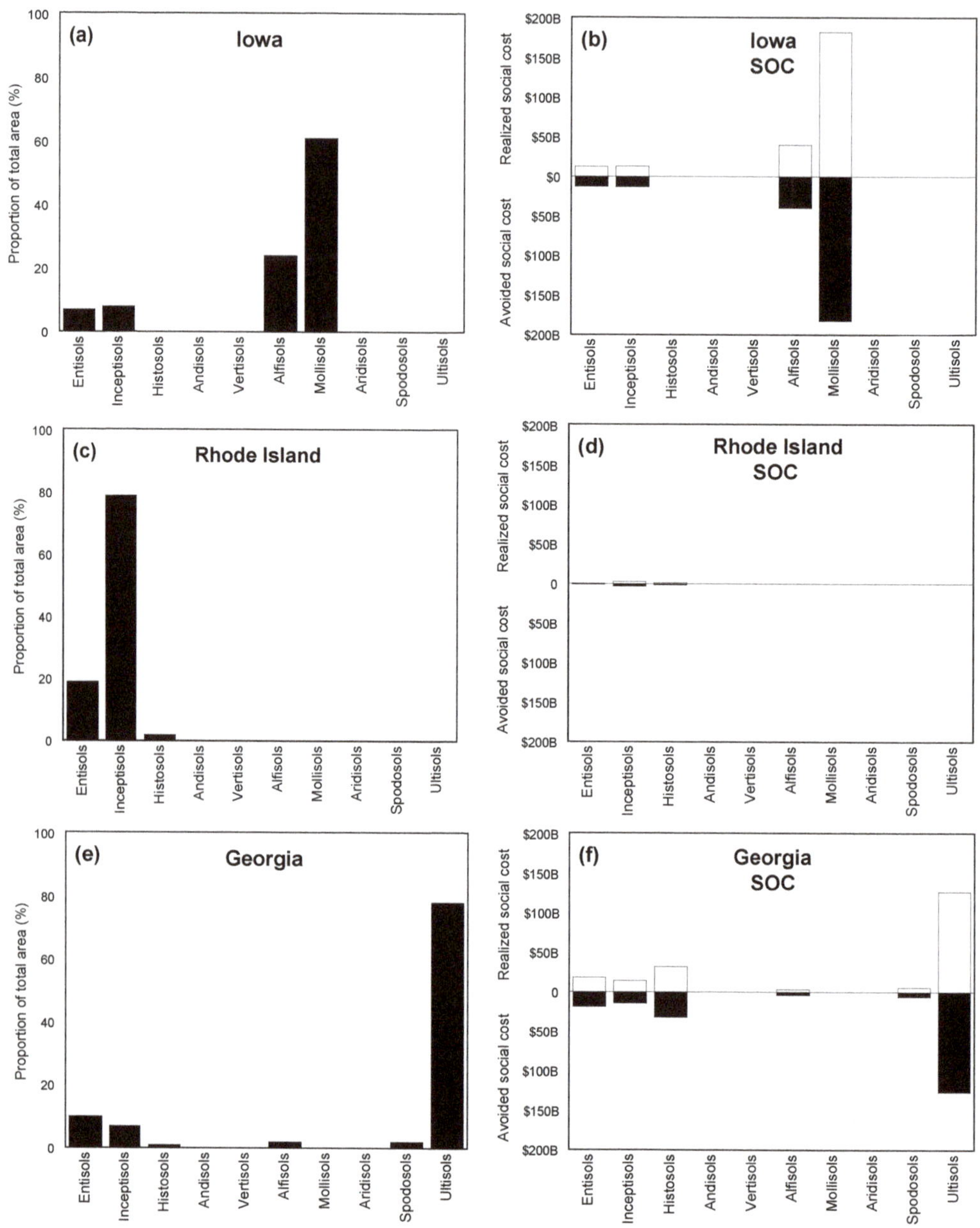

Figure 12. Diagram showing how the "portfolio effect" and the "distribution effect" of pedodiversity can vary by state: (**a**,**c**,**e**) pedodiversity by soil order area; (**b**,**d**,**f**) value of soil organic carbon (SOC) storage in the upper 2 m depth based on avoided or realized social cost of CO_2 (SC-CO_2) of $42 (USD) per metric ton of CO_2 [35,51] by soil order. Note: B = billion = 10^9.

This focus on the direct use value, with little or no regard to the passive and intrinsic use-values, may lead to unsustainable use of pedodiversity (Table 3 and Figure 10). The amount available to replace the lost value of pedodiversity, through insurance value (Table 3) may not be possible. For example, replacement of some soil nutrients (e.g., phosphorus and potassium) may not be economically feasible since their mineral supply is very limited in the world [81,82]). Estimates of social costs can be performed based on taxonomic pedodiversity (biophysical accounts) (Table 14) and using administrative (boundary-based) accounts (Table 15).

Table 14. Degree of soil development and area-normalized midpoint values of soil organic carbon (SOC) storage in the upper 2 m depth within the contiguous United States (U.S.), based on midpoint SOC numbers from Guo et al., 2006 [39] and a social cost of carbon (SC-CO_2) of $42 (USD) per metric ton of CO_2 [51].

Slight <—— Degree of Weathering and Soil Development ——> Strong					
Slight Weathering		**Intermediate Weathering**		**Strong Weathering**	
Soil Order	**Midpoint SOC Value per Area ($ m^{-2})**	**Soil Order**	**Midpoint SOC Value per Area ($ m^{-2})**	**Soil Order**	**Midpoint SOC Value per Area ($ m^{-2})**
Entisols	1.23	Aridisols	0.62	Spodosols	1.89
Inceptisols	1.37	Vertisols	2.26	Ultisols	1.09
Histosols	21.58	Alfisols	1.16	Oxisols	-
Gelisols	-	Mollisols	2.08		

Biocapacity and ecological footprint are commonly used in environmental carrying capacity (ECC) assessments (e.g., urban areas) [83], and clearly, pedocapacity, or the capacity of the soil to provide various ES, should be a part of these calculations as well. The value of pedocapacity should include both intrinsic (e.g., avoided social costs) and extrinsic (e.g., realized social costs) estimates. Extrinsic realized social costs may be impossible to estimate because their impacts extend beyond the pedosphere boundary, such as in the case of realized social costs of carbon (SC-CO_2) (Figure 13). Limited biocapacity (including pedocapacity) in urban areas often results in urban areas exceeding their ECC, sometimes crossing into other countries [83].

Table 15. Integration of biophysical accounts (science-based) and administrative accounts (boundary-based). Degree of soil development and area-normalized midpoint values of soil organic carbon (SOC) storage in the upper 2 m depth within the contiguous United States (U.S.), based on midpoint SOC numbers from Guo et al., 2006 [39] and a social cost of carbon (SC-CO_2) of $42 (USD) per metric ton of CO_2 [51].

Slight ←—— Degree of Weathering and Soil Development ——→ Strong					
Slight Weathering		**Intermediate Weathering**		**Strong Weathering**	
State (Region)	**Midpoint SOC Value per Area ($ m^{-2})**	**State (Region)**	**Midpoint SOC Value per Area ($ m^{-2})**	**State (Region)**	**Midpoint SOC Value per Area ($ m^{-2})**
Connecticut	2.42	Iowa	3.16	Alabama	1.42
Delaware	4.10	Illinois	1.96	Florida	5.44
Maryland	2.06	Indiana	2.16	Georgia	2.08
Maine	2.54	Michigan	3.71	Kentucky	1.12
New Hampshire	2.40	Minnesota	3.99	Mississippi	1.60
New Jersey	2.56	Missouri	1.36	North Carolina	3.42
New York	2.08	Ohio	1.57	South Carolina	2.77
Pennsylvania	0.91	Wisconsin	3.17	Tennessee	1.09
Rhode Island	2.70	**(Midwest)**	**2.73**	Virginia	1.23
Vermont	2.23			**(Southeast)**	**2.31**
West Virginia	0.74				
(East)	**1.82**				

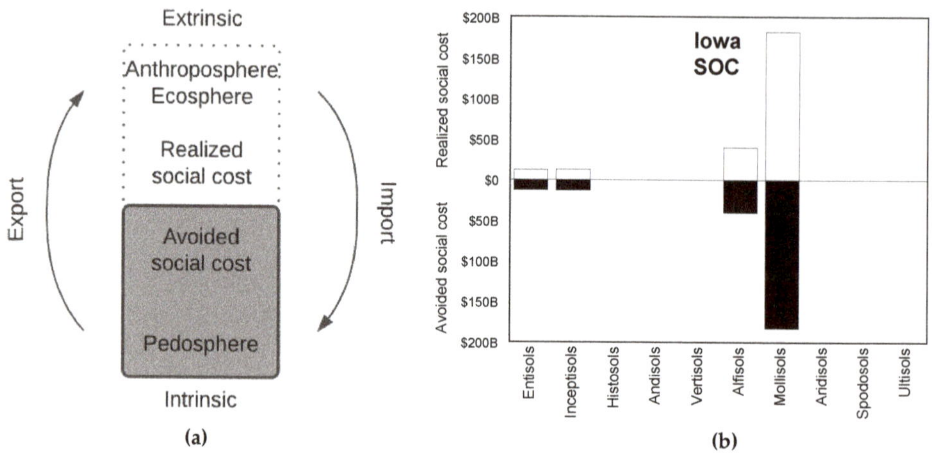

Figure 13. Relationship between intrinsic (e.g., avoided social costs) and extrinsic (e.g., realized social costs) estimates of social costs associated with pedosphere in general (**a**), and using the state of Iowa and soil organic carbon (SOC) as an example (**b**) based on midpoint SOC numbers from Guo et al., 2006 [41] and a social cost of carbon (SC-CO_2) of $42 (USD) per metric ton of CO_2 [51].

Most research efforts are focused on documenting biodiversity loss, but pedodiversity loss can be of catastrophic consequence to humanity; therefore, it is important to understand the extinction patterns and their underlying processes [84]. Global warming has various impacts on the soil, especially on soil organic matter (SOM) decomposition, which is an oxidation process accompanied by oxygen consumption and CO_2 release [85]:

$$\text{R-(C, 4H)} + 2O_2 \rightarrow CO_2\uparrow + 2H_2O + \text{energy } (478 \text{ kJ mol}^{-1} \text{ C}) \quad (1)$$

The decomposition of SOM, which is accompanied by the release of CO_2 and other gases, accelerates in the presence of increased heat (e.g., global warming) and can be compared to a "fire triangle." Analogous to the "fire triangle," the "SOM decomposition triangle" represents the three items (soil organic matter, oxygen, and heat) that feed SOM decomposition emissions of CO_2 and other invisible gases that fuel global warming (Figure 14). Unlike a regular fire, which is visible, the invisible greenhouse gases are like an invisible "fire" that can only be prevented and minimized by identifying the location of "fuel loading" (soil organic matter) throughout the landscape.

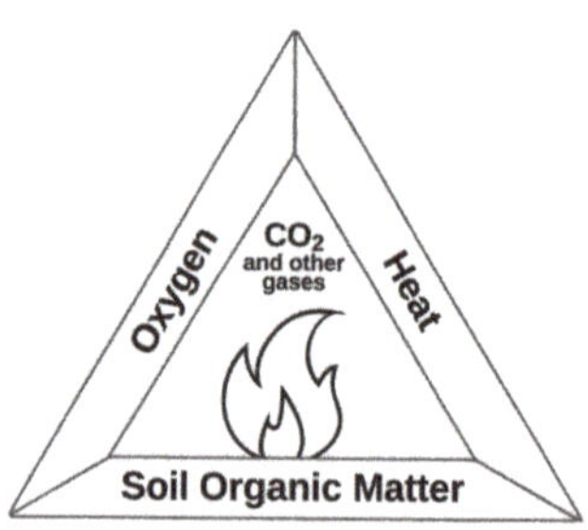

Figure 14. The "SOM decomposition triangle" represents the three items (soil organic matter, oxygen, and heat) that feed soil decomposition emissions of CO_2 and other gases that fuel global warming.

The Earth's regions and soils with high SOM levels (e.g., Histosols, Gelisols, Alfisols, Mollisols, and Vertisols) tend to be more susceptible to greenhouse gas emissions with increasing global temperatures. Histosols and Gelisols are of particular concern because they are threatened by draining (Histosols) and thawing (Gelisols), which can cause soil degradation with global consequences [86,87]. For example, Pastick et al. (2015) [86] reported that 16 to 24% (out of 38%) of near-surface permafrost will disappear by the end of the 21st century. States and regions with a higher proportion of their area occupied by high-risk soils ("hotspots") [37] are experiencing the highest losses in ES (especially in provisioning), which is often caused by the demand for ES (e.g., provisioning) outside their boundaries. According to Hansjurgens et al. (2018) [88], pedodiversity distribution around the world poses an important question about "fairness" not only in the provisioning of ES but also in the associated and past ED costs. Administrative accounts (e.g., states and regions) in combination with pedodiversity concepts can provide information to develop cost-effective policy options to manage benefits (ES) and risks (ED) from pedodiversity. These benefits and risks often extend beyond the boundaries of individual states and regions (e.g., greenhouse emissions), therefore creating a need for a long-term coordinated vision, collaboration, and monitoring. It should be noted that both the ES framework and its valuation measures are human-centric, bias, and focused on short-term human scale interests instead of treating and valuing pedodiversity at a long-term geologic time scale [12,20]. According to Table 3 and Figure 10, pedodiversity tangibility values tend to decrease from "actual use" values to "intrinsic" values (benefits to nature). Soil series are often associated with these monetary "actual use" values (e.g., provisioning: food, etc.) because they represent soil properties within property boundaries in contrast to soil orders, which are often associated with large spatial extents which cross multiple property boundaries representing "intrinsic" values and social costs (Figure 15). According to Guerry et al. (2015) [89], "perhaps the most difficult challenge in the path of success is removing the fundamental asymmetry at the heart of economic systems, which rewards the production of marketed commodities but not the provision of nonmarketed ecosystem services or the sustainable use of natural capital that supports these services."

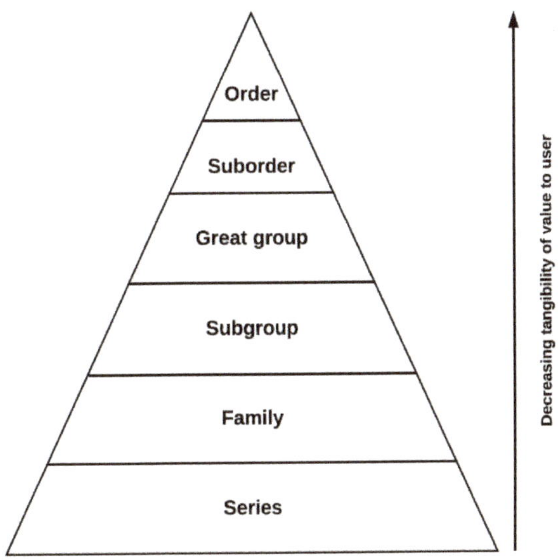

Figure 15. Taxonomic pedodiversity and tangibility of value to the user (adapted from Soil Survey Staff, 1999 [15]).

Market transformations of pedodiversity can result not only in welfare but damages as well [90], which can pose a threat to soil security, national security, food security, infrastructure, and human life [91–93]. Pedodiversity can be both valuable and problematic to human well-being, depending on the point of view. The value of pedodiversity is that it is a human construct, which is used to "categorize" the soil continuum in a discrete way [93] and can be applied to ES/ED within administrative boundaries for socio-economic analysis. The problem with the discretization of both soils and related ES/ED is that it can oversimplify the complex nature of pedodiversity, which is a product of the interaction between the Earth's various spheres and their diversities (Figure 1). For example, Bach et al. (2020) [94] discusses the contribution of soil biodiversity to ES, which varies by soil type (taxonomic pedodiversity) and would require integration of pedodiversity with soil biodiversity for sustainable soil management. Human activity (e.g., agriculture and urbanization) can erode soil pedodiversity by converting soils to more uniform human-altered soils (Anthrosols) with a reduction in soil ES [95]. The perception of pedodiversity [96] and its contribution to ES/ED depends on the human "behavioral dimensions" ("human nature"), which are less understood in both perceived ES benefits and ED, especially with regards to regulating ES/ED (e.g., greenhouse gas emission) which tend to be of global significance [97].

5. Conclusions

This study examined the application of soil diversity (pedodiversity) concepts (taxonomic, genetic, parametric, and functional) and its measures to value ES/ED with examples based on the contiguous United States (U.S.), its administrative units, and the systems of soil classification (e.g., U.S. Department of Agriculture (USDA) Soil Taxonomy, Soil Survey Geographic (SSURGO) Database). Pedodiversity provides an important context (e.g., "portfolio effect", "distribution effect", and "evenness effect") for analyzing, interpreting, and reporting ES/ED within the ES framework for business applications. Taxonomic pedodiversity in the contiguous U.S. exhibits high soil diversity, which is not evenly distributed within administrative units. Pedodiversity distribution around the country poses an important question about "fairness" not only in the provisioning of ES but also in the associated and past ED costs. Pedodiversity in the U.S. is under various threats, including land cover change (urbanization, agriculture, deforestation) and climate change (existential threat to the soil order of Gelisols). Pedodiversity losses are especially high in agriculturally productive and important soils (e.g., Alfisols, Mollisols) and regions (e.g., Midwest, Northern Plains, South Central) with some of the lowest proportions of U.S. total population. There is a mismatch between "potential" and "realized" supply/demand of flow-dependent ES/ED. With over 80% of the U.S. population living in urban environments, there is an increase in demand for ES, which is not always supplied by local soil resources and requires soil ecosystem goods and services to be "imported" from other geographic areas. The flow of ecosystem goods and services is often accompanied by the expansion of agricultural areas based on available soil resources. Low-fertility soils and other extrinsic factors (e.g., low precipitation) may limit the flow of ecosystem goods and services. Climate change will have a direct impact on pedodiversity and the classification of soils, with some soil types disappearing and others changing in both extent, and properties. Administrative accounts (e.g., states and regions) in combination with pedodiversity concepts can provide information to develop cost-effective policy options to manage benefits (ES) and risks (ED) from pedodiversity. These benefits and risks often extend beyond the boundaries of individual states and regions (e.g., greenhouse emissions), creating a need for a long-term coordinated vision, collaboration, and monitoring.

Author Contributions: Conceptualization, E.A.M.; methodology, E.A.M., M.A.S., and H.A.Z.; formal analysis, E.A.M.; writing—original draft preparation, E.A.M.; writing—review and editing, E.A.M., C.J.P., G.C.P., and M.A.S.; visualization, H.A.Z. All authors have read and agreed to the published version of the manuscript.

Funding: This research received no external funding.

Acknowledgments: We would like to thank the reviewers for their constructive comments and suggestions.

Conflicts of Interest: The authors declare no conflict of interest.

Abbreviations

ED	Ecosystem disservices
ES	Ecosystem services
EPA	Environmental Protection Agency
SC-CO_2	Social cost of carbon emissions
SDGs	Sustainable Development Goals
SOC	Soil organic carbon
SIC	Soil inorganic carbon
SOM	Soil organic matter
SSURGO	Soil Survey Geographic Database
TEV	Total economic value
TSV	Total system value
USDA	United States Department of Agriculture
US	United States

References

1. Jenny, H. *Factors of Soil Formation*; McGraw Hill: New York, NY, USA, 1941.
2. Amundson, R.; Guo, Y.; Gong, P. Soil diversity and land use in the United States. *Ecosystems* **2003**, *6*, 470–482. [CrossRef]
3. Mattson, S. The constitution of the pedosphere. *Ann. Agric. Coll. Swed.* **1938**, *5*, 261–279.
4. Odeh, L.O.A. In Discussion of: Ibáñez, J.J.; De-Alba, S.; Lobo, A.; Zucarello, V. Pedodiversity and global soil pattern at coarse scales. *Geoderma* **1998**, *83*, 203–205.
5. Ibáñez, J.J.; De-Alba, S.; Lobo, A.; Zucarello, V. Pedodiversity and global soil patterns at coarse scales (with Discussion). *Geoderma* **1998**, *83*, 171–214. [CrossRef]
6. Phillips, J.D. The relative importance of intrinsic and extrinsic factors in pedodiversity. *Ann. Am. Assoc. Geogr.* **2001**, *91*, 609–621. [CrossRef]
7. Dobrovolskii, G.V. Dokuchaev's language as a reflection of his broad vision and literary talent. *Eurasian Soil Sci.* **2007**, *40*, 1008–1015. [CrossRef]
8. Fridland, V.M. *Pattern of the Soil Cover*; John Wiley & Sons: Hoboken, NJ, USA, 1977; ISBN 13-978-0470991671.
9. Hole, F.D.; Campbell, J.B. *Soil Landscape Analysis*; Rowman and Littlefield: Lanham, MD, USA, 1985; 214p, ISBN 13 978-0865981409.
10. McBratney, A.B. On variation, uncertainty and informatics in environmental soil management. *Aust. J. Soil Res.* **1992**, *30*, 913–935. [CrossRef]
11. Ibáñez, J.J.; De-Alba, S.; Bermúdez, F.F.; García-Álvarez, A. Pedodiversity: Concepts and measures. *Catena* **1995**, *24*, 215–232. [CrossRef]
12. Mikhailova, E.A.; Post, C.J.; Schlautman, M.A.; Post, G.C.; Zurqani, H.A. The business side of ecosystem services of soil systems. *Earth* **2020**, *1*, 2. [CrossRef]
13. Ibáñez, J.J.; Saldaña, A.; Olivera, D. Biodiversity and pedodiversity: A matter of coincidence? *SJSS* **2012**, *2*, 8–12. [CrossRef]
14. Guo, Y.; Amundson, R.; Gong, P.; Ahrens, R. Taxonomic structure, distribution, and abundance of the soils in the USA. *SSSAJ* **2003**, *67*, 1507–1516. [CrossRef]
15. Soil Survey Staff. A Basic System of Soil Classification for Making and Interpreting Soil Surveys. In *Soil Taxonomy*; Agricultural Handbook 436; US Department of Agriculture, Natural Resources Conservation Service: Washington, DC, USA, 1999.
16. Adhikari, K.; Hartemink, A.E. Linking soils to ecosystem services—A global review. *Geoderma* **2016**, *262*, 101–111. [CrossRef]
17. Comerford, N.B.; Franzlubbers, A.J.; Stromberger, M.E.; Morris, L.; Markewitz, D.; Moore, R. Assessment and evaluation of soil ecosystem services. *Soil Horiz.* **2013**, *54*, 1–14. [CrossRef]
18. Baveye, P.C.; Baveye, J.; Gowdy, J. Soil "ecosystem" services and natural capital: Critical appraisal of research on uncertain ground. *Front. Environ. Sci.* **2016**, *4*, 41. [CrossRef]
19. Millennium Ecosystem Assessment (MEA). *Ecosystems and Human Well-Being: Synthesis*; Island Press: Washington, DC, USA, 2005.
20. Bartkowski, B.; Bartke, S.; Helming, K.; Paul, C.; Techen, A.; Hansjürgens, B. Potential of the economic valuation of soil-based ecosystem services to inform sustainable soil management and policy. *Peer J.* **2020**, *8*, e8749. [CrossRef]
21. Bartkowski, B. Are diverse ecosystems more valuable? Economic value of biodiversity as result of uncertainty and spatial interactions in ecosystem service provision. *Ecosyst. Serv.* **2017**, *24*, 50–57. [CrossRef]
22. De Groot, R.; Jax, K.; Harrison, P. Links between biodiversity and ecosystem services. In *OpenNESS Ecosystem Services Reference Book*; Potschin, M., Jax, K., Eds.; EC FP7 Grant Agreement No. 308428; 2016; Available online: http://www.openness-project.eu/library/reference-book (accessed on 10 October 2020).

23. Schnediders, A.; Van Daele, T.; Van Landuyt, W.; Van Reeth, W. Biodiversity and ecosystem services: Complementary approaches for ecosystem management? *Ecol. Indic.* **2012**, *21*, 123–133. [CrossRef]
24. Cardinale, B.; Duffy, J.; Gonzalez, A.; Hooper, D.; Perrings, C.; Venail, P.; Narwani, A.; Mace, G.; Tilman, D.; Naeem, S.; et al. Biodiversity loss and its impact on humanity. *Nature* **2012**, *486*, 59–67. [CrossRef]
25. Stephenson, J. Business, biodiversity and ecosystem services: Policies priorities for engaging business to improve health of ecosystems and conserve biodiversity. In Proceedings of the 28th Round Table on Sustainable Development, Telangana, India, 16 October 2012.
26. Chandler, R.D.; Mikhailova, E.A.; Post, C.J.; Moysey, S.M.J.; Schlautman, M.A.; Sharp, J.L.; Motallebi, M. Integrating soil analyses with frameworks for ecosystem services and organizational hierarchy of soil systems. *Commun. Soil Sci. Plant Anal.* **2018**, *49*, 1835–1843. [CrossRef]
27. Soil Survey Staff, Natural Resources Conservation Service, United States Department of Agriculture. Soil Survey Geographic (SSURGO) Database. Available online: https://www.nrcs.usda.gov/wps/portal/nrcs/detail/soils/survey/?cid=nrcs142p2_053627 (accessed on 10 September 2020).
28. Clarivate Analytics. Web of Science. Subscription-Based Website. 2020. Available online: https://clarivate.com/tag/web-of-science/ (accessed on 21 September 2020).
29. Pavan, A.L.R.; Ometto, A.R. Ecosystem services in life cycle assessment: A novel conceptual framework for soil. *Sci. Total Environ.* **2018**, *643*, 1337–1347. [CrossRef] [PubMed]
30. Groshans, G.R.; Mikhailova, E.A.; Post, C.J.; Schlautman, M.A. Accounting for soil inorganic carbon in the ecosystem services framework for the United Nations sustainable development goals. *Geoderma* **2018**, *324*, 37–46. [CrossRef]
31. Nimmo-Bell. *MAF Biosecurity New Zealand. TEV for Biodiversity*; Nimmo-Bell: Wellington, New Zealand, 2011; Available online: http://www.nimmo-bell.co.nz/pdf/ManualRev29411.pdf (accessed on 10 October 2020).
32. Van Zyl, S.; Au, J. *The Start of a Conversation on the Value of New Zealand's Natural Capital*; Living Standards Series: Discussion Paper 18/03; Office of the Chief Economic Advisor: Wellington, New Zealand, 2018. Available online: https://www.treasury.govt.nz/sites/default/files/2018-02/dp18-03.pdf (accessed on 10 October 2020).
33. Soil Survey Staff; Natural Resources Conservation Service; United States Department of Agriculture. Web Soil Survey. Available online: http://websoilsurvey.sc.egov.usda.gov/ (accessed on 16 October 2020).
34. Soil Science Society of America. Penistaja New Mexico State Soil. State Soil Booklets. Available online: https://www.soils4teachers.org/files/s4t/k12outreach/nm-state-soil-booklet.pdf (accessed on 14 October 2020).
35. Mikhailova, E.A.; Groshans, G.R.; Post, C.J.; Schlautman, M.A.; Post, G.C. Valuation of soil organic carbon stocks in the contiguous United States based on the avoided social cost of carbon emissions. *Resources* **2019**, *8*, 153. [CrossRef]
36. Mikhailova, E.A.; Groshans, G.R.; Post, C.J.; Schlautman, M.A.; Post, G.C. Valuation of total soil carbon stocks in the contiguous United States based on the avoided social cost of carbon emissions. *Resources* **2019**, *8*, 157. [CrossRef]
37. Mikhailova, E.A.; Post, C.J.; Schlautman, M.A.; Post, G.C.; Zurqani, H.A. Determining farm-scale site-specific monetary values of "soil carbon hotspots" based on avoided social costs of CO_2 emissions. *Cogent Environ. Sci.* **2020**, *6:1*, 1817289. [CrossRef]
38. Brevik, E.C.; Hartemink, A.E. Soil maps of the United States of America. *Soil Sci. Soc. Am.* **2013**, *77*, 1117–1132. [CrossRef]
39. Guo, Y.; Amundson, R.; Gong, P.; Yu, Q. Quantity and spatial variability of soil carbon in the conterminous United States. *Soil Sci. Soc. Am. J.* **2006**, *70*, 590–600. [CrossRef]
40. Hartemink, A.E.; Zhang, Y.; Bockheim, J.G.; Curi, N.; Silva, S.H.G.; Grauer-Gray, J.; Lowe, D.J.; Krasilnikov, P. Soil Horizon Variation: A review. *Adv. Agron.* **2020**, *160*. [CrossRef]
41. Mikhailova, E.A.; Bryant, R.B.; Vassenev, I.I.; Schwager, S.J.; Post, C.J. Cultivation effects on soil organic carbon and total nitrogen at depth in the Russian Chernozem. *Soil Sci. Soc. Am. J.* **2000**, *64*, 738–745. [CrossRef]
42. Bullock, C.H.; Collier, M.J.; Convery, F. Peatlands, their economic value and priorities for their future management—The example of Ireland. *Land Use Policy* **2012**, *29*, 921–928. [CrossRef]
43. Anisimov, O.A. Potential feedback of thawing permafrost to the global climate system through methane emission. *Environ. Res. Lett.* **2007**, *2*, 045016. [CrossRef]
44. Singh, B.; Schulze, D.G. Soil minerals and plant nutrition. *Nat. Educ. Knowl.* **2015**, *6*, 1–10.
45. Zurqani, H.A.; Mikhailova, E.A.; Post, C.J.; Schlautman, M.A.; Elhawej, A.R. A review of Libyan soil databases for use within an ecosystem services framework. *Land* **2019**, *8*, 82. [CrossRef]
46. Groshans, G.R.; Mikhailova, E.A.; Post, C.J.; Schlautman, M.A.; Zhang, L. Determining the value of soil inorganic carbon stocks in the contiguous United States based on the avoided social cost of carbon emissions. *Resources* **2019**, *8*, 119. [CrossRef]
47. Mikhailova, E.A.; Post, C.J.; Gerard, P.D.; Schlautman, M.A.; Cope, M.P.; Groshans, G.R.; Stiglitz, R.Y.; Zurqani, H.A.; Galbraith, J.M. Comparing field sampling and soil survey database for spatial heterogeneity in surface soil granulometry: Implications for the ecosystem services assessment. *Front. Environ. Sci.* **2019**, *7*, 128. [CrossRef]
48. Oliver, M.A.; Gregory, P.J. Soil, food security and human health: A review. *Eur. J. Soil Sci.* **2015**, *66*, 257–276. [CrossRef]
49. Merrill, D.; Leatherby, L. Here's how America uses its land. *Bloomberg*. 2018. Available online: https://www.bloomberg.com/graphics/2018-us-land-use/ (accessed on 14 October 2020).
50. Schlesinger, W.H.; Amundson, R. Managing for soil carbon sequestration: Let's get realistic. *Glob. Chang. Biol.* **2019**, *25*, 386–389. [CrossRef]

51. EPA. The Social Cost of Carbon. EPA Fact Sheet. 2016. Available online: https://19january2017snapshot.epa.gov/climatechange/social-cost-carbon_.html (accessed on 15 March 2019).
52. Mikhailova, E.A.; Zurqani, H.A.; Post, C.J.; Schlautman, M.A. Assessing ecosystem services of atmospheric calcium and magnesium deposition for potential soil inorganic carbon sequestration. *Geosciences* **2020**, *10*, 200. [CrossRef]
53. Duncombe, J. The ticking time bomb of Arctic permafrost. *Eos* **2020**, *101*. [CrossRef]
54. Restuccia, F.; Huang, X.; Rein, G. Self-ignition of natural fuels: Can wildfires of carbon-rich soil start by self-heating? *Fire Saf. J.* **2017**, *91*, 828–834. [CrossRef]
55. Borrelli, P.; Robinson, D.A.; Panagos, P.; Lugato, E.; Yang, J.E.; Alewell, C.; Wuepper, D.; Montarella, L.; Ballabio, C. Land use and climate change impacts on global soil erosion by water (2015–2070). *PNAS* **2020**, *117*, 21994–22001. [CrossRef]
56. Pavao-Zuckerman, M.A. The nature of urban soils and their role in ecological restoration in cities. *Restor. Ecol.* **2008**, *16*, 642–649. [CrossRef]
57. Vasenev, V.I.; Van Oudenhoven, A.P.E.; Romzaykina, O.N.; Hajiaghaeva, R.A. The ecological functions and ecosystem services of urban and technogenic soils: From theory to practice (A review). *Eurasian Soil Sci.* **2018**, *51*, 1119–1132. [CrossRef]
58. Grunewald, K.; Bastian, O. Special issue: Maintaining ecosystem services to support urban needs. *Sustainability* **2017**, *9*, 1647. [CrossRef]
59. Groshans, G.R.; Mikhailova, E.A.; Post, C.J.; Schlautman, M.A.; Zurqani, H.A.; Zhang, L. Assessing the value of soil inorganic carbon for ecosystem services in the contiguous United States based on liming replacement costs. *Land* **2018**, *7*, 149. [CrossRef]
60. Wikipedia. List of States and Territories of the United States by Population. Available online: https://en.wikipedia.org/wiki/List_of_states_and_territories_of_the_United_States_by_population (accessed on 22 October 2020).
61. United States Summary. *2010 Census of Population and Housing, Population and Housing Unit Counts*; CPH-2-5; U.S. Government Printing Office, U.S. Census Bureau: Washington, DC, USA, 2012; p. 42. Available online: https://www2.census.gov/library/publications/decennial/2010/cph-2/cph-2-1.pdf (accessed on 10 October 2020).
62. Goldenberg, R.; Kalantari, Z.; Cvetkovic, V.; Mörtberg, U.; Deal, B.; Destouni, G. Distinction, quantification and mapping of potential and realized supply-demand of flow-dependent ecosystem services. *Sci. Total Environ.* **2017**, *593–594*, 599–609. [CrossRef]
63. Hewes, L. *The Suitcase Farming Frontier: A study in the Historical Geography of the Central Great Plains*; University of Nebraska Press: Linkoln, NE, USA, 1974; 281p.
64. Lee, J.A.; Gill, T.E. Multiple causes of wind erosion in the Dust Bowl. *Aeolian Res.* **2015**, *19*, 15–36. [CrossRef]
65. Wentland, S.A.; Ancona, Z.H.; Bagstad, K.J.; Boyd, J.; Hass, J.L.; Gindelsky, M.; Moulton, J.G. Accounting for land in the United States: Integrating physical land cover, land use, and monetary valuation. *Ecosyst. Serv.* **2020**, *46*, 101178. [CrossRef]
66. Lu, M.; Zhou, X.; Yang, Q.; Li, H.; Luo, Y.; Fang, C.; Chen, J.; Yang, X.; Li, B. Responses of ecosystem carbon cycle to experimental warming: A meta-analysis. *Ecology* **2013**, *94*, 726–738. [CrossRef] [PubMed]
67. Nearing, M.A.; Pruski, F.F.; O'Neal, M.R. Expected climate change impacts on soil erosion rates: A review. *J. Soil Water Conserv.* **2004**, *59*, 43–50.
68. Chen, C.; McCarl, B.; Chang, C. Climate change, sea level rise and rice: Global market implications. *Clim. Chang.* **2012**, *110*, 543–560. [CrossRef]
69. Walthall, C.L.; Hatfield, J.; Backlund, P.; Lengnick, L.; Marshall, E.; Walsh, M.; Adkins, S.; Aillery, M.; Ainsworth, E.A.; Ammann, C.; et al. *Climate Change and Agriculture in the United States: Effects and Adaptation*; USDA Technical Bulletin 1935; USDA: Washington, DC, USA, 2012; 186p.
70. Reilly, J.; Tubiello, F.; McCarl, B.; Abler, D.; Darwin, R.; Fuglie, K.; Hollinger, S.; Izarralde, C.; Jagtap, S.; Jones, J.; et al. Agriculture and climate change: New results. *Clim. Chang.* **2003**, *57*, 43–69. [CrossRef]
71. Amundson, R.; Berhe, A.A.; Hopmans, J.W.; Olson, C.; Sztein, A.E.; Sparks, D.L. Soil and human security in the 21st century. *Science* **2015**, *348*. [CrossRef] [PubMed]
72. Pascual, U.; Termansen, M.; Hedlund, K.; Brussaard, L.; Faber, J.H.; Foudi, S.; Lemanceau, P.; Jørgensen, S.L. On the value of soil biodiversity and ecosystem services. *Ecosyst. Serv.* **2015**, *15*, 11–18. [CrossRef]
73. Mace, G.M.; Norris, K.; Fitter, A.H. Biodiversity and ecosystem services: A multilayered relationship. *Trends Ecol. Evol.* **2012**, *27*, 19–26. [CrossRef]
74. Gerasimova, M.I. Chinese Soil Taxonomy: Between the American and the International classification systems. *Eurasian J. Soil Sci.* **2010**, *43*, 945–949. [CrossRef]
75. Tilman, D.; Lehman, C.L.; Bristow, C.E. Diversity-stability relationships: Statistical inevitability or ecological consequence? *Am. Nat.* **1998**, *151*, 277–282. [CrossRef] [PubMed]
76. Gantioler, S.; Rayment, M.; Bassi, S.; Kettunen, M.; McConville, A.; Landgrebe, R.; Gerdes, H.; ten Brink, P. *Costs and Socio-Economic Benefits Associated with the Natura 2000 Network*; Final report to the European Commission; DG Environment on Contract ENV.B.2/SER/2008/0038; Institute for European Environmental Policy/GHK /Ecologic: Brussels, Belgium, 2010.
77. Wall, D.H.; Nielsen, U.N. Biodiversity and ecosystem services: Is it the same below ground? *Nat. Educ. Knowl.* **2012**, *3*, 8.
78. Vos, C.C.; Grashof-Bokdam, C.J.; Opdam, P.F.M. *Biodiversity and Ecosystem Services: Does Species Diversity Enhance Effectiveness and Reliability? A Systematic Literature Review*; WOT-Technical Report 25; Statutory Research Tasks Unit for Nature and the Environment (WOT Natuur and Milieu): Wageningen, The Netherlands, 2014; 64p, ISSN 2352-2739.
79. Lal, R. Soils and sustainable agriculture. A review. *Agron. Sustain. Dev.* **2008**, *28*, 57–64. [CrossRef]
80. Power, A.G. Ecosystem services and agriculture: Tradeoffs and synergies. *Phil. Trans. R. Soc. B* **2010**, *365*, 2959–2971. [CrossRef]

81. Cordell, D.; Drangert, J.O.; White, S. The story of phosphorus: Global food security and food for thought. *Glob. Environ. Chang.* **2009**, *19*, 292–305. [CrossRef]
82. Mikhailova, E.A.; Post, G.C.; Cope, M.P.; Post, C.J.; Schlautman, M.A.; Zhang, L. Quantifying and mapping atmospheric potassium deposition for soil ecosystem services assessment in the United States. *Front. Environ. Sci.* **2019**, *7*, 74. [CrossRef]
83. Świader, M.; Lin, D.; Szewrański, S.; Kazak, J.K.; Iha, K.; van Hoof, J.; Belčáková, I.; Altiok, S. The application of ecological footprint and biocapacity for environmental carrying capacity assessment: A new approach for European cities. *Environ. Sci. Policy* **2020**, *105*, 56–74. [CrossRef]
84. Raffaelli, D. How extinction patterns affect ecosystems. *Science* **2004**, *306*, 1141–1142. [CrossRef] [PubMed]
85. Brady, N.C. ; Weil, R.R. *The Nature and Properties of Soils*, 13rd ed.; Pearson Education: London, UK, 2002.
86. Pastick, N.J.; Torre Jorgenson, M.; Wylie, B.K.; Nield, S.J.; Johnson, K.D.; Finley, A.O. Distribution of near-surface permafrost in Alaska: Estimates of present and future conditions. *Remote Sens. Environ.* **2015**, *168*, 301–315. [CrossRef]
87. Leifeld, J.; Menichetti, L. The underappreciated potential of peatlands in global climate change mitigation strategies. *Nat. Commun.* **2018**, *9*, 1071. [CrossRef] [PubMed]
88. Hansjürgens, B.; Lienkamp, A.; Möckel, S. Justifying soil protection and sustainable soil management: Creation-ethical, legal and economic considerations. *Sustainability* **2018**, *10*, 3807. [CrossRef]
89. Guerry, A.D.; Polasky, S.; Lubchenco, J.; Chaplin-Kramer, R.; Daily, G.C.; Griffin, R.; Ruckelshaus, M.; Bateman, I.J.; Duraiappah, A.; Elmqvist, T.; et al. Natural capital and ecosystem services informing decisions: From promise to practice. *PNAS* **2015**, *112*, 7348–7355. [CrossRef]
90. Jones, C.A.; DiPinto, L. The role of ecosystem services in USA natural resources liability litigation. *Ecosyst. Serv.* **2018**, *29*, 333–351. [CrossRef]
91. Zhu, Y.; Meharg, A. Protecting global soil resources for ecosystem services. *Ecosyst. Health Sustain.* **2015**, *1*, 11. [CrossRef]
92. McBratney, A.; Field, D.J.; Koch, A. The dimensions of soil security. *Geoderma* **2014**, *213*, 203–213. [CrossRef]
93. Ibáñez, J.J. *Diversity of Soils*; Oxford University Press: Oxford, UK, 2017. [CrossRef]
94. Bach, E.M.; Ramirez, K.S.; Fraser, T.D.; Wall, D.H. Soil biodiversity integrates solutions for a sustainable future. *Sustainability* **2020**, *12*, 2662. [CrossRef]
95. Dazzi, C.; Papa, G.L. Soil genetic erosion: New conceptual developments in soil security. *Int. Soil Water Conserv. Res.* **2019**, *7*, 317–324. [CrossRef]
96. Chen, J.; Zhang, X.-L.; Gong, Z.-T.; Wang, J. Pedodiversity: A controversial concept. *J. Geogr. Sci.* **2001**, *11*, 110–116. [CrossRef]
97. Asah, S.T.; Guerry, A.D.; Blahna, D.J.; Lawler, J.J. Perception, acquisition and use of ecosystem services: Human behavior, and ecosystem management and policy implications. *Ecosyst. Serv.* **2014**, *10*, 180–186. [CrossRef]

land

MDPI

Article

Soil Carbon Regulating Ecosystem Services in the State of South Carolina, USA

Elena A. Mikhailova [1,*], Hamdi A. Zurqani [1,2], Christopher J. Post [1], Mark A. Schlautman [3], Gregory C. Post [4], Lili Lin [5] and Zhenbang Hao [5]

1 Department of Forestry and Environmental Conservation, Clemson University, Clemson, SC 29634, USA; hzurqan@clemson.edu (H.A.Z.); cpost@clemson.edu (C.J.P.)
2 Department of Soil and Water Sciences, University of Tripoli, Tripoli 13538, Libya
3 Department of Environmental Engineering and Earth Sciences, Clemson University, Anderson, SC 29625, USA; mschlau@clemson.edu
4 Economics Department, Reed College, Portland, OR 97202, USA; grpost@reed.edu
5 University Key Lab for Geomatics Technology and Optimized Resources Utilization in Fujian Province, No. 15 Shangxiadian Road, Fuzhou 350002, China; lilil@g.clemson.edu (L.L.); zhenbanghao@fafu.edu.cn (Z.H.)
* Correspondence: eleanam@clemson.edu

Abstract: Sustainable management of soil carbon (C) at the state level requires valuation of soil C regulating ecosystem services (ES) and disservices (ED). The objective of this study was to assess the value of regulating ES from soil organic carbon (SOC), soil inorganic carbon (SIC), and total soil carbon (TSC) stocks, based on the concept of the avoided social cost of carbon dioxide (CO_2) emissions for the state of South Carolina (SC) in the United States of America (U.S.A.) by soil order, soil depth (0–200 cm), region and county using information from the State Soil Geographic (STATSGO) database. The total estimated monetary mid-point value for TSC in the state of South Carolina was $124.36B (i.e., $124.36 billion U.S. dollars, where B = billion = 10^9), $107.14B for SOC, and $17.22B for SIC. Soil orders with the highest midpoint value for SOC were: Ultisols ($64.35B), Histosols ($11.22B), and Inceptisols ($10.31B). Soil orders with the highest midpoint value for SIC were: Inceptisols ($5.91B), Entisols ($5.53B), and Alfisols ($5.0B). Soil orders with the highest midpoint value for TSC were: Ultisols ($64.35B), Inceptisols ($16.22B), and Entisols ($14.65B). The regions with the highest midpoint SOC values were: Pee Dee ($34.24B), Low Country ($32.17B), and Midlands ($29.24B). The regions with the highest midpoint SIC values were: Low Country ($5.69B), Midlands ($5.55B), and Pee Dee ($4.67B). The regions with the highest midpoint TSC values were: Low Country ($37.86B), Pee Dee ($36.91B), and Midlands ($34.79B). The counties with the highest midpoint SOC values were Colleton ($5.44B), Horry ($5.37B), and Berkeley ($4.12B). The counties with the highest midpoint SIC values were Charleston ($1.46B), Georgetown ($852.81M, where M = million = 10^6), and Horry ($843.18M). The counties with the highest midpoint TSC values were Horry ($6.22B), Colleton ($6.02B), and Georgetown ($4.87B). Administrative areas (e.g., counties, regions) combined with pedodiversity concepts can provide useful information to design cost-efficient policies to manage soil carbon regulating ES at the state level.

Keywords: accounting; carbon emissions, CO_2; climate change; inorganic; organic; pedodiversity

check for updates

Citation: Mikhailova, E.A.; Zurqani, H.A.; Post, C.J.; Schlautman, M.A.; Post, G.C.; Lin, L.; Hao, Z. Soil Carbon Regulating Ecosystem Services in the State of South Carolina, USA. *Land* **2021**, *10*, 309. https://doi.org/10.3390/land10030309

Academic Editor: Chiara Piccini

Received: 1 March 2021
Accepted: 15 March 2021
Published: 17 March 2021

1. Introduction

Economic valuation of soil carbon is vital for achieving the United Nations (UN) Sustainable Development Goals (SDGs), especially SDG 13: "Take urgent action to combat climate change and its impacts on future climate" [1]. The ecosystem services (ES) framework is often used in connection with UN SDGs because it is focused on the economic valuation of benefits (ES) and/or disservices (ED) people obtain from nature [2]. The ES framework includes three general categories of services: provisioning, regulating/maintenance, and cultural supporting services [2]. Although TSC is composed of SOC and SIC, only SOC is currently included in the list of soil properties important for ES [3]. Soil organic carbon is derived from living matter and tends to be concentrated in the topsoil (Table 1). In a well-aerated soil, all of the organic compounds found in plant residue are subject to enzymatic oxidation. This reaction is accompanied by oxygen consumption and CO_2 release [4], which is often associated with ED in the form of realized social costs of carbon dioxide (CO_2) emissions [5]. Soil organic carbon is a fraction of soil organic matter (SOM) of <2 mm particle size fraction (Table 1). Soil databases provide SOM (%) and/or SOC (%) in their reports listed in the tables of soil physical properties. Soil organic matter contributes to numerous soil functions (e.g., nutrient and energy reserve, etc.), which are linked to ecosystem goods and services (e.g., nutrient storage and availability, gas regulation, etc.) [6,7]. The role of SOM in delivering these ecosystem goods and services varies with scales from local (e.g., fertility maintenance) to global (e.g., mitigation of carbon emissions) [6,7]. Soil inorganic carbon, which is found in different types of carbonates (e.g. calcium, magnesium), is also essential in various ES/ED (e.g., provisioning services as a liming material for food production). It is reported as calcium carbonate ($CaCO_3$, %) of <2 mm particle size fraction in the tables of soil chemical properties (Table 1).

Previous research on social costs of SOC and SIC in the U.S.A. was conducted at various scales using both biophysical (e.g., soil orders) and administrative accounts (e.g., states, regions, farm, etc.) [8–10]. These analyses allowed estimation of potential social costs of soil carbon, which is useful for decision-making at the national level using detailed tables and maps of social costs of C showing areas with high soil C content, which can become "soil carbon hotspots" upon disturbance [10]. At the national level, the analysis showed that states have different types of soils with various soil C types (e.g., Maryland is dominated by SOC, state of New Mexico is dominated by SIC) [11], which requires soil- and carbon-specific management strategies. Some states demonstrated more soil variability compared to others.

Table 1. Total soil carbon: soil organic matter (SOM), soil organic carbon (SOC), soil inorganic carbon (SIC), and carbon sequestration pathway (adapted from Mikhailova et al., 2019 [8]).

Total soil carbon, TSC (Biotic + Abiotic) = Soil organic carbon, SOC (Biotic) + Soil inorganic carbon, SIC (Abiotic)		
Biotic		**Abiotic**
Soil organic matter (SOM) of <2 mm particle size fraction	**Soil organic carbon (SOC)**	**Soil inorganic carbon (SIC)**
- Fresh residue, decomposing organic matter, stable organic matter (humus), and living organisms. or - "Continuum of organic material in all stages of transformation and decomposition or stabilization [12]."	- Carbon fraction of soil organic matter of <2 mm particle size fraction.	- Carbon fraction of calcium carbonate ($CaCO_3$) of <2 mm particle size fraction.
Conversion (using Van Bemmelen factor of 0.58 or 1.724): SOM (%) = SOC (%) × 1.724 or SOC (%) = SOM (%) × 0.58 [13]		**Conversion**: $CaCO_3$ (%) = SIC (%) × 100/12 or SIC (%) = $CaCO_3$ (%) × 0.12
Pathways to increased C sequestration: Additions of organic matter (e.g., compost additions, etc.); land/agricultural management (e.g., no-till operations, land conservation, etc.); afforestation, etc. [6,7].		**Pathways to increased C sequestration**: Additions of Ca^{2+} and Mg^{2+} cations outside the soil (e.g., atmospheric deposition, etc.) [14].

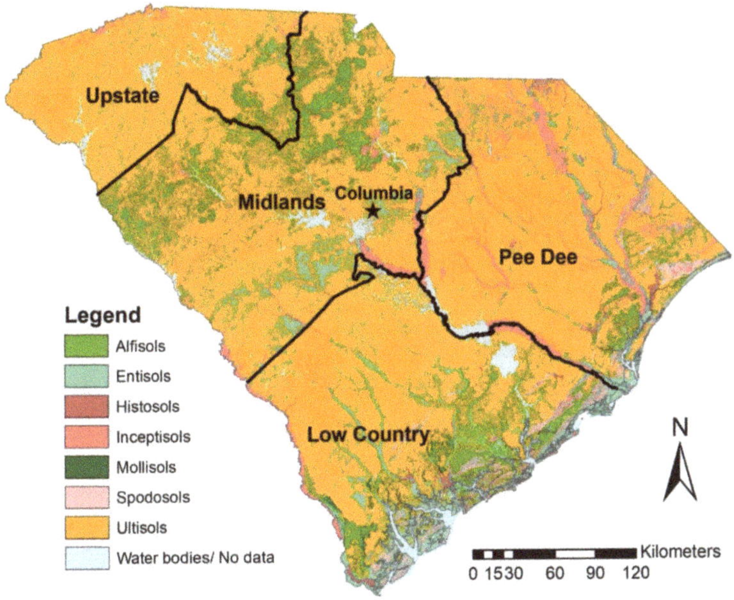

Figure 1. General soil map of South Carolina (U.S.A.) (33.8361° N, 81.1637° W) (adapted from [15]).

The ES framework is increasingly being used to address environmental concerns (e.g., global warming, climate change, etc.), but because of "the difficulty in relating soil properties to ES, soil ES are still not fully considered in the territorial planning decision process" [16]. According to Fossey et al., 2020 [16], soil databases play an essential role in assessing ES/ED in territorial planning. For sustainable soil C management decisions at the state level and its counties, it is critical to determine soil C and the distribution of its social costs within the state overall and by individual counties linked to biophysical units (e.g., soil orders). This type of analysis will allow prioritization of soil C management within the state based on this distribution. The hypothesis of this study is that pedodiversity concepts overlayed with administrative units (Figures 1 and 2) can be used to identify spatial patterns of soil carbon hotspots for sustainable management.

The specific objective of this study was to assess the value of SOC, SIC, and TSC in the state of South Carolina (U.S.A.) based on the social cost of carbon (SC–CO_2) and avoided emissions provided by carbon sequestration, which the U.S. Environmental Protection Agency (EPA) has determined to be $46 per metric ton of CO_2, which is applicable for the year 2025 based on 2007 U.S. dollars and an average discount rate of 3% [17]. This study provides the monetary values of SOC, SIC, and TSC for soil depth (0–200 cm) across the state and by considering different spatial aggregation levels (i.e., region, county) using State Soil Geographic (STATSGO) database, and information previously reported by Guo et al. (2006) [18].

2. Materials and Methods

The Accounting Framework

This study used both biophysical (science-based, Figure 1) and administrative (boundary-based, Figure 3) accounts to calculate monetary values for SOC, SIC, and TSC (Tables 2 and 3).

Table 2. A conceptual overview of the accounting framework used in this study (adapted from Groshans et al., 2018 [19]).

Biophysical Accounts (Science-Based)	Administrative Accounts (Boundary-Based)	Monetary Account(s)	Benefit(s)	Total Value
Soil extent:	Administrative extent:	Ecosystem good(s) and service(s):	Sector:	Types of value:
Separate constitute stock 1: Soil organic carbon (SOC)				
Separate constitute stock 2: Soil inorganic carbon (SIC)				
Composite (total) stock: Total soil carbon (TSC) = Soil organic carbon (SOC) + Soil inorganic carbon (SIC)				
			Environment:	The social cost of carbon (SC-CO_2) and avoided emissions:
- Soil order	- State - Region - County	- Regulating (e.g., carbon sequestration)	- Carbon sequestration	- $46 per metric ton of CO_2 (2007 U.S. dollars with an average discount rate of 3% [16])

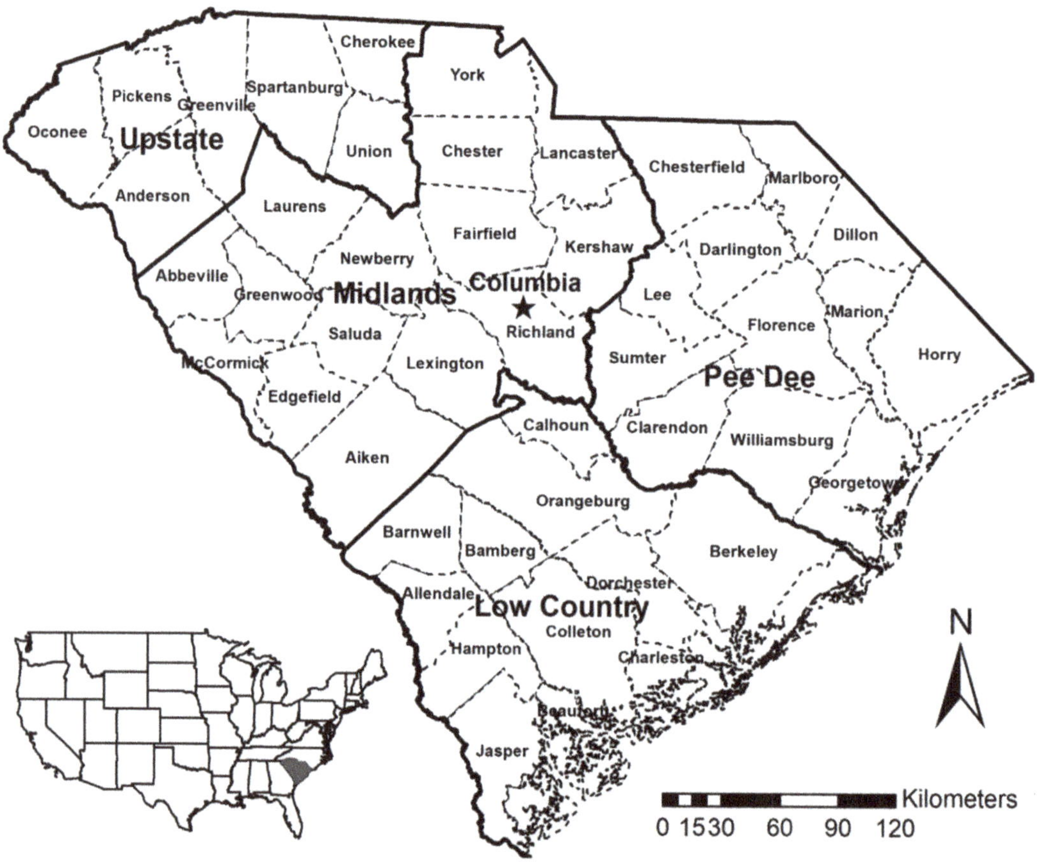

Figure 2. Administrative map of South Carolina (U.S.A.) (33.8361° N, 81.1637° W) with 46 counties and four regions [20].

Table 3. Soil diversity (pedodiversity) by soil order (taxonomic pedodiversity), region, and county in South Carolina (U.S.A.) based on Soil Survey Geographic (SSURGO) Database (2020) [15].

County (Region)	Total Area (km^2) (Rank)	Slight <— Degree of Weathering and Soil Development —> Strong						
		Slightly Weathered			Moderately Weathered		Strongly Weathered	
		Entisols	Inceptisols	Histosols	Alfisols	Mollisols	Spodosols	Ultisols
		Area (km^2)						
Anderson	1841 (14)	102	0	0	0	0	0	1739
Cherokee	1012 (44)	118	94	0	100	0	0	700
Greenville	1916 (11)	108	166	0	0	0	0	1642
Oconee	1620 (23)	37	56	0	0	0	0	1527
Pickens	1257 (35)	39	55	0	0	0	0	1163
Spartanburg	1881 (12)	8	161	0	77	0	0	1635
Union	1322 (32)	74	61	0	364	0	0	823
(Upstate)	**10,849 (4)**	**486**	**593**	**0**	**541**	**0**	**0**	**9229**
Abbeville	1269 (34)	52	38	0	401	0	0	778
Aiken	2758 (4)	445	213	11	5	0	0	2084
Chester	1493 (25)	60	82	0	570	6	0	775
Edgefield	1289 (33)	101	56	0	61	0	0	1071
Fairfield	1683 (20)	1	175	0	585	0	0	922
Greenwood	1171 (38)	0	116	0	335	0	0	720
Kershaw	1871 (13)	383	161	20	47	0	0	1260
Lancaster	1409 (30)	53	151	0	62	0	0	1143
Laurens	1837 (15)	125	19	0	351	0	0	1342
Lexington	1756 (17)	454	86	0	36	0	10	1170
Newberry	1621 (22)	70	64	0	278	0	0	1209
Richland	1827 (16)	163	360	8	19	0	0	1277
Saluda	1170 (39)	19	83	0	77	0	0	991
York	1753 (18)	5	134	0	577	0	0	1037
(Midlands)	**22,899 (1)**	**1931**	**1738**	**31**	**3404**	**6**	**10**	**15,779**
Chesterfield	2053 (9)	173	655	0	23	0	0	1202
Clarendon	1566 (24)	39	192	6	0	0	0	1329
Darlington	1442 (28)	36	258	9	0	0	1	1138
Dillon	1040 (42)	91	128	8	0	0	19	794
Florence	2046 (10)	97	224	0	0	0	2	1723
Georgetown	2064 (8)	351	274	57	409	0	115	858
Horry	2888 (1)	252	431	64	287	0	330	1524
Lee	1058 (40)	29	131	0	1	0	0	897
Marion	1241 (36)	107	286	27	0	0	49	772
Marlboro	1230 (37)	75	269	81	17	0	2	786
Sumter	1694 (19)	9	350	0	4	0	2	1329
Williamsburg	2400 (6)	25	209	0	0	0	3	2163
(Pee Dee)	**20,722 (3)**	**1284**	**3407**	**252**	**741**	**0**	**523**	**14,515**
Allendale	1055 (41)	25	101	6	0	0	0	923
Bamberg	1018 (44)	126	1	0	41	0	3	847
Barnwell	1416 (29)	78	138	0	0	0	0	1200
Beaufort	1402 (31)	698	34	6	40	23	210	391
Berkeley	2809 (3)	145	208	23	409	0	137	1887
Calhoun	748 (46)	66	26	0	0	0	0	656
Charleston	2317 (7)	765	332	0	727	0	273	220
Colleton	2677 (5)	280	49	88	140	85	109	1,926
Dorchester	1455 (26)	274	17	1	280	0	30	853
Hampton	1443 (27)	136	87	3	101	0	40	1076
Jasper	1669 (21)	318	58	57	246	116	24	850
McCormick	921 (45)	64	65	0	247	0	0	545
Orangeburg	2844 (2)	73	18	0	25	0	1	2,727
(Low Country)	**21,774 (2)**	**3048**	**1134**	**184**	**2256**	**224**	**827**	**14,101**
Totals	**76,252**	**6749**	**6872**	**475**	**6942**	**230**	**1360**	**53,624**

The present study is based on the SOC [21], SIC [21], TSC estimated values for the SOC, SIC, and TSC storage (in Mg or metric tons) and content (in kg m^{-2}) in the contiguous U.S. from Guo et al. (2006) [18]. A monetary valuation for TSC was calculated using the social cost of carbon (SC-CO_2) of $46 per metric ton of CO_2, which is applicable for 2025 based on 2007 U.S. dollars and an average discount rate of 3% [17]. According to the EPA, the SC-CO_2 is intended to be a comprehensive estimate of climate change damages. Still, it can underestimate the true damages and cost of CO_2 emissions due to the exclusion of various important climate change impacts recognized in the literature [17]. Soil carbon (SC) storage and content numbers were then converted to U.S. dollars and dollars per square meter in Microsoft Excel using the following equations, with a social cost of carbon of $46/Mg CO_2:

$$\$ = (\text{SC Storage, Mg}) \times \frac{44\ \text{Mg CO}_2}{12\ \text{Mg TSC}} \times \frac{\$46}{\text{Mg CO}_2} \quad (1)$$

$$\frac{\$}{\text{m}^2} = \left(\text{SC Content}, \frac{\text{kg}}{\text{m}^2}\right) \times \frac{1\ \text{Mg}}{10^3\ \text{kg}} \times \frac{44\ \text{Mg CO}_2}{12\ \text{Mg TSC}} \times \frac{\$46}{\text{Mg CO}_2} \quad (2)$$

Table 4 presents area-normalized content (kg m^{-2}) and monetary values ($ m^{-2}) of soil carbon, which were used to estimate total soil carbon storage and total soil carbon value by multiplying corresponding content (values) numbers by an area of a particular soil order within a county (region) (Table 3). For example, for the soil order of Entisols, Guo et al (2006) [18] reported an area-normalized midpoint SOC content number of 8.0 kg·m^{-2} in the upper 2 m (Table 4), which was used to calculate the total SOC storage in soil order by multiplying its area in particular county or region. Then, the reported area-normalized midpoint SOC content number of 8.0 kg·m^{-2} in the upper 2 m (Table 4) was converted to monetary values ($ m^{-2}) of soil organic carbon using a social cost of carbon (SC-CO_2) of $46 per metric ton of CO_2 (2007 U.S. dollars with an average discount rate of 3% [17]), which is $1.35 m^{-2} to calculate the total monetary value of SOC storage.

Table 4. Area-normalized content (kg m^{-2}) and monetary values ($ m^{-2}) of soil organic carbon (SOC), soil inorganic carbon (SIC), total soil carbon (TSC) by soil order based on numbers in the upper 2 m of the soil based on data from Guo et al., 2006 [18] and a social cost of carbon (SC-CO_2) of $46 per metric ton of CO_2 (2007 U.S. dollars with an average discount rate of 3% [17]).

Soil Order	SOC Content	SIC Content	TSC Content	SOC Value	SIC Value	TSC Value
	Minimum–Midpoint–Maximum Values			Midpoint Values		
	(kg m^{-2})	(kg m^{-2})	(kg m^{-2})	($ m^{-2})	($ m^{-2})	($ m^{-2})
Slightly Weathered						
Entisols	1.8–8.0–15.8	1.9–4.8–8.4	3.7–12.8–24.2	1.35	0.82	2.17
Inceptisols	2.8–8.9–17.4	2.5–5.1–8.4	5.3–14.0–25.8	1.50	0.86	2.36
Histosols	63.9–140.1–243.9	0.6–2.4–5.0	64.5–142.5–248.9	23.62	0.41	24.03
Moderately Weathered						
Alfisols	2.3–7.5–14.1	1.3–4.3–8.1	3.6–11.8–22.2	1.27	0.72	1.99
Mollisols	5.9–13.5–22.8	4.9–11.5–19.7	10.8–25.0–42.5	2.28	1.93	4.21
Strongly Weathered						
Spodosols	2.9–12.3–25.5	0.2–0.6–1.1	3.1–12.9–26.6	2.07	0.10	2.17
Ultisols	1.9–7.1–13.9	0.0–0.0–0.0	1.9–7.1–13.9	1.20	0.00	1.20

Note: TSC = SOC + SIC.

3. Results

The total estimated monetary mid-point value for TSC in the state of South Carolina was \$124.36B (i.e., \$124.36 billion U.S. dollars, where B = billion = 10^9), \$107.14B for SOC, and \$17.22B for SIC. The state of South Carolina ranked 31st for TSC, 25th for SOC, and 32nd for SIC. Figure 3 shows the distribution of soil carbon by South Carolina regions.

3.1. Storage and Value of SOC by County, Region, and Soil Order for the State of South Carolina (U.S.A.)

Soil orders with the highest midpoint storage and value for SOC were: Ultisols (\$64.35B), Histosols (\$11.22B), and Inceptisols (\$10.31B) (Tables 5 and 6). The regions with the highest midpoint storage and SOC values were: Pee Dee (\$34.24B), Low Country (\$32.17B), and Midlands (\$29.24B) (Tables 5 and 6). The counties with the highest midpoint SOC storage and values were Colleton (\$5.44B), Horry (\$5.37B), and Berkeley (\$4.12B) (Tables 5 and 6).

3.2. Storage and Value of SIC by County, Region, and Soil Order for the State of South Carolina (U.S.A.)

Soil orders with the highest midpoint storage and value for SIC were: Inceptisols (\$5.91B), Entisols (\$5.53B), and Alfisols (\$5.0B) (Tables 7 and 8). The regions with the highest midpoint SIC storage and values were: Low Country (\$5.69B), Midlands (\$5.55B), and Pee Dee (\$4.67B) (Tables 7 and 8). The counties with the highest midpoint SIC storage and values were Charleston (\$1.46B), Georgetown (\$852.81M), and Horry (\$843.18M) (Tables 7 and 8).

3.3. Storage and Value of TSC (SOC + SIC) by County, Region, and Soil Order for the State of South Carolina (U.S.A.)

Soil orders with the highest midpoint storage and value for TSC were: Ultisols (\$64.35B), Inceptisols (\$16.22B), and Entisols (\$14.65B) (Tables 9 and 10). The regions with the highest midpoint TSC storage and values were: Low Country (\$37.86B), Pee Dee (\$36.91B), and Midlands (\$34.79B) (Tables 9 and 10). The counties with the highest midpoint TSC storage and values were Horry (\$6.22B), Colleton (\$6.02B), and Georgetown (\$4.87B) (Tables 9 and 10).

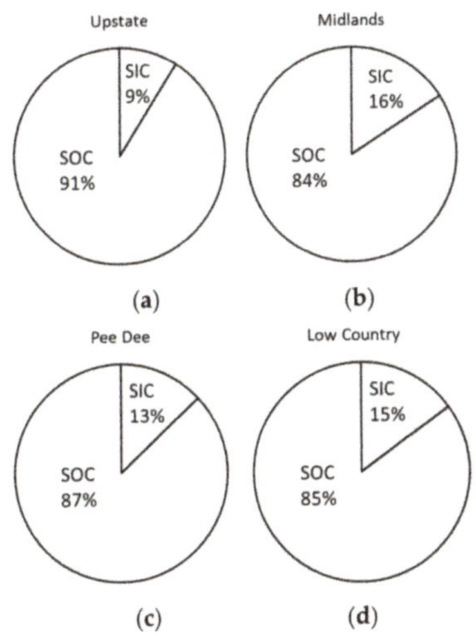

Figure 3. Distribution of soil carbon by region in the state of South Carolina: (**a**) Upstate, (**b**) Midlands, (**c**) Pee Dee, and (**d**) Low Country.

Table 5. Mid-point total soil organic carbon (SOC) storage values by county, region, and soil order for the state of South Carolina (U.S.A.), based on mid-point soil organic carbon (SOC) content numbers in the upper 2 m of the soil based on data from Guo et al., 2006 [18].

County (Region)	Total Storage (kg) (Rank)	Slight <—			Degree of Weathering and Soil Development			—> Strong
			Slightly Weathered		Moderately Weathered		Strongly Weathered	
		Entisols	Inceptisols	Histosols	Alfisols	Mollisols	Spodosols	Ultisols
					Total Storage (kg)			
Anderson	1.32×10^{10} (19)	8.16×10^{8}	0	0	0	0	0	1.23×10^{10}
Cherokee	7.50×10^{9} (43)	9.44×10^{8}	8.37×10^{8}	0	7.50×10^{8}	0	0	4.97×10^{9}
Greenville	1.40×10^{10} (15)	8.64×10^{8}	1.48×10^{9}	0	0	0	0	1.17×10^{10}
Oconee	1.16×10^{10} (28)	2.96×10^{8}	4.98×10^{8}	0	0	0	0	1.08×10^{10}
Pickens	9.06×10^{9} (37)	3.12×10^{8}	4.90×10^{8}	0	0	0	0	8.26×10^{9}
Spartanburg	1.37×10^{10} (16)	6.40×10^{7}	1.43×10^{9}	0	5.78×10^{8}	0	0	1.16×10^{10}
Union	9.71×10^{9} (34)	5.92×10^{8}	5.43×10^{8}	0	2.73×10^{9}	0	0	5.84×10^{9}
(Upstate)	$\mathbf{7.87 \times 10^{10}}$ **(4)**	$\mathbf{3.89 \times 10^{9}}$	$\mathbf{5.28 \times 10^{9}}$	**0**	$\mathbf{4.06 \times 10^{9}}$	**0**	**0**	$\mathbf{6.55 \times 10^{10}}$
Abbeville	9.29×10^{9} (36)	4.16×10^{8}	3.38×10^{8}	0	3.01×10^{9}	0	0	5.52×10^{9}
Aiken	2.18×10^{10} (5)	3.56×10^{9}	1.90×10^{9}	1.54×10^{9}	3.75×10^{7}	0	0	1.48×10^{10}
Chester	1.11×10^{10} (30)	4.80×10^{8}	7.30×10^{8}	0	4.28×10^{9}	8.10×10^{7}	0	5.50×10^{9}
Edgefield	9.37×10^{9} (35)	8.08×10^{8}	4.98×10^{8}	0	4.58×10^{8}	0	0	7.60×10^{9}
Fairfield	1.25×10^{10} (24)	8.00×10^{6}	1.56×10^{9}	0	4.39×10^{9}	0	0	6.55×10^{9}
Greenwood	8.66×10^{9} (39)	0	1.03×10^{9}	0	2.51×10^{9}	0	0	5.11×10^{9}
Kershaw	1.66×10^{10} (11)	3.06×10^{9}	1.43×10^{9}	2.80×10^{9}	3.53×10^{8}	0	0	8.95×10^{9}
Lancaster	1.03×10^{10} (33)	4.24×10^{8}	1.34×10^{9}	0	4.65×10^{8}	0	0	8.12×10^{9}
Laurens	1.33×10^{10} (17)	1.00×10^{9}	1.69×10^{8}	0	2.63×10^{9}	0	0	9.53×10^{9}
Lexington	1.31×10^{10} (20)	3.63×10^{9}	7.65×10^{8}	0	2.70×10^{8}	0	1.23×10^{8}	8.31×10^{9}
Newberry	1.18×10^{10} (27)	5.60×10^{8}	5.70×10^{8}	0	2.09×10^{9}	0	0	8.58×10^{9}
Richland	1.48×10^{10} (14)	1.30×10^{9}	3.20×10^{9}	1.12×10^{9}	1.43×10^{8}	0	0	9.07×10^{9}
Saluda	8.50×10^{9} (40)	1.52×10^{8}	7.39×10^{8}	0	5.78×10^{8}	0	0	7.04×10^{9}
York	1.29×10^{10} (21)	4.00×10^{7}	1.19×10^{9}	0	4.33×10^{9}	0	0	7.36×10^{9}
(Midlands)	$\mathbf{1.73 \times 10^{11}}$ **(3)**	$\mathbf{1.54 \times 10^{10}}$	$\mathbf{1.55 \times 10^{10}}$	$\mathbf{4.34 \times 10^{9}}$	$\mathbf{2.55 \times 10^{10}}$	$\mathbf{8.10 \times 10^{7}}$	$\mathbf{1.23 \times 10^{8}}$	$\mathbf{1.12 \times 10^{11}}$
Chesterfield	1.59×10^{10} (12)	1.38×10^{9}	5.83×10^{9}	0	1.73×10^{8}	0	0	8.53×10^{9}
Clarendon	1.23×10^{10} (25)	3.12×10^{8}	1.71×10^{9}	8.41×10^{8}	0	0	0	9.44×10^{9}
Darlington	1.19×10^{10} (26)	2.88×10^{8}	2.30×10^{9}	1.26×10^{9}	0	0	1.23×10^{7}	8.08×10^{9}
Dillon	8.86×10^{9} (38)	7.28×10^{8}	1.14×10^{9}	1.12×10^{9}	0	0	2.34×10^{8}	5.64×10^{9}
Florence	1.50×10^{10} (13)	7.76×10^{8}	1.99×10^{9}	0	0	0	2.46×10^{7}	1.22×10^{10}
Georgetown	2.38×10^{10} (4)	2.81×10^{9}	2.44×10^{9}	7.99×10^{9}	3.07×10^{9}	0	1.41×10^{9}	6.09×10^{9}
Horry	3.19×10^{10} (2)	2.02×10^{9}	3.84×10^{9}	8.97×10^{9}	2.15×10^{9}	0	4.06×10^{9}	1.08×10^{10}
Lee	7.77×10^{9} (42)	2.32×10^{8}	1.17×10^{9}	0	7.50×10^{6}	0	0	6.37×10^{9}
Marion	1.33×10^{10} (18)	8.56×10^{8}	2.55×10^{9}	3.78×10^{9}	0	0	6.03×10^{8}	5.48×10^{9}
Marlboro	2.01×10^{10} (8)	6.00×10^{8}	2.39×10^{9}	1.13×10^{10}	1.28×10^{8}	0	2.46×10^{7}	5.58×10^{9}
Sumter	1.27×10^{10} (23)	7.20×10^{7}	3.12×10^{9}	0	3.00×10^{7}	0	2.46×10^{7}	9.44×10^{9}
Williamsburg	1.75×10^{10} (10)	2.00×10^{8}	1.86×10^{9}	0	0	0	3.69×10^{7}	1.54×10^{10}
(Pee Dee)	$\mathbf{1.91 \times 10^{11}}$ **(1)**	$\mathbf{1.03 \times 10^{10}}$	$\mathbf{3.03 \times 10^{10}}$	$\mathbf{3.53 \times 10^{10}}$	$\mathbf{5.56 \times 10^{9}}$	**0**	$\mathbf{6.43 \times 10^{9}}$	$\mathbf{1.03 \times 10^{11}}$
Allendale	8.49×10^{9} (41)	2.00×10^{8}	8.99×10^{8}	8.41×10^{8}	0	0	0	6.55×10^{9}
Bamberg	7.38×10^{9} (44)	1.01×10^{9}	8.90×10^{6}	0	3.08×10^{8}	0	3.69×10^{7}	6.01×10^{9}
Barnwell	1.04×10^{10} (32)	6.24×10^{8}	1.23×10^{9}	0	0	0	0	8.52×10^{9}
Beaufort	1.27×10^{10} (22)	5.58×10^{9}	3.03×10^{8}	8.41×10^{8}	3.00×10^{8}	3.11×10^{8}	2.58×10^{9}	2.78×10^{9}
Berkeley	2.44×10^{10} (3)	1.16×10^{9}	1.85×10^{9}	3.22×10^{9}	3.07×10^{9}	0	1.69×10^{9}	1.34×10^{10}
Calhoun	5.42×10^{9} (46)	5.28×10^{8}	2.31×10^{8}	0	0	0	0	4.66×10^{9}
Charleston	1.94×10^{10} (9)	6.12×10^{9}	2.95×10^{9}	0	5.45×10^{9}	0	3.36×10^{9}	1.56×10^{9}
Colleton	3.22×10^{10} (1)	2.24×10^{9}	4.36×10^{8}	1.23×10^{10}	1.05×10^{9}	1.15×10^{9}	1.34×10^{9}	1.37×10^{10}
Dorchester	1.10×10^{10} (31)	2.19×10^{9}	1.51×10^{8}	1.40×10^{8}	2.10×10^{9}	0	3.69×10^{8}	6.06×10^{9}
Hampton	1.12×10^{10} (29)	1.09×10^{9}	7.74×10^{8}	4.20×10^{8}	7.58×10^{8}	0	4.92×10^{8}	7.64×10^{9}
Jasper	2.08×10^{10} (6)	2.54×10^{9}	5.16×10^{8}	7.99×10^{9}	1.85×10^{9}	1.57×10^{9}	2.95×10^{8}	6.04×10^{9}
McCormick	6.81×10^{9} (45)	5.12×10^{8}	5.79×10^{8}	0	1.85×10^{9}	0	0	3.87×10^{9}
Orangeburg	2.03×10^{10} (7)	5.84×10^{8}	1.60×10^{8}	0	1.88×10^{8}	0	1.23×10^{7}	1.94×10^{10}
(Low Country)	$\mathbf{1.90 \times 10^{11}}$ **(2)**	$\mathbf{2.44 \times 10^{10}}$	$\mathbf{1.01 \times 10^{10}}$	$\mathbf{2.58 \times 10^{10}}$	$\mathbf{1.69 \times 10^{10}}$	$\mathbf{3.02 \times 10^{9}}$	$\mathbf{1.02 \times 10^{10}}$	$\mathbf{1.00 \times 10^{11}}$
Totals (kg)	$\mathbf{6.34 \times 10^{11}}$	$\mathbf{5.40 \times 10^{10}}$	$\mathbf{6.12 \times 10^{10}}$	$\mathbf{6.65 \times 10^{10}}$	$\mathbf{5.21 \times 10^{10}}$	$\mathbf{3.11 \times 10^{9}}$	$\mathbf{1.67 \times 10^{10}}$	$\mathbf{3.81 \times 10^{11}}$

Table 6. The total dollar value of soil organic carbon (SOC) by county, region, and soil order for the state of South Carolina (U.S.A.), based on mid-point soil organic carbon (SOC) numbers for the upper 2 m from Guo et al. 2006 [18] and a social cost of carbon (SC-CO_2) of $46 per metric ton of CO_2 (2007 U.S. dollars with an average discount rate of 3% [17]).

County (Region)	Total Value ($) (Rank)	Slight <— Degree of Weathering and Soil Development —> Strong						
		Slightly Weathered			Moderately Weathered		Strongly Weathered	
		Entisols	Inceptisols	Histosols	Alfisols	Mollisols	Spodosols	Ultisols
		Value ($)						
Anderson	2.22×10^9 (19)	1.38×10^8	0	0	0	0	0	2.09×10^9
Cherokee	1.27×10^9 (43)	1.59×10^8	1.41×10^8	0	1.27×10^8	0	0	8.40×10^8
Greenville	2.37×10^9 (15)	1.46×10^8	2.49×10^8	0	0	0	0	1.97×10^9
Oconee	1.97×10^9 (28)	5.00×10^7	8.40×10^7	0	0	0	0	1.83×10^9
Pickens	1.53×10^9 (37)	5.27×10^7	8.25×10^7	0	0	0	0	1.40×10^9
Spartanburg	2.31×10^9 (16)	1.08×10^7	2.42×10^8	0	9.78×10^7	0	0	1.96×10^9
Union	1.64×10^9 (34)	9.99×10^7	9.15×10^7	0	4.62×10^8	0	0	9.88×10^8
(Upstate)	**1.33×10^{10} (4)**	**6.56×10^8**	**8.90×10^8**	**0**	**6.87×10^8**	**0**	**0**	**1.11×10^{10}**
Abbeville	1.57×10^9 (36)	7.02×10^7	5.70×10^7	0	5.09×10^8	0	0	9.34×10^8
Aiken	3.69×10^9 (5)	6.01×10^8	3.20×10^8	2.60×10^8	6.35×10^6	0	0	2.50×10^9
Chester	1.87×10^9 (30)	8.10×10^7	1.23×10^8	0	7.24×10^8	1.37×10^7	0	9.30×10^8
Edgefield	1.58×10^9 (35)	1.36×10^8	8.40×10^7	0	7.75×10^7	0	0	1.29×10^9
Fairfield	2.11×10^9 (24)	1.35×10^6	2.63×10^8	0	7.43×10^8	0	0	1.11×10^9
Greenwood	1.46×10^9 (39)	0	1.74×10^8	0	4.25×10^8	0	0	8.64×10^8
Kershaw	2.80×10^9 (11)	5.17×10^8	2.42×10^8	4.72×10^8	5.97×10^7	0	0	1.51×10^9
Lancaster	1.75×10^9 (33)	7.16×10^7	2.27×10^8	0	7.87×10^7	0	0	1.37×10^9
Laurens	2.25×10^9 (17)	1.69×10^8	2.85×10^7	0	4.46×10^8	0	0	1.61×10^9
Lexington	2.21×10^9 (20)	6.13×10^8	1.29×10^8	0	4.57×10^7	0	2.07×10^7	1.40×10^9
Newberry	1.99×10^9 (27)	9.45×10^7	9.60×10^7	0	3.53×10^8	0	0	1.45×10^9
Richland	2.51×10^9 (14)	2.20×10^8	5.40×10^8	1.89×10^8	2.41×10^7	0	0	1.53×10^9
Saluda	1.44×10^9 (40)	2.57×10^7	1.25×10^8	0	9.78×10^7	0	0	1.19×10^9
York	2.18×10^9 (21)	6.75×10^6	2.01×10^8	0	7.33×10^8	0	0	1.24×10^9
(Midlands)	**2.92×10^{10} (3)**	**2.61×10^9**	**2.61×10^9**	**7.32×10^8**	**4.32×10^9**	**1.37×10^7**	**2.07×10^7**	**1.89×10^{10}**
Chesterfield	2.69×10^9 (12)	2.34×10^8	9.83×10^8	0	2.92×10^7	0	0	1.44×10^9
Clarendon	2.08×10^9 (25)	5.27×10^7	2.88×10^8	1.42×10^8	0	0	0	1.59×10^9
Darlington	2.02×10^9 (26)	4.86×10^7	3.87×10^8	2.13×10^8	0	0	2.07×10^6	1.37×10^9
Dillon	1.50×10^9 (38)	1.23×10^8	1.92×10^8	1.89×10^8	0	0	3.93×10^7	9.53×10^8
Florence	2.54×10^9 (13)	1.31×10^8	3.36×10^8	0	0	0	4.14×10^6	2.07×10^9
Georgetown	4.02×10^9 (4)	4.74×10^8	4.11×10^8	1.35×10^9	5.19×10^8	0	2.38×10^8	1.03×10^9
Horry	5.37×10^9 (2)	3.40×10^8	6.47×10^8	1.51×10^9	3.64×10^8	0	6.83×10^8	1.83×10^9
Lee	1.31×10^9 (42)	3.92×10^7	1.97×10^8	0	1.27×10^6	0	0	1.08×10^9
Marion	2.24×10^9 (18)	1.44×10^8	4.29×10^8	6.38×10^8	0	0	1.01×10^8	9.26×10^8
Marlboro	3.39×10^9 (8)	1.01×10^8	4.04×10^8	1.91×10^9	2.16×10^7	0	4.14×10^6	9.43×10^8
Sumter	2.14×10^9 (23)	1.22×10^7	5.25×10^8	0	5.08×10^6	0	4.14×10^6	1.59×10^9
Williamsburg	2.95×10^9 (10)	3.38×10^7	3.14×10^8	0	0	0	6.21×10^6	2.60×10^9
(Pee Dee)	**3.22×10^{10} (1)**	**1.73×10^9**	**5.11×10^9**	**5.95×10^9**	**9.41×10^8**	**0**	**1.08×10^9**	**1.74×10^{10}**
Allendale	1.43×10^9 (41)	3.38×10^7	1.52×10^8	1.42×10^8	0	0	0	1.11×10^9
Bamberg	1.25×10^9 (44)	1.70×10^8	1.50×10^6	0	5.21×10^7	0	6.21×10^6	1.02×10^9
Barnwell	1.75×10^9 (32)	1.05×10^8	2.07×10^8	0	0	0	0	1.44×10^9
Beaufort	2.14×10^9 (22)	9.42×10^8	5.10×10^7	1.42×10^8	5.08×10^7	5.24×10^7	4.35×10^8	4.69×10^8
Berkeley	4.12×10^9 (3)	1.96×10^8	3.12×10^8	5.43×10^8	5.19×10^8	0	2.84×10^8	2.26×10^9
Calhoun	9.15×10^8 (46)	8.91×10^7	3.90×10^7	0	0	0	0	7.87×10^8
Charleston	3.28×10^9 (9)	1.03×10^9	4.98×10^8	0	9.23×10^8	0	5.65×10^8	2.64×10^8
Colleton	5.44×10^9 (1)	3.78×10^8	7.35×10^7	2.08×10^9	1.78×10^8	1.94×10^8	2.26×10^8	2.31×10^9
Dorchester	1.86×10^9 (31)	3.70×10^8	2.55×10^7	2.36×10^7	3.56×10^8	0	6.21×10^7	1.02×10^9
Hampton	1.89×10^9 (29)	1.84×10^8	1.31×10^8	7.09×10^7	1.28×10^8	0	8.28×10^7	1.29×10^9
Jasper	3.51×10^9 (6)	4.29×10^8	8.70×10^7	1.35×10^9	3.12×10^8	2.64×10^8	4.97×10^7	1.02×10^9
McCormick	1.15×10^9 (45)	8.64×10^7	9.75×10^7	0	3.14×10^8	0	0	6.54×10^8
Orangeburg	3.43×10^9 (7)	9.86×10^7	2.70×10^7	0	3.18×10^7	0	2.07×10^6	3.27×10^9
(Low Country)	**3.22×10^{10} (2)**	**4.11×10^9**	**1.70×10^9**	**4.35×10^9**	**2.87×10^9**	**5.11×10^8**	**1.71×10^9**	**1.69×10^{10}**
Totals ($)	**1.07×10^{11}**	**9.11×10^9**	**1.03×10^{10}**	**1.12×10^{10}**	**8.82×10^9**	**5.24×10^8**	**2.82×10^9**	**6.43×10^{10}**

Table 7. Mid-point total soil inorganic carbon (SIC) storage by county, region, and soil order for the state of South Carolina (U.S.A.), based on mid-point soil inorganic carbon (SIC) contents in the upper 2 m based on data from Guo et al., 2006 [18].

County (Region)	Total Storage (kg) (Rank)	Slight <— Degree of Weathering and Soil Development —> Strong						
		Slightly Weathered			Moderately Weathered		Strongly Weathered	
		Entisols	Inceptisols	Histosols	Alfisols	Mollisols	Spodosols	Ultisols
		Total Storage (kg)						
Anderson	4.90×10^8 (43)	4.90×10^8	0	0	0	0	0	0
Cherokee	1.48×10^9 (29)	5.66×10^8	4.79×10^8	0	4.30×10^8	0	0	0
Greenville	1.37×10^9 (30)	5.18×10^8	8.47×10^8	0	0	0	0	0
Oconee	4.63×10^8 (45)	1.78×10^8	2.86×10^8	0	0	0	0	0
Pickens	4.68×10^8 (44)	1.87×10^8	2.81×10^8	0	0	0	0	0
Spartanburg	1.19×10^9 (32)	3.84×10^7	8.21×10^8	0	3.31×10^8	0	0	0
Union	2.23×10^9 (17)	3.55×10^8	3.11×10^8	0	1.57×10^9	0	0	0
(Upstate)	**7.68×10^9 (4)**	**2.33×10^9**	**3.02×10^9**	**0**	**2.33×10^9**	**0**	**0**	**0**
Abbeville	2.17×10^9 (19)	2.50×10^8	1.94×10^8	0	1.72×10^9	0	0	0
Aiken	3.27×10^9 (10)	2.14×10^9	1.09×10^9	2.64×10^7	2.15×10^7	0	0	0
Chester	3.23×10^9 (11)	2.88×10^8	4.18×10^8	0	2.45×10^9	6.90×10^7	0	0
Edgefield	1.03×10^9 (37)	4.85×10^8	2.86×10^8	0	2.62×10^8	0	0	0
Fairfield	3.41×10^9 (9)	4.80×10^6	8.93×10^8	0	2.52×10^9	0	0	0
Greenwood	2.03×10^9 (21)	0	5.92×10^8	0	1.44×10^9	0	0	0
Kershaw	2.91×10^9 (13)	1.84×10^9	8.21×10^8	4.80×10^7	2.02×10^8	0	0	0
Lancaster	1.29×10^9 (31)	2.54×10^8	7.70×10^8	0	2.67×10^8	0	0	0
Laurens	2.21×10^9 (18)	6.00×10^8	9.69×10^7	0	1.51×10^9	0	0	0
Lexington	2.78×10^9 (14)	2.18×10^9	4.39×10^8	0	1.55×10^8	0	6.00×10^6	0
Newberry	1.86×10^9 (23)	3.36×10^8	3.26×10^8	0	1.20×10^9	0	0	0
Richland	2.72×10^9 (15)	7.82×10^8	1.84×10^9	1.92×10^7	8.17×10^7	0	0	0
Saluda	8.46×10^8 (38)	9.12×10^7	4.23×10^8	0	3.31×10^8	0	0	0
York	3.19×10^9 (12)	2.40×10^7	6.83×10^8	0	2.48×10^9	0	0	0
(Midlands)	**3.29×10^{10} (2)**	**9.27×10^9**	**8.86×10^9**	**7.44×10^7**	**1.46×10^{10}**	**6.90×10^7**	**6.00×10^6**	**0**
Chesterfield	4.27×10^9 (5)	8.30×10^8	3.34×10^9	0	9.89×10^7	0	0	0
Clarendon	1.18×10^9 (34)	1.87×10^8	9.79×10^8	1.44×10^7	0	0	0	0
Darlington	1.51×10^9 (28)	1.73×10^8	1.32×10^9	2.16×10^7	0	0	6.00×10^5	0
Dillon	1.12×10^9 (35)	4.37×10^8	6.53×10^8	1.92×10^7	0	0	1.14×10^7	0
Florence	1.61×10^9 (26)	4.66×10^8	1.14×10^9	0	0	0	1.20×10^6	0
Georgetown	5.05×10^9 (2)	1.68×10^9	1.40×10^9	1.37×10^8	1.76×10^9	0	6.90×10^7	0
Horry	4.99×10^9 (3)	1.21×10^9	2.20×10^9	1.54×10^8	1.23×10^9	0	1.98×10^8	0
Lee	8.12×10^8 (39)	1.39×10^8	6.68×10^8	0	4.30×10^6	0	0	0
Marion	2.07×10^9 (20)	5.14×10^8	1.46×10^9	6.48×10^7	0	0	2.94×10^7	0
Marlboro	2.00×10^9 (22)	3.60×10^8	1.37×10^9	1.94×10^8	7.31×10^7	0	1.20×10^6	0
Sumter	1.85×10^9 (24)	4.32×10^7	1.79×10^9	0	1.72×10^7	0	1.20×10^6	0
Williamsburg	1.19×10^9 (33)	1.20×10^8	1.07×10^9	0	0	0	1.80×10^6	0
(Pee Dee)	**2.76×10^{10} (3)**	**6.16×10^9**	**1.74×10^{10}**	**6.05×10^8**	**3.19×10^9**	**0**	**3.14×10^8**	**0**
Allendale	6.50×10^8 (41)	1.20×10^8	5.15×10^8	1.44×10^7	0	0	0	0
Bamberg	7.88×10^8 (40)	6.05×10^8	5.10×10^6	0	1.76×10^8	0	1.80×10^6	0
Barnwell	1.08×10^9 (36)	3.74×10^8	7.04×10^8	0	0	0	0	0
Beaufort	4.10×10^9 (6)	3.35×10^9	1.73×10^8	1.44×10^7	1.72×10^8	2.65×10^8	1.26×10^8	0
Berkeley	3.65×10^9 (7)	6.96×10^8	1.06×10^9	5.52×10^7	1.76×10^9	0	8.22×10^7	0
Calhoun	4.49×10^8 (46)	3.17×10^8	1.33×10^8	0	0	0	0	0
Charleston	8.66×10^9 (1)	3.67×10^9	1.69×10^9	0	3.13×10^9	0	1.64×10^8	0
Colleton	3.45×10^9 (8)	1.34×10^9	2.50×10^8	2.11×10^8	6.02×10^8	9.78×10^8	6.54×10^7	0
Dorchester	2.63×10^9 (16)	1.32×10^9	8.67×10^7	2.40×10^6	1.20×10^9	0	1.80×10^7	0
Hampton	1.56×10^9 (27)	6.53×10^8	4.44×10^8	7.20×10^6	4.34×10^8	0	2.40×10^7	0
Jasper	4.37×10^9 (4)	1.53×10^9	2.96×10^8	1.37×10^8	1.06×10^9	1.33×10^9	1.44×10^7	0
McCormick	1.70×10^9 (25)	3.07×10^8	3.32×10^8	0	1.06×10^9	0	0	0
Orangeburg	5.50×10^8 (42)	3.50×10^8	9.18×10^7	0	1.08×10^8	0	6.00×10^5	0
(Low Country)	**3.36×10^{10} (1)**	**1.46×10^{10}**	**5.78×10^9**	**4.42×10^8**	**9.70×10^9**	**2.58×10^9**	**4.96×10^8**	**0**
Totals (kg)	**1.02×10^{11}**	**3.24×10^{10}**	**3.50×10^{10}**	**1.14×10^9**	**2.99×10^{10}**	**2.65×10^9**	**8.16×10^8**	**0**

Table 8. The total dollar value of soil inorganic carbon (SIC) by county, region, and soil order for the state of South Carolina (U.S.A.), based on mid-point soil inorganic carbon (SIC) numbers for the upper 2 m from Guo et al. 2006 [18] and a social cost of carbon (SC-CO_2) of $46 per metric ton of CO_2 (2007 U.S. dollars with an average discount rate of 3% [17]).

County (Region)	Total Value ($) (Rank)	Slight <— Degree of Weathering and Soil Development —> Strong						
		Slightly Weathered			Moderately Weathered		Strongly Weathered	
		Entisols	Inceptisols	Histosols	Alfisols	Mollisols	Spodosols	Ultisols
		Value ($)						
Anderson	8.36×10^7 (43)	8.36×10^7	0	0	0	0	0	0
Cherokee	2.50×10^8 (29)	9.68×10^7	8.08×10^7	0	7.20×10^7	0	0	0
Greenville	2.31×10^8 (30)	8.86×10^7	1.43×10^8	0	0	0	0	0
Oconee	7.85×10^7 (45)	3.03×10^7	4.82×10^7	0	0	0	0	0
Pickens	7.93×10^7 (44)	3.20×10^7	4.73×10^7	0	0	0	0	0
Spartanburg	2.00×10^8 (32)	6.56×10^6	1.38×10^8	0	5.54×10^7	0	0	0
Union	3.75×10^8 (17)	6.07×10^7	5.25×10^7	0	2.62×10^8	0	0	0
(Upstate)	**1.30×10^9 (4)**	**3.99×10^8**	**5.10×10^8**	**0**	**3.90×10^8**	**0**	**0**	**0**
Abbeville	3.64×10^8 (19)	4.26×10^7	3.27×10^7	0	2.89×10^8	0	0	0
Aiken	5.56×10^8 (10)	3.65×10^8	1.83×10^8	4.51×10^6	3.60×10^6	0	0	0
Chester	5.42×10^8 (11)	4.92×10^7	7.05×10^7	0	4.10×10^8	1.16×10^7	0	0
Edgefield	1.75×10^8 (37)	8.28×10^7	4.82×10^7	0	4.39×10^7	0	0	0
Fairfield	5.73×10^8 (9)	8.20×10^5	1.51×10^8	0	4.21×10^8	0	0	0
Greenwood	3.41×10^8 (21)	0	9.98×10^7	0	2.41×10^8	0	0	0
Kershaw	4.95×10^8 (13)	3.14×10^8	1.38×10^8	8.20×10^6	3.38×10^7	0	0	0
Lancaster	2.18×10^8 (31)	4.35×10^7	1.30×10^8	0	4.46×10^7	0	0	0
Laurens	3.72×10^8 (18)	1.03×10^8	1.63×10^7	0	2.53×10^8	0	0	0
Lexington	4.73×10^8 (14)	3.72×10^8	7.40×10^7	0	2.59×10^7	0	1.00×10^6	0
Newberry	3.13×10^8 (23)	5.74×10^7	5.50×10^7	0	2.00×10^8	0	0	0
Richland	4.60×10^8 (15)	1.34×10^8	3.10×10^8	3.28×10^6	1.37×10^7	0	0	0
Saluda	1.42×10^8 (38)	1.56×10^7	7.14×10^7	0	5.54×10^7	0	0	0
York	5.35×10^8 (12)	4.10×10^6	1.15×10^8	0	4.15×10^8	0	0	0
(Midlands)	**5.55×10^9 (2)**	**1.58×10^9**	**1.49×10^9**	**1.27×10^7**	**2.45×10^9**	**1.16×10^7**	**1.00×10^6**	**0**
Chesterfield	7.22×10^8 (5)	1.42×10^8	5.63×10^8	0	1.66×10^7	0	0	0
Clarendon	2.00×10^8 (34)	3.20×10^7	1.65×10^8	2.46×10^6	0	0	0	0
Darlington	2.55×10^8 (28)	2.95×10^7	2.22×10^8	3.69×10^6	0	0	1.00×10^5	0
Dillon	1.90×10^8 (35)	7.46×10^7	1.10×10^8	3.28×10^6	0	0	1.90×10^6	0
Florence	2.72×10^8 (26)	7.95×10^7	1.93×10^8	0	0	0	2.00×10^5	0
Georgetown	8.53×10^8 (2)	2.88×10^8	2.36×10^8	2.34×10^7	2.94×10^8	0	1.15×10^7	0
Horry	8.43×10^8 (3)	2.07×10^8	3.71×10^8	2.62×10^7	2.07×10^8	0	3.30×10^7	0
Lee	1.37×10^8 (39)	2.38×10^7	1.13×10^8	0	7.20×10^5	0	0	0
Marion	3.50×10^8 (20)	8.77×10^7	2.46×10^8	1.11×10^7	0	0	4.90×10^6	0
Marlboro	3.38×10^8 (22)	6.15×10^7	2.31×10^8	3.32×10^7	1.22×10^7	0	2.00×10^5	0
Sumter	3.11×10^8 (24)	7.38×10^6	3.01×10^8	0	2.88×10^6	0	2.00×10^5	0
Williamsburg	2.01×10^8 (33)	2.05×10^7	1.80×10^8	0	0	0	3.00×10^5	0
(Pee Dee)	**4.67×10^9 (3)**	**1.05×10^9**	**2.93×10^9**	**1.03×10^8**	**5.34×10^8**	**0**	**5.23×10^7**	**0**
Allendale	1.10×10^8 (41)	2.05×10^7	8.69×10^7	2.46×10^6	0	0	0	0
Bamberg	1.34×10^8 (40)	1.03×10^8	8.60×10^5	0	2.95×10^7	0	3.00×10^5	0
Barnwell	1.83×10^8 (36)	6.40×10^7	1.19×10^8	0	0	0	0	0
Beaufort	6.98×10^8 (6)	5.72×10^8	2.92×10^7	2.46×10^6	2.88×10^7	4.44×10^7	2.10×10^7	0
Berkeley	6.15×10^8 (7)	1.19×10^8	1.79×10^8	9.43×10^6	2.94×10^8	0	1.37×10^7	0
Calhoun	7.65×10^7 (46)	5.41×10^7	2.24×10^7	0	0	0	0	0
Charleston	1.46×10^9 (1)	6.27×10^8	2.86×10^8	0	5.23×10^8	0	2.73×10^7	0
Colleton	5.84×10^8 (8)	2.30×10^8	4.21×10^7	3.61×10^7	1.01×10^8	1.64×10^8	1.09×10^7	0
Dorchester	4.44×10^8 (16)	2.25×10^8	1.46×10^7	4.10×10^5	2.02×10^8	0	3.00×10^6	0
Hampton	2.64×10^8 (27)	1.12×10^8	7.48×10^7	1.23×10^6	7.27×10^7	0	4.00×10^6	0
Jasper	7.37×10^8 (4)	2.61×10^8	4.99×10^7	2.34×10^7	1.77×10^8	2.24×10^8	2.40×10^6	0
McCormick	2.86×10^8 (25)	5.25×10^7	5.59×10^7	0	1.78×10^8	0	0	0
Orangeburg	9.34×10^7 (42)	5.99×10^7	1.55×10^7	0	1.80×10^7	0	1.00×10^5	0
(Low Country)	**5.69×10^9 (1)**	**2.50×10^9**	**9.75×10^8**	**7.54×10^7**	**1.62×10^9**	**4.32×10^8**	**8.27×10^7**	**0**
Totals ($)	**1.72×10^{10}**	**5.53×10^9**	**5.91×10^9**	**1.95×10^8**	**5.00×10^9**	**4.44×10^8**	**1.36×10^8**	**0**

Table 9. Mid-point total soil carbon (TSC) storage by county, region, and soil order for the state of South Carolina (U.S.A.), based on mid-point (TSC) contents in the upper 2 m based on data from Guo et al. 2006 [18].

County (Region)	Total Storage (kg) (Rank)	Slight <— Degree of Weathering and Soil Development —> Strong						
			Slightly Weathered		Moderately Weathered		Strongly Weathered	
		Entisols	Inceptisols	Histosols	Alfisols	Mollisols	Spodosols	Ultisols
		Total Storage (kg)						
Anderson	1.37×10^{10} (26)	1.31×10^{9}	0	0	0	0	0	1.23×10^{10}
Cherokee	8.98×10^{9} (42)	1.51×10^{9}	1.32×10^{9}	0	1.18×10^{9}	0	0	4.97×10^{9}
Greenville	1.54×10^{10} (20)	1.38×10^{9}	2.32×10^{9}	0	0	0	0	1.17×10^{10}
Oconee	1.21×10^{10} (31)	4.74×10^{8}	7.84×10^{8}	0	0	0	0	1.08×10^{10}
Pickens	9.53×10^{9} (39)	4.99×10^{8}	7.70×10^{8}	0	0	0	0	8.26×10^{9}
Spartanburg	1.49×10^{10} (22)	1.02×10^{8}	2.25×10^{9}	0	9.09×10^{8}	0	0	1.16×10^{10}
Union	1.19×10^{10} (32)	9.47×10^{8}	8.54×10^{8}	0	4.30×10^{9}	0	0	5.84×10^{9}
(Upstate)	**8.64×10^{10} (4)**	**6.22×10^{9}**	**8.30×10^{9}**	**0**	**6.38×10^{9}**	**0**	**0**	**6.55×10^{10}**
Abbeville	1.15×10^{10} (34)	6.66×10^{8}	5.32×10^{8}	0	4.73×10^{9}	0	0	5.52×10^{9}
Aiken	2.51×10^{10} (7)	5.70×10^{9}	2.98×10^{9}	1.57×10^{9}	5.90×10^{7}	0	0	1.48×10^{10}
Chester	1.43×10^{10} (24)	7.68×10^{8}	1.15×10^{9}	0	6.73×10^{9}	1.50×10^{8}	0	5.50×10^{9}
Edgefield	1.04×10^{10} (37)	1.29×10^{9}	7.84×10^{8}	0	7.20×10^{8}	0	0	7.60×10^{9}
Fairfield	1.59×10^{10} (17)	1.28×10^{7}	2.45×10^{9}	0	6.90×10^{9}	0	0	6.55×10^{9}
Greenwood	1.07×10^{10} (36)	0	1.62×10^{9}	0	3.95×10^{9}	0	0	5.11×10^{9}
Kershaw	1.95×10^{10} (11)	4.90×10^{9}	2.25×10^{9}	2.85×10^{9}	5.55×10^{8}	0	0	8.95×10^{9}
Lancaster	1.16×10^{10} (33)	6.78×10^{8}	2.11×10^{9}	0	7.32×10^{8}	0	0	8.12×10^{9}
Laurens	1.55×10^{10} (19)	1.60×10^{9}	2.66×10^{8}	0	4.14×10^{9}	0	0	9.53×10^{9}
Lexington	1.59×10^{10} (18)	5.81×10^{9}	1.20×10^{9}	0	4.25×10^{8}	0	1.29×10^{8}	8.31×10^{9}
Newberry	1.37×10^{10} (25)	8.96×10^{8}	8.96×10^{8}	0	3.28×10^{9}	0	0	8.58×10^{9}
Richland	1.76×10^{10} (13)	2.09×10^{9}	5.04×10^{9}	1.14×10^{9}	2.24×10^{8}	0	0	9.07×10^{9}
Saluda	9.35×10^{9} (40)	2.43×10^{8}	1.16×10^{9}	0	9.09×10^{8}	0	0	7.04×10^{9}
York	1.61×10^{10} (16)	6.40×10^{7}	1.88×10^{9}	0	6.81×10^{9}	0	0	7.36×10^{9}
(Midlands)	**2.06×10^{11} (3)**	**2.47×10^{10}**	**2.43×10^{10}**	**4.42×10^{9}**	**4.02×10^{10}**	**1.50×10^{8}**	**1.29×10^{8}**	**1.12×10^{11}**
Chesterfield	2.02×10^{10} (10)	2.21×10^{9}	9.17×10^{9}	0	2.71×10^{8}	0	0	8.53×10^{9}
Clarendon	1.35×10^{10} (28)	4.99×10^{8}	2.69×10^{9}	8.55×10^{8}	0	0	0	9.44×10^{9}
Darlington	1.34×10^{10} (29)	4.61×10^{8}	3.61×10^{9}	1.28×10^{9}	0	0	1.29×10^{7}	8.08×10^{9}
Dillon	9.98×10^{9} (38)	1.16×10^{9}	1.79×10^{9}	1.14×10^{9}	0	0	2.45×10^{8}	5.64×10^{9}
Florence	1.66×10^{10} (15)	1.24×10^{9}	3.14×10^{9}	0	0	0	2.58×10^{7}	1.22×10^{10}
Georgetown	2.89×10^{10} (3)	4.49×10^{9}	3.84×10^{9}	8.12×10^{9}	4.83×10^{9}	0	1.48×10^{9}	6.09×10^{9}
Horry	3.68×10^{10} (1)	3.23×10^{9}	6.03×10^{9}	9.12×10^{9}	3.39×10^{9}	0	4.26×10^{9}	1.08×10^{10}
Lee	8.59×10^{9} (43)	3.71×10^{8}	1.83×10^{9}	0	1.18×10^{7}	0	0	6.37×10^{9}
Marion	1.53×10^{10} (21)	1.37×10^{9}	4.00×10^{9}	3.85×10^{9}	0	0	6.32×10^{8}	5.48×10^{9}
Marlboro	2.21×10^{10} (8)	9.60×10^{8}	3.77×10^{9}	1.15×10^{10}	2.01×10^{8}	0	2.58×10^{7}	5.58×10^{9}
Sumter	1.45×10^{10} (23)	1.15×10^{8}	4.90×10^{9}	0	4.72×10^{7}	0	2.58×10^{7}	9.44×10^{9}
Williamsburg	1.86×10^{10} (12)	3.20×10^{8}	2.93×10^{9}	0	0	0	3.87×10^{7}	1.54×10^{10}
(Pee Dee)	**2.19×10^{11} (2)**	**1.64×10^{10}**	**4.77×10^{10}**	**3.59×10^{10}**	**8.74×10^{9}**	**0**	**6.75×10^{9}**	**1.03×10^{11}**
Allendale	9.14×10^{9} (41)	3.20×10^{8}	1.41×10^{9}	8.55×10^{8}	0	0	0	6.55×10^{9}
Bamberg	8.16×10^{9} (45)	1.61×10^{9}	1.40×10^{7}	0	4.84×10^{8}	0	3.87×10^{7}	6.01×10^{9}
Barnwell	1.15×10^{10} (35)	9.98×10^{8}	1.93×10^{9}	0	0	0	0	8.52×10^{9}
Beaufort	1.68×10^{10} (14)	8.93×10^{9}	4.76×10^{8}	8.55×10^{8}	4.72×10^{8}	5.75×10^{8}	2.71×10^{9}	2.78×10^{9}
Berkeley	2.80×10^{10} (5)	1.86×10^{9}	2.91×10^{9}	3.28×10^{9}	4.83×10^{9}	0	1.77×10^{9}	1.34×10^{10}
Calhoun	5.87×10^{9} (46)	8.45×10^{8}	3.64×10^{8}	0	0	0	0	4.66×10^{9}
Charleston	2.81×10^{10} (4)	9.79×10^{9}	4.65×10^{9}	0	8.58×10^{9}	0	3.52×10^{9}	1.56×10^{9}
Colleton	3.57×10^{10} (2)	3.58×10^{9}	6.86×10^{8}	1.25×10^{10}	1.65×10^{9}	2.13×10^{9}	1.41×10^{9}	1.37×10^{10}
Dorchester	1.36×10^{10} (27)	3.51×10^{9}	2.38×10^{8}	1.43×10^{8}	3.30×10^{9}	0	3.87×10^{8}	6.06×10^{9}
Hampton	1.27×10^{10} (30)	1.74×10^{9}	1.22×10^{9}	4.28×10^{8}	1.19×10^{9}	0	5.16×10^{8}	7.64×10^{9}
Jasper	2.52×10^{10} (6)	4.07×10^{9}	8.12×10^{8}	8.12×10^{9}	2.90×10^{9}	2.90×10^{9}	3.10×10^{8}	6.04×10^{9}
McCormick	8.51×10^{9} (44)	8.19×10^{8}	9.10×10^{8}	0	2.91×10^{9}	0	0	3.87×10^{9}
Orangeburg	2.09×10^{10} (9)	9.34×10^{8}	2.52×10^{8}	0	2.95×10^{8}	0	1.29×10^{7}	1.94×10^{10}
(Low Country)	**2.24×10^{11} (1)**	**3.90×10^{10}**	**1.59×10^{10}**	**2.62×10^{10}**	**2.66×10^{10}**	**5.60×10^{9}**	**1.07×10^{10}**	**1.00×10^{11}**
Totals (kg)	**7.36×10^{11}**	**8.64×10^{10}**	**9.62×10^{10}**	**6.77×10^{10}**	**8.19×10^{10}**	**5.75×10^{9}**	**1.75×10^{10}**	**3.81×10^{11}**

Table 10. Total soil carbon (TSC) values by county, region, and soil order for the state of South Carolina (U.S.A.), based on mid-point total soil carbon (TSC) numbers for the upper 2 m from Guo et al. 2006 [18] and a social cost of carbon (SC-CO_2) of $46 per metric ton of CO_2 (2007 U.S. dollars with an average discount rate of 3% [17]).

County (Region)	Total Value ($) (Rank)	Slight <— Degree of Weathering and Soil Development —> Strong						
		Slightly Weathered		Moderately Weathered			Strongly Weathered	
		Entisols	Inceptisols	Histosols	Alfisols	Mollisols	Spodosols	Ultisols
		Value ($)						
Anderson	2.31×10^9 (26)	2.21×10^8	0	0	0	0	0	2.09×10^9
Cherokee	1.52×10^9 (42)	2.56×10^8	2.22×10^8	0	1.99×10^8	0	0	8.40×10^8
Greenville	2.60×10^9 (20)	2.34×10^8	3.92×10^8	0	0	0	0	1.97×10^9
Oconee	2.04×10^9 (31)	8.03×10^7	1.32×10^8	0	0	0	0	1.83×10^9
Pickens	1.61×10^9 (39)	8.46×10^7	1.30×10^8	0	0	0	0	1.40×10^9
Spartanburg	2.51×10^9 (22)	1.74×10^7	3.80×10^8	0	1.53×10^8	0	0	1.96×10^9
Union	2.02×10^9 (32)	1.61×10^8	1.44×10^8	0	7.24×10^8	0	0	9.88×10^8
(Upstate)	**1.46×10^{10} (4)**	**1.05×10^9**	**1.40×10^9**	**0**	**1.08×10^9**	**0**	**0**	**1.11×10^{10}**
Abbeville	1.93×10^9 (34)	1.13×10^8	8.97×10^7	0	7.98×10^8	0	0	9.34×10^8
Aiken	4.24×10^9 (7)	9.66×10^8	5.03×10^8	2.64×10^8	9.95×10^6	0	0	2.50×10^9
Chester	2.41×10^9 (24)	1.30×10^8	1.94×10^8	0	1.13×10^9	2.53×10^7	0	9.30×10^8
Edgefield	1.76×10^9 (37)	2.19×10^8	1.32×10^8	0	1.21×10^8	0	0	1.29×10^9
Fairfield	2.69×10^9 (17)	2.17×10^6	4.13×10^8	0	1.16×10^9	0	0	1.11×10^9
Greenwood	1.80×10^9 (36)	0	2.74×10^8	0	6.67×10^8	0	0	8.64×10^8
Kershaw	3.30×10^9 (11)	8.31×10^8	3.80×10^8	4.81×10^8	9.35×10^7	0	0	1.51×10^9
Lancaster	1.97×10^9 (33)	1.15×10^8	3.56×10^8	0	1.23×10^8	0	0	1.37×10^9
Laurens	2.62×10^9 (19)	2.71×10^8	4.48×10^7	0	6.98×10^8	0	0	1.61×10^9
Lexington	2.69×10^9 (18)	9.85×10^8	2.03×10^8	0	7.16×10^7	0	2.17×10^7	1.40×10^9
Newberry	2.31×10^9 (25)	1.52×10^8	1.51×10^8	0	5.53×10^8	0	0	1.45×10^9
Richland	2.97×10^9 (13)	3.54×10^8	8.50×10^8	1.92×10^8	3.78×10^7	0	0	1.53×10^9
Saluda	1.58×10^9 (40)	4.12×10^7	1.96×10^8	0	1.53×10^8	0	0	1.19×10^9
York	2.72×10^9 (16)	1.09×10^7	3.16×10^8	0	1.15×10^9	0	0	1.24×10^9
(Midlands)	**3.48×10^{10} (3)**	**4.19×10^9**	**4.10×10^9**	**7.45×10^8**	**6.77×10^9**	**2.53×10^7**	**2.17×10^7**	**1.89×10^{10}**
Chesterfield	3.41×10^9 (10)	3.75×10^8	1.55×10^9	0	4.58×10^7	0	0	1.44×10^9
Clarendon	2.28×10^9 (28)	8.46×10^7	4.53×10^8	1.44×10^8	0	0	0	1.59×10^9
Darlington	2.27×10^9 (29)	7.81×10^7	6.09×10^8	2.16×10^8	0	0	2.17×10^6	1.37×10^9
Dillon	1.69×10^9 (38)	1.97×10^8	3.02×10^8	1.92×10^8	0	0	4.12×10^7	9.53×10^8
Florence	2.81×10^9 (15)	2.10×10^8	5.29×10^8	0	0	0	4.34×10^6	2.07×10^9
Georgetown	4.87×10^9 (3)	7.62×10^8	6.47×10^8	1.37×10^9	8.14×10^8	0	2.50×10^8	1.03×10^9
Horry	6.22×10^9 (1)	5.47×10^8	1.02×10^9	1.54×10^9	5.71×10^8	0	7.16×10^8	1.83×10^9
Lee	1.45×10^9 (43)	6.29×10^7	3.09×10^8	0	1.99×10^6	0	0	1.08×10^9
Marion	2.59×10^9 (21)	2.32×10^8	6.75×10^8	6.49×10^8	0	0	1.06×10^8	9.26×10^8
Marlboro	3.73×10^9 (8)	1.63×10^8	6.35×10^8	1.95×10^9	3.38×10^7	0	4.34×10^6	9.43×10^8
Sumter	2.45×10^9 (23)	1.95×10^7	8.26×10^8	0	7.96×10^6	0	4.34×10^6	1.59×10^9
Williamsburg	3.15×10^9 (12)	5.43×10^7	4.93×10^8	0	0	0	6.51×10^6	2.60×10^9
(Pee Dee)	**3.69×10^{10} (2)**	**2.79×10^9**	**8.04×10^9**	**6.06×10^9**	**1.47×10^9**	**0**	**1.13×10^9**	**1.74×10^{10}**
Allendale	1.54×10^9 (41)	5.43×10^7	2.38×10^8	1.44×10^8	0	0	0	1.11×10^9
Bamberg	1.38×10^9 (45)	2.73×10^8	2.36×10^6	0	8.16×10^7	0	6.51×10^6	1.02×10^9
Barnwell	1.93×10^9 (35)	1.69×10^8	3.26×10^8	0	0	0	0	1.44×10^9
Beaufort	2.84×10^9 (14)	1.51×10^9	8.02×10^7	1.44×10^8	7.96×10^7	9.68×10^7	4.56×10^8	4.69×10^8
Berkeley	4.73×10^9 (5)	3.15×10^8	4.91×10^8	5.53×10^8	8.14×10^8	0	2.97×10^8	2.26×10^9
Calhoun	9.92×10^8 (46)	1.43×10^8	6.14×10^7	0	0	0	0	7.87×10^8
Charleston	4.75×10^9 (4)	1.66×10^9	7.84×10^8	0	1.45×10^9	0	5.92×10^8	2.64×10^8
Colleton	6.02×10^9 (2)	6.08×10^8	1.16×10^8	2.11×10^9	2.79×10^8	3.58×10^8	2.37×10^8	2.31×10^9
Dorchester	2.30×10^9 (27)	5.95×10^8	4.01×10^7	2.40×10^7	5.57×10^8	0	6.51×10^7	1.02×10^9
Hampton	2.15×10^9 (30)	2.95×10^8	2.05×10^8	7.21×10^7	2.01×10^8	0	8.68×10^7	1.29×10^9
Jasper	4.25×10^9 (6)	6.90×10^8	1.37×10^8	1.37×10^9	4.90×10^8	4.88×10^8	5.21×10^7	1.02×10^9
McCormick	1.44×10^9 (44)	1.39×10^8	1.53×10^8	0	4.92×10^8	0	0	6.54×10^8
Orangeburg	3.53×10^9 (9)	1.58×10^8	4.25×10^7	0	4.98×10^7	0	2.17×10^6	3.27×10^9
(Low Country)	**3.79×10^{10} (1)**	**6.61×10^9**	**2.68×10^9**	**4.42×10^9**	**4.49×10^9**	**9.43×10^8**	**1.79×10^9**	**1.69×10^{10}**
Totals ($)	**1.24×10^{11}**	**1.46×10^{10}**	**1.62×10^{10}**	**1.14×10^{10}**	**1.38×10^{10}**	**9.68×10^8**	**2.95×10^9**	**6.43×10^{10}**

4. Discussion

Pedodiversity (soil diversity) in South Carolina is a source of various ES goods, services, and disservices (ED). This study demonstrates the value of regulating ES/ED in the state and its regions and counties. According to Mikhailova et al. (2021) [22], taxonomic pedodiversity (e.g., soil order) "provides a general description of the stock, its type, and spatial distribution," which is often rereferred to as a "portfolio" to describe the link between pedodiversity and its stocks. South Carolina soil "portfolio" is composed of seven soil orders: Entisols (9% of the total state area), Inceptisols (9%), Histosols (1%), Alfisols (9%), Mollisols (0%), Spodosols (2%), and Ultisols (70%) (Figure 4, Table 11). Highly weathered Ultisols have the highest proportion of the total area of the state (Figure 4a), which contributes to the highest SOC and TSC storage and their associated social costs of carbon. The contribution of SIC to associated social costs of carbon is small at the state level and primarily associated with Inceptisols, Entisols, and Alfisols.

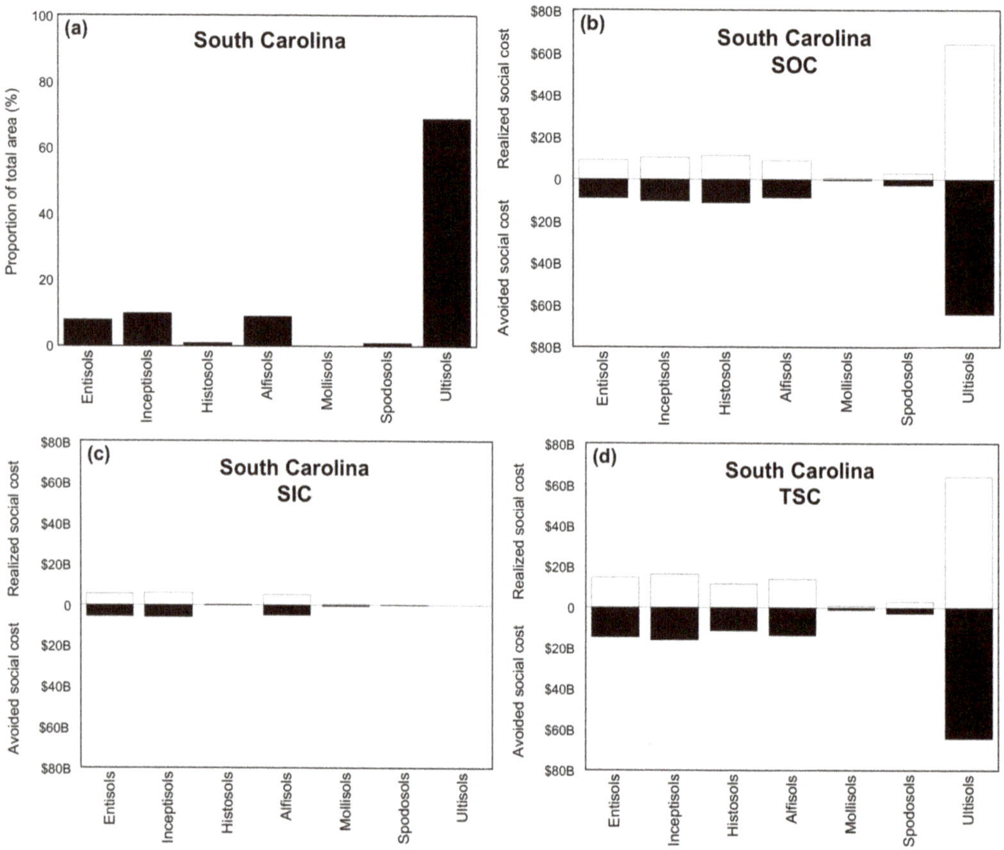

Figure 4. Diagram showing how the "portfolio-effect" and "distribution-effect" of pedodiversity can vary within the state: (**a**) pedodiversity by soil order area; (**b**) value of soil organic carbon (SOC) storage, (**c**) value of soil inorganic carbon (SIC) storage, (**d**) value of total soil carbon (TSC) storage in the upper 2-m depth based on avoided or realized the social cost of CO_2 (SC-CO_2) of $46 (USD) per metric ton of CO_2 [17] by soil order. Note: B = billion = 10^9.

Soil "portfolio" differs within each county, and Figure 5 illustrates this concept using three counties from different regions: Anderson (Upstate), Newberry (Midlands), and Colleton (Low Country).

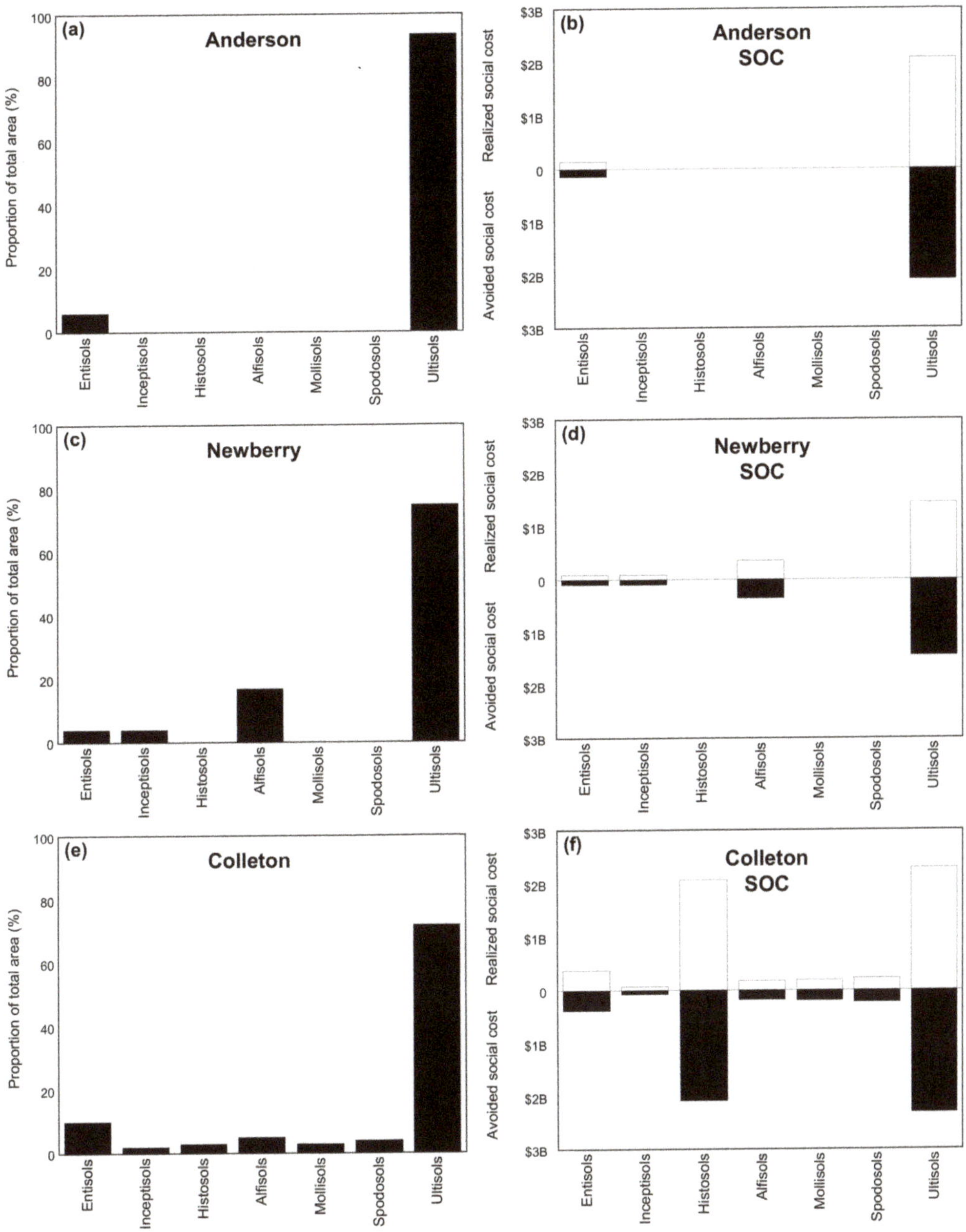

Figure 5. Diagram showing how the "portfolio-effect" and "distribution-effect" of pedodiversity can vary by county: (**a**,**c**,**e**) pedodiversity by soil order area; (**b**,**d**,**f**) value of soil organic carbon (SOC) storage in the upper 2-m depth based on avoided or realized the social cost of CO_2 (SC-CO_2) of $46 (USD) per metric ton of CO_2 [17] by soil order. Note: B = billion = 10^9.

In all three cases, Ultisols occupy the largest proportion of the area in each county. The type of soil order influences the value of SOC storage. In Colleton County, the soil order of Histosols contributes to the social costs of C as much as the Ultisols even though its area is much smaller (Figure 5) because of high SOC content of 142.5 kg m^{-2}. Figures 4 and 5 represent social costs of soil C from different point of views: "avoided" versus "realized" social costs. Soil carbon stored in the soil represents the "avoided social cost" of soil C if not converted to CO_2 and released into the atmosphere. When CO_2 is released into the atmosphere, it becomes the "realized social cost" because of the damages from global warming. In South Carolina, Histosols and Alfisols are particularly sensitive to climate change because of relatively high soil C content, which is most likely to experience higher decomposition rates due to increases in temperature and precipitation. All soils in the state of South Carolina have low recarbonization potential.

Table 11. Distribution of soil carbon regulating ecosystem services in the state of South Carolina (U.S.A.) by soil order (photos courtesy of USDA/NRCS [23]) in the upper 2-m depth based on avoided or realized the social cost of CO_2 (SC-CO_2) of $46 (USD) per metric ton of CO_2 [17].

Soil Regulating Ecosystem Services in the State of South Carolina						
Slight <—		**Degree of Weathering and Soil Development**				**—> Strong**
	Slightly Weathered 18%			**Moderately Weathered 9%**		**Strongly Weathered 72%**
Entisols 9%	**Inceptisols 9%**	**Histosols 1%**	**Alfisols 9%**	**Mollisols 0%**	**Spodosols 2%**	**Ultisols 70%**

The social cost of soil organic carbon (SOC) in USD: $107.14B						
$9.11B	$10.30B	$11.20B	$8.82B	$524.00M	$2.82B	$64.30B
9%	10%	10%	8%	0%	3%	60%
The social cost of soil inorganic carbon (SIC) in USD: $17.22B						
$5.53B	$5.91B	$195.00M	$5.00B	$444.00M	$136.00M	$0
32%	34%	1%	29%	3%	1%	0%
The social cost of total soil carbon (TSC) in USD: $124.36B						
$14.60B	$16.20B	$11.40B	$13.80B	$968.00M	$2.95B	$64.30B
12%	13%	9%	11%	1%	2%	52%
Sensitivity to climate change						
Low	Low	High	High	High	Low	Low
Soil organic and inorganic carbon sequestration (recarbonization) potential						
Low	Low	Low	Low	Low	Low	Low

Note: Entisols, Inceptisols, Alfisols, Mollisols, Spodosols, Ultisols are mineral soils. Histosols are mostly organic soils. M = million = 10^6; B = billion = 10^9.

Amelung et al. (2020) [24] proposed linking soil C sequestration to food security using soil- and site-specific potentials and opportunities for soil C sequestration. In this respect, the state of South Carolina faces serious limitations in both soil- (dominated by highly-weathered soil order, Ultisols) and site-specific (high demand for soil C due to rapid urbanization and population growth; rapid changes in coastal areas, etc.) potentials. Soil order Histosols (which often contains organic soils) is located in the coastal areas of the state and can be drained for agriculture and urbanization, leading to high losses of soil C into the atmosphere [24]. Recarbonization of soils in the state of South Carolina may not be economically feasible due to past excessive levels of soil degradation [25], high fertilization and liming costs (including transportation) associated with increasing soil C in mostly highly-weathered and acid soils in the state. It should be noted that the reported soil survey-based C values may be an overestimate of actual soil C measured in the field, but the overall trends for the soil orders should be similar [10]. Soil C should be regularly monitored to quantify soil contributions to ES and its flows [26,27].

5. Conclusions

This study examined the application of soil diversity (pedodiversity) concepts (taxonomic) and its measures to value soil C regulating ES/ED in the state of South Carolina (U.S.A.), its administrative units (regions, counties), and the systems of soil classification (e.g., U.S. Department of Agriculture (USDA) Soil Taxonomy, Soil Survey Geographic (SSURGO) Database) to be considered in territorial planning. Pedodiversity provides a critical context (e.g., "portfolio-effect," "distribution-effect," "evenness-effect," etc.) for analyzing, interpreting, and reporting ES/ED within the ES framework for sustainable management of soil carbon within the state. Taxonomic pedodiversity in South Carolina exhibits high soil diversity (7 soil orders: Entisols, Inceptisols, Histosols, Alfisols, Mollisols, Spodosols, and Ultisols), which is not evenly distributed within the state, regions, and counties. In general, pedodiversity tends to increase from the Upstate to Low Country, where three counties (Beaufort, Colleton, and Jasper) have all seven orders. Similarly, soil carbon storage and its associated social costs tend to increase in a similar geographic direction. Ultisols occupy the highest proportion of the state area (70%) and have the highest SOC storage and related social costs of carbon ($64.30B). The contribution of SIC to associated social costs of carbon is small ($17.22B) at the state level and primarily associated with Inceptisols ($5.91B), Entisols ($5.53B), and Alfisols ($5.00B). In the state of South Carolina, Histosols and Alfisols are particularly sensitive to climate change because of relatively high soil C content, which is most likely experience higher rates of decomposition due to increases in temperature and precipitation. All soils in the state of South Carolina have low recarbonization potential. Administrative areas (e.g., counties, regions) combined with pedodiversity concepts can provide useful information to design cost-efficient policies to manage soil carbon regulating ES at the state level.

Author Contributions: Conceptualization, E.A.M.; methodology, E.A.M., M.A.S. and H.A.Z.; formal analysis, E.A.M.; writing—original draft preparation, E.A.M.; writing—review and editing, E.A.M., C.J.P., G.C.P. and M.A.S.; visualization, H.A.Z., L.L. and Z.H. All authors have read and agreed to the published version of the manuscript.

Funding: This research received no external funding.

Data Availability Statement: Not applicable.

Acknowledgments: We would like to thank the reviewers for their constructive comments and suggestions.

Conflicts of Interest: The authors declare no conflict of interest.

Abbreviations

ED	Ecosystem disservices
ES	Ecosystem services
EPA	Environmental Protection Agency
SC-CO_2	Social cost of carbon emissions
SDGs	Sustainable Development Goals
SOC	Soil organic carbon
SIC	Soil inorganic carbon
SOM	Soil organic matter
SSURGO	Soil Survey Geographic Database
TSC	Total soil carbon
USDA	United States Department of Agriculture
U.S.A.	United States of America

References

1. Keestra, S.D.; Bouma, J.; Wallinga, J.; Tittonell, P.; Smith, P.; Cerda, A.; Montanarella, L.; Quinton, J.N.; Pachepsky, Y.; Van der Putten, W.H.; et al. The significance of soils and soil science towards realization of the United Nations Sustainable Development Goals. *Soil* **2016**, *2*, 111–128. [CrossRef]
2. Wood, S.L.; Jones, S.K.; Johnson, J.A.; Brauman, K.A.; Chaplin-Kramer, R.; Fremier, A.; Girvetz, E.; Gordon, L.J.; Kappel, C.V.; Mandle, L.; et al. Distilling the role of ecosystem services in the Sustainable Development Goals. *Ecosyt. Serv.* **2017**, *29*, 701–782. [CrossRef]
3. Adhikari, K.; Hartemink, A.E. Linking soils to ecosystem services—A global review. *Geoderma* **2016**, *262*, 101–111. [CrossRef]
4. Plaster, E.J. *Soil Science and Management*, 4th ed.; Delmar Learning, a Division of Thomson Learning, Inc.: Clifton Park, NY, USA, 2003; ISBN 0766839362.
5. Mikhailova, E.A.; Post, C.J.; Schlautman, M.A.; Post, G.C.; Zurqani, H.A. The business side of ecosystem services of soil systems. *Earth* **2020**, *1*, 15–34. [CrossRef]
6. Feller, C.; Manlay, R.J.; Swift, M.J.; Bernoux, M. Functions, services and value of soil organic matter for human societies and the environment: A historical perspective. In *Functions of Soils for Human Societies and the Environment*, 2nd ed.; Frossard, E., Blum, W.E.H., Warkentin, B.P., Eds.; Geological Society of London: London, UK, 2006; Volume 206, pp. 9–22.
7. Sheikh, M.A.; Kumar, M.; Todaria, N.P.; Pandey, R. Biomass and soil carbon along altitudinal gradients in temperate *Cedrus deodara* forests in Central Himalaya, India: Implications for climate change mitigation. *Ecol. Indic.* **2020**, *111*, 106025. [CrossRef]
8. Mikhailova, E.A.; Groshans, G.R.; Post, C.J.; Schlautman, M.A.; Post, G.C. Valuation of soil organic carbon stocks in the contiguous United States based on the avoided social cost of carbon emissions. *Resources* **2019**, *8*, 153. [CrossRef]
9. Groshans, G.R.; Mikhailova, E.A.; Post, C.J.; Schlautman, M.A.; Zhang, L. Determining the value of soil inorganic carbon stocks in the contiguous United States based on the avoided social cost of carbon emissions. *Resources* **2019**, *8*, 119. [CrossRef]
10. Mikhailova, E.A.; Post, C.J.; Schlautman, M.A.; Post, C.J.; Zurqani, H.A. Determining farm-scale site-specific monetary values of "soil carbon hotspots" based on avoided social costs of CO_2 emissions. *Cogent Environ. Sci.* **2020**, *6*, 1–1817289. [CrossRef]
11. Mikhailova, E.A.; Groshans, G.R.; Post, C.J.; Schlautman, M.A.; Post, C.J. Valuation of total soil carbon stocks in the contiguous United States based on the avoided social cost of carbon emissions. *Resources* **2019**, *8*, 157. [CrossRef]
12. Lehmann, J.; Kleber, M. The contentious nature of soil organic matter. *Nature* **2015**, *528*, 60–68. [CrossRef] [PubMed]
13. Heaton, L.; Fullen, M.A.; Bhattacharyya, R. Critical analysis of the van Bemmelen conversion factor used to convert soil organic matter data to soil organic carbon data: Comparative analyses in a UK loamy sand soil. *Espaço Aberto* **2016**, *6*, 35–44. [CrossRef]
14. Mikhailova, E.A.; Goddard, M.A.; Post, C.J.; Schlautman, M.A.; Galbraith, J.M. Potential contribution of combined atmospheric Ca^{2+} and Mg^{2+} wet deposition within the continental U.S. to soil inorganic carbon sequestration. *Pedosphere* **2013**, *23*, 808–814. [CrossRef]
15. Soil Survey Staff, Natural Resources Conservation Service, United States Department of Agriculture. Soil Survey Geographic (SSURGO) Database. Available online: https://nrcs.app.box.com/v/soils (accessed on 10 September 2020).
16. Fossey, M.; Angers, D.; Bustany, C.; Cudennec, C.; Durand, P.; Gascuel-Odoux, C.; Jaffrezic, A.; Pérès, G.; Besse, C.; Walter, C. A framework to consider soil ecosystem services in territorial planning. *Front. Environ. Sci.* **2020**, *8*. Article 28. [CrossRef]
17. EPA. The Social Cost of Carbon. EPA Fact Sheet. 2016. Available online: https://19january2017snapshot.epa.gov/climatechange/social-cost-carbon_.html (accessed on 15 March 2019).
18. Guo, Y.; Amundson, R.; Gong, P.; Yu, Q. Quantity and spatial variability of soil carbon in the conterminous United States. *Soil Sci. Soc. Am. J.* **2006**, *70*, 590–600. [CrossRef]
19. Groshans, G.R.; Mikhailova, E.A.; Post, C.J.; Schlautman, M.A. Accounting for soil inorganic carbon in the ecosystem services framework for the United Nations sustainable development goals. *Geoderma* **2018**, *324*, 37–46. [CrossRef]
20. The United States Census Bureau, 2018 TIGER/Line Boundary Shapefiles. Available online: https://www.census.gov/geographies/mapping-files/time-series/geo/tiger-line-file.2018.html (accessed on 10 October 2020).

21. Burt, R. Soil Survey Laboratory Methods Manual. Soil Survey Investigations Report No. 42 Version 3. U.S. Department of Agriculture, Natural Resources Conservation Service. 1996. Available online: https://www.nrcs.usda.gov/Internet/FSE_DOCUMENTS/nrcseprd1026806.pdf (accessed on 10 October 2020).
22. Mikhailova, E.A.; Zurqani, H.A.; Post, C.J.; Schlautman, M.A.; Post, C.J. Soil diversity (pedodiversity) and ecosystem services. *Land* **2021**, *10*, 288. [CrossRef]
23. Soil Survey Staff, Natural Resources Conservation Service, United States Department of Agriculture. Photos of soil orders. Available online: https://www.nrcs.usda.gov/wps/portal/nrcs/detail/soils/edu/?cid=nrcs142p2_053588 (accessed on 20 February 2021).
24. Amelung, W.; Bossio, D.; de Vries, W.; Kögel-Knabner, I.; Lehmann, J.; Amundson, R.; Bol, R.; Collins, C.; Lal, R.; Leifeld, J.; et al. Towards a global-scale soil climate mitigation strategy. *Nat. Commun.* **2020**, *11*, 5427. [CrossRef] [PubMed]
25. Galang, M.A.; Markewitz, D.; Morris, L.A.; Bussell, P. Land use change and gully erosion in the Piedmont region of South Carolina. *J. Soil Water Conserv.* **2007**, *62*, 122–129.
26. Kroeger, T.; Casey, F. An assessment of market-based approaches to providing ecosystem services on agricultural lands. *Ecol. Econ.* **2007**, 321–332. [CrossRef]
27. Greainer, L.; Keller, A.; Grêt-Regamey, A.; Papritz, A. Soil function assessment: Review of methods for quantifying the contributions of soils to ecosystem services. *Land Use Policy* **2017**, *69*, 224–237. [CrossRef]

land

MDPI

Article

Impacts of Agricultural Land Reclamation on Soil Nutrient Contents, Pools, Stoichiometry, and Their Relationship to Oat Growth on the East China Coast

Xuefeng Xie [1,2,†], Qi Xiang [1,†], Tao Wu [1], Ming Zhu [2,3], Fei Xu [4], Yan Xu [5] and Lijie Pu [2,3,*]

1 College of Geography and Environmental Sciences, Zhejiang Normal University, Jinhua 321004, China; xiexuefeng@zjnu.cn (X.X.); xiangqi@zjnu.edu.cn (Q.X.); twu@zjnu.cn (T.W.)
2 Key Laboratory of the Coastal Zone Exploitation and Protection, Ministry of Natural Resources, Nanjing 210023, China; zhuming@nju.edu.cn
3 School of Geography and Ocean Science, Nanjing University, Nanjing 210023, China
4 Institute of Land and Urban-Rural Development, Zhejiang University of Finance & Economics, Hangzhou 310018, China; xufei16@zufe.edu.cn
5 School of Environmental Science and Engineering, Suzhou University of Science and Technology, Suzhou 215009, China; yanxu@usts.edu.cn
* Correspondence: ljpu@nju.edu.cn
† These two authors contributed equally to this study as co-first author.

Abstract: Agricultural land reclamation of coastal tidal land (CTL) with organic amendments may modulate the soil properties, and therefore promote crop growth. However, the linkages between soil nutrient contents, pools, stoichiometry, and crop growth under the supplement of organic amendments in CTL is limited. In this study, six treatments including the control (CK), organic manure (OM), polyacrylamide plus organic manure (PAM + OM), straw mulching plus organic manure (SM + OM), buried straw plus organic manure (BS + OM), and bio-organic manure plus organic manure (BM + OM) were conducted to explore these linkages in newly reclaimed CTL in Jiangsu Province, eastern China. The results showed that the application of different soil reclamation treatments increased soil nutrient contents, pools, and modulated their stoichiometric ratio, which thus promoted the growth of oat. Soil under all reclamation treatments increased the contents of surface soil organic carbon (SOC), total nitrogen (TN), and total phosphorus (TP), and the BM + OM treatment had the highest increase, which increased by 11.7–182.4%, 24.3–85.7%, 3.2–29.4%, respectively. The highest soil C pools were observed in the oat heading stage (36.67–41.34 Mg C ha^{-1}), whereas the soil N and P pools were more stable during the oat growth period. Similarly, the highest surface soil C/N and C/P were observed in the oat heading stage (11.23–14.67 and 8.97–14.21), whereas the N/P in surface soil increased compared with the CK treatment during the oat growth period, with the exception of the filling stage. Land reclamation treatments significantly promoted oat growth by changing soil C, N, and P contents, pools, and stoichiometry, among which soil SOC, TN, TP, C/P, and N/P are more closely related to oat growth ($p < 0.05$).

Keywords: land reclamation; ecological stoichiometry; redundancy analysis; coastal tidal land

Citation: Xie, X.; Xiang, Q.; Wu, T.; Zhu, M.; Xu, F.; Xu, Y.; Pu, L. Impacts of Agricultural Land Reclamation on Soil Nutrient Contents, Pools, Stoichiometry, and Their Relationship to Oat Growth on the East China Coast. *Land* **2021**, *10*, 355. https://doi.org/10.3390/land10040355

Academic Editors: Chiara Piccini and Rosa Francaviglia

Received: 26 February 2021
Accepted: 23 March 2021
Published: 1 April 2021

1. Introduction

With an increasingly prominent contradiction between human and land resources, the agricultural reclamation of coastal tidal land (CTL) has become an important approach to increase the cultivated land, as well as improve agricultural productivity and ensure food security [1–3]. However, soil salinization seriously limits the soil quality and inhibits the growth and yield of crops in newly reclaimed coastal tidal land [4]. For instance, the increase of the salt ions can lead to physiological water shortage of plants and inhibit nutrient absorption, thus resulting in dysplasia of plants and reduction in crop yields [5–7]. Earlier study indicated that the increase of Na^{+} and Mg^{2+} ions may cause the structural damage and photosynthesis disorder of plant cells, and thus inhibit the production of chlorophyll [8]. Therefore, the physical (deep ploughing, straw/film mulching, etc.), chemical (macromolecular polymer, organic/inorganic fertilizer, biochar, gypsum, etc.), biological (bio-organic fertilizer, salt-tolerant plants, etc.), and engineering (irrigation and drainage system) improvement have been widely applied to reclaim the saline soil [9–12]. Many studies have confirmed that land reclamation apparently affected the contents of soil C, N, P, and their pools. For instance, straw returning can improve soil physical properties, inhibit soil salinity, and increase the content of soil organic carbon and total nitrogen [9,13]. Previous studies have indicated that application of organic and inorganic fertilizer can reduce soil salinity, improve soil nutrient content and pools, and promote crop yields [14,15]. For example, application of chemical fertilizer can accelerate the consumption of soil organic carbon, whereas the straw returning can offset the mineralization of organic carbon and increase the soil C pool [16]. Moreover, appropriate application of polyacrylamide (PAM) can improve the soil nutrient retention capacity [17]. Besides, the planting of salt-tolerant plants can improve the physicochemical properties, reduce the soil salinity, and increase the soil nutrient content of CTL [18]. For instance, oat (*Avena sativa* L.) cultivation is considered as an efficient reclamation approach to improve CTL due to its high capacity to accumulate salt ions in straw biomass [18].

Ecological stoichiometry deals with the balance of multiple chemical elements (mainly C, N, and P) in the process of ecologic interaction [19], which is used to track the changes of ecosystem structure and nutrient cycling [20]. Soil nutrient directly affects the growth and productivity of plant communities, and soil C/N/P stoichiometry is considered an important indicator of soil nutrient characteristics [21]. Therefore, the study of soil C/N/P stoichiometry can indicate soil nutrient status, which is conducive to a better understanding of soil limiting elements, and scientifically adjusts the fertilization type, so as to promote plant growth and improve crop productivity [22,23]. Large numbers of studies have demonstrated that land reclamation can affect the soil C/N/P stoichiometry. For example, intensive fertilization in farmland has led to a decrease in C/N and C/P, and the N/P was more sensitive to nitrogen addition [24]. Besides, deep plowing broke the nutrient fixation status, and significantly increased the soil C/N and reduced the C/P [25]. Straw mulching directly affected the rate of mineralization and decomposition of nutrients by the adjusted soil temperature and water content, which in turn caused the changes in soil C/N/P [26].

Although the impact of different land reclamation treatments on soil nutrients have been fully revealed, the linkages between soil nutrient contents, pools, stoichiometry, and crop growth under the supplement of organic amendments in CTL is limited. Therefore, we hypothesized that different soil reclamation treatments can increase soil nutrient content, pools, and modulate their stoichiometric ratio, thus promoting the growth of oat. Specifically, the objectives of this study were to: (1) identify the effect of different land reclamation treatments on C, N, and P contents, pools, stoichiometry, and oat growth; and (2) explore the linkages between soil C, N, and P contents, pools, stoichiometry, and oat growth parameters following the reclamation of CTL.

2. Materials and Methods

2.1. Study Area

This experiment was carried out in Tongzhou Bay (32°11′ N, 121°22′ E), Nantong City, Jiangsu Province, eastern China (Figure 1). The region has a subtropical monsoon climate, with an average annual temperature of about 14–15 °C. The average annual rainfall is about 1000–1080 mm, which is relatively concentrated from June to September. The area was reclaimed for marine aquaculture in 2008, and the experiment field was established in 2016. The groundwater depth is 1.2–1.8 m. The soil is characterized by a sandy loam texture, high bulk density, salinity, and sodicity, and has low nutrients (Table 1).

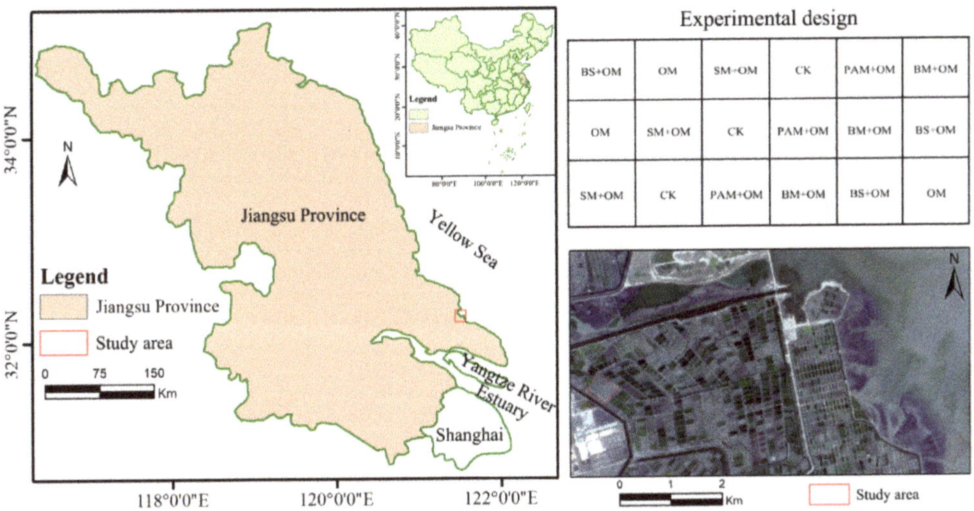

Figure 1. Location of the study area.

Table 1. Soil properties before field experiment.

Soil Depth (cm)	Sand (%)	Silt (%)	Clay (%)	BD (g cm^{-3})	$EC_{1:5}$ (dS m^{-1})	$pH_{1:2.5}$	SOC (g kg^{-1})	TN (g kg^{-1})	TP (g kg^{-1})
0–10	79.35	17.45	3.19	1.49	1.82	8.06	4.13	0.55	0.73
10–20	79.01	17.72	3.26	1.53	1.43	8.16	4.04	0.54	0.69
20–40	80.54	16.37	3.08	1.52	2.32	7.98	3.69	0.47	0.67

Note: BD, bulk density; EC, electrical conductivity; SOC, soil organic carbon; TN, total nitrogen; TP, total phosphorus.

2.2. Experimental Design

In this experiment, 18 plots (3 × 2 m) were insulated by double-layer plastic sheets buried to 60 cm deep and 50 cm wide to the soil surface to reduce interference between the plots. Due to the high soil bulk density, all plots were plowed to 20 cm deep before the experiment. The following 6 reclamation treatments were applied: (1) control (CK); (2) organic manure (OM); (3) polyacrylamide plus organic manure (PAM + OM); (4) straw mulching plus organic manure (SM + OM); (5) buried straw plus organic manure (BS + OM); and (6) bio-organic manure plus organic manure (BM + OM). All treatments were randomly designed and repeated three times, and the specific measures of each treatments are shown in Table 2. The physicochemical properties of all applied amendments are presented in Table 3. All treatments were conducted in September 2016. Oat seed was sown in drill (drill spacing 60 cm) with 90 kg ha^{-1} on 3 November 2016, and the urea (46% N) was sprayed with 180 kg ha^{-1} on the soil surface at the jointing stage (March 2017). The field management practices were consistent with local farmers, and the crop was harvested on 2 June 2017. The oat growing period can be divided into the seedling stage (0–60 days), jointing stage (60–90 days), heading stage (90–120 days), filling stage (120–150 days), and maturation stage (150–210 days).

Table 2. Experimental treatment design and specific measures.

Treatment	Specific Measures	References
CK	No application of amendments.	
OM	Chicken manure was evenly applied at soil surface at 15 ton ha^{-1}, then the plot was ploughed and harrowed to a depth of 10–15 cm with a physically acceptable evenness and mellowness.	[27]
PAM + OM	Both nonionic polyacrylamide (5%, approximately 2 ton ha^{-1}) and chicken manure (15 ton ha^{-1}) were evenly applied at soil surface, then the plot was ploughed and harrowed to a depth of 10–15 cm with a physically acceptable evenness and mellowness.	[28]
SM + OM	Chicken manure (15 ton ha^{-1}) was evenly applied at soil surface, and the plot was ploughed and harrowed to a depth of 10–15 cm with a physically acceptable evenness and mellowness, then the wheat straw (15 ton ha^{-1}) was cut to 10 cm long and evenly mulched.	[29]
BS + OM	Wheat straw (15 ton ha^{-1}) was cut to 10 cm long and evenly buried nearly 20 cm underground after removal of soil, followed by the addition of chicken manure (15 ton ha^{-1}); thereafter, the plot was ploughed and harrowed to a depth of 10–15 cm with a physically acceptable evenness and mellowness.	[30]
BM + OM	Both Jiahua (a compound bio-organic manure made from cow dung and crushed corn straw by deep fermentation and decomposition of *Bacillus* and *Saccharomyces*, containing approximately 2.0×10^8 CFU of viable bacteria g^{-1}) and chicken manure were evenly applied at soil surface at a rate of 15 ton ha^{-1}, and the plot was ploughed and harrowed to a depth of 10–15 cm with a physically acceptable evenness and mellowness.	[28]

Note: CK, control; OM, organic manure; PAM + OM, polyacrylamide plus organic manure; SM + OM, straw mulching plus organic manure; BS + OM, buried straw plus organic manure; BM + OM, bio-organic manure plus organic manure.

Table 3. Physicochemical properties of soil amendments.

Amendment	TOC (%)	TN (%)	TP (%)
Chicken manure	13.14	1.42	0.87
Polyacrylamide	–	0.07	–
Wheat straw	16.53	0.62	0.23
Jiahua bio-organic manure	27.10	4.58	3.63

Note: TOC, total organic carbon; TN, total nitrogen; TP, total phosphorus; "–", not determined.

2.3. Soil Sampling and Determination

After the oats were sown, a composite soil sample was randomly collected at 0–10 cm (surface layer), 10–20 cm (subsurface layer), and 20–40 cm (deep layer) in each plot with five replicates at 30-day intervals. A total of 432 soil samples were collected during the whole oat growing period. All samples were stored in polyethylene bags and brought back to the laboratory. After removing all visible plant roots, stones, and organisms, soil samples were naturally air-dried and passed through a 0.149 mm sieve to measure physicochemical properties. All methods applied for measuring the soil physicochemical properties have been described in detail by Lu [31]. Briefly, soil bulk density (BD) was determined by oven drying to constant mass at 105 °C for 48 h; soil organic carbon (SOC) was determined by potassium dichromate oxidation-spectrophotometry; soil total nitrogen (TN) was determined by the Kjeldahl method; soil total phosphorus (TP) was determined by the colorimetric method after digestion with hydrofluoric and perchloric acid. The pools of SOC, TN, and TP were calculated using the following equation:

$$Y_P = \sum_{i=1}^{n} X_i \times BD_i \times D_i \times 0.1 \quad (1)$$

where Y_P is the pools of SOC, TN, and TP; X_i is the concentration of SOC, TN, and TP in the ith layer; BD_i is the bulk density of the ith layer; D_i is the depth interval of ith layer; and 0.1 is the conversion factor from g cm^{-2} to kg m^{-2}.

2.4. Determination of Oat Growth Parameters

During the oat growing period, 5 oat plants were randomly selected from each plot, and the plant height and stem diameter were recorded every 30 days. The plant height was determined by measuring the absolute height from the ground to the highest position of the main stem with a steel tape, and the stem diameter was measured by vernier caliper from the internode position at the base of the main stem.

2.5. Statistical Analysis

The measured soil properties and oat growth parameters were analyzed with one–way ANOVA to test the significant differences among the different reclamation treatments, and the means comparisons were separated using the Fisher's least significant difference (LSD) test at $p = 0.05$. Redundancy analysis (RDA) was applied to clarify the relationship between oat growth, soil C, N, and P content, pools, and stoichiometry. All data analyses were carried out in SPSS 20.0 for Windows software package and Canoco 4.5 for Windows software package.

3. Results

3.1. Soil C, N, and P Content

During the oat growing period, the content of SOC under different reclamation treatments in all soil layers showed a trend of first increasing and then decreasing with the highest content observed in the heading stage (Figure 2). In the surface layer, compared with the CK treatment, the SOC content of each reclamation treatment gradually increased, especially under the BM + OM treatment, which increased by 11.7–182.4%. However, compared with the CK treatment, the content of SOC in the subsurface layer increased by 0.0–40.0% during the entire growing season, except for the SM + OM treatment. Additionally, no significant differences were observed in SOC content between different treatments during the entire growing season in the subsurface layer, except for the heading stage; whereas no significant differences were found in the SOC content in the seedling, heading, and maturation stage among different treatments in the deep layer. The content of TN in all soil layers remained relatively stable under different treatments during the oat growing season, except for the BM + OM treatment (Figure 2). In the surface layer, the content of TN under the PAM + OM treatment was slightly lower than that of the CK treatment (decreased by 9.1%) during the filling stage, whereas soils under the BM + OM, BS + OM, SM + OM, and OM treatments were higher than that of the CK treatment, and increased by 24.3–85.7%, 9.1–47.2%, 12.1–25.0%, and 1.2–22.9%, respectively. Besides, there were no significant differences in TN content between different reclamation treatments in subsurface and deep layers during the oat growing season, except for the heading stage in the subsurface layer and the filling stage in the deep layer, respectively. The dynamics of TP content in all soil layers were similar to that of TN content (Figure 2). Throughout the oat growing season, BM + OM, BS + OM, OM, SM + OM, and PAM + OM treatments increased surface layer TP content by 3.2–29.4%, 0.5–17.4%, 3.8–14.8%, 4.9–10.4%, and −0.5–13.6% compared with the CK treatment. In general, there were no significant differences in TP content between different reclamation treatments during the jointing, heading, and filling stages in the surface layer, whereas no significant differences were observed in TP content during the middle and later stages of oat growth in subsurface and deep layers.

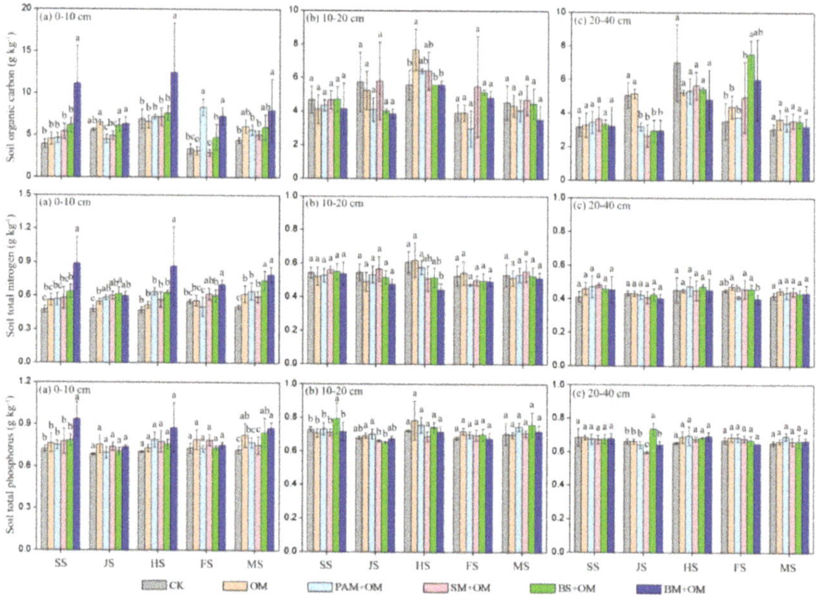

Figure 2. Dynamics of soil organic carbon, total nitrogen, and total phosphorus contents under different oat growth stages in 0–10 cm (**a**), 10–20 cm (**b**), and 20–40 cm (**c**). SS: seedling stage; JS: jointing stage; HS: heading stage; FS: filling stage; MS: maturation stage. Values are means of three replicates ± SD; error bars refer to standard deviation; values having different lowercase letters on the bars indicate significant differences among different treatments (least significant difference (LSD), $p < 0.05$).

3.2. Soil C, N, and P Pools

During the oat growing period, the soil organic carbon pools (SOCP) under different reclamation treatments in the 0–40 cm soil layer showed a trend of first increasing and then decreasing (Figure 3). Except for the BS + OM treatment (filling stage), the highest SOCP under different treatments were found in the heading stage. Compared with the CK treatment, the SOCP increased to different degrees under different reclamation treatments in seedling, filling, and maturation stages. Among them, the BM + OM and BS + OM treatments showed significant differences in SOCP in seedling and filling stages, respectively, whereas there were no significant differences between the treatments in the heading and maturation stage. In the jointing stage, the SOCP in OM treatment was significantly higher than other treatments.

Generally, soil total nitrogen pool (TNP) under the BS + OM treatment was slightly higher than that of other treatments (Figure 3). In the seedling and maturation stages, the TNP under different reclamation treatments increased by 6.5–19.9% and 6.0–11.2%, respectively, compared with the CK treatment. However, no significant differences were observed in TNP under different treatments at jointing and heading stages. Besides, the TNP under the PAM + OM treatment was significantly lower than CK, OM, SM + OM, and BS + OM treatments.

During the oat growing season, soil total phosphorus pool (TPP) remained relatively stable under different reclamation treatments, and no significant differences were observed between different treatments in seedling and heading stages (Figure 3). The TPP in the jointing stage was similar to the maturation stage, with the highest value in the BS + OM treatment, whereas the lowest value was in the SM + OM treatment. Additionally, TPP under the SM + OM treatment in the filling stage was significantly lower than that of CK, OM, PAM + OM, and BS + OM treatments.

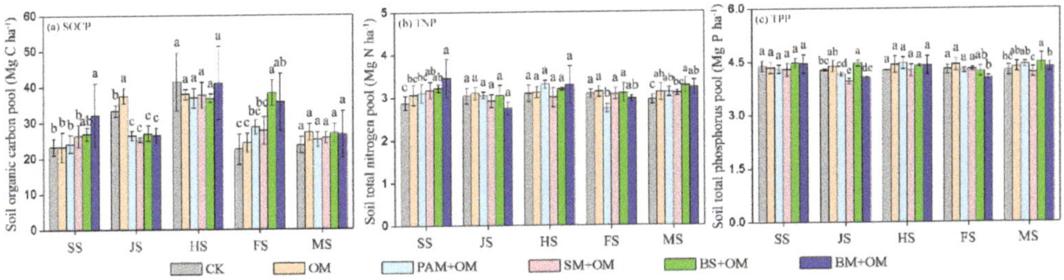

Figure 3. Dynamics of SOCP (**a**), TNP (**b**), and TPP (**c**) under different oat growth stages in 0–40 cm. SOCP: soil organic carbon; TNP: total nitrogen; TPP: total phosphorus pools; SS: seedling stage; JS: jointing stage; HS: heading stage; FS: filling stage; MS: maturation stage. Values are means of three replicates ± SD; error bars refer to standard deviation; values having different lowercase letters on the bars indicate significant differences among different treatments (LSD, $p < 0.05$).

3.3. Soil C, N, and P Stoichiometry

Soil C/N roughly increased first and then decreased with the growth of oat under different treatments and soil layers, and reached the highest value at the heading stage (Figure 4). In the surface layer, significant differences were found in soil C/N between different treatments in the seedling, jointing, heading, and filling stage, whereas no significant difference was found in the maturation stage. Among them, soil C/N of the BM + OM treatment was significantly higher than that of other treatments at the seedling stage, and the PAM + OM treatment had the lowest soil C/N at jointing and heading stages and the highest C/N at the filling stage. The C/N of the subsurface layer did not differ significantly under different treatments. Similarly, in the deep layer, there was no significant difference in C/N at seedling, heading, and maturation stages. The C/N of the deep layer under CK and OM treatments were apparently higher than that of other treatments at the jointing stage, whereas BS + OM and BM + OM treatments at the filling stage were notably higher than that of CK, OM, and PAM + OM.

The dynamic of soil C/P during the oat growing period was similar to that of C/N, with the highest value appearing at the heading stage (Figure 4). Overall, the BM + OM treatment had a significant impact on the C/P of different soil layers. In the surface layer, compared with the CK treatment, soil C/P under the BS + OM and BM + OM treatments increased by 1.0–44.5% and 28.0–126.3%, respectively, during the whole oat growth period. There was no significant difference between OM and CK treatments. Additionally, PAM + OM and SM + OM treatments were significantly lower than the CK treatment at the jointing stage, whereas they were significantly higher than the CK treatment at the filling stage and seedling stage, respectively. The reclamation treatments have little effect on the C/P in the subsurface layer. Except that the BM + OM treatment was significantly higher than other treatments at the heading stage, and there was no significant difference in C/P between different treatments at other growth stages. Soil C/P under different treatments in the deep layer is not significantly different at the heading stage. However, in the seedling and maturation stages, the C/P under BM + OM treatment was significantly higher than other treatments.

Soil N/P fluctuated slightly under different treatments during the oat growing season (Figure 4). In the surface layer, soil N/P under BM + OM and BS + OM (except the filling stage) treatments was significantly higher than the CK treatment throughout the oat growing season. Compared with the CK treatment, the N/P of the subsurface layer under BS + OM, PAM + OM, and OM treatments, respectively, decreased by 1.2–17.5%, 3.6–12.9%, 1.1–12.0%, and no significant differences were found between jointing and filling stages. Moreover, no significant differences were observed between reclamation treatments throughout the growth period in the deep layer.

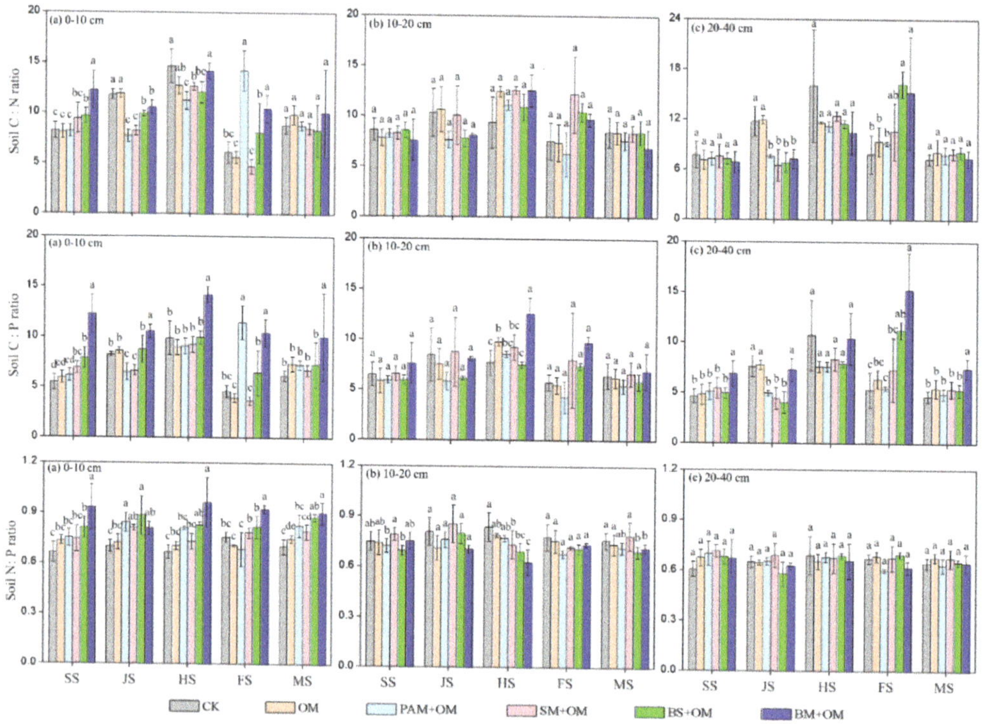

Figure 4. Dynamics of soil C/N, C/P, and N/P under different oat growth stages in 0–10 cm (**a**), 10–20 cm (**b**), and 20–40 cm (**c**). SS: seedling stage; JS: jointing stage; HS: heading stage; FS: filling stage; MS: maturation stage. Values are means of three replicates ± SD; error bars refer to standard deviation; values having different lowercase letters on the bars indicate significant differences among different treatments (LSD, $p < 0.05$).

3.4. Oat Growth Parameters

During the growth period of oat, all reclamation treatments can increase the stem diameter and plant height of oat in varying degrees (Figure 5). The stem diameter and plant height of oat under the BM + OM treatment was much higher than that of other treatments. As the growing season progressed, there were no significant differences in stem diameter under BS + OM, SM + OM, and PAM + OM treatments except for the maturation stage. Similarly, the plant height under PAM + OM, SM + OM, and BS + OM treatments were significantly higher than that of the OM treatment in the heading, filling, and maturation stage.

3.5. Redundancy Analysis

In this study, RDA was performed to explore the relationship between oat growth parameters and soil C, N, and P indicators at each oat growth period. As shown in Figure 6, the first two axes in the seedling, jointing, heading, filling, and maturation stages cumulatively explained 75.4%, 96.9%, 98.8%, 76.9%, and 63.0% of the variation of oat growth, which indicates that the first two axes can fully explain the relationship between oat growth and soil C, N, and P indicators. During the oat grown season, soil C, N, and P contents and their stoichiometry are positively correlated with oat growth parameters, whereas soil N and P pool in jointing and filling stages are negatively correlated (Figure 6).

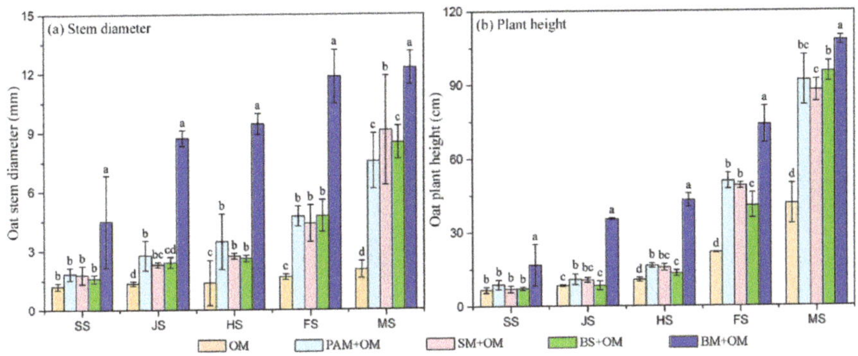

Figure 5. Dynamics of oat stem diameter (**a**) and plant height (**b**) under different growth stages. SS: seedling stage; JS: jointing stage; HS: heading stage; FS: filling stage; MS: maturation stage. Values are means of three replicates ± SD; error bars refer to standard deviation; values having different lowercase letters on the bars indicate significant differences among different treatments (LSD, $p < 0.05$).

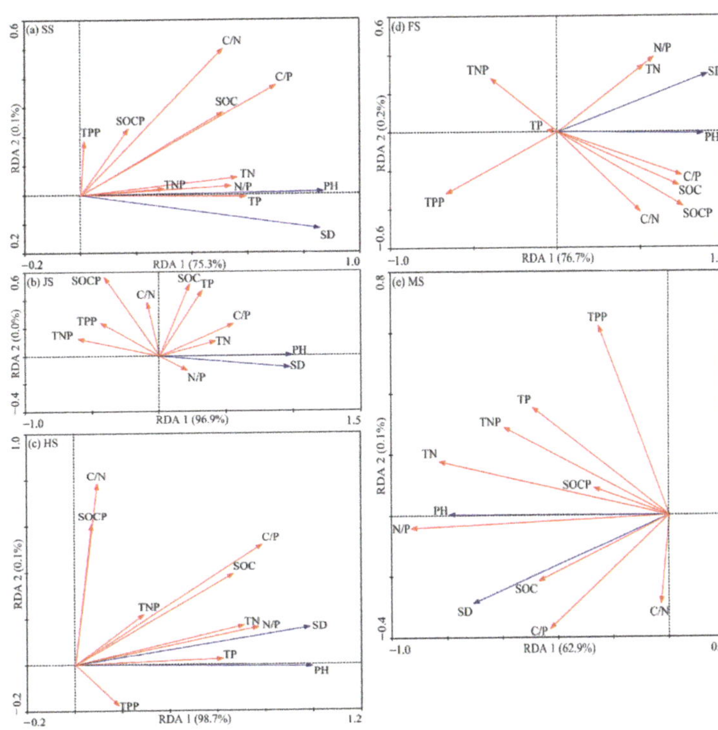

Figure 6. Coordination biplots of redundancy analysis (RDA) displaying the relationship between oat growth parameters and surface soil C, N, and P indicators in SS (**a**), JS (**b**), HS (**c**), FS (**d**), and MS (**e**). Oat growth parameters are response variables and soil C, N, and P indicators are explanatory variables. The positive and negative correlation between two soil properties depends on the same or opposite direction of arrows, and the correlation is determined by the projection length of the arrows of two soil properties. PH: plant height; SD: stem diameter; SOC: soil organic carbon; TN: total nitrogen; TP: total phosphorus; SOCP: soil organic carbon pool; TNP: total nitrogen pool; TPP: total phosphorus pool; SS: seedling stage; JS: jointing stage; HS: heading stage; FS: filling stage; MS: maturation stage.

4. Discussion

4.1. Response of Soil Nutrient Contents, Pools, and Stoichiometry Following Agricultural Land Reclamation

Land reclamation of CTL with organic and inorganic amendment may alter the mineralization and decomposition processes of soil nutrients [12,32]. In this experiment, soil C, N, and P contents, pools, and stoichiometry under different land reclamation treatments generally decreased with the increase of soil depth. This might be attributed to the various amendments that are mainly carried out in the upper soil layer, and the litter of oat plants is also concentrated on the soil surface. Earlier studies have indicated that organic manure was an efficient way to supplement the soil nutrients [11,33]. For example, the organic manure compost increased the cation exchange capacity, available macro-nutrient contents, and biological activities, whereas it decreased the soil salinity [11,14]. In the present experiment, application of chicken manure has increased the content of SOC, TN, and TP in the surface soil. Polyacrylamide changes the soil structure and improves soil water retention and corrosion resistance, which is conducive to maintaining soil moisture and fertility [34,35]. In this experiment, the content of SOC, TN, and TP in the surface layer under PAM + OM treatment basically increased during the oat growing period (Figure 1). Earlier studies have also shown that PAM as a structural modifier can effectively change the structure of soil aggregates, retaining soil water content and fertility, and reducing the loss of soil N, P, and other nutrient elements, and improving the stability of agricultural production [17]. Numerous studies have shown that straw is rich in organic matter and nutrient elements such as nitrogen, phosphorus, and potassium, and straw returning can increase the number and activities of microorganisms and promotes the decomposition of organic nutrients, resulting in the increase of soil nutrients [30,36]. In our study, the SOC, TN, and TP contents of the soil under treatments of SM + OM and BS + OM roughly increased, and the BS + OM is more conducive to the enhancement of deep soil fertility than that of SM + OM, which is consistent with Zhao et al. [13]. Many studies have shown that the bio-organic fertilizer provided a large amount of organic matter for microorganisms [12,37]. Similarly, the SOC, TN, and TP contents of the surface soil under the BM + OM treatment increased significantly in our experiment, which might be due to the bio-organic fertilizer used in this experiment containing the organic matter $\geq$45% and total nutrients (N + P + K) $\geq$12%, as well as containing a large number of microbial communities, which are beneficial to the release of available nutrients [28].

Application of organic amendment may also change the soil C, N, and P pools. In our study, the SOCP under different land reclamation treatments roughly increased from the seedling to heading stage, and then decreased from the heading to maturation stage, which is similar to the results of a previous study [21]. In addition, SOCP at the seedling, filling, and maturation stages increased significantly under different reclamation treatments, whereas at the jointing and heading stages it decreased (Figure 3). This may be owing to the gradual decomposition of organic amendment in the early oat growth period providing enough carbon source, while the oat grows rapidly at the heading stage, and the enhancement of photosynthesis improves the carbon fixation capacity of oat [38]. In this experiment, soil TNP and TPP basically remained stable throughout the growth period, which was not consistent with former studies that confirmed that the application of organic fertilizers can increase soil carbon and nitrogen pools [39]. It may be because nitrogen is usually present in organic matter in the form of organic nitrogen, which makes the changes in soil carbon and nitrogen more synchronized [40]. The decline in the phosphorus pool may be caused by the reduction in the fixation of inorganic phosphorus after the application of organic fertilizers, and part of the organic phosphorus in organic fertilizers is easily decomposed [41]. In general, fertilizers gradually dissolved and were slowly absorbed by the soil under different reclamation treatments. Therefore, the soil C, N, and P contents and pools increased in the early stage of oat growth, and the application of urea at the jointing stage further improved the soil N content and pool. After that, the oat enters the rapid growth stage and needs to absorb a large amount of soil nutrients to maintain growth, resulting in the decline of soil C, N, and P content and pools.

Land reclamation remarkably altered soil C/N/P stoichiometry, ascribing to the disproportionate increase of C, N, and P content. In our study, soil C/N, C/P, and N/P under different reclamation treatments were between 4.73–17.23, 3.69–15.22, and 0.58–0.96, respectively. SOC is a key factor to adjust soil C/N and C/P changes under different reclamation treatments, whereas soil N/P changes are mainly controlled by TN in this study. Generally, the soil C/N ratio is inversely proportional to its decomposition rate, and soil with a lower C/N ratio has faster mineralization [23]. The soil C/P ratio is considered an important indicator for assessing the mineralization ability of soil phosphorus, which can measure the potential of soil organic matter mineralization to release phosphorus or absorb and retain phosphorus [22,23]. In this experiment, soil C/N and C/P of each soil layer under different treatments generally increased first and then decreased, which might be due to the relatively high temperature in the seedling, filling, and maturation stages, and enhanced the soil microbial activities and accelerated the decomposition of organic matter, resulting from the decrease of SOC [42]. Simultaneously, the increase in precipitation can increase the mineralization rate of soil nitrogen, and ultimately lead to a decrease in soil C/N, which is consistent with the results of Yan [43]. The soil N/P ratio has been suggested to be useful for assessing N or P limitations [19,22]. Soil N/P in this study was much lower than the average level of Chinese national wetlands (13.6), which suggest that N is the main limiting element in this area [20].

4.2. Linkages between Soil Nutrient Contents, Pools, Stoichiometry, and Oat Growth Following Agricultural Land Reclamation

Land reclamation with organic amendments altered soil C, N, and P content, pools, and stoichiometry, thereby promoting the growth of oat. Our experiment indicated that the stem diameter and plant height of oat under BM + OM, BS + OM, SM + OM, and PAM + OM treatments were significantly higher than that of the OM treatment, while oat cannot germinate under the CK treatment due to high salinity. This is consistent with previous studies which demonstrated that the application of organic amendments in saline soils can improve soil structure, reduce soil salinity, increase nutrient contents, and thus promote the plant growth and crop yield [44,45]. However, the growth of oat in the middle and later stages of the OM treatment was slower than that of other treatments, which, due to the increase of soil salinity, inhibited the oat growth [28]. PAM modifier plays an important role in reducing nutrient loss [34] and can significantly promote oat growth (Figure 5). Straw returning to the field has been demonstrated to reduce soil water evaporation, regulate soil temperature, release a large amount of organic matter and nutrient elements during the process of decay, and promote crop growth and yield [9,13]. The effects of SM + OM and BS + OM treatments on oat growth are more obvious in the filling stage and maturation stage. This might be attributed to the slow decomposing rate of straw due to lower temperatures in the early stage (winter), and the accelerated decomposition of straw as the temperature rises in the later stage (spring) to release a large amount of nutrients, which ensures nutrient supply and enables rapid plant growth [46]. In this experiment, the BM + OM treatment significantly promoted the oat growth, which can be ascribed to the bio-organic fertilizer-enhanced soil microbial activity, and which continuously provides nutrients for plant growth [47].

Redundancy analysis revealed that soil SOC, TN, TP, C/P, and N/P is highly correlated with oat growth throughout the growth period of oat, indicating that soil C, N, and P content and their stoichiometric relationships are important factors affecting the growth of oat. Previous studies have shown that C, N, and P are essential nutrients for crop growth, and appropriate N and P content are beneficial to the increase of vegetation height, density, and biomass [48]. The phosphorus content in the study area was relatively low and stable under different soil reclamation treatments (Figure 3), and the fluctuation of C and N content led to changes in soil C/P and N/P, which ultimately affected the growth of oat. In this experiment, the correlation between SOCP, TNP, TPP, and oat growth are weak or even negatively correlated during oat growing season, which ascribes to the oat absorbing a large amount of N and P elements in the jointing and filling stages to meet the growth of oat, resulting in the decrease of N and P pools.

5. Conclusions

Our study revealed that the applied land reclamation treatments can be considered as an efficient approach to increase surface soil nutrients and pools. During the oat growth period, the BM + OM treatment significantly increased SOC, TN, and TP content, with the increasing rate of 11.7–182.4%, 24.3–85.7%, and 3.2–29.4%, respectively. The highest SOCP was observed in the oat heading stage (36.67–41.34 Mg C ha^{-1}), whereas the differences in TNP and TPP under all land reclamation treatments were not significant. The C/N and C/P ratio under different reclamation treatments showed a trend of increasing first and then decreasing, with the highest value in the oat heading stage (11.23–14.67 and 8.97–14.21), whereas the N/P fluctuates with the growth of oat. Simultaneously, the C/N/P ratio of all treatments indicated that the study area was regarded as N limited. Moreover, land reclamation treatments promoted the growth of oat, among which the highest stem diameter and plant height of oat were observed in the BM + OM treatment (12.27 mm and 108.06 cm). Furthermore, we observed soil C, N, and P contents and stoichiometry ($p < 0.05$) were more closely related to the oat growth compared with their pools. This study suggested that BM + OM can be recommended as priority agricultural management for reclamation of CTL.

Author Contributions: Conceptualization, X.X. and L.P.; Methodology, X.X. and T.W.; Writing—original draft, X.X. and Q.X.; Supervision, M.Z., F.X., and Y.X.; Project administration, L.P.; Funding acquisition, X.X. and L.P. All authors have read and agreed to the published version of the manuscript.

Funding: This research was funded by the National Natural Science Foundation of China (41871083, 41230751, 41701609, 41701618), the Open Fund of Key Laboratory of Coastal Zone Exploitation and Protection, Ministry of Natural Resources (2019CZEPK09), and the Natural Science Foundation of Zhejiang Province, China (LQ21D010007, LY21D010008).

Institutional Review Board Statement: Not applicable.

Informed Consent Statement: Not applicable.

Data Availability Statement: The data presented in this study are available on request from the corresponding author.

Conflicts of Interest: The authors declare no conflict of interest.

References

1. Xu, Y.; Pu, L.J.; Zhang, R.S.; Zhu, M.; Zhang, M.; Bu, X.G.; Xie, X.F.; Wang, Y. Effects of agricultural reclamation on soil physicochemical properties in the mid-eastern coastal area of China. *Land* **2021**, *10*, 142. [CrossRef]
2. Xie, X.F.; Wu, T.; Zhu, M.; Jiang, J.J.; Xu, Y.; Wang, X.H.; Pu, L.J. Comparison of random forest and multiple linear regression models for estimation of soil extracellular enzyme activities in agricultural reclaimed coastal saline land. *Ecol. Indic.* **2021**, *120*, 106925. [CrossRef]
3. Xie, X.F.; Pu, L.J.; Zhu, M.; Meadows, M.; Sun, L.C.; Wu, T.; Bu, X.G.; Xu, Y. Differential effects of various reclamation treatments on soil characteristics: An experimental study of newly reclaimed tidal mudflats on the east China coast. *Sci. Total Environ.* **2021**, *768*, 144996. [CrossRef] [PubMed]
4. Xie, X.F.; Pu, L.J.; Zhu, M.; Xu, Y.; Wang, X.H. Linkage between soil salinization indicators and physicochemical properties in a long-term intensive agricultural coastal reclamation area, Eastern China. *J. Soils Sediment.* **2019**, *19*, 1–9. [CrossRef]
5. Awad, A.S.; Nair, N.G. Salt tolerance of agaricus-bisporus in relation to water-stress and toxicity of sodium-ions. *Ann. Appl. Biol.* **1989**, *115*, 215–220. [CrossRef]
6. Nawaz, F.; Shehzad, M.A.; Majeed, S.; Ahmad, K.S.; Aqib, M.; Usmani, M.M.; Shabbir, R.N. Role of Mineral Nutrition in Improving Drought and Salinity Tolerance in Field Crops. In *Agronomic Crops*; Hassanuzzaman, M., Ed.; Springer: Singapore, 2020; pp. 129–147.
7. Castiglione, S.; Oliva, G.; Vigliotta, G.; Novello, G.; Gamalero, E.; Lingua, G.; Cicatelli, A.; Guarino, F. Effects of Compost Amendment on Glycophyte and Halophyte Crops Grown on Saline Soils: Isolation and Characterization of Rhizobacteria with Plant Growth Promoting Features and High Salt Resistance. *Appl. Sci.* **2021**, *11*, 2125. [CrossRef]

8. Kingsbury, R.W.; Epstein, E. Salt sensitivity in wheat–a case for specific ion toxicity. *Plant Physiol.* **1986**, *80*, 651–654. [CrossRef]
9. Akhtar, K.; Wang, W.Y.; Ren, G.X.; Khan, A.; Feng, Y.Z.; Yang, G.H.; Wang, H.Y. Integrated use of straw mulch with nitrogen fertilizer improves soil functionality and soybean production. *Environ. Int.* **2019**, *132*, 105092. [CrossRef]
10. Hussain, A.; Zahir, Z.A.; Ditta, A.; Tahir, M.U.; Ahmad, M.; Mumtaz, M.Z.; Hayat, K.; Hussain, S. Production and implication of bio-activated organic fertilizer enriched with zinc-solubilizing bacteria to boost up maize (*Zea mays* L.) production and biofortification under two cropping seasons. *Agronomy* **2020**, *10*, 39. [CrossRef]
11. Gunarathne, V.; Senadeera, A.; Gunarathne, U.; Biswas, J.K.; Almaroai, Y.A.; Vithanage, M. Potential of biochar and organic amendments for reclamation of coastal acidic-salt affected soil. *Biochar* **2020**, *2*, 107–120. [CrossRef]
12. Abo El-Ezz, S.F.; El-Hadidi, E.M.; El-Sherpiny, M.A.; Mahmoud, S.E. Land Reclamation Using Compost, Agricultural Gypsum and Sugar Beet Mud. *J. Soil Sci. Agri. Eng.* **2020**, *11*, 503–511. [CrossRef]
13. Zhao, Y.G.; Pang, H.C.; Wang, J.; Huo, L.; Li, Y.Y. Effects of straw mulch and buried straw on soil moisture and salinity in relation to sunflower growth and yield. *Field Crop. Res.* **2014**, *161*, 16–25. [CrossRef]
14. Yang, L.; Bian, X.G.; Yang, R.P.; Zhou, C.L.; Tang, B.P. Assessment of organic amendments for improving coastal saline soil. *Land Degrad. Dev.* **2018**, *29*, 3204–3211. [CrossRef]
15. Zhang, Y.; Sun, C.; Chen, Z.; Zhang, G.; Chen, L.; Wu, Z. Stoichiometric analyses of soil nutrients and enzymes in a Cambisol soil treated with inorganic fertilizers or manures for 26years. *Geoderma* **2019**, *353*, 382–390. [CrossRef]
16. Fan, R.Q.; Du, J.J.; Liang, A.Z.; Lou, J.; Li, J.Y. Carbon sequestration in aggregates from native and cultivated soils as affected by soil stoichiometry. *Biol. Fert. Soils* **2020**, *56*, 1109–1120. [CrossRef]
17. Zhou, J.J.; Liang, X.Q.; Shan, S.D.; Yan, D.W.; Chen, Y.F.; Yang, C.K.; Lu, Y.Y.; Niyungeko, C.; Tian, G.M. Nutrient retention by different substrates from an improved low impact development system. *J. Environ. Manag.* **2019**, *238*, 331–340. [CrossRef]
18. Han, L.P.; Liu, H.T.; Yu, S.H.; Wang, W.H.; Liu, J.T. Potential application of oat for phytoremediation of salt ions in coastal saline-alkali soil. *Ecol. Eng.* **2013**, *61*, 274–281. [CrossRef]
19. Cleveland, C.C.; Liptzin, D. C:N:P stoichiometry in soil: Is there a "redfield ratio" for the microbial biomass? *Biogeochemistry* **2007**, *85*, 235–252. [CrossRef]
20. Zhang, J.H.; Zhao, N.; Liu, C.C.; Yang, H.; Li, M.L.; Yu, G.R.; Wilcox, K.; Yu, Q.; He, N.P. C:N:P stoichiometry in China's forests: From organs to ecosystems. *Funct. Ecol.* **2018**, *31*, 50–60. [CrossRef]
21. He, H.; Xia, G.T.; Yang, W.J.; Zhu, Y.P.; Wang, G.D.; Shen, W.B. Response of soil C:N:P stoichiometry, organic carbon stock, and release to wetland grasslandification in Mu Us Desert. *J. Soils Sediment.* **2019**, *19*, 3954–3968. [CrossRef]
22. Yang, Y.; Liu, B.R.; An, S.S. Ecological stoichiometry in leaves, roots, litters and soil among different plant communities in a desertified region of Northern China. *Catena* **2018**, *166*, 328–338. [CrossRef]
23. Xu, C.Y.; Pu, L.J.; Li, J.G.; Zhu, M. Effect of reclamation on C, N, and P stoichiometry in soil and soil aggregates of a coastal wetland in eastern China. *J. Soils Sediment.* **2019**, *19*, 1215–1225. [CrossRef]
24. Jiang, Y.F.; Guo, X. Stoichiometric patterns of soil carbon, nitrogen, and phosphorus in farmland of the Poyang Lake region in Southern China. *J. Soils Sediment.* **2019**, *19*, 3476–3488. [CrossRef]
25. He, J.X.; Du, L.; Zhai, C.; Guan, Y.P.; Wang, J.; Zhang, Z.H.; Wu, S.; Ogundeji, O.A.; Gu, S.Y. Physicochemical properties and stoichiometry of Mollisols in responses to tillage and fertilizer management. *Arch. Agron. Soil Sci.* **2020**. [CrossRef]
26. Hou, H.Z.; Zhang, X.C.; Wang, J.; Yin, J.D.; Fang, Y.J.; Yu, X.F.; Wang, H.L.; Ma, Y.F. Plastic-soil mulching increases the photosynthetic rate by relieving nutrient limitations in the soil and flag leaves of spring wheat in a semiarid area. *J. Soils Sediment.* **2020**, *20*, 3158–3170. [CrossRef]
27. Zhang, J.B.; Yang, J.S.; Yao, R.J.; Yu, S.P.; Li, F.R.; Hou, X.J. The effects of farmyard manure and mulch on soil physical properties in a reclaimed coastal tidal flat salt-affected soil. *J. Integr. Agric.* **2014**, *13*, 1782–1790. [CrossRef]
28. Xie, X.F.; Pu, L.J.; Shen, H.Y.; Wang, X.H.; Zhu, M.; Ge, Y.; Sun, L.C. Effects of soil reclamation on the oat cultivation in the newly reclaimed coastal land, eastern China. *Ecol. Eng.* **2019**, *129*, 115–122. [CrossRef]
29. Liang, J.C.; Shi, H.B.; Yang, S.Q.; Liu, R.M.; Zhou, J.; Li, L.X.; Wang, L.R. The effects of straw mulching on soil water, soil salinity and grain yield of a salty sunflower field. *Chin. J. Soil Sci.* **2014**, *45*, 1202–1206. (In Chinese)
30. Liu, E.K.; Yan, C.R.; Mei, X.R.; He, W.Q.; Bing, S.H.; Ding, L.P.; Liu, Q.; Liu, S.A.; Fan, T.L. Long-term effect of chemical fertilizer, straw, and manure on soil chemical and biological properties in Northwest China. *Geoderma* **2010**, *158*, 173–180. [CrossRef]
31. Lu, R.K. *Chemical Analysis of Agricultural Soils*; China Agricultural Science and Technology Press: Beijing, China, 1999. (In Chinese)
32. Xie, X.F.; Pu, L.J.; Wang, Q.Q.; Zhu, M.; Xu, Y.; Zhang, M. Response of soil physicochemical properties and enzyme activities to long-term reclamation of coastal saline soil, Eastern China. *Sci. Total Environ.* **2017**, *607–608*, 1419–1427. [CrossRef]
33. Xie, X.F.; Pu, L.J.; Zhu, M.; Wu, T.; Xu, Y.; Wang, X.H. Effect of long-term reclamation on soil quality in agricultural reclaimed coastal saline soil, Eastern China. *J. Soils Sediment.* **2020**, *20*, 3909–3920. [CrossRef]
34. Xu, S.T.; Zhang, L.; Zhou, L.; Mi, J.Z.; McLaughlin, N.B.; Liu, J.H. Effect of synthetic and natural water absorbing soil amendments on soil microbiological parameters under potato production in a semi-arid region. *Eur. J. Soil Biol.* **2016**, *75*, 8–14. [CrossRef]
35. Tian, X.M.; Fan, H.; Wang, J.Q.; Ippolitoc, J.; Li, Y.B.; Feng, S.S.; An, M.J.; Zhang, F.H.; Wang, K.Y. Effect of polymer materials on soil structure and organic carbon under drip irrigation. *Geoderma* **2019**, *340*, 94–103. [CrossRef]
36. He, H.; Zhang, Y.T.; Wei, C.Z.; Li, J.H. Characteristics of decomposition and nutrient release of corn straw under different organic fertilizer replacement rates. *Appl. Ecol. Env. Res.* **2019**, *17*, 13455–13472.

37. Fallah Nosratabad, A.R.; Etesami, H.; Shariati, S. Integrated use of organic fertilizer and bacterial inoculant improves phosphorus use efficiency in wheat (*Triticum aestivum* L.) fertilized with triple superphosphate. *Rhizosphere* **2017**, *3*, 109–111. [CrossRef]
38. Pawlowski, L.; Pawlowska, M.; Cel, W.; Wang, L.; Li, C.; Mei, T.T. Characteristic of carbon dioxide absorption by cereals in Poland and China. *Gospod. Surowcami Min.* **2019**, *35*, 165–176.
39. De Almeida, R.F.; Mikhael, J.E.R.; Franco, F.O.; Santana, L.M.F.; Wendling, B. Measuring the labile and recalcitrant pools of carbon and nitrogen in forested and agricultural soils: A study under tropical conditions. *Forests* **2019**, *10*, 544. [CrossRef]
40. Ren, T.; Wang, J.G.; Chen, Q.; Zhang, F.S.; Lu, S.C. The effects of manure and nitrogen fertilizer applications on soil organic carbon and nitrogen in a high-input cropping system. *PLoS ONE* **2014**, *9*, e97732. [CrossRef] [PubMed]
41. González Jiménez, J.L.; Healy, M.G.; Daly, K. Effects of fertiliser on phosphorus pools in soils with contrasting organic matter content: A fractionation and path analysis study. *Geoderma* **2018**, *338*, 128–135. [CrossRef]
42. Siebers, N.; Sumann, M.; Kaiser, K.; Amelung, W. Climatic effects on phosphorus fractions of native and cultivated North American grassland soils. *Soil Sci. Soc. Am. J.* **2017**, *81*, 299–309. [CrossRef]
43. Yan, G.Y.; Xing, Y.J.; Han, S.J.; Zhang, J.H.; Wang, Q.G.; Mu, C.C. Long-time precipitation reduction and nitrogen deposition increase alter soil nitrogen dynamic by influencing soil bacterial communities and functional groups. *Pedosphere* **2020**, *30*, 363–377. [CrossRef]
44. Samson, M.E.; Menasseri-Aubry, S.; Chantigny, M.H.; Angers, D.A.; Royer, I.; Vanasse, A. Crop response to soil management practices is driven by interactions among practices, crop species and soil type. *Field Crop. Res.* **2019**, *243*, 107623. [CrossRef]
45. Li, Z.Q.; Zhang, X.; Xu, J.; Cao, K.; Wang, J.H.; Xu, C.X.; Cao, W.D. Green manure incorporation with reductions in chemical fertilizer inputs improves rice yield and soil organic matter accumulation. *J. Soils Sediment.* **2020**, *20*, 2784–2793. [CrossRef]
46. Zhou, G.X.; Zhang, J.B.; Chen, L.; Zhang, C.Z.; Yu, Z.H. Temperature and straw quality regulate the microbial phospholipid fatty acid composition associated with straw decomposition. *Pedosphere* **2016**, *26*, 386–398. [CrossRef]
47. Tinna, D.; Garg, N.; Sharma, S.; Pandove, G.; Chawla, N. Utilization of plant growth promoting rhizobacteria as root dipping of seedlings for improving bulb yield and curtailing mineral fertilizer use in onion under field conditions. *Sci. Hortic.* **2020**, *270*, 109432. [CrossRef]
48. Chrysargyris, A.; Panayiotou, C.; Tzortzakis, N. Nitrogen and phosphorus levels affected plant growth, essential oil composition and antioxidant status of lavender plant (*Lavandula angustifolia* Mill.). *Ind. Crop. Prod.* **2016**, *83*, 577–586. [CrossRef]

land

Article

Effects of Land-Use Change on Soil Functionality and Biodiversity: Toward Sustainable Planning of New Vineyards

Elena Gagnarli [1], Giuseppe Valboa [2], Nadia Vignozzi [2], Donatella Goggioli [1], Silvia Guidi [1], Franca Tarchi [1], Lorenzo Corino [3] and Sauro Simoni [1,*]

1 CREA—Research Centre for Plant Protection and Certification, 50125 Florence, Italy; elena.gagnarli@crea.gov.it (E.G.); donatella.goggioli@crea.gov.it (D.G.); silvia.guidi@crea.gov.it (S.G.); franca.tarchi@crea.gov.it (F.T.)
2 CREA—Research Centre for Agriculture and Environment, 50125 Florence, Italy; giuseppe.valboa@crea.gov.it (G.V.); nadia.vignozzi@crea.gov.it (N.V.)
3 Fattoria 'La Maliosa' Loc. Podere Monte Cavallo, 58014 Saturnia, Italy; lorenzo.corino@gmail.com
* Correspondence: sauro.simoni@crea.gov.it; Tel.: +39-055-2492229

Abstract: Sustainable agriculture largely depends on soil biodiversity and requires efficient methods to assess the effectiveness of agronomic planning. Knowledge of the landscape and relative pedosite is enriched by data on the soil microarthropod community, which represent useful bio-indicators for early soil-quality detection in land-use change (LUC). In the hilly Maremma region of Grosseto, Italy, two areas, a >10ys meadow converted into a vineyard and an old biodynamic vineyard (no-LUC), were selected for evaluating the LUC effect. For maintaining soil vitality and ecosystem services by meadow, the vineyard was planted and cultivated using criteria of the patented "Corino method". The aim was to evaluate the LUC impact, within one year, by assessing parameters characterizing soil properties and soil microarthropod communities after the vineyard was planted. The adopted preservative method in the new vineyards did not show a detrimental impact on the biodiversity of soil microarthropods, and in particular, additional mulching contributed to a quick recovery from soil stress due to working the plantation. In the short term, the adopted agricultural context confirmed that the targeted objectives preserved the soil quality and functionality.

Keywords: sustainability; vineyards; best agronomic practices; Collembola; Acari

Citation: Gagnarli, E.; Valboa, G.; Vignozzi, N.; Goggioli, D.; Guidi, S.; Tarchi, F.; Corino, L.; Simoni, S. Effects of Land-Use Change on Soil Functionality and Biodiversity: Toward Sustainable Planning of New Vineyards. *Land* **2021**, *10*, 358. https://doi.org/10.3390/land10040358

Academic Editor: Nigussie Haregeweyn

Received: 28 February 2021
Accepted: 24 March 2021
Published: 1 April 2021

1. Introduction

In terms of soil functionality maintenance, high-quality soils have ensured the integration of soil productivity with other ecosystem services. During the last few years, European policies have enhanced compliance and rules to avoid land degradation [1]. Sustainable development goals for soil management address efforts of rural development and, simultaneously, protection of soil functionality [2]. To support short-term needs and long-term (global) goals, the conventional practices for the new planting of vineyards should be reviewed.

The global agriculture challenge is to increase the output from available land while reducing the negative effects of its use [3]. The traditional agricultural landscape is disappearing due to land-cover changes, and these modifications in vegetation impact regional climate, carbon sequestration, and biodiversity [4]. During the last few years, concern for the environment and sustainability has compelled many governments to adjust land-use policies to balance multiple uses of land resources [4] by increasing expectations that productive agricultural landscapes should be managed by coupling preservation or enhancement of biodiversity [5]. At the various trophic levels in the food chain, the interactions between the communities of soil can be altered according to different strategies of soil management: farming increase and agronomic practices (i.e., land-leveling soil and deep tillage) impact belowground biodiversity [5]. Furthermore, the conversion of natural habitats to agriculture or other intensive human land uses leads to biodiversity loss [6].

Changes in land use have mainly been studied regarding their consequences on productivity and human well-being [7], while their effects on the environment have been poorly investigated [8]. Referring to land-use change (LUC), only a few current European monitoring systems have focused on the status and/or trends recorded in soil functions [9,10]. High activity in physicochemical processes and richness of organisms is recognized in the upper soil layer (from 0 to 20 cm); however, at the same time, this is the layer most vulnerable to erosion and degradation [3]. Usually, the conversion of natural habitat to agricultural land results in the reduction of the edaphic species' richness, along with lower genetic variability and the loss of functional groups/ecosystem functions [11]. Microtopographical changes occurring during and after the planting of vineyards induce soil structural changes, which directly affect ecosystem services and biodiversity for a potentially long time lag [12]. However, little is known about how soil structural changes occur during and after the planting of vineyards and which key factors and processes play a major role in soil degradation due to cultivation works. In viticulture, deep earthworks performed before the plantation of vine plants severely affect the properties of the soil profile, vine phenology, and grape yield by altering the ecosystem functioning for years [13,14]. After deep tillage, soil organism communities are simplified and often need several years to recover [14,15] (Figure 1a). Deep ploughing may not be beneficial for soil types high in clay, as it can simply reseal the clay bank [16]. Conventional ploughing (≥30 cm depth) hinders soil aggregate formation and depletes soil organic matter, thus returning soils to early stages of ecological succession and stimulating soil erosion with the loss of the nutrient-rich upper soil layer [14]. Furthermore, the economic issue must also be considered. In hilly Italian viticultural areas, the cost of a new vineyard, including mechanization and labor, amounts to approximately EUR 20,000/ha [17].

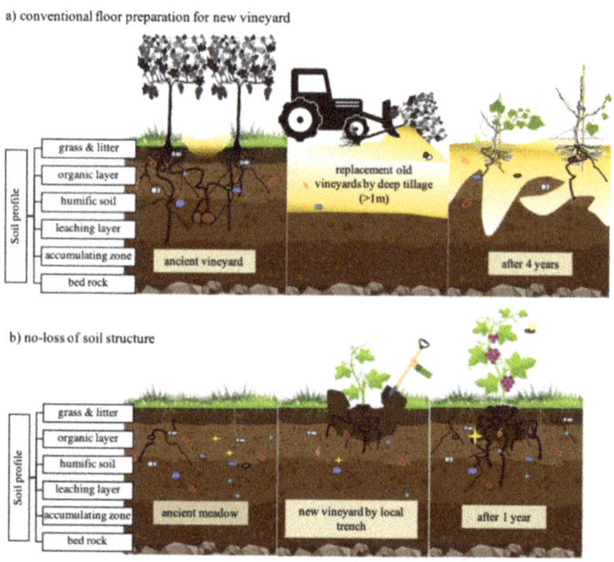

Figure 1. A cross-section of the soil profile in two different examples of management for starting new vineyards in hilly areas: (**a**) no land-use change, deep earthworks, and substitution of an ancient vineyard by implying a recovery time for soil functions and grape production >4 years [10]; (**b**) land-use change with the "Corino method" by maintaining the soil "heritage" using perennial meadows as the biological potential ecosystem service described in this study case (pre- and post-LUC).

The area selection of a new vineyard can be the starting point for a well-prepared soil bed: some sites are suited to low-mowed row middles of native vegetation, while others may require annual planting of winter grasses [16]. Recently, the culture, lifestyle, landscape patrimony, longevity of vineyards, and asset value of the land, as well as farmers' profit margins, were jointly considered in the "Corino method", aiming at increasing the vitality of the soil and the health of the environment, producers, and consumers [18] (Figure 1b). The benefits provided by minimal soil disturbance are multiple: they rely on physical (i.e., erosion reduction, increase of water retention, temperature), chemical (organic carbon storage, nutrient availability, pH), and biological (diversity of organisms, soil quality) properties of soils [19].

Under reduced mechanical disturbance, maintaining the soil profile results in a positive effect on the stability of soil aggregates and mycorrhizal associations. The development of grass cover protects soils from erosion and extreme temperatures [8].

Soils with a good structure allow air, water, and nutrients to move freely through pores within and between the aggregates, thereby influencing the water and nutrient reservoir for vine growth [20]. The content of organic matter and other chemical parameters of soil impact nutrient availability and, indirectly, crop plant growth. Concerning soil organic matter in the soil agro-ecosystems, the more considered living soil components are plant and microorganism contributions [21,22]. Studies focusing on the mesofauna community have not been provided to assess the effects of land use on soil biodiversity. Several ecosystem functions are ascribable to mesofauna and strictly related to soil fertility and agricultural production (i.e., the decomposition of the organic matter and nutrient cycling) [23,24]. A more diverse and abundant soil community provides better soil functions [25], efficiently returning ecosystem processes [26,27]. The biomass and density of the microarthropod population closely reflect the resource availability [25], promoting organic matter breakdown and the recycling of essential nutrients for plant growth [8,28].

Assessments of soil biodiversity can be highly indicative to estimate the impact of human activity and soil biological quality. To quickly assess soil disturbance, the presence of most adapted forms of hypogeal life and assemblage of the edaphic arthropod fauna community can represent a useful tool [2,29,30]. By evaluating the microarthropods' level of adaptation to the soil, the multitaxon indication by the index of Biological Soil Quality (BSQar) can provide efficient information [29]. Several studies have been carried out in vineyards for the evaluation of soil biodiversity and variability among management systems [31], the influence of soil physical and chemical characteristics on the edaphic community [15], and comparison of different ecological indices, e.g., the Shannon diversity index, etc. [30,32,33]. Considering the richness and abundance of soil arthropods as biotic factors, to be incorporated in landscape modeling, their use may implement, at a low cost, the evaluation of short-term conservation in viticulture.

This study aimed to estimate the effects on the short-term change in soil biodiversity for a pluriannual meadow after its conversion into a vineyard, by following rules in the cited Corino method. This purpose was pursued by evaluating if the entire soil-beneficial "inheritance" passes on from the meadow to the vineyard. The approach is based on the possible role, through LUC, of the previous natural habitat (meadow) not as a competitor—i.e., for water availability—but rather as valuable and functional in maintaining ecosystem services, in addition to being a resource involved in assuring natural mulching.

2. Materials and Methods

2.1. Study Area

The study area is in the central part of the Maremma region (Grosseto province, Tuscany, Italy) (Figure 2), and is characterized by hills between 300 and 600 m above sea level, dotted with sulfur-rich sources of water (such as those of nearby the Saturnia-Springs). Soils are shallow but rich in substances useful for the vine plant. The climate is mild, typically Mediterranean, with a constant wind all year round and a dry summer period. Viticulture is the primary activity in the local agricultural economy; its ancestral link with grapes has been strong since the time of the Etruscans, who settled in this area between the sixth and the first century B.C.

Figure 2. Two study areas (satellite image source: Google 2019): the LUC area (5100 m^2, Coord. X: 42.639210, Y: 11.539317) and no-LUC area (8200 m^2; coord. X: 42.619682, Y: 11.537605) delineated by the dotted white circles and 4 different plots (VV area: 4200 m^2; VM area: 4000 m^2; VN area: 3000 m^2; MC area: 2100 m^2) delineated by the red boxes. The schematic representation of the experimental design describes the steps during the conversion process (LUC) from meadows to new vineyards by conservative practices in the upper soil layer (from 0 to 20 cm).

The area extends to approximately 1.8 ha and consists of two areas, distant 2 km, selected as part of a vegetable-based biodynamic farm (La Maliosa Farm, Saturnia, Italy) (Figure 2). The studied vineyards lie on south-/southwest-facing slopes, about 300 m above sea level, in a complex mosaic landscape characterized by multicultivar vineyards surrounded by natural elements such as natural boundaries (i.e., trees and high fences). Farm management was based on the Corino method (IT Patent approved in 2019, IT201700005484A1), a set of agricultural practices developed by the farm owner and focused on soil vitality and environmental health [16]. The method makes use of good protection against erosion and the improvement of self-fertility by exploiting the role of green manure and natural mulching to improve the soil structure. The strengthening of the living organisms, helped by a gas exchange of oxygen/CO_2, will provide sustainable vitality in soil and permanent benefit for grapevines.

Here, vineyards have been rewarded by adopting Tuscan Maremma's historical native vine varieties (mainly Ciliegiolo, Sangiovese, Procanico, and Cannonau grigio). The vine plants were reclaimed from a >50-year-old and semiabandoned vineyard; this choice exploited the wealth of grapevine germplasm, both for the red and white vines selected and retrieved within the farm. For vine disease containment, powdered sulfur of 80 kg/ha and copper metal of less than 3 kg/ha/year were applied. The vine vegetation was arranged on stakes without shoot topping to prolong the foliar activity until late in the season, and pruning mixed with arch, spurred cordon, and sapling.

Considering the soil as a living organism to be preserved in its functions, the vineyard location and grape varieties were chosen to minimize preplanting earthworks and to maintain the natural grassland bed.

The vine rows were not oriented along the maximum gradient of the land, but instead where natural terraces allowed the mitigation of soil erosion, to save the value of the landscape and to maximize the physiological functions of the young plants [16]. In all vineyards, the inter-row spaces were kept under natural grass cover throughout the year. The grass was periodically mowed (two times/year), shredded together with plant residues, and spread on the soil surface as a source of organic matter and to avoid possible plant competition. The contemporary adoption of intercropping with mulches, green manure, and periodic cultivation equipment with manual management (no mould-board plough use) was aimed at enhancing vital soil processes both in the short and long term.

Four vineyards were selected in the two different areas (Figure 2): Vigna Nuova (VN) and Monte Cavallo (MC), in the LUC area, established in March 2014 after land-use conversion from grassland; and Vigna Vecchia (VV) and Vigna Maliosa (VM), in the no-LUC area, with pluridecennial vine plants.

A first soil sampling was carried out in June 2013 to gain information about microarthropod communities and soil texture in all sites. A few months later, in November 2014, two experimental subplots were selected within each area: (1) one managed by straw mulching (mu) between vine plants, in order to reduce soil erosion, control weed development, and improve soil moisture content; and (2) a second one kept as a control plot without any mulching treatment. Different soil-sampling procedures were planned according to the specific analyses to be performed. In each vineyard, three soil cores (7×5 cm, 10 cm depth) were collected from the intrarow space, at 20 cm from the vine plant, for zoological analysis. Close to these soil cores, three subsamples were collected by auger to 20 cm depth for chemical (total organic C, total N, total $CaCO_3$, pH, and electrical conductivity) and physical (particle-size distribution) analyses.

2.2. Soil Properties and Microarthropod Communities

Soil texture was determined using the SediGraph method [34] and the USDA classification [35]. The extraction of micro-arthropods was carried out by Berlese-Tullgren selectors for 5 days; specimens were collected in jars with 80% ethanol solution and were counted at a stereomicroscope (10×–60×). Mean microarthropod density was calculated by year (pre- and post-LUC), area (ancient vineyards, grassland, and new vineyards) and plot (VV, VM, NV, and MC). To determine the effects of LUC, differences in the population structure were analyzed among arthropod densities of three main abundant groups: Acari, Collembola, and "other arthropods". In order to assess the biological soil quality (BSQar), the microarthropods were separated into biological form (BF) morphotypes (see Parisi et al. [29]) according to their degree of morphological adaptation to soil life. Each BF was associated with a score, ranging from 1 (surface-living organisms) to 20 (deep-living organisms). Generally, the soil was considered to have a good "biological quality" when the soil fauna community was abundant and diversified in well-adapted forms to an edaphic environment. For estimating the complexity, stability, and thus general health of soil ecosystem, the following diversity indices were also calculated: taxa richness (S), the Margalef index [36], the Shannon diversity index (H') [37], Buzas and Gibson's evenness index (E') [38], Simpson's index (1-D) [39], and the Berger–Parker index [40].

2.3. Mulching Effect on Soil

After LUC, each vineyard was split into two subplots: (1) added straw mulch (mu) and (2) control without mulching (no-mulch). The soil chemical parameters in both subplots were subjected to a Pearson's correlation coefficient. The material for chemical analysis was sampled by 3 topsoil (0–20 cm) cores randomly in each subplot using a hand-auger. The samples were air-dried and sieved through a 2 mm mesh before analysis. For C and N determination, a representative fraction from each sample was ground and homogenized to 0.5 mm. TOC and TN were measured by dry combustion on a Thermo Flash 2000 CN soil analyzer. Then, 70 mg soil was weighed into an Sn-foil capsule to analyze the total C (organic C + mineral C) and N contents. Separately, 20 to 40 mg of soil was weighed into an Ag-foil capsule, pretreated with 10% Cl until complete removal of carbonates, and then analyzed for total C content (corresponding to the TOC content). The total equivalent $CaCO_3$ content was calculated from the difference between the total C measured before and after the HCl treatment [41]. Soil pH was measured potentiometrically in a 1:2.5 soil:water suspension. Electrical conductivity was measured in a 1:2 soil:water extract after 2 h of shaking, overnight standing, and filtration.

2.4. Data Analysis

The effect of LUC on arthropod density was assessed by comparing, within the same year, VV, VM (no-LUC area), MC, and VN (LUC area) by means of one-way ANOVA followed by Tukey's post-hoc test. Richness (S), Shannon (H'), Simpson (1-D), evenness (E), Margalef, and Berger–Parker were calculated and compared by the bootstrap method. The effects of LUC were evaluated by mean BSQ_{ar} values (Mann–Whitney test; $p < 0.05$). The impact of mulching on soil chemical properties was assessed through one-way ANOVAs within each vineyard. The relationships between soil parameters and the abundance of microarthropod groups (Acari, Collembola and other arthropods) were evaluated by correlation analysis (Pearson's "r" coefficient, $p < 0.05$). All analyses were performed using standard methods with PAST software [42].

3. Results

3.1. Soil Properties and Microarthropod Communities

In the no-LUC area, the soil texture was silty-clay (SIC), and in the pre-LUC area it was clay-loam (CL) (Table 1). After LUC, no significant variations in the fine soil-particle distribution were registered, and the textural class was unchanged. In the no-LUC vineyards, in the second year, the sand percentage increased, probably due to light farming interventions to prevent soil compaction.

Table 1. Soil textural classification by individual size-groups (%) of mineral particles.

	2013				2014			
	Sand (%)	**Clay (%)**	**Silt (%)**	**USDA Class**	**Sand (%)**	**Clay (%)**	**Silt (%)**	**USDA Class**
Vigna Vecchia (VV)	15.66	43.79	40.55	silty_clay	21.56	39.19	39.25	clay_loam
Vigna Maliosa (VM)	10.97	43.25	45.78	silty_clay	35.75	37.09	32.15	clay_loam
Monte Cavallo (MC)	28.40	34.90	36.6	clay_loam	27.44	35.42	37.14	clay_loam
Vigna Nuova (VN)	30.88	34.73	34.39	clay_loam	32.68	34.23	33.09	clay_loam

On the whole, 6647 microarthropods were collected. The most abundant group was Acari (61%), followed by Collembola (29%). The other microarthropod group was composed of 21 biological forms (BFs). Araneida and Palpigrada were present only in grasslands; and Coleoptera, Isopoda, and Embioptera disappeared in plots after planting vineyards (post-LUC area). The soil dwellers (i.e., Protura, Diplura, Pseudoscorpiona, Diplopoda, Pauropoda, and Symphyla) were sporadic but present.

Regarding abundance, no substantial difference was registered between LUC (meadow/ vineyard) and no-LUC (vineyard) areas, in the two years considered (2013: $F_{1,11} = 2.988$; $P = 0.146$; 2014: $F_{1,11} = 3.097$; $P = 0.109$). Only light differences in total microarthropod density were due to the plot ($F_{3,11} = 9.4$; $P < 0.01$), in 2013, with the lower density in VM; however, this value of abundance was similar to that registered in MC, the long-standing meadow; in 2014, no difference was detected ($F_{3,11} = 1.380$; $P = 0.317$).

The BSQ_{ar} index showed the highest value in meadows (Figure 3). The second-year evaluation (Y2) showed that the vineyard plantation did not affect soil quality despite the soil perturbation, and the BSQ_{ar} values were similar to the ancient vineyards (no-LUC) (Mann–Whitney test, not significant at 5% level) (Figure 3). The decrease of BSQ_{ar} value after LUC was associated with a loss of six arthropod groups, especially euedaphic forms in the no-mulch vineyards (Appendix A Table A1).

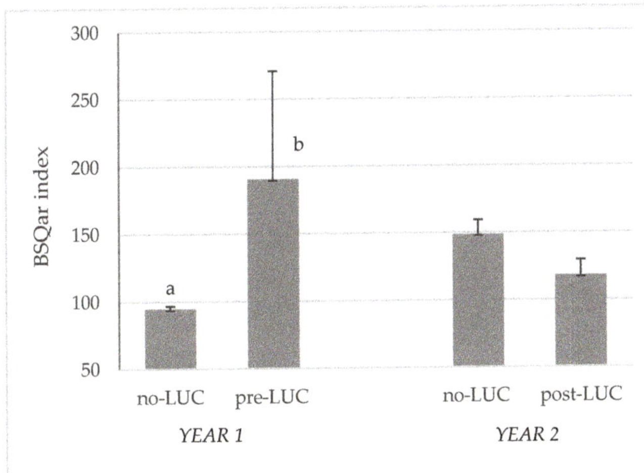

Figure 3. Biological soil quality by BSQar index in the experimental areas. Data were expressed as mean ± SD and 95% confidence interval by different letters. Significance was evaluated within the year (Mann–Whitney test, $p < 0.02$).

By referring to vineyards, after LUC, the diversity richness of soil arthropods became similar between areas (Table 2); the loss of richness in ex-meadow (post-LUC area) affected the Shannon and Simpson's indices, showing a decrease of relative frequencies of soil dominant groups.

Table 2. Biodiversity indices calculated in the no-LUC area and LUC area: S (richness), N (total abundance), H'(Shannon), E (Evenness), 1-D (Simpson's). Significant differences are in bold (Monte Carlo permutation test [42]).

	2013			2014		
Diversity Index	**No-LUC Area**	**Pre-LUC Area**	**p (eq)**	**No-LUC Area**	**Pre-LUC Area**	**p (eq)**
S	11	19		13	12	
N	550	1132		1241	293	
H'	1.05	1.07	0.72	**1.64**	**1.28**	**0.00**
E	**0.26**	**0.15**	**0.00**	**0.40**	**0.30**	**0.04**
1-D	0.47	0.50	0.43	**0.73**	**0.55**	**0.00**
Margalef	**1.60**	**2.60**	**0.00**	1.67	1.98	0.42
Berger- Parker	0.71	0.67	0.16	**0.43**	**0.64**	**0.00**

Concerning biodiversity, the meadow area showed high values of taxa richness of microarthropods; furthermore, different groups were well represented in their natural soil habitat (Figure 4). At the same time, the H', 1-D, and Berger–Parker indices registered in the old vineyard (no-LUC area) were similar to those of the meadow, probably due to similar habitat conditions (inter-row long-term management within permanent cover grass) (Table 2).

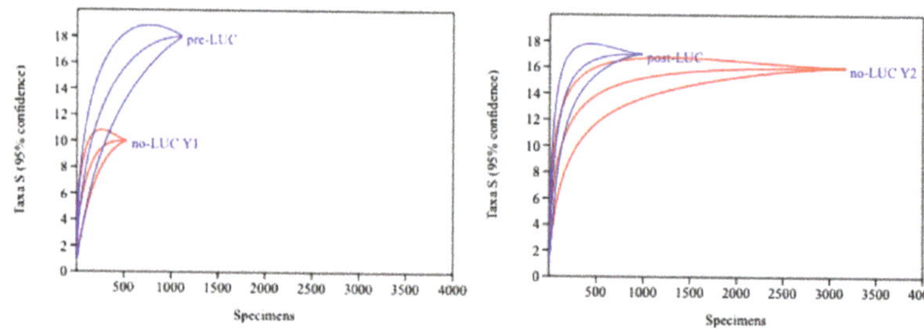

Figure 4. Individual-based taxon rarefaction curves, by year and area, that differentially estimate the relative importance of taxa richness change in composition of the arthropod community. Delimited area around the curves indicates 95% confidence interval.

By including small and rare taxa, the rarefaction curve begins to level off at a new plateau: pre-LUC meadows showed the highest diversity (Figure 4); however, after LUC, the diversity-rarefaction curve denoted changes in taxa richness, independently of the reduction of specimens (Figure 4).

3.2. Post-LUC Mulching Effect on Soil

The BSQ_{ar} values registered in the second year were all ≥110 (Table A1), and no significant decrease was registered between pre- and post-LUC areas (Mann–Whitney test, P = 0.334). High values of biodiversity were obtained where the mulch was added. Nevertheless, the application of a mulch layer closely around the vine plants changed the arthropod assemblages in soils, promoting the presence of epe- and hemiedaphic forms in both areas (Figure 5).

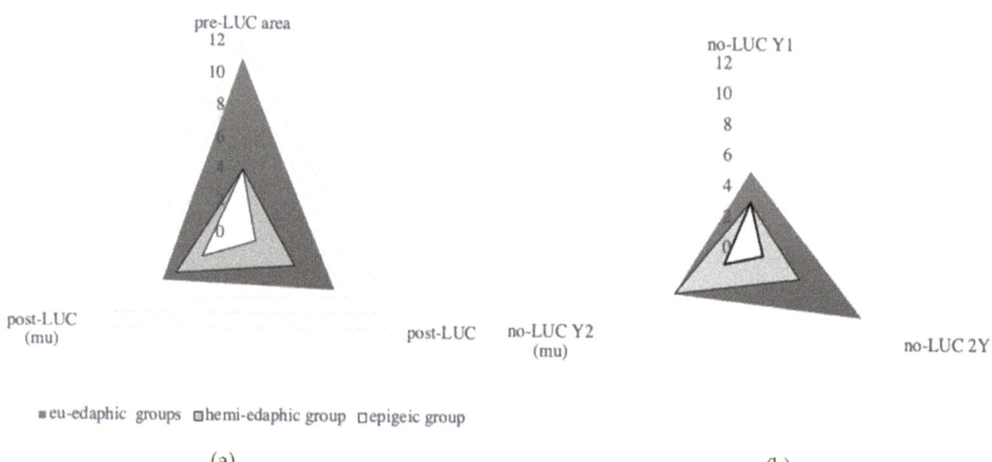

Figure 5. Composition of microarthropod communities by three morpho-functional levels in the different areas: (**a**) LUC area and (**b**) no-LUC area, by mulching (mu) effect.

The effect of mulching was different in two areas (Table 3). In old vineyards, the abundance and richness were similar, independent of the mulching addition; while the other biodiversity indices were higher in plots where straw mulch was not added. A positive effect of mulching on biodiversity was registered in new vineyards by determining differences in S, H′, and 1-D.

Table 3. Diversity indices of soil arthropods between different inter-row managements (mulch; no mulch plots) in 2014 according to richness (S), total abundance of arthropods (N), Shannon (H′), Evenness (E), Simpson's (1-D), and Margalef. Significant differences are in bold (Monte Carlo permutation test [42]).

	Old Vineyard			New Vineyard		
Diversity Index	**Mulch**	**No Mulch**	**p (eq)**	**Mulch**	**No Mulch**	**p (eq)**
S	13	13	n.s.	**16**	**12**	**0.0254**
N	1946	1241		709	293	
H′	**1.13**	**1.64**	**0.0001**	**1.51**	**1.28**	**0.0212**
E	**0.23**	**0.40**	**0.0001**	0.28	0.30	0.6038
1-D	**0.58**	**0.73**	**0.0001**	**0.64**	**0.55**	**0.0044**
Margalef	1.58	1.69	0.6725	2.29	1.94	0.2107

Soil electrical conductivity (EC) increased with mulching in the new vineyard ($F_{3,20} = 16.7$; $p < 0.001$), whereas it did not differ in the old vineyard (Figure 6). Soil TOC and TN contents were generally low and did not significantly change related to the floor management, except for a slight increasing trend under mulching (P = 0.68 and P = 0.81, respectively).

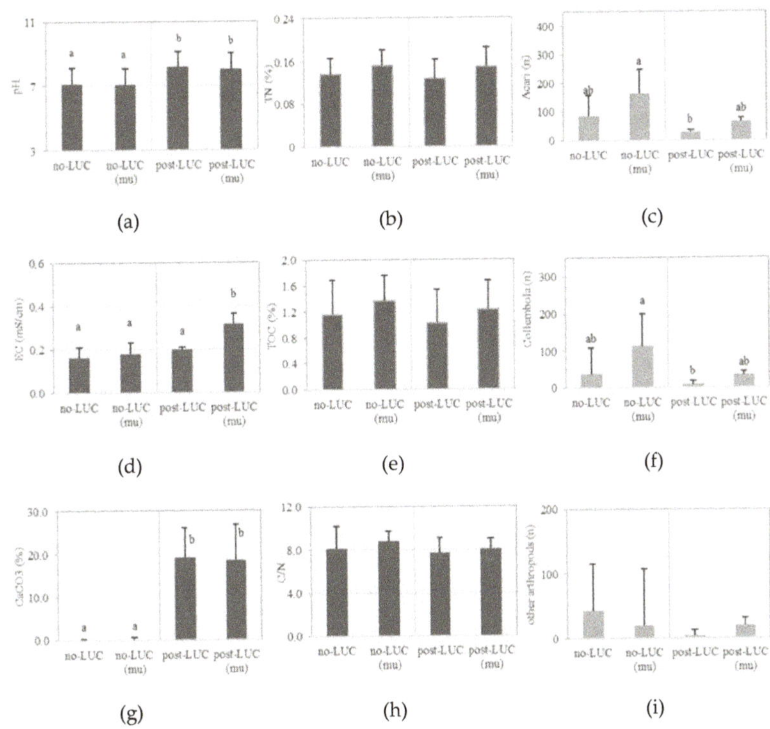

Figure 6. The evaluation of post-LUC mulching effects (mu) by separate ANOVAs on selected soil chemical properties ((**a**) pH, (**b**) TN, (**d**) EC, (**e**) TOC, (**g**) $CaCO_3$ (**h**) C/N) (TOC = total organic carbon; TN = total nitrogen; EC = electrical conductivity; $CaCO_3$ = total equivalent Ca carbonate; C/N = Carbon-to-Nitrogen ratio) and the average of abundance of three main edaphic animal groups, (**c**) Acari, (**f**) Collembola, (**i**) other arthropods in four different management areas (no-LUC; no-LUC (mu); post-LUC; post-LUC (mu)). Error bars indicate the mean standard error and different letters show statistically significant differences between variables (ANOVA; Tukey's test, $p < 0.05$).

The correlation analysis performed on soil properties showed a strong relationship between TOC and TN ($R^2 = 0.97$, $p < 0.01$) (Table 4).

Table 4. Pearson's correlation coefficients (significance level, two-tailed test: $p < 0.05$ *; $p < 0.01$ **) showing interactions among chemical properties of superficial soil samples and mean abundance of arthropod groups in different vineyards.

	pH	EC	TOC	TN	$CaCO_3$	Acari	Collembola	Other Arthr.
pH								
EC	0.58 (**)							
TOC	−0.15	0.33						
TN	−0.1	0.39	0.97 (**)					
$CaCO_3$	0.96 (**)	0.56 (**)	−0.3	−0.26				
Acari	−0.39	−0.16	0.16	0.18	−0.44 (*)			
Coll.	−0.24	−0.06	0.34	0.34	−0.28	0.84 (**)		
other arthr.	−0.21	−0.26	−0.21	−0.2	−0.22	0.52 (**)	0.28	

Moreover, the abundance of Collembola and other arthropods were positively related to the Acari ($R^2 = 0.84$ **; $R^2 = 0.52$ ** respectively), while the mites appeared to be negatively affected by $CaCO_3$ ($R^2 = -0.44$ *).

4. Discussion

In the case study, soil arthropod biodiversity was used as an indicator to assess the impact of LUC when planting vineyards in a meadow. In the Mediterranean region, where susceptible land suffers the most degradation because of topographical and climate characteristics [43], the habitat transformation should be carefully chosen according to the soil and environmental specificities [44,45]. Differently from deep soil working [14,45], the Corino method indicates preserving the top layer during a new vineyard planting to protect the soil floor heritage. As soon as LUC was adopted, the complexity of microarthropod communities indicated a short-term restored soil biological diversity [18]. A high number of arthropods, belonging to different taxa, was recorded in the meadow, with Acari and Collembola as dominating groups. Usually, identifying diversity richness is an important indication of the management and preservation of biofunctionality of soil [23].

The entire soil arthropod community was promptly able to react to soil perturbation, probably due to the maintenance of physical and chemical properties in the soil. Furthermore, according to Wong et al. [46], grapevine planting can cause a partial dissolution of soil carbonates, strongly characterizing the mineral phase of the soil and subsequently increasing the soluble salt concentration of the soil solution. It is extremely difficult for most plants to survive in soil whose structure has been destroyed, leading to the clay particles clogging the pore spaces [47,48]; also in vineyards [13]. In this study, the soil textural group did not change post-LUC, remaining moderately fine (clay-loam class texture). Vineyard age and vine age can represent a key issue for soil biota [31]. Among the tested vineyard plots, after LUC, the total abundances of soil microarthropods, independent of the sampling time and arthropod life cycles, were similar to those in the no-LUC areas. The method allowed the preservation of several patterns of euedaphic groups: Acari; Collembola; and some smaller Symphyla, Pauropoda, Diplura, and Hymenoptera Formicidae have been surviving in the soil, and a few months after the planting, they recolonized areas. The spatiotemporal patterns of Acari and Collembola may be due to changes in microclimatic soil properties, adaptive phenological characteristics of the organisms themselves, or pressure from a combination of different anthropogenic environmental change drivers [49,50].

The ANOVA showed only a slight difference in total microarthropod density due to the plot ($F_{3,11} = 9.9$; $P < 0.01$) in 2013, with the lower density in VM; however, this figure was similar to that registered in MC, the long-standing meadow. In 2014, concerning plot or areas, no difference was detected.

An excessive reduction in biological soil components and the loss of microarthropod species with unique functions in nutrient cycles may lead to degradation of soil and loss of agricultural productive capacity [51,52]. This aspect, not even adequately considered, is now assuming importance in providing the basic information required for assessment of sustainable ecosystem services in grape production [53–55]. Studies suggest that organically managed fields contain greater abundance and diversity of arthropods than conventionally managed ones [31,56], but evidence among different strategies in starting new vineyards is not available. This study confirmed, in post-LUC areas, that some groups are very sensitive to recent soil disturbance, such as pseudoscorpions or diplurans, and that their presence is highly associated with environment and soil-specific parameters [57].

Under added mulching, there was a minor increase in the average OC and total N contents. However, overall, the chemical parameters were similar. Only the electrical conductivity (EC) in the new vineyard was significantly higher compared to sites with no-mulching management, supporting soil mineral composition and interactions with soil organic matter and microbial activity [44]. Whereas mulching can provide immediate effects in terms of soil erosion reduction, soil temperature, and moisture control, its contribution to soil organic carbon enrichment also may require longer periods, especially in fine-textured or clayey soils [58]. Field screening performed in experimental sites indicated facilitated growth of new plants, probably favored by easy rooting and availability of rich oligo-elements [59]. In the present study, the scarce accumulation of organic matter in the upper soil layers seemed to have no influence on the abundance of microarthropods and might not necessarily be a limiting factor for the qualitative performance of the vineyard [60].

According to Decaëns et al. [61], the most abundant microarthropod groups were the soil-dwelling organisms: Acari (more than 50%) and Collembola (about 30%). Considering Acari, the highest presence of oribatids, living in dense clusters in the decomposing litter of the upper soil layers, is favored by thick organic horizons, acidic conditions, and recalcitrant litter materials [62].

On the whole, the complexity of the microarthropod population structure did not show significant differences, although the number of euedaphon groups was quite high, as evidenced in all plots by the BSQar values. The soil biological quality index proved to be a good indicator of soil-stress conditions at different levels. Protura, Diplura, and Pauropoda, even if they affect soil processes less compared to soil-dwelling organisms [27], are highly sensitive to soil-stress conditions, and can be relevant for biomonitoring purposes [29,63]. Taxa richness and other ecological indicators, such as the Shannon and Simpson's diversity indexes, confirmed the evidence showed by the BSQ_{ar} index, where the grassland is the habitat with the highest biodiversity [64]. According to Gope and Ray [65], the dynamics of microarthropods were probably dependent on the combined effect of vegetation cover and soil characteristics. Not all groups responded to the same extent: soil microarthropods with a larger body size appeared to be primarily affected by short-term consequences of LUC (disturbance, loss of habitat) [60], and after LUC, some functional groups, as the predators Palpigrada and Araneidae, disappeared. Nevertheless, the application of a mulch layer significantly increased the abundance of different arthropod predators [66], especially predator mites. Overall, a more diverse and abundant soil microarthropod community seems to provide better soil functions by reflecting the resource availability in the soil ecosystem [64].

5. Conclusions

Monitoring soil biodiversity enables the detection of biodiversity hot spots, as well as areas susceptible to changes, and helps to achieve successful implementation of ecosystem management. According to Novara et al. [43], the high eco-mosaic complexity of landscape significantly contributes to the ecosystem resilience. Despite the short time elapsed from LUC, the agronomic strategy employed in planting and managing new vineyards shows a great potential regarding landscape preservation. The strategy provides significant support to address and harmonize changes that are brought about by social, economic, and environmental processes. Based on the FAO input [3], new approaches, inspired by traditional agricultural management, can limit the problems caused by the continuous simplification of agro-ecosystems and the "one-size-fits-all" approach to vineyard management to obtain economic benefits [43]. The economic benefits of adopting a targeted approach rather than a conventional one for vineyard management can improve outputs in crop yield and fruit quality and/or reduced inputs [47].

Our research will continue in the future with the aim of monitoring/determining long-term effects of LUC based on the selected soil biological quality indicators. Monitoring soil quality means improving soil management so that it functions optimally now and is not degraded for future use.

6. Patents

Farming was done according to the Corino method (IT Patent approved in 2019, IT201700005484A1), a completely vegetal-based, closed-cycle agricultural method.

The object of the patent for the Corino method is a process to produce grapes that comprises several phases, such as the use of native vines. As described in the patent, the Corino method represents a humanized system with minimal environmental impact, and constitutes a significant step in technological development, as well as a fascinating return to origins, quality, and excellence.

In addition to the Corino method, the patent covers two products; namely, the grapes and the wine obtained through the Corino method. The main essential characteristics of the wine described in the patent are the absence of added sulfur dioxide, and the absence of additional chemical and microbiological interventions during its production.

Author Contributions: S.S. and L.C. conceived the assessment; S.S., N.V., G.V. and E.G. designed and performed the experiment; E.G., D.G., S.G., F.T., L.C. and S.S. obtained the soil samples; E.G., D.G., S.G., F.T., N.V. performed the lab analysis; E.G., G.V. and S.S. analyzed data; E.G. and S.S. wrote the paper; S.S., D.G. and G.V. revised the paper. All authors have read and agreed to the published version of the manuscript.

Funding: This research received no external funding.

Institutional Review Board Statement: Not applicable.

Informed Consent Statement: Not applicable.

Data Availability Statement: The data presented in this study are available on request from the corresponding author.

Acknowledgments: The authors are grateful to Antonella Manuli, owner of "La Maliosa" farm, for her encouragement and assistance at all stages of this research/work. The authors also wish to thank Kathleen Collins Tostanoski, English language native teacher, for the mindful English revision.

Conflicts of Interest: The authors declare no conflict of interest.

Appendix A

Table A1. EMI values for biological forms (BF) and BSQar indexes calculated by three morpho-functional levels [29] in the first (Y1) and second (Y2) years, in the different areas (no-LUC, LUC), considering mulching (mu) effect.

	Y1				Y2							
	No-LUC Area		Pre-LUC Area		No-LUC Area				Post-LUC Area			
BF	VV	VM	Pre-VN	Pre-MC	VV (mu)	VV	VM (mu)	VM	VN (mu)	VN	MC (mu)	MC
Acari	20	20	20	20	20	20	20	20	20	20	20	20
Collembola	20	20	20	20	20	20	20	20	20	20	20	20
Diplura	20	20	20		20					20	20	
Pauropoda			20	20	20	20	20	20	20	20	20	
Protura			20			20			20	20		
Pseudoscorpionida			20					20	20			
Embioptera	20		20									
Palpigrada			20									

Table A1. *Cont.*

	Y1				Y2							
	No-LUC Area		Pre-LUC Area		No-LUC Area				Post-LUC Area			
BF	VV	VM	Pre-VN	Pre-MC	VV (mu)	VV	VM (mu)	VM	VN (mu)	VN	MC (mu)	MC
Chilopoda			20		10	10	10	10	10	10	10	10
Coleoptera				6	20	20	20	6	1		6	
Diplopoda			20	20	10	20	20		20		20	20
Symphyla		20	20	20		20			20		20	20
Isopoda			10	10	10	10	10	10			10	
Diptera *larvae*		10	10	10	10	10	10	10	10	10	10	10
Coleoptera *larvae*	10				10	10	10		10		10	10
Hymenoptera	5				5	5	5	5		5		
Araneidae			5	5								
Psocoptera			1	1					1	1		
Hemiptera		1		1	1						1	
Thysanoptera		1	1				1					
Diptera	1	1		1					1			
BSQ$_{ar}$ *	96	93	247	134	156	185	146	141	173	126	167	110

References

1. Egidi, G.; Zambon, I.; Tombolin, I.; Salvati, L.; Cividino, S.; Seifollahi-Aghmiuni, S.; Kalantari, Z. Unraveling Latent Aspects of Urban Expansion: Desertification Risk Reveals More. *Int. J. Environ. Res. Public Health* **2020**, *17*, 4001. [CrossRef]
2. Keesstra, S.; Mol, G.; de Leeuw, J.; Okx, J.; de Cleen, M.; Visser, S. Soil-related sustainable development goals: Four concepts to make land degradation neutrality and restoration work. *Land* **2018**, *7*, 133. [CrossRef]
3. FAO. The State of Food and Agriculture Trends and Challenges. 2017. Available online: http://www.fao.org/3/a-i6583e.pdf (accessed on 16 March 2021).
4. Kanianska, R. Agriculture and Its Impact on Land-Use, Environment, and Ecosystem Services, Landscape Ecology. In *Landscape Ecology-The Influences of Land Use and Anthropogenic Impacts of Landscape Creation*; Almusaed, A., Ed.; InTech: Rijeka, Croatia, 2016; pp. 1–138. [CrossRef]
5. FAO-ITPS. *Protocol for the Assessment of Sustainable Soil Management*; FAO: Rome, Italy, 2020; Available online: http://www.fao.org/fileadmin/user_upload//GSP/SSM/SSM_Protocol_EN_006.pdf (accessed on 9 March 2021).
6. Hansen, A.J.; DeFries, R.S.; Turner, W. Land use change and biodiversity. *Land Chang. Sci.* **2004**, *6*, 277–299.
7. Verburg, P.H.; Crossman, N.; Ellis, E.C.; Heinimann, A.; Hostert, P.; Mertz, O.; Nagendra, H.; Sikor, T.; Erb, K.-H.; Golubiewski, N.; et al. Land system science and sustainable development of the earth system: A global land project perspective. *Anthropocene* **2015**, *12*, 29–41. [CrossRef]
8. Conti, D.F. Conservation agriculture and soil fauna: Only benefits or also potential threats? A review. *ECronicon Agric.* **2015**, *2.5*, 473–482.
9. Van Leeuwen, J.P.; Saby, N.P.A.; Jones, A.; Louwagie, G.; Micheli, E.; Rutgers, M.; Schulte, R.P.O.; Spiegel, H.; Toth, G.; Creamer, R.E. Gap assessment in current soil monitoring networks across Europe for measuring soil functions. *Environ. Res. Lett.* **2017**, *12*, 124007. [CrossRef]
10. Nejadhashemi, A.; Wardynski, B.; Munoz, J. Evaluating the impacts of land use changes on hydrologic responses in the agricultural regions of Michigan and Wisconsin. *Hydrol. Earth Sys. Sci. Discuss* **2011**, *8*, 3421–3468. [CrossRef]
11. Vandermeer, J.; Van Noordwijk, M.; Anderson, J.; Ong, C.; Perfecto, I. Global change and multi-species agroecosystems: Concepts and issues. *Agric. Ecosyst. Environ.* **1998**, *67*, 1–22. [CrossRef]
12. De Groot, G.A.; Jagers op Akkerhuis, G.A.J.M.; Dimmers, W.J.; Charrier, X.; Faber, J.H. Biomass and diversity of soil mite functional groups respond to extensification of land management, potentially affecting soil Ecosystem Services. *Front. Environ. Sci.* **2016**, *4*, 15. [CrossRef]
13. Costantini, E.A.; Agnelli, A.E.; Fabiani, A.; Gagnarli, E.; Mocali, S.; Priori, S.; Simoni, S.; Valboa, G. Short term recovery of soil physical, chemical, micro- and mesobiological functions in a new vineyard under organic farming. *Soil* **2015**, *1*, 443–457. [CrossRef]
14. Costantini, E.A.; Valboa, G.; Gagnarli, E.; Mocali, S.; Fabiani, A.; Priori, S.; Simoni, S.; Storchi, P.; Perria, R.; Vignozzi, N.; et al. Soil Resilience and Yield Performance in a Vineyard Established after Intense Pre-Planting Earthworks, PICO presentation in Session SSS10.7. In Proceedings of the European Geosciences Union General Assembly, Wien, Austria, 23–28 April 2017.

15. Mania, E.; Piazzi, M.; Gangemi, L.; Rossi, A.E.; Cassi, F.; Isocrono, D.; Pedullà, M.; Guidoni, S. The Soil Biodiversity as a Support to Environmental Sustainability in Vineyard. In Proceedings of the XI International Terroir Congress, Willamette Valley, OR, USA, 10–14 July 2016; pp. 316–320.
16. White, R.E. *Soils for Fine Wine*, 2nd ed.; Winetitles; Oxford University Press: New York, NY, USA, 2003; pp. 1–279.
17. Morando, A.; Lavezzaro, S.; Corradi, C. Costi d'impianto e produzione del vigneto. *Vitenda* **2018**, *23*, 81–94.
18. Corino, L. *The Essence of Wine and Natural Viticulture*; Quintadicopertina: Genova, Italy, 2018; pp. 1–114.
19. Hobbs, P.R.; Sayre, K.; Gupta, R. The role of conservation agriculture in sustainable agriculture. *Philos. Trans. R. Soc. B* **2008**, *363*, 543–555. [CrossRef] [PubMed]
20. Longbottom, M. *Managing Grapevine Nutrition in a Changing Environment*; Research to Practice Manual Australian Wine Research Institute: Adelaide, Australia, 2009.
21. Balser, T.C.; Gutknecht, J.L.M.; Liang, C.F. How Will Climate Change Impact Soil Microbial Communities? In *Soil Microbiology and Sustainable Crop Production*; Dixon, G.R., Tilston, E.L., Eds.; University of Reading Press: Reading, UK, 2010; pp. 373–397.
22. Bardgett, R.D.; Hobbs, P.J.; Frostegard, A. Changes in soil fungal: Bacterial biomass ratios following reductions in the intensity of management of an upland grassland. *Biol. Fertil. Soils* **1996**, *22*, 261–264. [CrossRef]
23. Brussaard, L.; Behan-Pelletier, V.; Bignell, D.E.; Brown, V.K.; Didden, W.; Folgarait, P.; Fragoso, C.; Freckman, D.W.; Gupta, V.V.S.R.; Hattori, T. Biodiversity and ecosystem functioning in soil. *Ambio* **1997**, *26*, 563–570.
24. Wall, D.; Nielsen, U.N. Biodiversity and Ecosystem services: Is it the same below ground? *Nat. Educ. Knowl.* **2012**, *3*, 8.
25. Nakamoto, T.; Jamagishi, J.; Miura, F. Effect of reduced tillage on weeds and soil organisms in winter wheat and summer maize cropping on *Humic Andosols* in Central Japan. *Soil Till. Res.* **2006**, *85*, 94–106. [CrossRef]
26. Lavelle, P.; Decaëns, T.; Aubert, M.; Barot, S.; Blouin, M.; Bureau, F.; Margerie, P.; Mora, P.; Rossi, J.P. Soil invertebrates and ecosystem services. *Eur. J. Soil Biol.* **2006**, *42*, S3–S15. [CrossRef]
27. Barrios, E. Soil biota ecosystem services and land productivity. *Ecol. Econ.* **2007**, *64*, 269–285. [CrossRef]
28. Wardle, D.A.; Bardgett, R.D.; Klironomos, J.N.; Setälä, W.H.; van der Putten, W.D.H. Ecological Linkages between aboveground and belowground biota. *Science* **2004**, *304*, 1629–1633. [CrossRef] [PubMed]
29. Parisi, V.; Menta, C.; Gardi, C.; Jacomini, C.; Mozzanica, E. micro-arthropod communities as a tool to assess soil quality and biodiversity: A new approach in Italy. *Agric. Ecosyst. Environ.* **2005**, *105*, 323–333. [CrossRef]
30. Menta, C.; Conti, F.D.; Pinto, S.; Bodini, A. Soil Biological Quality index (BSQ—ar): 15 years of application at global scale. *Ecol. Indic.* **2017**, *85*, 773–780. [CrossRef]
31. Gagnarli, E.; Goggioli, D.; Tarchi, F.; Guidi, S.; Nannelli, R.; Vignozzi, N.; Valboa, G.; Lottero, M.R.; Corino, L.; Simoni, S. Study case of micro-arthropod communities to assess soil quality in different managed vineyards. *Soil* **2015**, *1*, 527–536. [CrossRef]
32. Galli, L.; Capurro, M.; Menta, C.; Rellini, I. Is the QBS—ar index a good tool to detect the soil quality in Mediterranean areas? A cork tree *Quercus suber* L. (Fagaceae) wood as a case of study. *Ital. J. Zool.* **2014**, *81*, 126–135. [CrossRef]
33. Karlen, D.L.; Mausbach, M.J.; Doran, J.W.; Cline, R.G.; Harris, R.F.; Schuman, G.E. Soil Quality: A Concept, Definition, and Framework for Evaluation. *Soil Sci. Soc. Am. J.* **1997**, *61*, 4–10. [CrossRef]
34. Andrenelli, M.C.; Fiori, V.; Pellegrini, S. Soil particle-size analysis up to 250μm by X-ray granulometer: Device set-up and regressions for data conversion into pipette-equivalent values. *Geoderma* **2013**, *192*, 380–393. [CrossRef]
35. USDA ARS/NRCS. Soil Quality Test Kit Guide, Section II. Background & Interpretive Guide for Individual Tests. Available online: https://www.nrcs.usda.gov/Internet/FSE_DOCUMENTS/nrcs142p2_052490.pdf (accessed on 9 March 2021).
36. Margalef, R. Information theory in ecology. *Gen. Syst.* **1958**, *3*, 36–71.
37. Shannon, C.A. Mathematical theory of communication. *Bell Syst. Tech. J.* **1948**, *27*, 379–423. [CrossRef]
38. Buzas, I.A.; Gibson, T.G. Species diversity: Benthonic Forminifera in western North Atlantic. *Science* **1969**, *163*, 72–75. [CrossRef]
39. Simpson, E.H. Measurement of diversity. *Nature* **1949**, *163*, 688. [CrossRef]
40. Berger, W.H.; Parker, F.L. Diversity of planktonic foraminifera in deep-sea sediments. *Science* **1970**, *168*, 1345–1347. [CrossRef]
41. Sequi, P.; De Nobili, M. Frazionamento del Carbonio Organico. In *Metodi di Analisi Chimica del Suolo*; Violante, P., Ed.; Franco Angeli: Milano, Italy, 1970; pp. 1–13.
42. Hammer, Ø.; Harper, D.A.T.; Ryan, P.D. PAST: Paleontological statistics software package for education and data analysis. *Palaeon. Electr.* **2001**, *4*, 9.
43. Novara, A.; Cerdà, A.; Gristina, L. Sustainable vineyard floor management: An equilibrium between water consumption and soil conservation. *Environ. Sci. Health* **2018**, *5*, 33–37. [CrossRef]
44. Chou, M.Y.; Vanden Heuvel, J.; Bell, T.H.; Panke-Buisse, K.; Kao-Kniffin, J. Vineyard under—vine floor management alters soil microbial composition while the fruit microbiome shows no corresponding shifts. *Sci. Rep.* **2018**, *8*, 11039. [CrossRef]
45. Gay, S.H.; Louwagie, G.; Sammeth, F.; Ratinger, T.; Cristoiu, A.; Marechal, B.; Prosperi, P.; Rusco, E.; Terres, J.; Adhikari, K.; et al. Addressing soil degradation in EU agriculture: Relevant processes practices and policies. In *Technical Report JSC on the Project 'Sustainable Agriculture and Soil Conservation (SoCo)*; (JRC Working Papers JRC50424); Louwagie, G., Gay, S.H., Burrell, A., Eds.; European Commission: Luxembourg, 2009. [CrossRef]
46. Wong, V.N.L.; Dalal, R.C.; Greene, R.S.B. Carbon dynamics of sodic and saline soils following gypsum and organic material additions: A laboratory incubation. *Appl. Soil Ecol.* **2009**, *41*, 29–40. [CrossRef]
47. Proffitt, T.; Bramley, R.; Lamb, D.; Winter, E. *Precision Viticulture–A New Era in Vineyard Management and Wine Production*; Winetitles Pty Ltd.: Ashford, Australia, 2006; pp. 1–90.

48. Capello, G.; Biddoccu, M.; Ferraris, S.; Cavallo, E. Effects of Tractor Passes on Hydrological and Soil Erosion Processes in Tilled and Grassed Vineyards. *Water* **2019**, *11*, 2118. [CrossRef]
49. Vignozzi, N.; Agnelli, A.E.; Brandi, G.; Gagnarli, E.; Lagomarsino, A.; Pellegrini, S.; Simoncini, S.; Simoni, S.; Valboa, G.; Caruso, G.; et al. Soil ecosystem functions in a high—density olive orchard managed by different soil conservation practices. *Appl. Soil Ecol.* **2019**, *134*, 64–76. [CrossRef]
50. Taylor, A.R.; Pflug, A.; Schroeter, D.; Wolter, V. Impact of micro-arthropod biomass on the composition of the soil fauna community and ecosystem processes. *Eur. J. Soil Biol.* **2010**, *46*, 80–86. [CrossRef]
51. Rana, N.; Rana, S.A.; Khan, H.A.; Sohail, M.J.I. Assessment of handicaps owing to high input (hip) farming on the soil macro—invertebrates diversity in sugarcane field. *Pak. J. Agric. Sci.* **2010**, *47*, 271–278.
52. Wolters, V.; Silver, W.L.; Bignell, D.E.; Coleman, D.C.; Lavelle, P.; van der Putten, W.H.; de Ruiter, P.; Rusek, J.; Wall, D.H.; Wardle, D.A.; et al. Effects of global changes on above—And belowground biodiversity in terrestrial ecosystems: Implications for ecosystem functioning. *BioScience* **2000**, *50*, 1089–1098. [CrossRef]
53. Cheeke, T.E.; Cruzan, M.B.; Rosenstiel, T.N. A field evaluation of arbuscular mycorrhizal fungal colonization in multiple lines of Bt and non-Bt maize. *App. Environ. Microbiol.* **2013**, *79*, 4078–4086. [CrossRef]
54. Pizzigallo, A.; Granai, C.; Borsa, S. The joint use of LCA and emergy evaluation for the analysis of two Italian wine farms. *J. Environ. Manag.* **2008**, *86*, 396–406. [CrossRef] [PubMed]
55. Heller, M. Food Product Environmental Footprint Literature Summary: Wine. In *Monographic Report by: Center for Sustainable Systems University of Michigan*; State of Oregon, Dept Environmental Quality: Portland, Oregon, USA, 2017; pp. 1–17.
56. Hole, D.G.; Perkins, A.J.; Wilson, J.D.; Alexander, I.H.; Grice, P.V.; Evans, A.D. Does organic farming benefit biodiversity? *Biol. Conserv.* **2005**, *122*, 113–130. [CrossRef]
57. Ruiz, N.; Lavelle, P.; Jimenez, J. *Soil Macrofauna Field Manual: Technical Level*; Food and Agriculture Organization of The United Nations (FAO): Rome, Italy, 2008; p. 100.
58. Prosdocimi, M.; Jordán, A.; Tarolli, P.; Keesstra, S.; Novara, A.; Cerdà, A. The Immediate Effectiveness of Barley Straw Mulch in Reducing Soil Erodibility and Surface Runoff Generation in Mediterranean Vineyards. *Sci. Total Environ.* **2016**, *547*, 323–330. [CrossRef] [PubMed]
59. Rawson, G.A. The Influence of Geology and Soil Characteristics on the Fruit Composition of Winegrape (Vitis Vinifera cv. Shiraz) Hunter Valley New South Wales: Implications for Regionality in the Australian Wine Industry. Ph.D. Thesis, University of Newcastle, Callaghan, Australia, 2002.
60. Postma-Blaauw, M.B.; Goede, R.G.M.; Bloem, J.; Faber, J.H.; Brussaard, L. Soil biota community structure and abundance under agricultural intensification and extensification. *Ecology* **2010**, *91*, 460–473. [CrossRef] [PubMed]
61. Decaëns, T.; Jiménez, J.J.; Gioia, C.; Measey, G.J.; Lavelle, P. The values of soil animals for conservation biology. *Eur. J. Soil Biol.* **2006**, *42*, S23–S38. [CrossRef]
62. Maraun, M.; Scheu, S. The structure of oribatid mite communities (Acari Oribatida): Patterns mechanisms and implications for future research. *Ecography* **2000**, *23*, 374–382. [CrossRef]
63. Menta, C. Soil Fauna Diversity–Function Soil Degradation Biological Indices Soil Restoration. In *Agricultural and Biological Sciences "Biodiversity Conservation and Utilization in a Diverse World"*; Lameed, G.A., Ed.; InTech: Rijeka, Croatia, 2012. Available online: https://www.intechopen.com/books/biodiversity-conservation-and-utilization-in-a-diverse-world/soil-fauna-diversity-function-soil-degradation-biological-indices-soil-restoration (accessed on 7 March 2021).
64. Menta, C.; Leoni, A.; Gardi, C.; Conti, F.D. Are grasslands important habitats for soil micro-arthropod conservation? *Biodivers. Conserv.* **2011**, *20*, 1073–1087. [CrossRef]
65. Gope, R.; Ray, D.C. Ecological studies on soil microarthropods in Banana (*Musa* sp.) Plantation of Cachar district (Assam). *Indian J. Environ. Ecoplan.* **2006**, *12*, 105–109.
66. Miura, F.; Nakamoto, T.; Kaneda, S.; Okano, S.; Nakajima, M.; Murakami, T. Dynamics of soil biota at different depths under two contrasting tillage practices. *Soil Biol. Biochem.* **2008**, *40*, 406–414. [CrossRef]

land

MDPI

Article

Responses of Soil Infiltration to Water Retention Characteristics, Initial Conditions, and Boundary Conditions

Lesheng An [1], Kaihua Liao [2,*] and Chun Liu [1]

[1] School of Resources and Environment, Anqing Normal University, Anqing 246133, China; als00316@163.com (L.A.); yixiang0302@126.com (C.L.)
[2] Key Laboratory of Watershed Geographic Sciences, Nanjing Institute of Geography and Limnology, Chinese Academy of Sciences, Nanjing 210008, China
* Correspondence: khliao@niglas.ac.cn; Tel.: +86-25-86882139

Abstract: (1) Background: Simulation of soil water infiltration process and analysis of its influencing factors are important for water resources management. (2) Methods: In this study, the relative contributions of the soil water retention characteristics (SWRC) estimation, initial water content, and constant pressure head at upper boundary to the cumulative infiltration under various soil conditions were quantified based on the 1-D Richards' equation and 900 scenarios. Scenario simulations were performed for two SWRC estimation methods (Jensen method and Rosetta); three different initial water contents (0.15, 0.20, and 0.25 cm^3/cm^3); five different constant pressure heads (0.5, 1, 2, 4, and 8 cm); and thirty soil samples with varying texture and bulk density. (3) Results: Rosetta representing the drying branch of the SWRC yielded higher simulated cumulative infiltration compared with the Jensen method representing the wetting branch of the SWRC. However, the Jensen method–predicted cumulative infiltration fluxes matched well with the measured values with a low RMSE of 0.80 cm. (4) Conclusions: The relative contribution of the SWRC estimation method to cumulative infiltration (19.1–72.2%) was compared to that of constant pressure head (14.0–65.5%), and generally greater than that of initial water content (2.2–29.9%). Findings of this study have practical significance for investigating the transport of water, nutrients, and contaminants in the unsaturated zone.

Keywords: soil infiltration; Jensen method; Rosetta; 1-D Richards' equation

check for updates

Citation: An, L.; Liao, K.; Liu, C. Responses of Soil Infiltration to Water Retention Characteristics, Initial Conditions, and Boundary Conditions. *Land* **2021**, *10*, 361. https://doi.org/10.3390/land10040361

Academic Editor: Chiara Piccini

Received: 27 February 2021
Accepted: 23 March 2021
Published: 1 April 2021

1. Introduction

Knowledge of the soil water infiltration process is important for the management of water resources across spatio-temporal scale [1]. For example, urban development causes water losses. Some of them can have a substantial effect on the catchment, including soil drought. To estimate water losses due to urbanization, we need to know the process of water infiltration [2–4]. In addition, water infiltration has a substantial influence on availability of water and nutrients for plants, microbial activity, and chemical weathering [5,6]. In the 20th century, numerous models were established to study the process of soil infiltration, such as Richards' equation [7], the Green–Ampt model [8], the Philip model, and the Horton model [9,10]. Richards' equation had rigorous physical basis since this model was derived based on the Darcy–Buckingham Law and mass conservation for water movement in unsaturated soils [11]. Therefore, it has been often applied as a reference to test the accuracy of the other infiltration models [12–14].

Analytical solutions of Richards' equation require the soil water retention curve (SWRC), which relates pressure head and soil water content. Soil water infiltration has been widely reported to be influenced by the SWRC, which is closely related to basic soil properties (e.g., texture and bulk density) [15]. However, direct measurement of the SWRC is time consuming and laborious [16]. Pedotransfer functions (PTFs) (e.g., Rosetta) were often used to predict the SWRC from easily measurable soil properties [17]. For example, Liao et al. [18] applied PTFs to assess the SWRCs and their spatial variability in

Qingdao City, China. Minasny et al. [19] used artificial neural networks (ANNs) coupled with bootstrap aggregation to predict the SWRC and hydraulic conductivities. However, the majority of PTFs were developed to estimate the drying branch of the SWRC and neglected the wetting branch of the curve. The hysteresis phenomenon was always found when measuring the SWRC in laboratory [20–22]. Recently, Jensen et al. [23] proposed a theoretical approach to predict the drying branch of the SWRC from soil texture.

Richards' equation can be solved analytically with simple geometric condition as well as initial and boundary conditions [24,25]. Previous studies have found that soil infiltration was influenced by the initial water content. For example, Hino et al. [26] indicated that the loss of infiltrated rainfall was significantly correlated with the initial water content when it does not exceed the infiltration rate of the soil. Leuther et al. [27] found that the infiltration front stability is dependent on the initial water content of the soil from two orchards in Israel. In addition, the boundary condition was also found to affect the water infiltration of soil in previous studies. As reported by Feng et al. [28], the higher water ponding depth at the soil surface induced a monotonic increase in infiltration rate for a water-repellent sand. A similar result was also observed by Hsu et al. [29] for prewetted sand columns. In this case, the relative impact of the SWRC estimation on the soil infiltration may rely on the settings of the initial and boundary conditions. Previous studies mostly considered only one or two factors affecting soil infiltration (e.g., Hsu et al. [29], Lassabatere et al. [30], and Bughici and Wallach [31]). Only a few studies assessed the coupling effects of multiple factors on soil infiltration. For example, Gong et al. [15] evaluated the coupling effects of surface charges, adsorbed counterions, and grain-size distribution on soil water infiltration. However, the response of soil infiltration to the interactions among the SWRC estimation, initial water content, and boundary condition has been rarely investigated.

The objectives of this study were to (i) compare the Jensen method– and Rosetta-predicted SWRC for simulating soil infiltration process based on an indoor downward water infiltration experiment in a soil column and (ii) quantify the relative contributions of the SWRC estimation, initial water content, and boundary condition to soil infiltration based on scenario analysis, with consideration of various soil conditions.

2. Materials and Methods

2.1. Soil Data Resources

Thirty soil samples were selected from the Unsaturated Soil Hydraulic Database (UNSODA) database [32] (Table 1). The soil particle-size distribution, organic matter content (OMC), and bulk density (BD) of the 30 soils were provided. The modified logistic growth (MLG) model was applied to obtain the full description of soil particle-size distribution [33]:

$$W = 1/(1 + a * \exp(-b * D^c)), \quad (1)$$

where W is the cumulative weight percentage (%) corresponding to particle diameter D (μm), and a, b and c are empirical parameters. Figure 1 shows that the predicted cumulative grain-size distribution by using the MLG model matched well with the measured one. The clay (<0.002 mm), silt (0.002–0.05 mm), fine sand (0.05–0.5 mm), and coarse sand contents (0.5–2 mm) of each soil sample were then determined. From Table 1, the BD, OMC, coarse sand, fine sand, silt, and clay contents ranged between 0.72–1.81 g/cm^3, 0.00–5.60%, 0.00–80.00%, 4.40–100.00%, 0.00–55.91%, and 0.00–62.00%, respectively, indicating a large discrepancy among different samples. In addition, it was also found that the content of organic matter in soil with low BD was generally higher [34].

Table 1. Basic soil properties of the 30 soil samples selected from the Unsaturated Soil Hydraulic Database UNSODA.

Soil Code	Bulk Density (g/cm^3)	Organic Matter (%)	Coarse Sand (%)	Fine Sand (%)	Silt (%)	Clay (%)
SC1010	1.64	0.01	7.90	75.10	14.00	3.00
SC1011	1.52	-	7.30	75.20	14.50	3.00
SC1012	1.40	-	9.50	73.00	15.00	2.50
SC1013	1.49	-	8.00	77.00	13.00	2.00
SC1014	1.53	-	8.40	78.60	11.00	2.00
SC1015	1.72	-	8.30	73.70	12.00	6.00
SC1020	1.61	-	80.00	10.00	5.50	4.50
SC1021	1.58	-	72.00	17.00	5.00	6.00
SC1022	1.60	-	76.30	13.00	6.40	4.30
SC1023	1.67	-	78.40	15.60	4.00	2.00
SC1024	1.68	-	72.30	22.70	3.00	2.00
SC1030	1.48	1.70	8.60	70.40	13.70	7.30
SC1031	1.48	0.20	9.30	69.70	13.20	7.80
SC1032	1.53	0.10	11.00	69.00	12.20	7.80
SC1041	1.51	0.78	7.50	85.50	5.00	2.00
SC1300	1.26	-	0.03	38.81	29.77	31.40
SC1301	1.27	-	0.03	38.81	29.77	31.40
SC1310	1.60	-	0.00	95.60	2.40	2.00
SC1410	1.41	-	0.00	100.00	0.00	0.00
SC2020	0.72	5.60	0.00	4.40	33.60	62.00
SC2021	0.89	5.60	0.00	4.40	33.60	62.00
SC2022	0.75	5.60	0.00	4.40	33.60	62.00
SC2310	1.71	-	9.99	89.29	0.72	0.01
SC3340	1.41	0.89	16.01	81.81	2.17	0.02
SC4690	1.32	-	0.06	23.31	54.83	21.80
SC4700	1.28	-	0.08	11.21	55.91	32.79
SC4710	1.28	-	0.01	50.69	37.79	11.51
SC4720	1.48	-	8.90	82.61	8.49	0.00
SC4940	1.76	-	38.02	35.11	0.57	26.30
SC4941	1.81	-	40.90	31.13	0.27	27.70

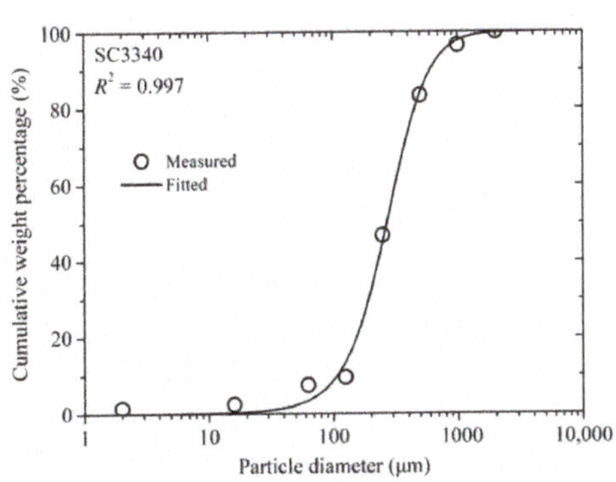

Figure 1. Measured and predicted cumulative grain-size distribution curve for soil sample with a code of 3340 (SC3340) by using the modified logistic growth model.

2.2. The Jensen Method

Jensen et al. [23] developed a theoretical approach to predict the drying branch of the SWRC from texture by scaling volumetric particle fractions to pore volume fractions. The reason is that there is a similar shape between the particle-size distribution and the SWRC [35]. They proposed five functions for water contents at pF 1.7, 2.0, 2.2, 2.7, and 3.0.

Recently, Liao et al. [36] further modified the Jensen method for estimating the wetting branch of the SWRC. For pF 1.7, 2.0, and 2.2, the functions of water contents are expressed as

$$\theta(\Psi) = \omega\, \theta_s\, (1 - (V_{CS} + \beta_1\, V_{FS})^{*}\, (V_{CS} + V_{FS})), \quad (2)$$

where ω is an empirical parameter, θ_s is the saturated water content, V_{CS} and V_{FS} are the relative volume coarse sand and fine sand, respectively, and β_1 is a parameter reflecting the water filling degree of the pores of the fine sand fraction. For pF 2.7 and 3.0, the functions of water contents are obtained:

$$\theta(\Psi) = \omega\, \theta_s\, (1 - (V_{CS} + \beta_1\, V_{FS} + \beta_2\, V_S)^{*}\, (V_{CS} + V_{FS} + V_S)), \quad (3)$$

where vs. is the relative volume silt, and β_2 is a parameter reflecting the water filling degree of the pores of the silt fraction. The ω and β_1 values ranged between 0.63–0.72 and 0.87–0.96 for the five pF values, respectively, while the β_2 value equaled to 1 at pF 2.7 and 3.0. The obtained five data pairs of pressure head and water content were then used to parameterize the van Genuchten [20] model to predict the continuous wetting branch of the SWRC. The van Genuchten model is given as

$$\theta(h) = \theta_r + (\theta_s - \theta_r)/[1 + (\alpha\, h)^{\,n}]^{\,(1 - 1/n)}, \quad (4)$$

where θ_r is the residual water content (cm^3/cm^3), and α and n are shape-defining parameters.

Rosetta implemented five hierarchical PTFs for the estimation of SWRC [17]. The hierarchical in PTFs allowed the estimation of van Genuchten model parameters using limited (textural classes only) to more extended (texture, BD, and one or two water retention points) input data. Rosetta was based on ANN analyses combined with the bootstrap method, thus allowing the program to provide uncertainty estimates of the predicted SWRC. In this study, for a comparison purpose, the soil water retention parameters were predicted with both the modified Jensen method [36] (representing the wetting branch of the SWRC) and the Rosetta software [17] (representing the drying branch of the SWRC) using particle-size distribution and bulk density.

2.3. Soil Water Infiltration Experiment

An indoor ponded infiltration experiment was performed in a single vertical column of uniformly packed loam passed through a 2 mm sieve. The soil BD, OMC, coarse sand, fine sand, silt, and clay contents were 1.40 g/cm^3, 2.04%, 0.50%, 45.11%, 44.80%, and 9.59%, respectively. θ_s was determined as 0.372 cm^3/cm^3 using the gravimetric method, while the saturated hydraulic conductivity (K_s) was measured as 0.057 cm/min in the laboratory by variable falling head method. Figure 2 shows the sketch map of the experiment equipment. The soil column had a size of 60 cm in height and 18.3 cm in width. There were gravel layer and drainage room under the soil column. After installing the soil column, the height of the Markov bottle was then fixed to ensure ponded infiltration under the constant pressure head of 2 cm. The Markov bottle readings were recorded in the process of soil infiltration. The experiment was over when the water was overflowed from the drainage. Due to the short duration of the experiment, the effect of evaporation on soil infiltration can be neglected.

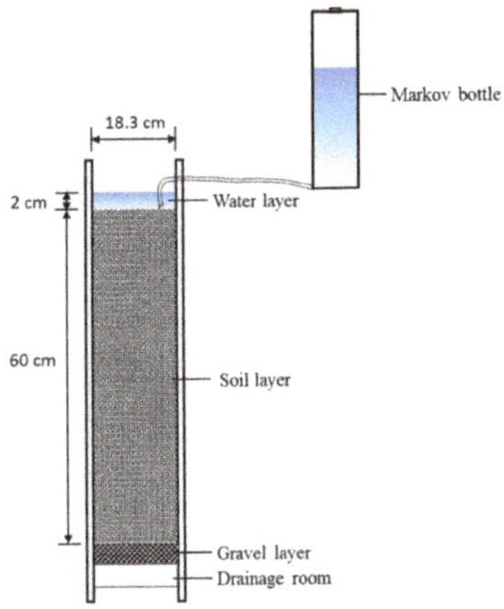

Figure 2. Sketch map of the soil water infiltration experiment equipment.

2.4. Numerical Method and Scenario Simulation

The 1-D Richards' equation was employed to simulate the soil infiltration process and can be formulated as

$$\partial\theta/\partial t = \partial/\partial z\,[K\,(\partial h/\partial z + 1)], \quad (5)$$

where t is time, z is the spatial coordinate, and K is the unsaturated hydraulic conductivity, which can be determined by the Mualem [37] model. The solution of Equation (5) requires knowledge of the initial soil profile water distribution:

$$\theta(z, t) = 0.08\ \mathrm{cm}^3/\mathrm{cm}^3\ t = 0, \quad (6)$$

The upper boundary condition is defined by the constant pressure head at the soil surface:

$$h(z, t) = 2\ \mathrm{cm}\ z = 0, t > 0, \quad (7)$$

While the lower boundary condition was defined as the free drainage at the depth of 60 cm:

$$\partial h/\partial z = 0\ z = 60\ \mathrm{cm}, t > 0, \quad (8)$$

The 1-D Richards' equation was solved by the Galerkin finite element method. In order to assess the coupling effects of the SWRC estimation, initial water content and upper boundary condition on soil infiltration with consideration of various soil conditions, a total of 900 scenarios were established: (i) 30 soil samples from the UNSODA database representing different soil conditions; (ii) two SWRC estimations (using the Jensen method and Rosetta software [17]) reflecting hysteresis impacts on soil infiltration; (iii) three initial water contents which are 0.15, 0.20, and 0.25 $\mathrm{cm}^3/\mathrm{cm}^3$, indicating the relative dry, intermediate and relatively wet conditions, respectively; and (IV) five constant pressure head values (0.5, 1, 2, 4, and 8 cm) at the upper boundary. For each soil sample, the K_s value was estimated with Rosetta using soil texture and bulk density. All scenario simulations were run with the same lower boundary condition (free drainage).

2.5. Evaluation Criteria and Contribution Rate Analysis

The performance of the model was evaluated by the coefficient of determination (R^2), the root mean squared error (RMSE). A good model will have a high R^2 and low RMSE. The multiple regression method was applied to quantify the relative contribution rates of different influencing factors to cumulative infiltration for each soil sample. The standardized regression equation can be expressed as

$$Y_{CI} = a_1 X_1 + a_2 X_2 + a_3 X_3, \tag{9}$$

where Y_{CI} is the standardized cumulative infiltration, a_1, a_2, and a_3 are the standardized regression coefficients, and X_1, X_2, and X_3 are the standardized values of the independent variables. The classical approach of the dummy variables was used for qualitative variable, i.e., the SWRC estimation method (1 for Jensen method and 0 for Rosetta software). The relative contribution rate of each factor to cumulative infiltration (CR_i) can be calculated as

$$CR_i = |a_i|/(|a_1| + |a_2| + |a_3|) \tag{10}$$

3. Results and Discussion

3.1. Test of the Jensen Method and Rosetta Software for Simulating Soil Water Infiltration

The Jensen method and Rosetta software were tested for simulating the indoor soil infiltration process. The θ_s, θ_r, α, and n values of the soil column obtained by the Jensen method were 0.372 cm^3/cm^3, 0, 0.0408 cm^{-1}, and 1.430, respectively, while those values predicted with Rosetta were 0.370 cm^3/cm^3, 0.041 cm^3/cm^3, 0.0109 cm^{-1}, and 1.517, respectively. As expected, the Jensen method–predicted α was larger than the Rosetta-estimated value. Figure 3a shows the predicted SWRCs of the soil column by using the Jensen method and Rosetta software. A substantial difference was found between the two SWRCs, showing that the Jensen method produced lower water contents than the Rosetta at the same pressure heads. In this case, the Rosetta produced higher simulated cumulative infiltration than the Jensen method using the 1-D Richards' equation (Figure 3b). The accuracy of the Rosetta was relatively low with a relatively high RMSE of 2.49 cm. However, the Jensen method–predicted values matched well with the measured ones with a low RMSE of 0.80 cm. The positions of the wetting front at different time for the Jensen method and Rosetta software are shown in Figure 4. The wetting front obtained by the Jensen method reached the depths of 18.0, 27.6, and 40.2 cm at time $t = 50$, 100, and 150 min, respectively, whereas those predicted with Rosetta reached the depths of 28.2, 41.4, and 52.8 cm, respectively. As expected, the simulated wetting front by the Rosetta software advanced more rapidly than that by the Jensen method during the infiltration process.

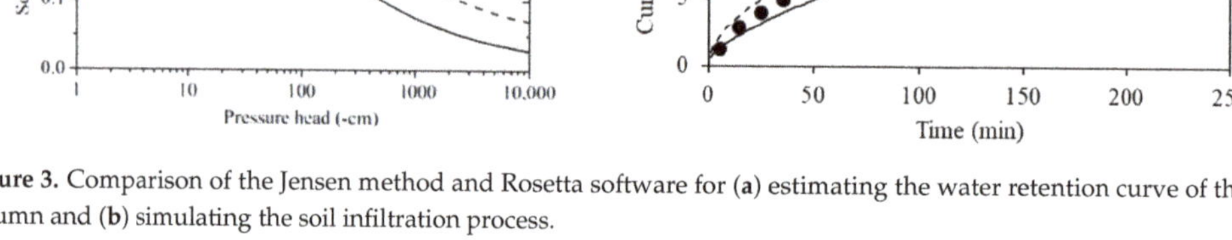

Figure 3. Comparison of the Jensen method and Rosetta software for (**a**) estimating the water retention curve of the soil column and (**b**) simulating the soil infiltration process.

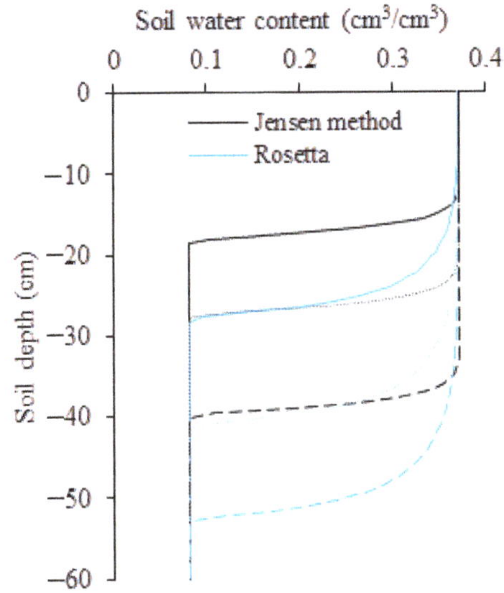

Figure 4. Comparison of the wetting front simulated by the Jensen method and Rosetta software at t = 50, 100, and 150 min after the infiltration.

Previous studies have been conducted to assess the hysteresis effect of SWRC on the water infiltration of various textured soils. Huang et al. [38] indicated that the simulation of the coarse-textured profile water content was improved during the infiltration process when hysteresis was taken into account. In the study by Abbasi et al. [39], a large difference was found between the measured fluxes and non-hysteresis-models-predicted values during the infiltration processes of the loamy and sandy loam soil. In our study, it is suggested that the Rosetta estimating the drying SWRC has to be used with caution when applying this method to simulate the infiltration process. In contrast, the Jensen method representing the wetting SWRC was demonstrated to be capable of predicting the water infiltration of the loam soil. However, this method was still needed to be validated for different textured soils.

3.2. Coupling Influences of the SWRC Estimation, Initial Water Content, and Upper Boundary Condition

Solving 1-D Richards' equation by different scenarios, 900 cumulative soil infiltration fluxes were obtained. For each soil sample, the average infiltration fluxes decreased as the increase in the initial water content (Figure 5). For example, the average of infiltration fluxes of SC1300 obtained by the Jensen method decreased from 5.62 cm to 5.20 cm as the initial water content increasing from 0.15 cm^3/cm^3 to 0.25 cm^3/cm^3. The relative contributions of various factors to cumulative soil infiltration were quantified by the regression method (Figure 6). Results indicated that initial water content had a generally lesser contribution (2.2–29.9%) to cumulative infiltration compared to the SWRC estimation method (19.1–72.2%) and constant pressure head (14.0–65.5%).

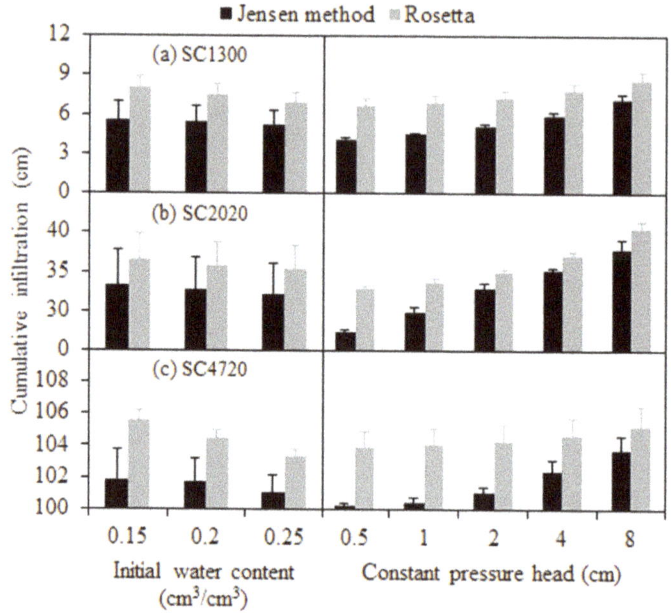

Figure 5. Average cumulative infiltration fluxes and their standard deviation (error bars) under three initial water contents and five constant pressure heads at upper boundary for soil samples with codes of (**a**) 1300, (**b**) 2020 and (**c**) 4720.

From Figure 5, the average infiltration fluxes decreased as increasing initial water content. This finding agreed with classical theory of infiltration that predicted cumulative infiltration to decrease as increases of initial water content [9]. For any SWRC prediction method, an increase in constant pressure head resulted in an increase in infiltration fluxes at any point in time. Therefore, the relationship between constant pressure head and cumulative infiltration was independent of initial water content. The above results are consistent with many previous studies [29,40,41]. In addition, SC4720 with a low OMC and clay content produced higher cumulative infiltration fluxes than SC1300 and SC2020 with a high OMC and clay content at the same initial water content and constant pressure head (Figure 5). Previous studies also observed that various soil conditions (e.g., texture, structure, and OMC) largely influenced soil infiltration and redistribution processes [42,43]. In the study by Franzluebbers [44], the stratification ratio of soil OMC (OMC at 0–3 cm depth divided by that at 6–12 cm depth) was found to control the water infiltration rate. Zhao et al. [42] indicated that a sandy loam soil can produce 23% higher infiltration than the clay and silt loam soils. Therefore, soil infiltration was synthetically affected by the static soil properties and dynamic factors (initial water content and constant pressure head). In previous studies, soil infiltration had also been demonstrated to be affected by soil properties, initial water content, and constant pressure head. For example, in the study by Camps-Roach et al. [45], the dynamic effect of the capillary pressure on water infiltration was related to the soil parameters (e.g., grain-size distributions). Hsu et al. [29] reported that the magnitude of the dynamic effect of capillary pressure depended primarily on the initial moisture content, not the constant pressure head. The results suggested that the effects of different factors on soil infiltration were dynamic and intertwined.

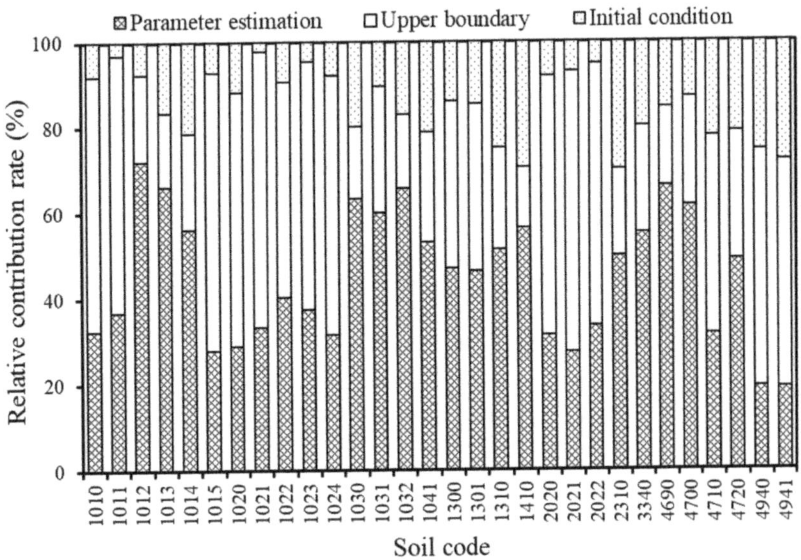

Figure 6. Relative contribution rates of different factors (parameter estimation methods, initial water contents, and constant pressure heads at upper boundary) to cumulative infiltration fluxes for the 30 soils investigated.

Influences of the SWRC estimation method, initial water content, and constant pressure head on the cumulative infiltration substantially varied under different soil conditions. Figure 7 shows the relationships between the relative contribution of the parameter estimation to soil infiltration and basic soil properties for the 30 soil samples. There was a bell-shape relationship between the relative contribution of the parameter estimation and BD, showing that the BD of 1.40 g/cm^3 corresponded to the maximum relative contribution rate. This is related to the similar relationship between the SMBE and BD. However, the relative contribution of the parameter estimation was negatively and positively correlated with the clay and fine sand contents at the 0.01 level of significance, respectively.

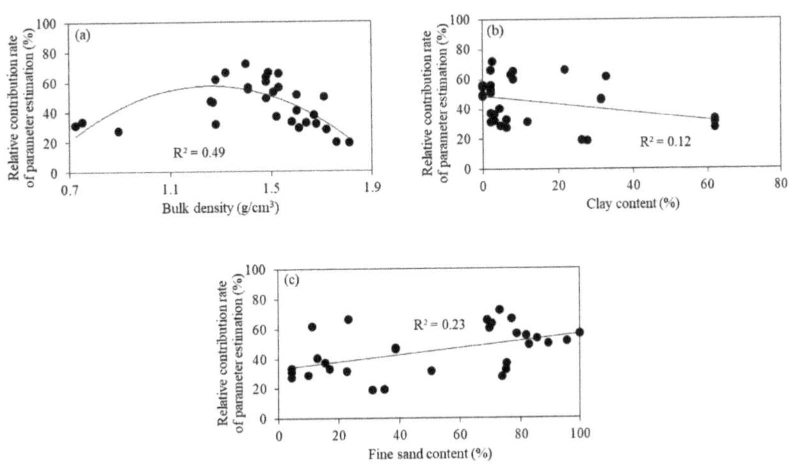

Figure 7. Relationships between the relative contribution of parameter estimation to cumulative infiltration fluxes and (**a**) bulk density, (**b**) clay, and (**c**) fine sand contents.

The relative contribution of the parameter estimation gradually increased with the soil texture from fine to coarse, yet the relative contribution of the initial and boundary conditions gradually decreased (figures not shown). The reason may be that there was a lager difference between the drying SWRC and wetting SWRC (hysteresis effect) for the coarser-textured soils than for the finer-textured soils [46]. Elmaloglou and Diamantopoulos [47] also found that the hysteresis decreased the soil leakage more in the loamy sand (6.4–10.3%) than in the silt loam (3.6–6.4%). Besides the SWRC estimation method, initial water content, and constant pressure head considered in this study, the vertical heterogeneity of soil texture greatly influenced soil infiltration as reported in the study of Zhu and Mohanty [48]. Huang et al. [38] found that the vertical change of soil texture led to the increase of water storage compared with homogeneous soil profile with similar texture. In addition, climate change and tillage conversion also affected the soil infiltration. For example, air temperature changed the soil evapotranspiration rate and soil moisture content, which in turn influenced soil infiltration [49]. Conversion from conventional tillage to conservation tillage has been demonstrated to increase saturated hydraulic conductivity and water infiltration due to greater soil organic carbon accumulation over time in conservation tillage [50,51]. Therefore, to comprehensively detect the mechanisms of the water infiltration, investigations on the coupling effects of multiple factors (e.g., vertical change of soil texture and hydraulic properties, climate change, and tillage conversion) on soil infiltration under different land use types (e.g., farmland and garden) are still needed for future research.

4. Conclusions

This study compared the derived theoretical functions and Rosetta software for estimating the wetting SWRC of the 30 soil samples from UNSODA and for simulating the infiltration process of a loam soil using a downward infiltration experiment. The Jensen method was found to have better performance in estimating the wetting SWRC and simulating the cumulative infiltration than the Rosetta software in terms of RMSE. The 1-D Richards' equation and scenario simulation were used to quantify the response of cumulative infiltration to the coupling interactions among SWRC estimation method, initial water content and constant pressure head.

As expected, an increase in constant pressure head resulted in an increase in the cumulative infiltration. The cumulative infiltration decreased as the initial water content increased. The SWRC estimation method and constant pressure head had a generally greater contribution to cumulative infiltration than initial water content. The relative contribution of the SWRC estimation to infiltration had a bell-shape relationship with the BD and a significant ($p < 0.01$) linear relationship with the clay and fine sand contents. This indicated that influences of the SWRC estimation method on the cumulative infiltration largely depended on the soil condition. The above results suggest that more efforts need to be made to obtain a high-precision SWRC and accurately describe the upper boundary conditions in the simulation of soil infiltration process.

Author Contributions: Conceptualization, L.A. and K.L.; methodology, L.A. and K.L.; software, L.A., C.L. and K.L.; formal analysis, L.A. and K.L.; data curation, L.A. and K.L.; writing—original draft preparation, L.A. and K.L.; writing—review and editing, L.A., C.L. and K.L.; funding acquisition, L.A. and K.L. All authors have read and agreed to the published version of the manuscript.

Funding: This study was funded by the National Natural Science Foundation of China, grant number 41771107; the Anhui Provincial Natural Science Foundation, grant number 1808085MD101; the Outstanding Young Talents Support Program in Universities of Anhui Province in 2020, grant number gxyq2020030; and the Youth Innovation Promotion Association, Chinese Academy of Sciences, grant number 2020317.

Institutional Review Board Statement: Not applicable.

Informed Consent Statement: Not applicable.

Data Availability Statement: Not applicable.

Conflicts of Interest: The authors declare no conflict of interest.

References

1. Herrada, M.A.; Gutiérrez-Martin, A.; Montanero, J.M. Modeling infiltration rates in a saturated/unsaturated soil under the free draining condition. *J. Hydrol.* **2014**, *515*, 10–15. [CrossRef]
2. Lepeška, T.; Wojkowski, J.; Wałęga, A.; Młyński, D.; Radecki-Pawlik, A.; Olah, B. Urbanization—Its hidden impact on water losses: Prądnik River Basin, Lesser Poland. *Water* **2020**, *12*, 1958. [CrossRef]
3. Su, W.; Gu, C.; Yang, G.; Chen, S.; Zhen, F. Measuring the impact of urban sprawl on natural landscape pattern of the Western Taihu Lake watershed, China. *Landsc. Urban Plan.* **2010**, *95*, 61–67. [CrossRef]
4. Su, W.; Ma, L.; Chen, S.; Yang, G. Conflict analysis and system optimization of urban ecological space. *J. Nat. Resour.* **2020**, *35*, 601–613.
5. Castellini, M.; Stellacci, A.M.; Sisto, D.; Iovino, M. The mechanical impact of water affected the soil physical quality of a loam soil under minimum tillage and no-tillage: An assessment using Beerkan multi-height runs and BEST-procedure. *Land* **2021**, *10*, 195. [CrossRef]
6. Suprayogo, D.; van Noordwijk, M.; Hairiah, K.; Meilasari, N.; Rabbani, A.L.; Ishaq, R.M.; Widiando, W. Infiltration-friendly agroforestry land uses on Volcanic slopes in the Rejoso Watershed, East Java, Indonesia. *Land* **2020**, *9*, 240. [CrossRef]
7. Ross, P.J. Efficient numerical methods for infiltration using Richards' equation. *Water Resour. Res.* **1990**, *26*, 279–290. [CrossRef]
8. Green, W.H.; Ampt, G.A. Studies on soil physics: 1. Flow of air and water through soils. *J. Agric. Sci.* **1911**, *4*, 1–24.
9. Philip, J.R. The theory of infiltration: 1, The infiltration equation and its solution. *Soil Sci.* **1957**, *83*, 345–357. [CrossRef]
10. Horton, R.E. The role of infiltration in the hydrologic cycle. *Trans. Am. Geophys. Union* **1933**, *14*, 446–460. [CrossRef]
11. Botros, F.E.; Onsoy, Y.S.; Ginn, T.R.; Harter, T. Richards equation-based modeling to estimate flow and nitrate transport in a deep Alluvial vadose zone. *Vadose Zone J.* **2012**, 11. [CrossRef]
12. Hsu, S.M.; Ni, C.F.; Hung, P.F. Assessment of three infiltration formulas based on model fitting and Richards equation. *J. Hydrol. Eng.* **2002**, *7*, 373–379. [CrossRef]
13. Smith, R.E. *Infiltration Theory for Hydrologic Application*; American Geophysical Union: Washington, DC, USA, 2002; Volume 15.
14. Su, N. Theory of infiltration: Infiltration into swelling soils in a material coordinate. *J. Hydrol.* **2010**, *395*, 103–108. [CrossRef]
15. Gong, Y.; Tian, R.; Li, H. Coupling effects of surface charges, adsorbed counterions and particle-size distribution on soil water infiltration and transport. *Eur. J. Soil Sci.* **2018**, *69*, 1008–1017. [CrossRef]
16. Liao, K.; Zhou, Z.; Li, Y.; Lai, X.; Zhu, Q.; Shan, N. Comparison of seven water retention functions used for modelling soil hydraulic conductivity due to film flow. *Soil Use Manag.* **2018**, *34*, 370–379. [CrossRef]
17. Schaap, M.G.; Leij, F.J.; van Genuchten, M.T. ROSETTA: A computer program for estimating soil hydraulic parameters with hierarchical pedotransfer functions. *J. Hydrol.* **2001**, *251*, 163–176. [CrossRef]
18. Liao, K.; Xu, S.; Wu, J.; Ji, S.; Lin, Q. Assessing soil water retention characteristics and their spatial variability using pedotransfer functions. *Pedosphere* **2011**, *21*, 413–422. [CrossRef]
19. Minasny, B.; Hopmans, J.W.; Harter, T.; Eching, S.O.; Tuli, A.; Denton, M.A. Neural networks prediction of soil hydraulic functions for alluvial soils using multistep outflow data. *Soil Sci. Soc. Am. J.* **2004**, *68*, 417–429. [CrossRef]
20. Van Genuchten, M.T. A closed-form equation for predicting the hydraulic conductivity of unsaturated soils. *Soil Sci. Soc. Am. J.* **1980**, *44*, 892–898. [CrossRef]
21. Braddock, R.D.; Parlange, J.Y.; Lee, H. Application of a soil water hysteresis model to simple water retention curves. *Transp. Porous Med.* **2001**, *44*, 407–420. [CrossRef]
22. Lourenço, S.D.N.; Jones, N.; Morley, C.; Doerr, S.H.; Bryant, R. Hysteresis in the soil water retention of a sand–clay mixture with contact angles lower than ninety degrees. *Vadose Zone J.* **2015**, *14*, 1–8. [CrossRef]
23. Jensen, D.K.; Tuller, M.; de Jonge, L.W.; Arthur, E.; Moldrup, P. A new two-stage approach to predicting the soil water characteristic from saturation to oven-dryness. *J. Hydrol.* **2015**, *521*, 498–507. [CrossRef]
24. Tracy, F.T. Clean two and three-dimensional analytical solutions of Richards' equation for testing numerical solvers. *Water Resour. Res.* **2006**, *42*, 1–11. [CrossRef]
25. Chen, X.; Liang, X.; Xia, J.; She, D. Impact of lower boundary condition of Richards' equation on water, energy, and soil carbon based on coupling land surface and biogeochemical models. *Pedosphere* **2018**, *28*, 497–510. [CrossRef]
26. Hino, M.; Odaka, Y.; Nadaoka, K.; Sato, A. Effect of initial soil moisture content on the vertical infiltration process—A guide to the problem of runoff-ratio and loss. *J. Hydrol.* **1988**, *102*, 267–284. [CrossRef]
27. Leuther, F.; Weller, U.; Wallach, R.; Vogel, H.J. Quantitative analysis of wetting front instabilities in soil caused by treated waste water irrigation. *Geoderma* **2018**, *319*, 132–141. [CrossRef]
28. Feng, G.L.; Letey, J.; Wu, L. Water ponding depths affect temporal infiltration rates in a water-repellent sand. *Soil Sci. Soc. Am. J.* **2001**, *65*, 315–320. [CrossRef]
29. Hsu, S.Y.; Huang, V.; Park, S.W.; Hilpert, M. Water infiltration into prewetted porous media: Dynamic capillary pressure and Green-Ampt modeling. *Adv. Water Resour.* **2017**, *106*, 60–67. [CrossRef]
30. Lassabatere, L.; Loizeau, S.; Angulo-Jaramillo, R.; Winiarski, T.; Rossier, Y.; Delolme, C.; Gaudet, J.P. Influence of the initial soil water content on Beerkan water infiltration experiments. *Geophys. Res. Abstr.* **2012**, *14*, 2278.

31. Bughici, T.; Wallach, R. Formation of soil–water repellency in olive orchards and its influence on infiltration pattern. *Geoderma* **2016**, *262*, 1–11. [CrossRef]
32. Nemes, A.; Schaap, M.G.; Leij, F.J.; Wösten, J.H.M. Description of the unsaturated soil hydraulic database UNSODA version 2.0. *J. Hydrol.* **2001**, *251*, 151–162. [CrossRef]
33. Thornley, J.H.M.; Johnson, I.R. *Plant and Crop Modelling: A Mathematical Approach to Plant and Crop Physiology*; Clarendon: Oxford, UK, 1990.
34. Alexander, E.B. Bulk densities of California soils in relation to other soil properties. *Soil Sci. Soc. Am. J.* **1980**, *44*, 689–692. [CrossRef]
35. Arya, L.M.; Paris, J.F. A physicoempirical model to predict the soil-moisture characteristic from particle-size distribution and bulk-density data. *Soil Sci. Soc. Am. J.* **1981**, *45*, 1023–1030. [CrossRef]
36. Liao, K.; Lai, X.; Jiang, S.; Zhu, Q. Estimating the wetting branch of the soil water retention curve from grain-size fractions. *Eur. J. Soil Sci.* **2020**, 1–6. [CrossRef]
37. Mualem, Y. A new model predicting the hydraulic conductivity of unsaturated porous media. *Water Resour. Res.* **1976**, *12*, 513–522. [CrossRef]
38. Huang, M.; Barbour, S.L.; Elshorbagy, A.; Zettl, J.D.; Si, B.C. Infiltration and drainage processes in multilayered coarse soils. *Can. J. Soil Sci.* **2011**, *91*, 169–183. [CrossRef]
39. Abbasi, F.; Javaux, M.; Vanclooster, M.; Feyen, J. Estimating hysteresis in the soil water retention curve from monolith experiments. *Geoderma* **2012**, *189*, 480–490. [CrossRef]
40. Warrick, A.W.; Zerihun, D.; Sanchez, C.A.; Furman, A. Infiltration under variable ponding depths of water. *J. Irrig. Drain. Eng.* **2005**, 131. [CrossRef]
41. Zhang, G.; Feng, G.; Li, X.; Xie, C.; Pi, X. Flood effect on groundwater recharge on a typical silt loam soil. *Water* **2017**, *9*, 523. [CrossRef]
42. Zhao, L.T.; Gray, D.M.; Toth, B. Influence of soil texture on snowmelt infiltration into frozen soils. *Can. J. Soil Sci.* **2002**, *82*, 75–83. [CrossRef]
43. Lai, X.; Liao, K.; Feng, H.; Zhu, Q. Responses of soil water percolation to dynamic interactions among rainfall, antecedent moisture and season in a forest site. *J. Hydrol.* **2016**, *540*, 565–573. [CrossRef]
44. Franzluebbers, A.J. Water infiltration and soil structure related to organic matter and its stratification with depth. *Soil Tillage Res.* **2002**, *66*, 197–205. [CrossRef]
45. Camps-Roach, G.; O'Carroll, D.M.; Newson, T.A.; Sakaki, T.; Illangasekare, T.H. Experimental investigation of dynamic effects in capillary pressure: Grain size dependency and upscaling. *Water Resour. Res.* **2010**, *46*, W08544. [CrossRef]
46. Topp, G.C. Soil water hysteresis in silt loam and clay loam soils. *Water Resour. Res.* **1971**, *7*, 914–920. [CrossRef]
47. Elmaloglou, S.; Diamantopoulos, E. Effects of hysteresis on redistribution of soil moisture and deep percolation at continuous and pulse drip irrigation. *Agric. Water Manag.* **2009**, *96*, 533–538. [CrossRef]
48. Zhu, J.; Mohanty, B.P. Spatial averaging of van Genuchten hydraulic parameters for steady state flow in heterogeneous soils. *Vadose Zone J.* **2002**, *1*, 261–272. [CrossRef]
49. Pruski, F.F.; Nearing, M.A. Climate-induced changes in erosion during the 21st century for eight U.S. locations. *Water Resour. Res.* **2002**, *38*, 1298. [CrossRef]
50. Liu, C.; Lu, M.; Cui, J.; Li, B.; Fang, C. Effects of straw carbon input on carbon dynamics in agricultural soils: A meta-analysis. *Glob. Chang. Biol. Bioenergy* **2014**, *20*, 1366–1381. [CrossRef]
51. Li, Y.; Li, Z.; Cui, S.; Jagadamma, S.; Zhang, Q. Residue retention and minimum tillage improve physical environment of the soil in croplands: A global meta-analysis. *Soil Tillage Res.* **2019**, *194*, 104292. [CrossRef]

land

MDPI

Article

Response of Gross Mineralization and Nitrification Rates to Banana Cultivation Sites Converted from Natural Forest in Subtropical China

Xinghua Qin [1], Cheng Yang [2], Lin Yang [1], Erdeng Ma [3], Lei Meng [1,*] and Tongbin Zhu [4]

1 College of Tropical Crops, Hainan University, Haikou 570228, China; 13627731088@163.com (X.Q.); y523528@163.com (L.Y.)
2 Geological Survey of Jiangsu Province, Nanjing 210018, China; yc384522@163.com
3 Yunnan Academy of Tobacco Agricultural Sciences, Kunming 650021, China; maerdeng@yntsti.com
4 Key Laboratory of Karst Dynamics, MRL, Institute of Karst Geology, CAGS, Guilin 541004, China; zhutongbin@mail.cgs.gov.cn or zhutongbin@gmail.com
* Correspondence: menglei@hainanu.edu.cn or menglei94@sohu.com; Tel./Fax: +86-898-6627-9014

Abstract: Evaluations of gross mineralization (M_{Norg}) and nitrification (O_{NH4}) can be used to evaluate the supply capacity of inorganic N, which is crucial in determining appropriate N fertilizer application. However, the relevant research for banana plantations to date is limited. In this study, natural forest and banana plantations with different cultivation ages (3, 7, 10, and 22 y) were chosen in a subtropical region, and the ^{15}N dilution technique was used to determine the gross M_{Norg} and O_{NH4} rates. The objective was to evaluate the effect of the conversion of natural forests to banana plantations on inorganic N supply capacity (M_{Norg} + O_{NH4}) and other relevant factors. Compared to other natural forests in tropical and subtropical regions reported on by previous studies, the natural forest in this study was characterized by a relatively low M_{Norg} rate and a high O_{NH4} rate in the soil, resulting in the presence of inorganic N dominated by nitrate. Compared to the natural forest, 3 y banana cultivation increased the M_{Norg} and O_{NH4} rates and inorganic N availability in the soil, but these rates were significantly reduced with prolonged banana cultivation. Furthermore, the mean residence times of ammonium and nitrate were shorter in the 3 y than in the 7, 10, and 22 y banana plantations, indicating a reduced turnover of ammonium and nitrate in soil subjected to long-term banana cultivation. In addition, the conversion of natural forest to banana plantation reduced the soil organic carbon (SOC), total N and calcium concentrations, as well as water holding capacity (WHC), cation exchangeable capacity (CEC), and pH, more obviously in soils subjected to long-term banana cultivation. The M_{Norg} and O_{NH4} rates were significantly and positively related to the SOC and TN concentrations, as well as the WHC and CEC, suggesting that the decline in soil quality after long-term banana cultivation could significantly inhibit M_{Norg} and O_{NH4} rates, thus reducing inorganic N supply and turnover. Increasing the amount of soil organic matter may be an effective measure for stimulating N cycling for long-term banana cultivation.

Keywords: banana plantation; ^{15}N tracing; mineralization; nitrification; inorganic N supply and turnover

Citation: Qin, X.; Yang, C.; Yang, L.; Ma, E.; Meng, L.; Zhu, T. Response of Gross Mineralization and Nitrification Rates to Banana Cultivation Sites Converted from Natural Forest in Subtropical China. *Land* **2021**, *10*, 376. https://doi.org/10.3390/land10040376

Academic Editors: Angelinus Franke and Chiara Piccini

Received: 8 February 2021
Accepted: 31 March 2021
Published: 4 April 2021

1. Introduction

Due to the high economic benefits it offers, the banana (*Musa nana*) has been widely cultivated as a food source of regional populations in subtropical and tropical regions around the world, occupying an extremely important position in local markets [1,2]. Banana plantation area and production have increased from approximately 4.55 million ha and 67.2 million tons in 2000 to 5.16 million ha and 117 million tons in 2019, respectively [3]. In China, banana plantation area and production amounted to approximately 0.36 million ha and 12.0 million tons in 2019 [3]. To increase banana growth and yield, a suitable nitrogen (N) management strategy is required [4], since N is the main element limiting crop growth [5]. The current recommended rate of N fertilizer application for bananas is between 250 and 600 kg N ha^{-1}, in the form of split application or basal application [6,7].

In soil, inorganic N, such as ammonium (NH_4^+) and nitrate (NO_3^-), are the main N forms available for crop uptake [8,9]. In previous studies, the net transformation method for determining changes in NH_4^+ and NO_3^- concentrations has been widely used to evaluate soil N availability and its environmental effects, but this method cannot identify the production process [9,10]. In soil, inorganic N is produced mainly through the conversion of organic N to NH_4^+ (i.e., mineralization) and the subsequent oxidation of NH_4^+ to NO_3^- (i.e., nitrification) [11–13]. Thus, the determination of gross mineralization and nitrification rates using the ^{15}N dilution technique can provide a better understanding of the process and intensity of inorganic N production, which has been widely conducted in various ecosystems (e.g., forest, agriculture, grass) [14–16]. However, the relevant information regarding changes in gross mineralization and nitrification is limited for banana plantations in tropical or subtropical regions. Considering the wide distribution of banana plantations around the world, the investigation of gross mineralization and nitrification rates as a means of evaluating inorganic N supply is crucial to guide N fertilizer application.

At present, the unreasonable rates of fertilization and tillage in banana plantations lead to low fertilizer use efficiency, reductions in yield and quality, etc. [2,17]. Moreover, in the past decade, large banana plantations have been abandoned due to the outbreak of banana wilt disease [18]. Consequently, excessive logging and forest clearing for new banana plantations have been enacted in order to meet the strong market demand for bananas [19]. Previous studies have found a high N retention capacity in the highly weathered soils of natural forests in subtropical or tropical regions, which exhibit high gross mineralization rates along with low gross nitrification rates [11,13]. During the conversion of forest to farmland, an input of N, preferentially provided in inorganic form, as well as frequent irrigation and tillage can greatly change the soil properties and biochemical environment (e.g., soil organic matter, water holding capacity and pH) [20–23], which may affect mineralization and nitrification rates, and N availability in soils. At the initial stage of banana cultivation via conversion from forests, tillage can increase the soil's porosity and subsequently the O_2 diffusion into the soil, which could increase microbial abundance and activity, thus accelerating the decomposition of soil organic matter [24], suggesting a possible increase in mineralization rate. In addition, the application of organic fertilizer can increase the active organic matter sufficiently to stimulate an increase in mineralization rate [25]. On the other hand, increases in soil aeration and N fertilizer application during agricultural cultivation can stimulate increases in the abundance and activity of nitrifying microorganisms, such as ammonia-oxidizing archaea (AOA) and ammonia-oxidizing bacteria (AOB) [26,27], thereafter possibly increasing the oxidation of NH_4^+ to NO_3^-. These results indicate that the short-term conversion of natural forests to banana plantations may increase the NO_3^- production rate in soils. In soil, NH_4^+ is lost to the atmosphere through ammonia volatilization, but NO_3^- is more easily lost through leaching, runoff, and the emission of nitrogenous gases due to denitrification [28,29], especially in subtropical and tropical regions with high rainfall [30]. This may ultimately lower the sustainable supply capacity of soil inorganic N. Noticeably, this stimulating effect of mineral N fertilizer application on gross mineralization and nitrification rates may gradually decrease with the prolonged cultivation of banana, possibly due to the decline in soil quality [15,31]. For example, long-term rubber or oil cultivation sites converted from forests cause significant reductions in the organic carbon (C) and total N (TN) concentrations in soils, as well as the macro-aggregate levels, but cause increases in micro-aggregate [32,33]. Due to the decline in soil organic carbon (SOC) and macro-aggregate levels, the soil can become hard and compacted [34]. If this holds true for long-term banana cultivation as well, the rates of mineralization and nitrification and the inorganic N supply may be reduced as the reduction in substrate, the occurrence of soil compaction, and the input of organic N fertilizer prevent the effective conversion of N to a form available for plant uptake. Thus, we hypothesized that (1) short-term banana plantations converted from natural forests cause increase mineralization and nitrification rates, thus stimulating inorganic N supply, and (2) long-term banana cultivation causes significant reductions in both rates, and reduce inorganic N supply.

To verify our hypotheses, soils were sampled from natural forests and banana plantations with different ages of cultivation (3, 7, 10, and 22 y) in the subtropical region of Southwestern China. The purpose of this study was to determine the mineralization and nitrification rates in soils using the ^{15}N tracing approach, and thus to evaluate the effects of the conversion of natural forests to banana plantations on the soil's inorganic N supply capacity.

2. Material and Methods

2.1. Site Description and Sample Collection

The studied sample sites were located in Gulinqing Nature Reserve, Maguan County, Southeastern Yunnan Province, China (103°54′ E, 22°43′ N) (Figure 1). This region is characterized by a typical subtropical monsoon climate. The annual average temperature is between 18.2 and 22.2 °C, and the annual average precipitation is 1700 mm, which mainly occurs from May to October. The main tree species of the natural forest are *Dipterocarpus tonkinensis*, *Pometia tomentosa*, *Altingia yunnanensis*, *Shorea chinensis* var. *kwangsiensis*, *Burretiodendron hsienmu*, *Castanopsis fabri*, *Caryota urens*, *Lithocarpus truncatus*, *Fagus longipetiolata*, *Alnus nepalensis*, *Cunninghamia lanceolata*, *Dendrocalamus strictus*, and *Arenga pinnata*. Four banana plantations with 3-, 7-, 10-, and 22-year cultivation ages were chosen, all of which were converted from natural forests. The slope (approximately 8°) and altitude (approximately 450–650 m) were relatively consistent between the natural forest and the four banana plantations. Approximately 2400 bananas ha^{-1} were planted, and commercial organic fertilizer was applied at a rate of 36,000 kg ha^{-1} y^{-1} as the base fertilizer. Inorganic fertilizers were applied five to six times each year in a circular trench approximately 20 cm away from the banana plants. According to the field investigation, these plantations were fertilized with approximately 320–380, 220–240, and 150–530 kg ha^{-1} y^{-1} of N, phosphorus (P), and potassium (K), respectively. The organic fertilizer contained 3.1 g N kg^{-1}, 3.0 g P kg^{-1}, and 2.1 g K kg^{-1}. The soil of this area is a mixed soil deriving from carbonate rock and basalt weathering, and is classified as Latosol (US Soil Taxonomy), containing 18.5% clay, 65.6% sand, and 15.9% silt.

In July, 2020, three natural forest sites and three sites for each banana plantation type were selected as spatial replicates. The distance between each site exceeded 300 m. Five plots (about 1 × 1 m) were randomly established at intervals of 20 m for each site. Due to the high exposure rate of carbonate rock (20%), the soil layer was relatively thin (<40 cm). After removing the litter layer, the soils were sampled at the banana plantations' cultivation horizon using a hand auger (5 cm diameter) to a 0–15 cm depth, and all subsamples were mixed to form one composite sample. The sampling method for the natural forest was identical to that of the banana plantations. Fresh soil was passed through a 2 mm sieve after removing litter, plant roots, stones, and other impurities. All soils were immediately placed in a covered cooler with ice for transport to the laboratory. Subsequently, the soils were divided into two constituent parts. A portion of fresh soil was used to determine gross mineralization and nitrification rates and bacterial or archaeal *amoA* abundance, and the other portion of soil was air-dried to determine its basic physicochemical properties.

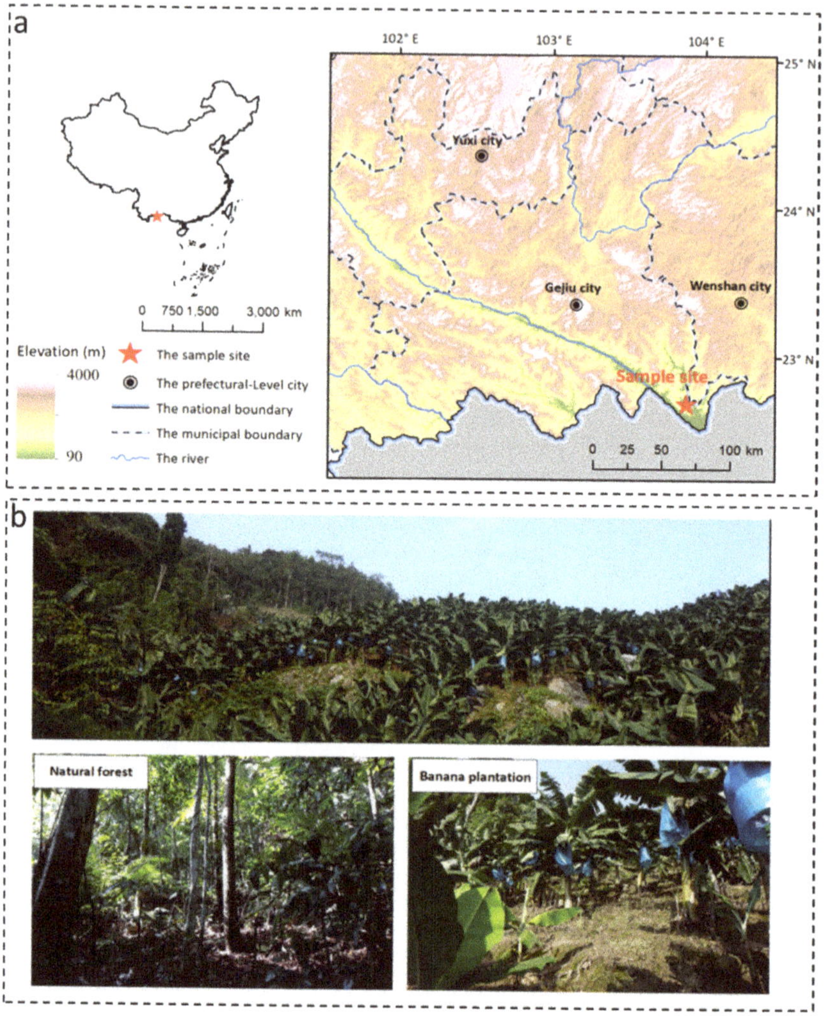

Figure 1. Location of the study site (**a**) in Gulinqing Nature Reserve, Maguan County, Yunnan Province, China, and an illustration of the land use types (**b**).

2.2. Gross Mineralization and Nitrification Rates

The gross mineralization and nitrification rates were quantified using ^{15}N pool dilution techniques [14,35,36]. A series of 30 g samples of fresh soil (oven-dried) were weighted into 250 mL Erlenmeyer flasks for each soil, and these were subsequently pre-incubated at 25 °C for 24 h. After the pre-incubation, 1 mL of $^{15}NH_4NO_3$ or $NH_4{}^{15}NO_3$ (10 atom% ^{15}N excess) solution, containing 1.5 mg $NH_4{}^+$–N kg^{-1} and 1.5 mg $NO_3{}^-$–N kg^{-1}, was evenly applied to the soil in each Erlenmeyer flask. Noticeably, the amount of applied NH_4NO_3 was relatively higher in this study compared to that previously reported [14,32]. Due to the rapid conversion of $NH_4{}^+$ to $NO_3{}^-$ in soils under natural forest conditions with a high pH [30,31], the application of a small amount of NH_4NO_3 can lead to a relatively low $NH_4{}^+$ content in the soil after 24 h of incubation, which cannot satisfy the determination criteria of ^{15}N. As such, NH_4NO_3 was applied at the relatively high rate of 50 mg $NH_4{}^+$-N kg^{-1} and 50 mg $NO_3{}^-$-N kg^{-1}. Consequently, the potential rather than actual gross nitrification rate was measured. Distilled water was added to adjust the soil's moisture to 60% water holding capacity (WHC). Then, the flasks were capped with plastic film with small holes in it and incubated for 24 h at 25 °C. The soil samples in the Erlenmeyer flasks were extracted with 150 mL 2M KCl solution at 0.5 and 24 h after NH_4NO_3 application in order to determine the NH_4^+ and NO_3^- concentrations and the respective ^{15}N atom% excess.

2.3. Analyses

Soil pH was determined using a SevenExcellence™ pH/mV detector. After carbonate removal using 1.0 M HCl, an elemental analyzer (Sercon Integra 2) was used to determine the TN and SOC concentrations. The neutral ammonium acetate exchange method was used to determine the soil's cation exchange capacity (CEC) [37]. Total X-ray fluorescence spectroscopy was used to quantify the total calcium (Ca), magnesium (Mg), P, and K concentrations in the soil. The soil's available K (AK) and available P (AP) were extracted with neutral NH_4OAc and NH_4F–HCl solutions, respectively, and subsequently determined via use of a flame photometer and a spectrophotometer. A continuous flow analyzer (Skalar, Breda, The Netherlands) was used to determine the NH_4^+ and NO_3^- concentrations in the extract. In addition, the KCl extracts were gradually distilled with magnesium oxide (MgO) and Devarda's alloy so as to separate the pools of NH_4^+ and NO_3^- for ^{15}N measurements [38]. In brief, 100 mL of KCl extract was steam-distilled with MgO to convert the NH_4^+ into ammonia (NH_3), and then Devarda's alloy was added, and it was distilled again to convert NO_3^- into NH_3 through the reduction of NO_3^- to NH_4^+. The NH_3 was trapped in a boric acid solution in a conical flask, acidified, and converted into ammonium sulfate ($(NH_4)_2SO_4$) using 0.02 M H_2SO_4. According to our preliminary experiment, the recovery ratios of NH_4^+ and NO_3^- using the distillation method were 98–102% and 96–98%, respectively. The H_2SO_4 solution containing NH_4^+ was then evaporated to dryness at 80 °C in order to analyze the ^{15}N atom% excess with an isotope mass spectrometer (Sercon Integra 2, SerCon Ltd., Crewe, UK).

2.4. Bacterial or Archaeal amoA Abundance Analysis

The FastDNA® Spin Kit for Soil (MP Biomedicals, OH, USA) was used to extract soil DNA, which was subsequently stored at −20 °C until use. A spectrophotometer (Nanodrop ND-2000, NanoDrop Technologies, DE, USA) was used to quantify soil DNA quantity and purity. The abundances of bacterial (AOB) and archaeal (AOA) *amoA* genes were determined via the quantitative PCR method on a real-time detection system (Bio-Rad CFX96, Laboratories Inc., Hercules, CA, USA). The primers of bacterial and archaeal *amoA* genes were *amoA*-1F/*amoA*-2R and Arch-*amoA*-F/Arch-*amoA*-R, respectively [39,40]. Detailed information about the quantitative PCR analysis can be obtained from Zhu et al. (2018) [41].

2.5. Data and Statistical Analyses

The gross mineralization and nitrification rates, expressed as mg N kg^{-1} d^{-1}, were calculated via the equation of Kirkham and Bartholomew (1954) [35], as follows:

$$\text{Mineralization } (M_{Norg}) = \frac{[NH_4^+]_0 - [NH_4^+]_t}{t} \times \frac{\log\left(\frac{APE_0}{APE_t}\right)}{\log\frac{[NH_4^+]_0}{[NH_4^+]_t}} \tag{1}$$

where t is the incubation time (day), $[NH_4^+]$ is the NH_4^+ concentration (mg N kg^{-1}), and APE is the ^{15}N atom% excess of NH_4^+.

$$\text{Nitrification } (O_{NH_4}) = \frac{[NO_3^-]_0 - [NO_3^-]_t}{t} \times \frac{\log\left(\frac{APE_0}{APE_t}\right)}{\log\frac{[NO_3^-]_0}{[NO_3^-]_t}} \tag{2}$$

where t is the incubation time (day), $[NO_3^-]$ is the NO_3^- concentration (mg N kg^{-1}), and APE is the ^{15}N atom% excess of NO_3^-.

The mean residence times of NH_4^+ (MRT NH_4^+) and NO_3^- (MRT NO_3^-), expressed as d, were calculated following the equation of Corre et al. (2007) [42].

$$\text{MRT NH}_4^+ = \frac{c(NH_4^+)}{M_{Norg}} \tag{3}$$

$$\text{MRT NO}_3 = \frac{c(NO_3^-)}{O_{NH4}} \tag{4}$$

where $c(NH_4^+)$ and $c(NO_3^-)$ are the initial NH_4^+ and NO_3^- concentrations (mg N kg^{-1}) in the studied soils, respectively. If the MRT value of a certain N pool is high, this indicates low turnover.

The supply capacity of inorganic N was calculated by M_{Norg} plus O_{Norg} [13]. SPSS 23 software (SPSS, Chicago, IL, USA) was used to analyze the relationships between soil properties, AOA and AOB abundances, and M_{Norg} and O_{Norg} rates. Analysis of variance (ANOVA) was used to compare the differences in soil properties, AOA and AOB abundances, and M_{Norg} and O_{Norg} rates between natural forests and banana plantations at the $p = 0.05$ level.

3. Results

3.1. Soil Physical and Chemical Properties, AOA and AOB Abundances

The conversion of natural forest to banana plantations reduced the SOC, TN, and CaO concentrations, as well as the WHC, CEC, and pH, more significantly as the cultivation ages increased (Table 1), but this process significantly increased the AK and AP concentrations. In all the studied soils, NO_3^- dominated the inorganic N pool with NO_3^-/NH_4^+ ratios of 2.31 (natural forest) and 3.47–8.15 (banana plantations). The difference in NH_4^+ concentration between natural forests and banana plantations was not significant due to the high variation; however, the highest NO_3^- concentration (70.9 mg N kg^{-1}) was found in soil under 10 y banana cultivation conditions. The other three banana cultivation conditions manifested concentrations of 17.1–46.2 mg N kg^{-1} ($p < 0.05$). The SOC and TN concentrations were significantly positively related to CaO, CEC, and WHC ($p < 0.05$), and a significant and positive relationship was also found between CaO and pH ($p < 0.05$) (Table 2).

Table 1. Physical and chemical properties of soils under natural forest and banana plantation conditions with different cultivation ages.

Parameter [i]	Natural Forest	3 y [ii]	7 y	10 y	22 y
SOC (g C kg^{-1})	34.6 ± 6.29 a	32.9±3.14 a	21.4±1.47 b	20.0±2.36 b	19.7±3.34 b
TN (g C kg^{-1})	3.25 ± 0.18 a	2.92 ± 0.18 a	2.06 ± 0.06 b	1.88 ± 0.09 b	1.99 ± 0.29 b
pH	6.75 ± 0.09 a	6.29 ± 0.08 b	5.05 ± 0.39 c	4.65 ± 0.22 cd	4.30 ± 0.16 d
WHC	0.92 ± 0.15 a	0.73 ± 0.02 b	0.69 ± 0.02 bc	0.67 ± 0.01 c	0.58 ± 0.03 d
CEC (cmol kg^{-1})	18.4 ± 1.25 a	15.6 ± 0.85 b	11.3 ± 0.07 c	10.6 ± 0.38 c	11.5 ± 0.49 c
CaO (%)	8.39 ± 3.13 a	3.94 ± 0.52 b	1.77 ± 0.48 c	1.45 ± 0.46 c	1.54 ± 0.53 c
AP (mg kg^{-1})	3.27 ± 0.51 c	173 ± 26.0 a	136 ± 10.3 b	197 ± 18.2 a	159 ± 19.6 a
AK (mg kg^{-1})	236 ± 103 b	783 ± 227 a	848 ± 331 a	800 ± 87.1 a	761 ± 74.8 a
NH_4^+ (mg N kg^{-1})	9.45 ± 0.36 a	6.35 ± 2.54 a	15.0 ± 8.41 a	10.2 ± 5.34 a	4.85 ± 0.7 a
NO_3^- (mg N kg^{-1})	21.9 ± 1.41 c	29.7 ± 6.79 c	46.2 ± 3.55 b	70.9 ± 12.7 a	17.1 ± 5.36 c
NO_3^-/NH_4^+	2.31 ± 0.07 b	5.08 ± 1.63 a	3.65 ± 1.53 ab	8.15 ± 4.07 a	3.47 ± 0.78 ab
AOA abundance $\times 10^7$ *amoA* gene copies (g dry soil)$^{-1}$	11.6 ± 1.23 a	14.4 ± 2.45 a	5.55 ± 3.23 b	3.66 ± 1.45 b	1.45 ± 0.29 b
AOB abundance $\times 10^5$ *amoA* gene copies (g dry soil)$^{-1}$	19.6 ± 3.20 a	24.6 ± 4.98 a	6.48 ± 3.13 b	11.6 ± 2.64 b	9.14 ± 3.78 b

[i] SOC, soil organic C; TN, total N; WHC, water holding capacity; CEC, cation exchange capacity; AP, available P; AK, available K; AOA, archaeal *amoA* gene; AOB, bacterial *amoA* gene. [ii] 3, 7, 10, and 22 y represent the cultivation durations of the banana plantations that were converted from natural forests. Identical letters for the same value indicate that there were no significant differences in the soils under natural forest conditions and in the four banana plantations in the same region at $p = 0.05$.

Table 2. The relationships between soil properties, gross mineralization (M_{Norg}) and nitrification (O_{Norg}) rates, and archaeal *amoA* gene (AOA) and bacterial *amoA* gene (AOB) abundances ($n = 12$), under banana plantation conditions.

	NH_4^+	NO_3^-	MNorg	ONH4	MRT NH_4^+	MRT NO_3^-
SOC	−0.09	−0.22	0.87 **	0.70 *	−0.39	−0.39
TN	−0.22	−0.34	0.91 **	0.72 **	−0.58	−0.51
pH	−0.21	−0.14	0.87 **	0.83 **	−0.53	−0.48
WHC	0.30	0.37	0.51	0.57	0.06	0.08
CEC	−0.34	−0.47	0.96 **	0.87 **	−0.64 *	−0.65 *
CaO	−0.10	−0.33	0.84 **	0.78 **	−0.33	−0.42
AP	−0.09	0.47	−0.14	0.04	0.18	0.10
AK	−0.19	0.23	−0.01	0.01	−0.18	−0.18
AOA	0.04	−0.15	0.83 **	0.79 **	−0.25	−0.29
AOB	−0.31	−0.19	0.80 **	0.87 **	−0.43	−0.54

*, $p < 0.05$; **, $p < 0.01$.

The AOA and AOB abundances in soils under 3 y banana cultivation conditions were 1.2 times higher than those in natural forests, but these values gradually decreased as banana cultivation time lengthened. Both the AOA and the AOB abundances were significantly related to SOC, TN, WHC, CEC, CaO, and pH ($p < 0.05$) (Table 2).

3.2. Gross Mineralization (M_{Norg}) and Nitrification (O_{Norg}) Rates

Compared to those under natural forest conditions (1.70 and 6.30 mg N kg^{-1} d^{-1}), the M_{Norg} and O_{Norg} rates were significantly increased to 3.23 and 11.9 mg N kg^{-1} d^{-1} in the soil under 3 y banana cultivation conditions, respectively, but decreased to 0.65–1.26 and 3.41–4.92 mg N kg^{-1} d^{-1} with the prolongation of banana cultivation (Figure 2). Contrastingly, the 3 y banana cultivation conditions lowered the residence times of NH_4^+ and NO_3^- to 2.07 and 2.88 d, respectively, compared to those in natural forests (5.57 and 3.62 d), while long-term banana cultivation (>3 y) increased the residence times of NH_4^+ (5.83–15.8 d) and NO_3^- (5.02–15.0 d) (Figure 3). Both the M_{Norg} and O_{Norg} rates in soils under banana cultivation conditions were significantly positively related to SOC, TN, WHC, CEC, and pH ($p < 0.05$) (Table 2). In addition, the O_{Norg} rate was significantly positively related to AOA and AOB abundances ($p < 0.05$) (Table 2).

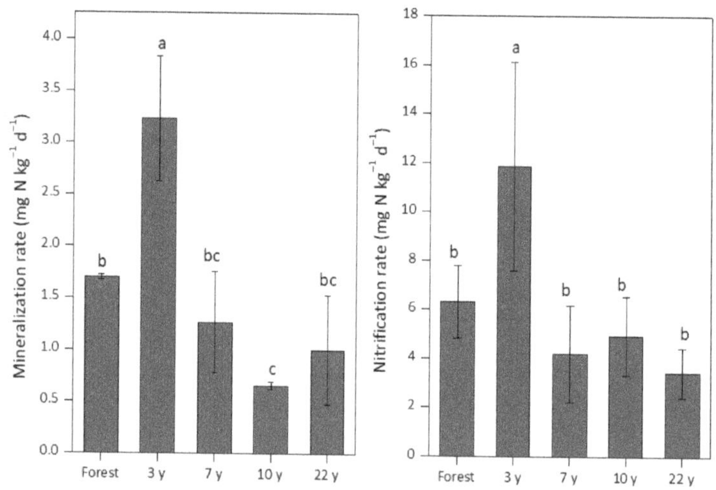

Figure 2. M_{Norg} and O_{NH4} rates in soils under natural forest conditions and in banana plantations with different cultivation durations. Identical letters for M_{Norg} and O_{NH4} indicate there were no significant differences in soils under natural forest conditions and in the four banana plantations in the same region, at $p = 0.05$. M_{Norg}, the mineralization of organic N to NH_4^+; O_{NH4}, the oxidation of NH_4^+ to NO_3^-.

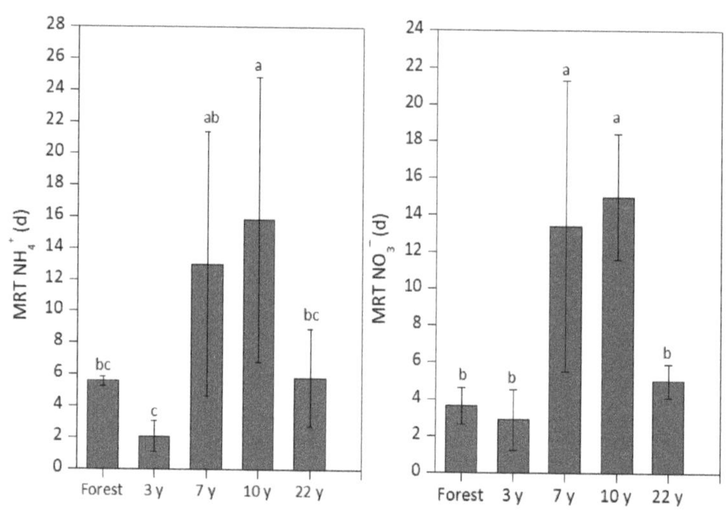

Figure 3. Mean residence times of NH_4^+ (MRT NH_4^+) and NO_3^- (MRT NO_3^-) in soils under natural forest conditions and in banana plantations with different cultivation durations. Identical letters for MRT NH_4^+ and MRT NO_3^- indicate there were no significant differences in soils under natural forest conditions and in the four banana plantations in the same region, at $p = 0.05$.

4. Discussion

4.1. Low Supply Capacity of Inorganic N in Soils under Natural Forest Conditions

The soil M_{Norg} rates under natural forest conditions reached 1.70 mg N kg^{-1} d^{-1}, which was significantly lower than the rates in other natural forests in tropical or subtropical regions, as reported by previous studies (2.29–9.20; average, 4.31 mg N kg^{-1} d^{-1}) [11–13], suggesting a low inorganic N supply capacity in our studied soils under natural forest conditions. However, higher O_{NH4} rates (6.30 mg N kg^{-1} d^{-1}) were found in our studied soils than in the highly weathered soils under natural forest conditions in subtropical or tropical regions (0.06–1.97, average 0.75 mg N kg^{-1} d^{-1}) [11–13], suggesting the rapid oxidation of NH_4^+ to NO_3^- and inorganic N dominated by NO_3^-. This was supported by the high NO_3^-/NH_4^+ ratio (2.31) found in soil under natural forest conditions. Considering the high rainfall in this region, the NO_3^- in soil can easily be lost through leaching and runoff, and thus inorganic N cannot be effectively conserved in soil. This result was inconsistent with those of previous studies conducted in natural forests in tropical or subtropical regions, in which high inorganic N supply and N retention capacities were found due to the high M_{Norg} and low O_{NH4} rates [11,12,28]. Our study, and other previous studies, have inferred a large variability in N dynamics under tropical or subtropical conditions.

The characteristics of the transformation of soil N in natural forests may be greatly related to differences in soil type. In this study, the soil was composed of a mixture of carbonate rock and basalt weathering, and was characterized by a relatively high Ca content. Calcium can react with organic matter to form stable calcium humate [43], which is more difficult to break down for soil organisms, thereby leading to a decline in M_{Norg} rate even when the content of soil organic matter (SOM) is high. Previous studies have found that the O_{NH4} rate is significantly positively related to pH in the soil [44,45]. Forest soil that has developed from basalt or granite in subtropical/tropical regions has a low pH (<4.5), which can inhibit the O_{NH4} rate [11,13]. In this study, however, the soils were characterized by a relatively high pH (5.42), which could increase the abundance and activity of nitrifying microorganisms [46,47]. This might explain why a high O_{NH4} rate was found in the studied soils under natural forest conditions. A high level of NO_3^- production through O_{NH4} in soil may increase the rate of denitrification in subtropical regions with high rainfall, thus subsequently increasing the nitrogenous gas emission potential [48]. Noticeably, the gross O_{NH4} rate may be overestimated in this study due to the application of $NH_4{}^{15}NO_3$ to the soils. Further in situ experiments must be conducted to elucidate the actual dynamic changes in the NO_3^- of soil under natural forest conditions.

4.2. Response of M_{Norg} and O_{Norg} Rates to Banana Cultivation

Previous studies found significantly positive relationships between SOC and TN concentrations and M_{Norg} rate [9,13,15], indicating that the SOM level is a critical driver of M_{Norg}. Although the conversion of natural forests to banana plantations reduces the SOC and total N concentrations (Table 1), large variations in M_{Norg} rate were found in banana plantations with different cultivation durations. Compared to natural forest conditions, short-term banana cultivation greatly increased the M_{Norg} rate, but this rate gradually decreased with prolonged banana cultivation. This result supports our hypothesis, and may be attributed to changes in SOM quality and quantity [25,32]. The conversion of natural forests to banana plantations significantly reduces soil organic matter, thus reducing the substrate of M_{Norg}. At the initial stage of banana cultivation, however, organic fertilizer application and the separation of stable calcium humate via reductions in Ca^{2+} protection can increase the soil's labile organic matter content [49,50], which could stimulate the mineralization of organic N to NH_4^+, and even a decline in TN concentration (Table 1). When the labile organic matter is gradually consumed through continuous banana cultivation, the stimulating effect on M_{Norg} may not offset the decline in M_{Norg} caused by the reduction in TN content. The mechanisms by which organic N quality affects M_{Norg} need investigating in the future.

Generally speaking, management measures (e.g., tillage and fertilization) can increase the soil O_{NH4} rate [12,13]. Indeed, 3 y banana cultivation increased the O_{NH4} rate to 11.9 mg N kg^{-1} d^{-1} compared to natural forest conditions, but this was not the case for the long-term (>3 y) banana cultivation assessed in this study, wherein the O_{NH4} rate was significantly reduced. This result supported Hypothesis 1. The different responses of O_{NH4} to various banana cultivation conditions may be related to changes in soil properties (e.g., pH, SOC, structure) [13,31]. Nitrogen fertilizer can increase the abundance and activity of nitrifying microorganisms, in order to increase the O_{NH4} rate in soils with high pH when natural forests are converted to banana plantations. However, due to the reductions in SOC and total N concentrations with the increasing duration of crop cultivation (Table 1), the reduction in macro-aggregates and the increase in micro-aggregates can cause the surface soil layer to become hard and compacted [34], which can reduce soil permeability and thus inhibit nitrifying microbial abundance and activity [31]. This inhibiting effect may be more obvious in soils composed of carbonate rock, which is characterized by a lower content of acid-insoluble matter and a heavy texture [51,52]. In addition, after long-term banana cultivation, the concentrations of SOC and TN, as well as the water holding capacity, CEC, and pH, are also reduced, thus deteriorating the soil condition, and all of these factors may have an adverse effect on the growth of nitrifying microorganisms and thus inhibit O_{NH4} [15,31]. Indeed, the O_{NH4} rate was found to be significantly positively related to the abundance of AOA and AOB, as well as the SOC, TN, CEC, WHC, and pH (Table 2), which supports the above speculation.

Due to the decline in the M_{Norg} and O_{NH4} rates, long-term banana cultivation significantly reduces inorganic N supply capacity, while increasing the resident time of inorganic N, compared to short-term banana cultivation. This result is consistent with Hypothesis 2, implying the reduced turnover of inorganic N in soils under long-term banana cultivation conditions in subtropical regions. According to previous studies [11,13,31], as well as our present result that M_{Norg} and O_{NH4} rates were positively related to SOC and TN concentrations, organic N fertilizer should be recommended over mineral N fertilizer to stimulate the supply and turnover of inorganic N in long-term banana plantations.

5. Conclusions

The present study highlights the fact that short-term (3 y) banana cultivation causes increased mineralization and nitrification rates, as well as increasing the turnover rate for inorganic N in the soil, but these rates are significantly reduced with the prolongation of banana cultivation. These results, combined with those of the previous studies, suggest the rapid reduction in soil inorganic N supply when natural forests are converted to economic crop plantations. In such cases, soil N cycling is blocked, and the applied N fertilizers cannot be effectively converted into a form of N that is available for banana uptake, suggesting low N use efficiency and high N loss potential. This is mainly attributed to the reduction in soil quality, i.e., reduced soil organic matter content, and high clay content. Considering the prevalence of banana cultivation worldwide, developing appropriate management measures in future will be necessary in order for banana plantations in specific regions to increase their inorganic N supply and turnover.

Author Contributions: Conceptualization, X.Q. and L.Y.; investigation, X.Q., C.Y., and L.Y.; formal analysis, C.Y. and E.M.; original draft preparation, X.Q. and L.M.; writing—review and editing, X.Q., L.M. and T.Z. All authors have read and agreed to the published version of the manuscript.

Funding: This work was supported by grants from the National Natural Science Foundation of China (41661051, 42067008) and Guangxi Natural Science Foundation of China (2017GXNSFBA198034; 2018GXNSFBA138042).

Institutional Review Board Statement: Not applicable.

Informed Consent Statement: Not applicable.

Data Availability Statement: Not applicable.

Conflicts of Interest: The authors declare no conflict of interest.

References

1. Zhang, J.Z.; Bei, S.K.; Li, B.S.; Zhang, J.L.; Christie, P.; Li, X.L. Organic fertilizer, but not heavy liming, enhances banana biomass, increases soil organic carbon and modifies soil microbiota. *Appl. Soil Ecol.* **2019**, *136*, 67–79. [CrossRef]
2. Meya, A.I.; Ndakidemi, P.A.; Mtei, K.M.; Swennen, R.; Merckx, R. Optimizing soil fertility management strategies to enhance banana production in volcanic soils of the northern highlands, Tanzania. *Agronomy* **2020**, *10*, 289. [CrossRef]

3. FAO. FAO Statistical Databases. 2020. Available online: http://www.fao.org/faostat/zh/?#data/QC (accessed on 1 January 2021).
4. Yao, L.X.; Li, G.L.; Yang, B.M.; Tu, S.H. Optimal fertilization of banana for high yield, quality, and nutrient use efficiency. *Better Crops Plant Food* **2009**, *93*, 10–11.
5. Vitousek, P.M.; Howarth, R.W. Nitrogen limitation on land and in the sea: How can it occur? *Biogeochemistry* **1991**, *13*, 87–115. [CrossRef]
6. Memon, N.U.N.; Memon, K.S.; Anwar, R.; Ahmad, S.; Nafees, M. Status and response to improved NPK fertilization practices in banana. *Pak. J. Bot.* **2010**, *42*, 2369–2381.
7. Gonçalves, A.L.; Kernaghan, J.R. *Banana Production Methods: A comparative Study*; Centro Ecológico: Ipê da Serra, Brazil, 2014; 40p.
8. Hobbie, E.A.; Högberg, P. Nitrogen isotopes link mycorrhizal fungi and plants to nitrogen dynamics. *New Phytol.* **2012**, *196*, 367–382. [CrossRef] [PubMed]
9. Schimel, J.P.; Bennett, J. Nitrogen mineralization: Challenges of a changing paradigm. *Ecology* **2004**, *85*, 591–602. [CrossRef]
10. Hart, S.C.; Nason, G.E.; Myrold, D.D.; Perry, D.A. Dynamics of gross nitrogen transformations in an old-growth forest: The carbon connection. *Ecology* **1994**, *75*, 880–891. [CrossRef]
11. Zhang, J.B.; Cai, Z.C.; Zhu, T.B.; Yang, W.Y.; Müller, C. Mechanisms for the retention of inorganic N in acidic forest soils of southern China. *Sci. Rep.* **2013**, *3*, 2342. [CrossRef] [PubMed]
12. Zhu, T.B.; Meng, T.C.; Zhang, J.B.; Yin, Y.; Cai, Z.C.; Yang, W.Y.; Zhong, W.H. Nitrogen mineralization, immobilization turnover, heterotrophic nitrification, and microbial groups in acid forest soils of subtropical China. *Biol. Fert. Soils* **2013**, *49*, 323–331. [CrossRef]
13. Xie, Y.; Yang, L.; Zhu, T.B.; Yang, H.; Zhang, J.B.; Yang, J.L.; Cao, J.H.; Bai, B.; Jiang, Z.C.; Liang, Y.M.; et al. Rapid recovery of nitrogen retention capacity in a subtropical acidic soil following afforestation. *Soil Biol. Biochem.* **2018**, *120*, 171–180. [CrossRef]
14. Wang, C.H.; Wang, N.N.; Zhu, J.X.; Liu, Y.; Xu, X.F.; Niu, S.L.; Yu, G.R.; Han, X.G.; He, N.P. Soil gross N ammonification and nitrification from tropical to temperate forests in eastern China. *Funct. Ecol.* **2018**, *32*, 83–94. [CrossRef]
15. Shan, Z.J.; Yin, Z.; Yang, H.; Zuo, C.Q.; Zhu, T.B. Long-term cultivation of fruit plantations decreases mineralization and nitrification rates in calcareous soil in the karst region in southwestern China. *Forests* **2020**, *11*, 1282. [CrossRef]
16. Knops, J.M.H.; Bradley, K.L.; Wedin, D.A. Mechanisms of plant species impacts on ecosystem nitrogen cycling. *Ecol. Lett.* **2002**, *5*, 454–466. [CrossRef]
17. Baijukya, F.P.; de Ridder, N.; Masuki, K.F.; Giller, K.E. Dynamics of banana-based farming systems in Bukoba District, Tanzania: Changes in land use, cropping and cattle keeping. *Agric. Ecosyst. Environ.* **2005**, *106*, 395–406. [CrossRef]
18. Dita, M.; Barquero, M.; Heck, D.; Mizubuti, E.S.G.; Staver, C.P. Fusarium wilt of banana: Current knowledge on epidemiology and research needs toward sustainable disease management. *Front. Plant Sci.* **2018**, *9*, 1468. [CrossRef] [PubMed]
19. Bradshaw, C.J.A. Little left to lose: Deforestation and forest degradation in Australia since European colonization. *J. Plant. Ecol.* **2012**, *5*, 109–120. [CrossRef]
20. Zhu, T.B.; Zhang, J.B.; Huang, P.; Suo, L.; Wang, C.; Ding, W.X.; Meng, L.; Zhou, K.X.; Hu, Z.W. N_2O emissions from banana plantations in tropical China as affected by the application rates of urea and a urease/nitrification inhibitor. *Biol. Fert. Soils* **2015**, *51*, 673–683. [CrossRef]
21. Grünzweig, J.M.; Sparrow, S.D.; Chapin, F.S. Impact of forest conversion to agriculture on carbon and nitrogen mineralization in subarctic Alaska. *Biogeochemistry* **2003**, *64*, 271–296. [CrossRef]
22. Zhu, T.B.; Zhang, J.B.; Meng, T.Z.; Zhang, Y.C.; Yang, J.J.; Müller, C.; Cai, Z.C. Tea plantation destroys soil retention of NO_3^- and increases N_2O emissions in subtropical China. *Soil Biol. Biochem.* **2014**, *73*, 106–114. [CrossRef]
23. Monkai, J.; Goldberg, S.D.; Hyde, K.D.; Harrison, R.D.; Mortimer, P.E.; Xu, J.C. Natural forests maintain a greater soil microbial diversity than that in rubber plantations in Southwest China. *Agr. Ecosyst. Environ.* **2018**, *265*, 190–197. [CrossRef]
24. Ryals, R.; Kaiser, M.; Torn, M.S.; Berhe, A.A.; Silver, W.L. Impacts of organic matter amendments on carbon and nitrogen dynamics in grassland soils. *Soil Biol. Biochem.* **2014**, *68*, 52–61. [CrossRef]
25. Zhang, J.B.; Cai, Z.C.; Yang, W.Y.; Zhu, T.B.; Yu, Y.J.; Yan, X.Y.; Jia, Z.J. Long-term field fertilization affects soil nitrogen transformations in a rice-wheat-rotation cropping system. *J. Plant Nutr. Soil Sci.* **2012**, *175*, 939–946. [CrossRef]
26. Hwang, S.; Hanaki, K. Effects of oxygen concentration and moisture content of refuse on nitrification, denitrification and nitrous oxide production. *Bioresour. Technol.* **2000**, *71*, 159–165. [CrossRef]
27. Bai, J.H.; Gao, H.F.; Deng, W.; Yang, Z.F.; Cui, B.S.; Xiao, R. Nitrification potential of marsh soils from two natural saline–alkaline wetlands. *Biol. Fert. Soils* **2010**, *46*, 525–529. [CrossRef]
28. Huygens, D.; Boeckx, P.; Templer, P.; Paulino, L.; Van Cleemput, O.V.; Oyarzún, C.; Müller, C.; Godoy, R. Mechanisms for retention of bioavailable nitrogen in volcanic rainforest soils. *Nat. Geosci.* **2008**, *1*, 543–548. [CrossRef]
29. Zhu, T.B.; Zhang, J.B.; Cai, Z.C. The contribution of nitrogen transformation processes to total N_2O emissions from soils used for intensive vegetable cultivation. *Plant Soil* **2011**, *343*, 313–327. [CrossRef]
30. Zhu, T.B.; Zeng, S.M.; Qin, H.L.; Zhou, K.X.; Yang, H.; Lan, F.N.; Huang, F.; Cao, J.H.; Müller, C. Low nitrate retention capacity in calcareous soil under woodland in the karst region of southwestern China. *Soil Biol. Biochem.* **2016**, *97*, 99–101. [CrossRef]
31. Garousi, F.; Shan, Z.J.; Ni, K.; Yang, H.; Shan, J.; Cao, J.H.; Jiang, Z.C.; Yang, J.L.; Zhu, T.B.; Müller, C. Decreased inorganic N supply capacity and turnover in calcareous soil under degraded rubber plantation in the tropical karst region. *Geoderma* **2021**, *381*, 114754. [CrossRef]

32. Allen, K.; Corre, M.D.; Tjoa, A.; Veldkamp, E. Soil nitrogen-cycling responses to conversion of lowland forests to oil palm and rubber plantations in Sumatra, Indonesia. *PLoS ONE* **2015**, *10*, e0133325. [CrossRef]
33. Li, H.M.; Ma, Y.X.; Liu, W.J.; Liu, W.J. Soil changes induced by rubber and tea plantation establishment: Comparison with tropical rain forest soil in Xishuangbanna, SW China. *Environ. Manag.* **2012**, *50*, 837–848. [CrossRef]
34. Nawaz, M.F.; Bourrié, G.; Trolard, F. Soil compaction impact and modelling. A review. *Agrono. Sustain. Dev.* **2013**, *33*, 291–309. [CrossRef]
35. Kirkham, D.; Bartholomew, W.V. Equations for following nutrient transformations in soil utilizing tracer data. *Soil Sci. Soc. Am. Proc.* **1954**, *18*, 33–34. [CrossRef]
36. Murphy, D.V.; Recous, S.; Stockdale, E.A.; Fillery, I.R.P.; Jensen, L.S.; Hatch, D.J.; Goulding, K.W.T. Gross nitrogen fluxes in soil: Theory, measurement and application of ^{15}N pool dilution techniques. *Adv. Agron.* **2003**, *79*, 69–118.
37. Ross, D.S.; Ketterings, Q. Recommended methods for determining soil cation exchange capacity. In *Cooperative Bulletin No. 493 Recommended Soil Testing Procedures for the Northeastern United States*; University of Delaware Newark: Newark, DE, USA, 2011; pp. 75–85.
38. Bremner, J.M.; Keeney, D.R. Determination and isotope-ratio analysis of different forms of nitrogen in soils: 3. Exchangeable ammonium, nitrate, and nitrite by extraction-distillation methods. *Soil Sci. Soc. Am. J.* **1966**, *30*, 577–582. [CrossRef]
39. Rotthauwe, J.H.; Witzel, K.P.; Liesack, W. The ammonia monooxygenase structural gene amoA as a functional marker: Molecular fine-scale analysis of natural ammonia-oxidizing populations. *Appl. Environ. Microb.* **1997**, *63*, 4704–4712. [CrossRef] [PubMed]
40. Francis, C.A.; Roberts, K.J.; Beman, J.M.; Santoro, A.E.; Oakley, B.B. Ubiquity and diversity of ammonia–oxidizing archaea in water columns and sediments of the ocean. *Proc. Natl. Acad. Sci. USA* **2005**, *102*, 14683–14688. [CrossRef] [PubMed]
41. Zhu, T.B.; Yang, C.; Wang, J.; Zeng, S.M.; Liu, M.Q.; Yang, J.L.; Bai, B.; Cao, J.H.; Chen, X.Y.; Müller, C. Bacterivore nematodes stimulate soil gross N transformation rates depending on their species. *Biolo. Fert. Soils* **2018**, *54*, 107–118. [CrossRef]
42. Corre, M.D.; Brumme, R.; Veldkamp, E.; Beese, F.O. Changes in nitrogen cycling and retention processes in soils under spruce forests along a nitrogen enrichment gradient in Germany. *Glob. Chang. Biol.* **2007**, *13*, 1509–1527. [CrossRef]
43. Rowley, M.C.; Grand, S.; Verrecchia, É.P. Calcium-mediated stabilization of soil organic carbon. *Biogeochemistry* **2018**, *137*, 27–49. [CrossRef]
44. Cheng, Y.; Wang, J.; Mary, B.; Zhang, J.B.; Cai, Z.C.; Chang, S.X. Soil pH has contrasting effects on gross and net nitrogen mineralizations in adjacent forest and grassland soils in central Alberta, Canada. *Soil Biol. Biochem.* **2013**, *57*, 848–857. [CrossRef]
45. Kemmitt, S.; Wright, D.; Goulding, K.; Jones, D. pH regulation of carbon and nitrogen dynamics in two agricultural soils. *Soil Biol Biochem.* **2006**, *38*, 898–911. [CrossRef]
46. He, J.Z.; Hang, W.H.; Zhi, H.X. Chapter six-ammonia-oxidizing archaea play a predominant role in acid soil nitrification. *Adv. Agron.* **2014**, *125*, 261–302.
47. Yao, H.Y.; Campbell, C.D.; Chapman, S.J.; Freitag, T.E.; Nicol, G.W.; Singh, B.K. Multi-factorial drivers of ammonia oxidizer communities: Evidence from a national soil survey. *Environ. Microbiol.* **2013**, *15*, 2545–2556. [CrossRef] [PubMed]
48. Yu, L.F.; Mulder, J.; Zhu, J.; Zhang, X.S.; Wang, Z.W.; Dörsch, P. Denitrification as a major nitrogen sink in forested monsoonal headwater catchments in the sub-tropics: Evidence from multi-site dual nitrate isotopes. *Glob. Chang. Biol.* **2019**, *25*, 1765–1778. [CrossRef]
49. Canfield, D.E.; Glazer, A.N.; Falkowski, P.G. The evolution and future of Earth's nitrogen cycle. *Science* **2010**, *330*, 192–196. [CrossRef] [PubMed]
50. Xiao, K.C.; He, T.G.; Chen, H.; Peng, W.X.; Song, T.Q.; Wang, K.L.; Li, D.J. Impacts of vegetation restoration strategies on soil organic carbon and nitrogen dynamics in a karst area, southwest China. *Ecol. Eng.* **2017**, *101*, 247–254. [CrossRef]
51. Wang, K.L.; Zhang, C.H.; Chen, H.S.; Yue, Y.M.; Zhang, W.; Zhang, M.Y.; Qi, X.K.; Fu, Z.Y. Karst landscapes of China: Patterns, ecosystem processes and services. *Landsc. Ecol.* **2019**, *34*, 2743–2763. [CrossRef]
52. Jiang, Z.H.; Liu, H.Y.; Wang, H.Y.; Peng, J.; Meersmans, J.; Green, S.M.; Quine, T.A.; Wu, X.C.; Song, Z.L. Bedrock geochemistry influences vegetation growth by regulating the regolith water holding capacity. *Nat. Commun.* **2020**, *11*, 2392. [CrossRef]

land

MDPI

Article

Cultivated Land Use Zoning Based on Soil Function Evaluation from the Perspective of Black Soil Protection

Rui Zhao [1,†], Junying Li [2,†], Kening Wu [1,3,4,*] and Long Kang [1]

1 School of Land Science and Technology, China University of Geosciences, Beijing 100083, China; zhaoruifighting@cugb.edu.cn (R.Z.); 2112200053@cugb.edu.cn (L.K.)
2 College of Resources and Environment, Shandong Agricultural University, Taian 271018, China; lijunying@sdau.edu.cn
3 Key Laboratory of Land Consolidation and Rehabilitation, Ministry of Natural Resources, Beijing 100035, China
4 Technology Innovation Center of Land Engineering, Ministry of Natural Resources, Beijing 100083, China
* Correspondence: wukening@cugb.edu.cn
† These authors contributed equally to this work.

Abstract: Given that cultivated land serves as a strategic resource to ensure national food security, blind emphasis on improvement of food production capacity can lead to soil overutilization and impair other soil functions. Therefore, we took Heilongjiang province as an example to conduct a multi-functional evaluation of soil at the provincial scale. A combination of soil, climate, topography, land use, and remote sensing data were used to evaluate the functions of primary productivity, provision and cycling of nutrients, provision of functional and intrinsic biodiversity, water purification and regulation, and carbon sequestration and regulation of cultivated land in 2018. We designed a soil function discriminant matrix, constructed the supply-demand ratio, and evaluated the current status of supply and demand of soil functions. Soil functions demonstrated a distribution pattern of high grade in the northeast and low grade in the southwest, mostly in second-level areas. The actual supply of primary productivity functions in 71.32% of the region cannot meet the current needs of the population. The dominant function of soil in 34.89% of the area is water purification and regulation, and most of the cultivated land belongs to the functional balance region. The results presented herein provide a theoretical basis for optimization of land patterns and improvement of cultivated land use management on a large scale, and is of great significance to the sustainable use of black soil resources and improvement of comprehensive benefits.

Keywords: agroecosystems; Heilongjiang province; supply and demand; soil multifunctionality; spatial scales

Citation: Zhao, R.; Li, J.; Wu, K.; Kang, L. Cultivated Land Use Zoning Based on Soil Function Evaluation from the Perspective of Black Soil Protection. *Land* **2021**, *10*, 605. https://doi.org/10.3390/land10060605

Academic Editors: Chiara Piccini and Rosa Francaviglia

Received: 28 April 2021
Accepted: 4 June 2021
Published: 7 June 2021

1. Introduction

Soil provides basic services for maintaining and guaranteeing agricultural production, plant growth, animal habitats, biodiversity, and environmental quality, and is one of the core elements linking the entire natural ecosystem [1]. Soil is the key foundation for cultivated land to perform its functions, a limited resource essential to the maintenance and sustainable use of land, and essential for cultivated land to have multifunctional roles [2]. Agriculture is facing the challenge of increasing primary productivity to meet the growing global demand for food security [3]. However, most of the soil resources in the world are in a barren or worse state, and one third of cultivated land is moderately or highly degraded [4]. The total amount of cultivated land resources in China is only 135 million ha, and the per capita cultivated land is only 0.30 ha, which is lower than the world's per capita level of 0.37 ha; thus, China is facing a serious shortage of cultivated land [5]. The cultivated land area of black soil in Northeast China accounts for 27% of China's total cultivated land [6], and the total grain output accounts for 25% of China's total output. However, commodity grain accounts for 33% of China's agricultural production. Because of long-term and high-intensity utilization, cultivated land resources in the black soils area have been overdrawn for a long time, and the excessive input of chemical fertilizers and pesticides has broken the original stable micro-ecological system of the black soils. The degradation of soil functions like soil biodiversity, nutrient maintenance, carbon storage, buffering, and water purification and regulation has become a shortcoming that restricts the improvement of regional food production capacity and sustainable agricultural development and poses a serious threat to national food security in Heilongjiang province [7]. In China, the Ministry of Natural Resources conducts surveys and evaluations of the quality of agricultural land and the Ministry of Agriculture and Rural Affairs evaluates the quality of cultivated land These agencies have formed a farmland resource evaluation system that takes the county as the project unit and the field as the evaluation unit and summarizes the results at the provincial and national scales, providing a solid foundation for the utilization and management of cultivated land. However, these tasks were restricted by the scientific methods and protection concepts of the time. This is because they ignored the general principle of scales in land quality evaluation. Specifically, there were problems of mixed scales, and it was difficult to describe scales accurately and quickly above the county level. This situation cannot meet the management needs of different levels. Additionally, the two sets of plans mainly characterize the production potential and serve cultivation. However, they have insufficient consideration of soil function and soil environmental conditions, and the research results have a single effect, which makes it difficult to serve the current multi-target soil health management and protection. As a result, a relatively new practice to fulfil this goal has emerged in which researchers have begun to calculate multiple soil functions to guide the sustainable use of cultivated land [8].

Soil function indicates a soil-based ecosystem service that consists of a series of soil processes that support the provision of ecosystem services and contribute to the production of goods and services that are beneficial to human social requirements and the environment [9]. The carbon, nitrogen, water, and biological reservoirs in soil make a significant contribution to the sustainability of the earth [10]. The core of ecosystem processes is the biogeochemical cycles of carbon, nitrogen, and water. Understanding and evaluating the natural capital savings of soil from the perspective of ecosystem service functions can improve the resource utilization efficiency of production activities [11]. The core of ecosystem processes is the biogeochemical cycle of carbon, nitrogen and water. Understanding and evaluating the natural capital savings of soil from the perspective of ecosystem service functions can improve the resource utilization efficiency of production activities [12], by increasing the ability of soil to retain water and fertilizer and preventing soil erosion, both of which increase and stabilize crop yields. In addition, increasing soil carbon storage helps slow climate change [13]. The nutrient cycle represented by the nitrogen cycle provides nutrient support for the production of biomass [14]. Ecosystem services related to the soil water cycle include food and water security supply services, soil storage water and purification water flow regulation services and support services [15]. Soil biodiversity is closely related to the formation of ecosystem services [16]. Soil ecological services comprise supporting services for biological production, habitat, species, and genetic diversity, services providing nutrients, water, and mineral raw materials, and services that regulate water, the carbon cycle, and greenhouse gas emissions, as well as services that preserve the soil landscape and cultural relics [17]. Therefore, the ability of soils to provide multiple ecosystem functions simultaneously is known as soil ecosystem multifunctionality [18,19], which provides a simple metric to assess the overall functioning of ecosystems or treatments [20]. However, there is still no consensus on the quantitative standards and methods for the versatility of ecosystems. This is mainly because of the structural complexity and functional diversity of ecosystems [21]. Soil function and its ecosystem service are used when evaluating soil quality for land use and management [9]. If one or more soil functions are restricted, soil quality may be threatened by compaction, erosion, loss of biodiversity and organic matter, salinization, pollution, or desertification, which influences the rational utilization and protection of the soil.

There is currently no soil function classification and evaluation system in the world. The European Commission's soil protection strategy was an important initiative that brought the concept of soil functions to the attention of the wider public and placed the concept on the political agenda. Seven soil functions were defined in the strategy: (i) production of food and biomass, (ii) storage, filtering, and transformation of compounds, (iii) habitats for living creatures and gene pools, (iv) the physical and cultural environment, (v) a source of raw materials, (vi) a carbon pool, and (vii) archives of geological and archaeological heritage [22]. Guerra et al. described and analyzed four soil ecosystem functions across soil macroecological studies and 17,186 sampling sites: (i) decomposition, (ii) soil respiration, (iii) nutrient cycling, and (iv) water infiltration [23]. Schulte et al. classified soil functions into five categories: primary productivity, water purification and regulation, carbon sequestration and regulation, provision of functional and intrinsic biodiversity, and nutrient supply and circulation [24]. In 1997, Zhao Qiguo et al. pointed out that soil quality evaluation in agricultural systems should take soil function as a central task, but mainly focus on the evaluation of soil production functions [5]. Based on the classification of ecosystem service function types, the classification of soil functions can be continuously improved, the characterization indicators of soil functions can be selected, and the indicators can be classified and interpreted [25,26]. Under the precondition of being in favor of protection of soil functions, most soil evaluations conducted to date have been based on individual or comprehensive evaluations of soil functions through reasonable selection of evaluation methods [27–30]. For example, Thoumazeau et al. proposed an integrated indicator set, Biofunctool®, to evaluate the impact of agricultural land management on soil functions. This set consists of 12 rapid and economical field indicators, including soil active organic carbon, soil basic respiration, earthworm activity, available nitrogen, and infiltration rate and stability of soil aggregates, to assess three dynamic soil functions; namely, carbon conversion, nutrient cycling, and structure maintenance [31,32]. In terms of the integration of multiple function evaluation results, Schulte et al. conceptually described the cooperative weighing of soil functions of various land use types such as cultivated land, forest land, and grassland based on the theory of functional soil management [24,33]. Land use types have different requirements for various soil functions. In terms of evaluation indicators and their weights, prior knowledge such as literature and expert experience is important for the selection of indicators and determination of weights [3]. The Delphi method and analytic hierarchy process commonly used in previous studies are typical applications of traditional expert systems. However, these methods are often criticized for being more subjective [34]. Poorly subjective methods can also be used in practical applications, such as meta-analysis or structural equation modeling [23,35]. The soil functions of cultivated land mainly involve five aspects that contribute to increased agricultural productivity and provide other regulatory and supportive ecosystem services: primary productivity, provision and cycling of nutrients, provision of functional and intrinsic biodiversity, water purification and regulation, and carbon sequestration and regulation [8]. In other countries, the theoretical framework, index selection, and standardization of soil function evaluation, and the mapping method of evaluation results have been relatively systematic [36,37], providing a reference for the development of soil multifunctionality evaluation and change analysis in China [38–44]. Existing studies on large scale soil function are still slightly inadequate in China. Therefore, we have taken Heilongjiang province as our research area, selecting evaluation indicators according to local conditions, establishing a soil function evaluation model to reveal regional differences in soil functions and providing references for the sustainable use of land resources and the formulation of cultivated land protection policies based on evaluation of the supply and demand of soil functions. The specific purposes of this article are to: (A) evaluate the soil function of paddy fields and dry land in Heilongjiang province and the current situation of supply and demand by using the multi-factor comprehensive evaluation method and analyzing the main restrictive factors of the soil function in each region; and (B) divide the cultivated land use function region guided by the dominant function of soil, combine this with analysis of soil function restriction and put forward corresponding utilization optimization and regulation suggestions for different areas. The research frame diagram is shown in Figure 1.

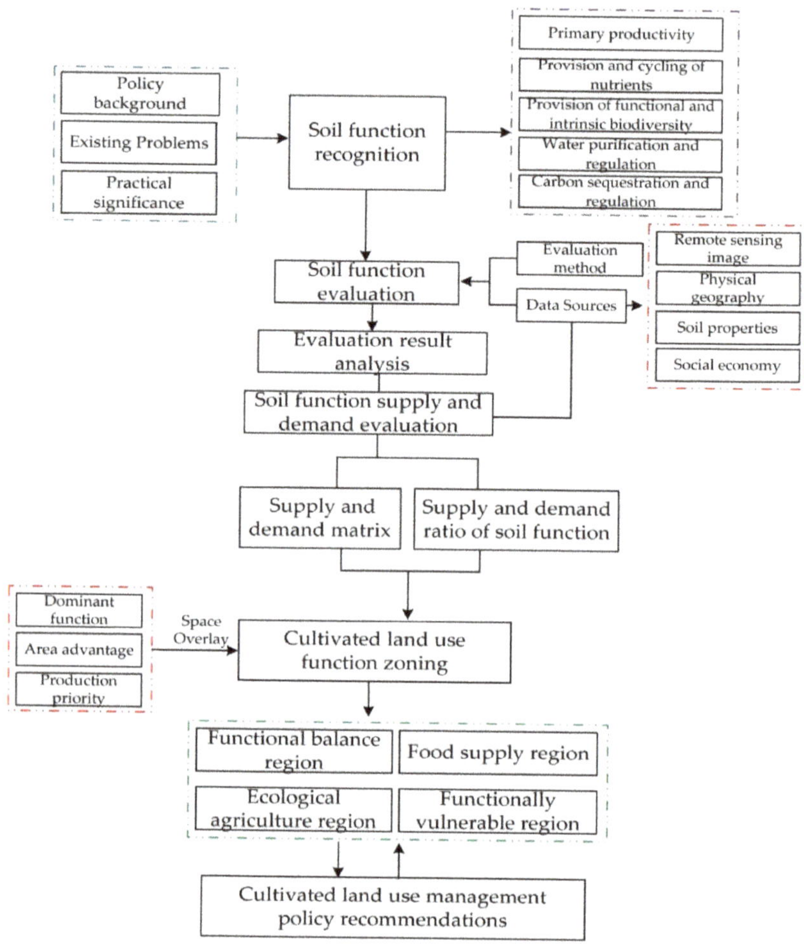

Figure 1. The research frame diagram.

2. Materials and Methods

2.1. Study Area

Being in the northernmost location and highest latitude in China (Figure 2), Heilongjiang province is a vast area, with a high biodiversity environment, diverse landform types, rich natural resources, and a large area of fertile black soil. The province has a total land area of 47.07 million ha, which accounts for 4.9% of China, making it the sixth largest province in the country. Because of its large area of cultivated land per capita, Heilongjiang province is an important agricultural province in China. Given that the plains are flat and open with high soil nutrient content, good agricultural cultivation conditions, and high concentrated and contiguous cultivated land, the region is suitable for large-scale mechanized operations [45]. The cultivated land quality in the Songnen Plain and Sanjiang Plain ranks first in China, with the black soil area accounting for about 33% [46]. Heilongjiang province contains the highest proportion of black soil in China, and nearly 30% of the high-quality cultivated land is located in the black soil area of Northeast China. It is primarily underlaid by Black soil, Meadow soil, Dark brown soil, Albic soil, Chernozem soil, and Chestnut soil. The black soil is loose and high in organic matter content, making it suitable for cultivation. Nevertheless, the recent rapid development of urbanization in Heilongjiang province has had adverse effects on land in the region, causing serious erosion and desertification [47]. Moreover, because of the vast area and the unreasonable structure of cultivated land utilization, the cultivated land use pattern urgently needs to be optimized.

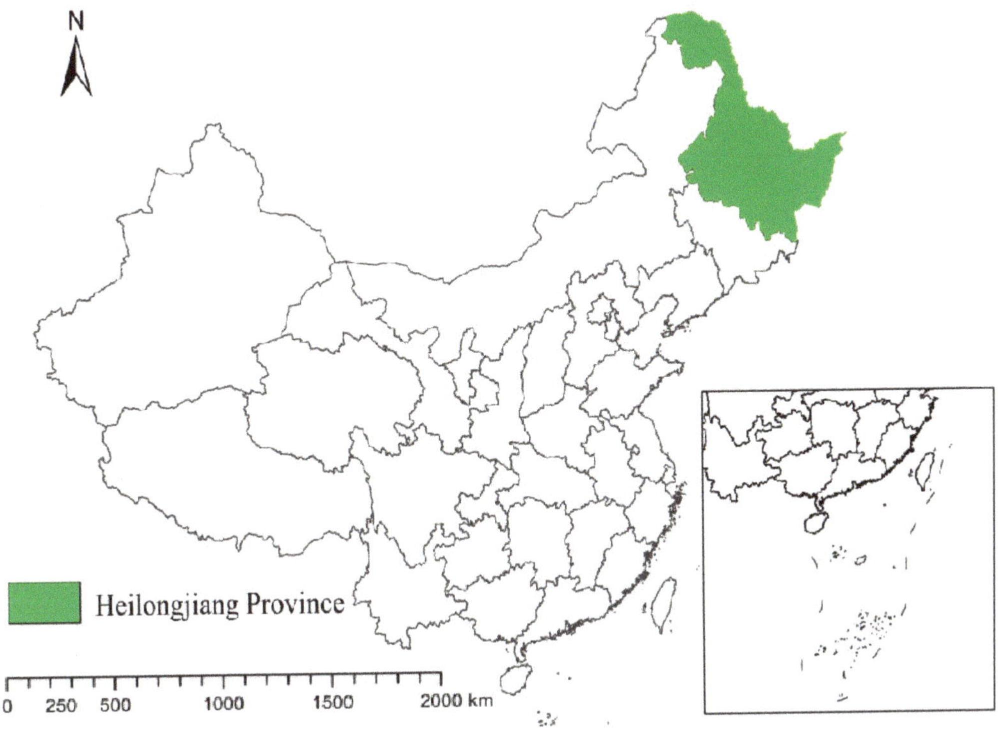

Figure 2. Location of the study area. Source: Drawn by the authors.

2.2. Construction of an Evaluation Method System of Soil Function

2.2.1. Construction of Evaluation Indicator System at the Provincial Scale

The primary productivity function is the ability of soil to produce plant biomass for human use and to provide food, feed, fiber, and fuel within the boundaries of natural or managed ecosystems [48]. The water purification and regulation function is the ability of the soil to absorb, store, and transport water for later use, prevent long-term drought, floods, and soil erosion, and remove harmful compounds from water [49]. The carbon sequestration and regulation function is the ability of soil to store carbon in an unstable form with the purpose of slowing down the increase in the concentration of carbon dioxide in the atmosphere [50]. The provision of functional and intrinsic biodiversity function includes numerous soil biological processes. Soil can provide habitats for animals, plants, and microorganisms, supporting their life activities and protecting biodiversity [51]. The function of provision and cycling of nutrients indicates the ability of soil to absorb and maintain nutrients, produce and retain the nutrients absorbed by crops, promote soil biochemical processes, and provide and retain nutrients [52]. We constructed a provincial-scale soil function evaluation indicator system for cultivated land (Table 1) based on existing large and medium-scale soil function evaluation indicator systems [53–55] that considered the availability of data and was founded on the principles of stability, dominance, spatial variability, regionality, and systemicity [56,57]. Due to difficulty obtaining soil biological data we indirectly characterized the large-scale soil biological diversity from the perspective of the suitability of the soil biological habitat and the suitable soil conditions for biological survival [58,59]. However, during practical evaluation and application, these indicators can be combined with necessary supplements or simplifications according to the actual data collection of the study area. It is important to note that the five soil function models show overlap in terms of data input. For example, soil organic carbon supports multiple soil ecosystem functions that are underpinned by processes such as C sequestration, N mineralization, aggregation, promotion of plant health, and compound retention [60]. Soil organic carbon is considered to be an important indicator for monitoring soil degradation and soil erosion because it affects soil aggregation and stability [61], and is the primary component influencing soil fertility [62]. Soils with higher organic carbon contents have better buffering and a stronger capacity to preserve soil water and fertilizer [63], and the amount of soil carbon sequestration is principally reflected in the soil organic carbon content [64]. In addition, increasing soil organic carbon in agricultural soils can enhance a myriad soil biological processes [60]. Therefore, soil organic carbon is input into all five function models. This creates interconnections between soil functions as observed in the field; however, the threshold values and decision rules used to assess the input attributes are unique for each model [65]. As several indicators use the same input variables, statistical correlations among indicators are expected.

Table 1. Evaluation indicator system for cultivated land soil function.

Soil Function	Indicator
Primary productivity	Soil texture Soil thickness/cm Available soil moisture/mm pH Gravel contents Soil organic carbon Bulk density Slope ≥10° Effective accumulated temperature
Provision and cycling of nutrients	Soil texture Soil thickness pH Soil organic carbon CEC Bulk density
Provision of functional and intrinsic biodiversity	Bulk density Soil organic carbon pH ≥10° Effective accumulated temperature Annual cumulative precipitation Soil thickness Available soil moisture Soil texture

Table 1. *Cont.*

Soil Function	Indicator
Water purification and regulation	Soil texture Soil thickness Available soil moisture Base saturation% Gravel contents Soil organic carbon CEC pH Bulk density
Carbon sequestration and regulation	Soil thickness Soil texture Bulk density Soil organic carbon Gravel contents

2.2.2. System for Evaluation of Soil Function Supply and Demand

Soil function is transferred from cultivated land ecosystems to social economic systems for utilization by humans, and the functional service flow is considered the actual supply. Human demand is formed by the consumption and use of products and services produced by soil functions. When there is small human demand or a high soil function, the soil function enables satisfaction of human demand (actual supply is greater than or equal to human demand). When human demand for the utilization of soil function is beyond its capacity, the actual supply will be less than or equal to human needs. The actual supply of soil functions can be employed by human society. To meet human requirements, however, the actual supply of soil functions may exceed its own capacity. A demand matrix was then assigned through the supply capacity of soil functions and human demand for soil functions to spatially support evaluation of the status of the supply and demand of soil functions, and select the indicator of "supply-demand ratio" to explain the surplus relationship between actual supply and demand, that is, whether the actual supply of soil functions enables us to meet the current requirements of human society [8].

2.3. *Collection and Processing of Basic Data*

Remote sensing data primarily consisted of Landsat satellite remote sensing data and the Normalized Differential Vegetation Index (NDVI). The remote sensing data of Heilongjiang land use stems from the geospatial data cloud, and collected Landsat 8 remote sensing data from September 2018. After acquiring the images, the ENVI 5.1 software was applied for splicing, radiometric calibration, atmospheric correction, geometric correction, mosaic cutting, remote sensing interpretation, and supervision classification of images to six types of land use: cultivated land, woodland, grassland, water area, urban and rural, industrial and mining, residential, and unused with the Kappa coefficients of 0.84 to meet the demand for accuracy [66]. NDVI data originated from NASA, and vegetation index data was processed in a linear manner to acquire the standard vegetation index. Meteorological data such as rainfall and accumulated temperature originated from the National Meteorological Science Data Sharing Service Platform, while soil data including the CEC, pH, gravel content, soil organic carbon, and bulk density were acquired from SoilGrids (https://soilgrids.org), while soil texture, soil thickness, available soil moisture, and base saturation were obtained from the Harmonized World Soil Database. Terrain data were mainly digital elevation data (DEM) derived from the geospatial data cloud. The grid method was applied to divide the evaluation unit in the research, and the cultivated land spot was extracted from the land use status map in 2018 after image interpretation, with a 1 × 1 km grid to sample the study area at equal intervals. ArcGIS 10.6 was employed to merge small map spots to generate 134,449 evaluation units. To facilitate subsequent evaluation and analysis, each unit was given a unique identification code.

2.4. Evaluation Indicator Gradation System

The optimal gradation and assignment of each indicator lies in a value of 100 points with the worst value being 60 points, and the others referring to the classification standard of cultivated land quality in the northeast reported in "Cultivated Land Quality Gradation (GB/T 33469-2016), "Agricultural Land Quality Gradation Regulations" (GB/T 28407-2012) and the gradation and assignment methods described in related studies [67–69]. The value was assigned in accordance with the degree of impact of indicators on soil functions and adjusted based on the foundation of the actual situation of the study area. The soil function supply matrix proposed by Coyle et al. represents the supply proportion (grid size) of five soil functions of different land use types [70], with the horizontal axis indicating land use type and the vertical axis five soil functions, white for primary productivity, purple for provision and cycling of nutrients, blue for water purification and regulation, green for provision of functional and intrinsic biodiversity, and dark gray for carbon sequestration and regulation (Figure 3). This matrix was used as a reference to determine the proportion of cultivated soil functions in this study.

Figure 3. The supply matrix of cultivated land soil function (Adapted with permission from ref. [70] Copyright 2016 Copyright Coyle, et al.).

The importance between two indicators was determined in combination with the opinions of five experts and the differences in accordance with the relevant literature. The solution to the article was to use a weighted average to assign the scoring value of each level of expert, and constructed the judgment matrix to ensure that the restrictions of each indicator and the total sorts and the single hierarchical arrangement passed the consistency test (random conformance rate, CR < 0.1) to gain the gradation and assignment and weight of evaluation indicators shown in Table 2.

Table 2. Grading assignment criteria and weights of cultivated land soil function evaluation indicators.

Soil Function	Indicator	Grading Assignment Criteria of Indicators					Indicator Weight	Data Source of Indicator	Weight of Function
		100	90	80	70	60			
Primary productivity	**Soil texture**	**Loam**	**Clay**		Sand		0.1052	Harmonized World Soil Database	0.6912
	Soil thickness/cm	≥100	(100, 80]	(80, 60]		<60	0.0540		
	Available soil moisture/mm	150	125	100	75	50	0.2100		
	pH	[5.5, 7.5)		[7.5, 8.5)		≥8.5 or <5.5	0.0879	SoilGrids	
	Gravel contents	≤4	(4, 6]	(6, 8]	(8, 11]	>11	0.0316		
	Organic carbon	≥90	(90, 70]	(70, 50]	(50, 40]	<40	0.2857		
	Bulk density g/cm^3	[1, 1.25)		[1.25, 1.45)		≥1.45 or <1	0.1566		
	Slope	<2	[2, 6)	[6, 15)		[15, 25)	0.0457	DEM data	
	≥10° Effective accumulated temperature	≥2600	(2600, 2480]	(2480, 2360]	(2360, 1995]	<1995	0.0232	Meteorological data	
Provision and cycling of nutrients	Soil texture	Loam	Clay		Sand		0.0350	Harmonized World Soil Database	0.2258
	Soil thickness	≥100	(100, 80]	(80, 60]		<60	0.1482		
	pH	[5.5, 7.5)		[7.5, 8.5)		≥8.5 or <5.5	0.2640		
	Organic carbon	≥90	(90, 70]	(70, 50]	(50, 40]	<40	0.4350	SoilGrids	
	CEC/cmol/kg	≥20	(20, 15.4]	(15.4, 10.5]	(10.5, 6.2]	<6.2	0.0363		
	Bulk density	[1, 1.25)		[1.25, 1.45)		≥1.45 or <1	0.0815		
Carbon sequestration and regulation	Soil thickness	≥100	(100, 80]	(80, 60]		<60	0.0756	Harmonized World Soil Database	0.0242
	Soil moisture	Loam	Clay		Sand		0.1427		
	Bulk density	[1, 1.25)		[1.25, 1.45)		≥1.45 or <1	0.2694	SoilGrids	
	Organic carbon	≥90	(90, 70]	(70, 50]	(50, 40]	<40	0.4690		
	Gravel contents	≤4	(4, 6]	(6, 8]	(8, 11]	>11	0.0434		
Provision of functional and intrinsic biodiversity	Bulk density	[1, 1.25)		[1.25, 1.45)		≥1.45 or <1	0.1091	SoilGrids	0.0242
	Organic carbon	≥90	(90, 70]	(70, 50]	(50, 40]	<40	0.3021		
	pH	[5.5, 7.5)		[7.5, 8.5)		≥8.5 or <5.5	0.2001		
	≥10° Effective accumulated temperature	≥2600	(2600, 2480]	(2480, 2360]	(2360, 1995]	<1995	0.0476	Meteorologic-al data	
	Annual cumulative precipitation	≥640	(640, 590]	(590, 550]	(550, 520]	<520	0.0476		
	Soil thickness	≥100	(100, 80]	(80, 60]		<60	0.0917	Harmonized World Soil Database	
	Available soil moisture	150	125	100	75	50	0.0476		
	Soil texture	Loam	Clay		Sand		0.1543		
Water purification and regulation	Soil texture	Loam	Clay		Sand		0.0487	Harmonized World Soil Database	0.0346
	Soil thickness	≥100	(100, 80]	(80, 60]		<60	0.0944		
	Available soil moisture	150	125	100	75	50	0.3037		
	Base Saturation/%	≥80		(80, 50]		<50	0.0252		
	Gravel contents	≤4	(4, 6]	(6, 8]	(8, 11]	>11	0.0487		
	Organic carbon	≥90	(90, 70]	(70, 50]	(50, 40]	<40	0.1913		
	CEC						0.0294	SoilGrids	
	pH	[5.5, 7.5)		[7.5, 8.5)		≥8.5 or <5.5	0.0944		
	Bulk density	[1, 1.25)		[1.25, 1.45)		≥1.45 or <1	0.1640		

2.5. Soil Function Evaluation Model

The multi-factor comprehensive evaluation method was employed to establish a soil function evaluation model of cultivated land:

$$C_i = \sum_{i=1}^{n} F_{ij} W_{ij} \quad (1)$$

where C_i indicates the comprehensive evaluation score of the i-th evaluation unit; F_{ij} is the score of the j-th evaluation indicator of the i-th evaluation unit; W_{ij} is the weight of the j-th evaluation factor of the i-th evaluation unit. The soil function evaluation results were divided into three levels using Jenks natural breaks classification, with the highest quality being the first-level function and the worst quality the third-level function.

2.6. Supply and Demand Ratio of Soil Function

The supply and demand matrix of soil function was established with reference to the ecosystem service matrix method, with the soil function evaluation results as the actual supply of soil functions and application of the following five-point system to the supply of paddy fields and dry land: 1 = low supply capacity, 2 = general supply capacity, 3 = medium supply capacity, 4 = high supply capacity, 5 = very high supply capacity.

First-level soil function supply is assigned 5 points, second-level 3 points, and third-level 1 point. By employing the ecosystem service valuation method and the Delphi method [71], a five-point system was assigned to the soil function requirements of the two types of cultivated land (dryland and paddy field) based on expert knowledge, targeting interviews, and statistical data, with 1 to 5 representing low to high demand. In accordance with the results of supply and demand assignments, a supply and demand evaluation matrix of cultivated soil function was then constructed (Figure 4).

Type of cultivated land	Soil Function									
	Primary productivity		Provision and cycling of nutrients		Provision of functional and intrinsic biodiversity		Water purification and regulation		Carbon sequestration and regulation	
Paddy field	1		1		1		1		1	
	3	5	3	4	3	3	3	5	3	3
	5		5		5		5		5	
Dry land	1		1		1		1		1	
	3	4	3	3	3	2	3	3	3	3
	5		5		5		5		5	

Figure 4. Soil function supply matrix of cultivated land in Heilongjiang province.

The supply-demand ratio refers to whether the total amount of soil function in a certain area can meet the demand, reflecting the balance between the actual supply and demand of soil function. The supply and demand of cultivated land function is classified into three types based on the supply-demand ratio: full satisfaction, general satisfaction, and dissatisfaction.

$$\text{The supply-demand ratio of soil function} = \text{human needs/actual supply} \quad (2)$$

A supply-demand ratio of soil function of less than 1 suggests a surplus state in which the supply of soil function can meet the demand, while a supply-demand ratio of 1 indicates a balanced condition in which the supply of soil function enables basic satisfaction of the demand, and a supply-demand ratio of more than 1 indicates a loss status in which the function of soil supply is unable to meet the demand.

3. Results

3.1. Analysis of Evaluation Results of Cultivated Land Soil Function

(1) The function of primary productivity

The second-level function dominated the paddy fields and drylands, accounting for 46.36% and 53.52%, respectively. As shown in Figure 5, there were second-level function areas scattered in the central part and some counties and cities in the south and north, first-level areas located in the cities of Hulin, Baoqing, Fujin, Raohe, Tongjiang, and Fuyuan County in the northeast, as well as parts of Suileng County, Hailun, and Bei'an in the central part, and third-level districts concentrated in the 15 counties and cities in the southwest and a small area of Ning'an and Dongning County in the southwest. The primary productivity was found to be principally influenced by indicators of slope, soil texture, soil thickness, and pH. Owing to the sticky soil texture, poor drainage, and excessive soil moisture in the southwestern region, the available nutrients are not easily released, resulting in low soil fertility and the accumulation of salt in the soil surface. This leads to intensified soil salinization, high pH, and alkaline soil [72]. There are more sandstorms and droughts in the western region, which causes serious soil erosion and a thin soil layer, resulting in soils not conducive to cultivated production [73]. The southeast is hilly, and most of the cultivated land has a slope of more than 5°, which is subject to sloping ridge-tillage and longitudinal ridge-tillage. This practice, coupled with excessive development and utilization by humans, has resulted in serious soil erosion in the southeast region, therefore, the soil is characterized by stripped topsoil, a thin soil layer, and descending fundamental fertility.

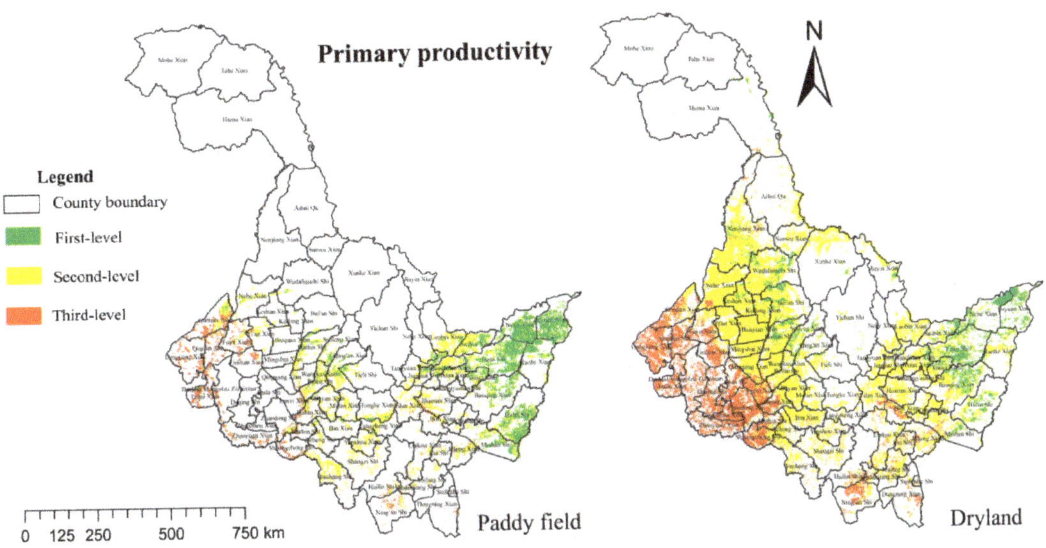

Figure 5. The grade of soil primary productivity function of paddy fields and drylands.

(2) The function of provision and cycling of nutrients

Most land in paddy fields and dry land were second-level function, followed by first-level function, and then third-level function. The third-level zone is distributed in the southwest, and the main influencing factor is CEC, that is, soil with high CEC enables the retention of more nutrients and exertion of better cushioning performance than that with low CEC (Figure 6). There were some third-level zones, especially in the area bordering Inner Mongolia, which is characterized by frequent windy weather year round, a dry climate, and low rainfall. In recent years, deforestation and overgrazing have resulte in serious soil desertification and low CEC in desertified areas, which has impacted th circulation and storage of soil nutrients [74].

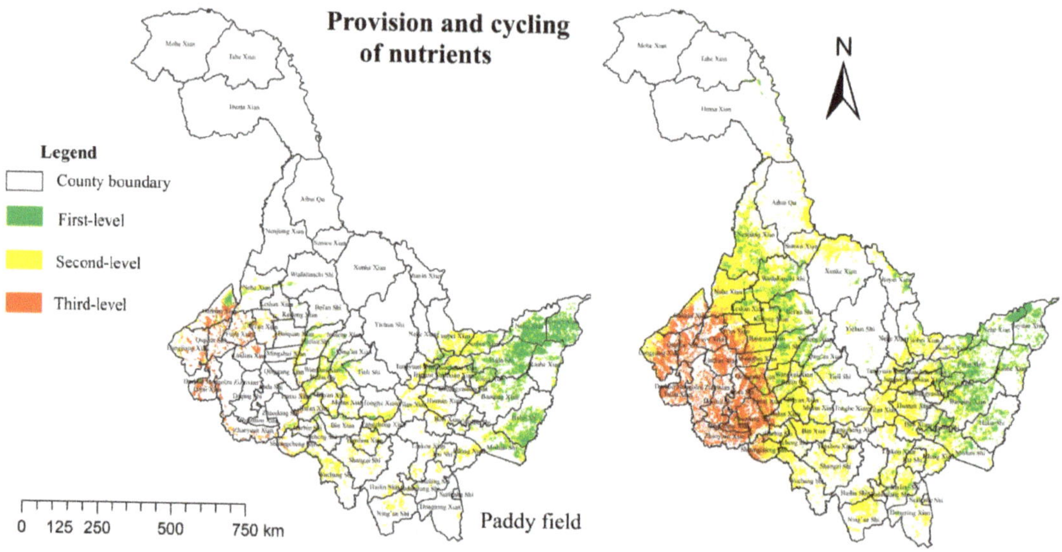

Figure 6. The grade of provision and cycling of nutrients function in paddy fields and drylands.

(3) The function of provision of functional and intrinsic biodiversity

The provision of functional and intrinsic biodiversity in paddy fields and drylands demonstrated a decreasing spatial distribution pattern from northeast to southwest (Figure 7). This function was primarily affected by pH and annual accumulated precipitation, and the suitable pH range is an imperative condition for the survival of animals and plants. The soil types in the southwestern part consisted of meadow chernozem, hydrochloride meadow soil, and hydrochloride chernozem, with the average pH ranging from 8 to 8.5, showing strong alkalinity and low nutrient content. Additionally, the low annual accumulated precipitation in the southwest is unable to meet the requirements of animal and plant survival. Land utilization methods influence the composition of soil biological communities [74], and paddy soil can provide soil organisms with a more stable source of nutrients such as water and organic carbon sources. In addition, paddy soil holds more soil biomass than dry land soil, which suggests that paddy soil biodiversity is slightly better than that of dry soil [75].

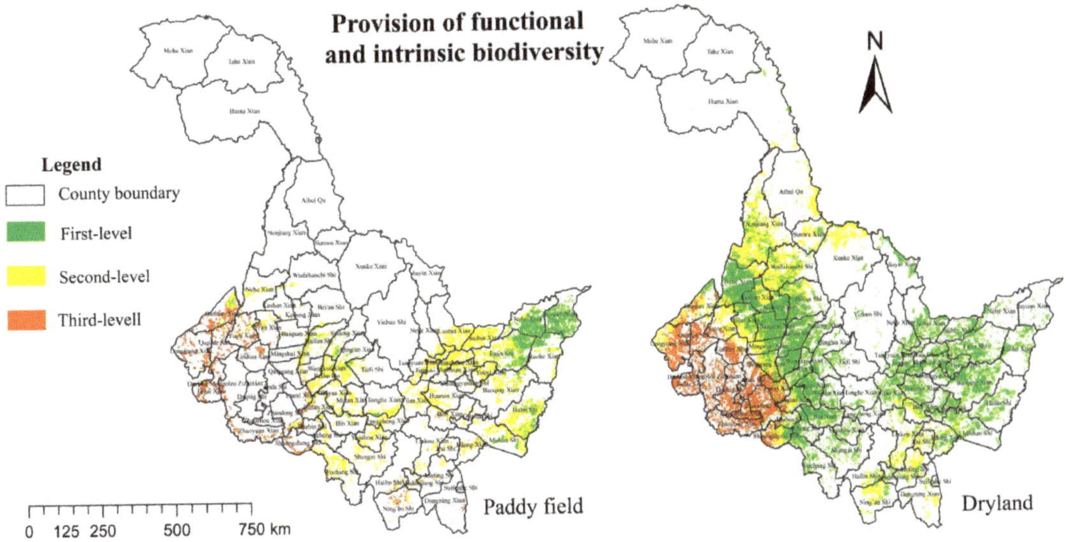

Figure 7. The grade of provision of functional and intrinsic biodiversity function in paddy fields and drylands.

(4) The function of water purification and regulation

Water purification and regulation functions in most areas of paddy fields and dry land are at a high level (Figure 8). The first-level area had the widest range, primarily distributed in the central, northeastern, and most of the northwestern areas of the province. The second and third-level area was mainly scattered in the north of Ning'an, the east of Shuangcheng, the middle of Boli County, the west of Longjiang County, and the northeast of Fuyu County. As the prime factor, soil organic carbon and available soil moisture were the main influencing factors, and higher organic carbon content contributes to the purification and buffering function [76]. When compared with dryland soil, paddy field soil has stronger carbon sequestration ability and organic carbon content, as well as more stable soil natural water content. The content of organic carbon and available soil moisture in the southwest was lower than the overall average level of the study area, and the unequal water and heat and seasonal rainfall have caused serious soil erosion, desertification, and thinning of the soil layer, which will affect the functional performance of water purification and regulation.

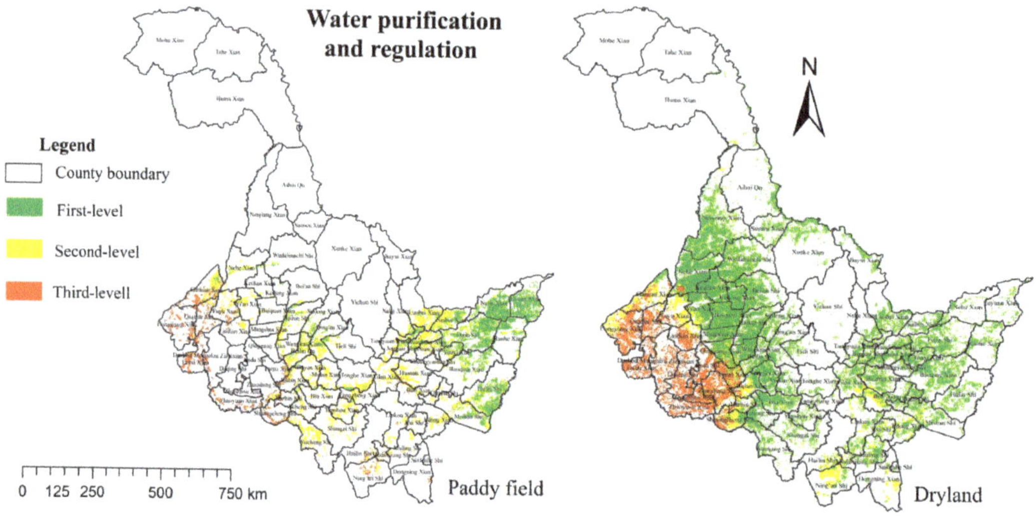

Figure 8. The grade of soil water purification and regulation function in paddy fields and drylands.

(5) The function of carbon sequestration and regulation

The first-level of carbon sequestration and regulation districts were principally distributed in the cities of Tongjiang, Fujin, Hulin, and Mishan, as well as Raohe County, Fuyuan County, and Baoqing County in the northeast, and part of the cities of Wudalianchi, Bei'an, and Hailun as well as Suileng County in the middle. The second-level area was concentrated in the middle, south, and north, while the third-level area includes most of the southwest and a small part of the south (Figure 9). Cultivated land utilization is an important factor of human activities that affects the carbon cycle of soil ecosystems, and there are significant differences in soil organic carbon content under different cultivated methods. Some studies have shown that rice has the best ability to absorb carbon and produce oxygen among the main crops per unit area, resulting in an obvious carbon sequestration effect of paddy soil and high organic carbon content [77,78], and the organic carbon content serves as a significant factor. Sanjiang Plain is located in the northeast, where the soil conditions are better than those in the southwest because there is a large area of meadow soil and black soil that is soft, fertile, and rich in organic matter. Organic matter with high content can improve the structure of soil aggregates, increase the exchange capacity of soil ions, and strengthen the ability of soil to fix carbon and release oxygen. There is serious soil erosion in the southwestern region, and the existing research demonstrates that the low organic matter content and clay particles of eroded soil lead to the 20% lower carbon sequestration capacity than that of high-quality soil. Therefore, the spatial distribution law gradually decreased from northeast to southwest.

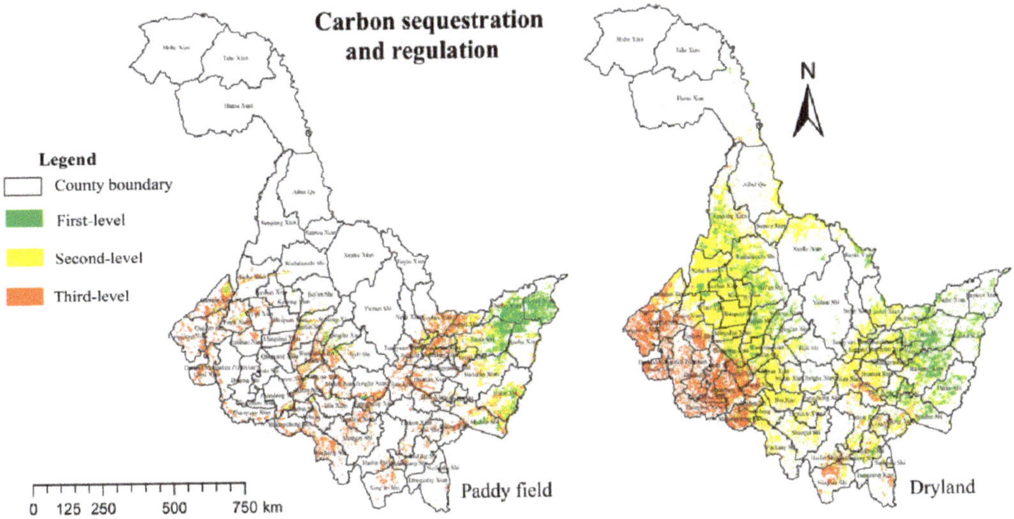

Figure 9. The grade of soil carbon sequestration and regulation function in paddy fields and drylands.

(6) Comprehensive Evaluation Results of Soil Function

The soil function gradation presented a spatial distribution pattern of high in the northeast and low in the southwest, with the third-level area distributed in a small portion in the south (Figure 10). This function predominated by the second-level area of soil function, which accounts for 46.59% of land area. The first-level area was mainly distributed in the east and northeast, and as well as in a small portion of the central part in strips, accounting for 28.56% with comprehensively unrestricted or low indicators and highly functional soil. The third-level area was scattered in 17 counties and cities in the southwest of Ning'an, the eastern part of Dongning County, and the intersecting part of the northern area of Boli County and the eastern part of Yilan County. The dry climate, low rainfall, and frequent windy weather year-round in the southwestern part has caused serious soil erosion and desertification. The soil types in this area include meadow chernozem, hydrochloride meadow soil, and hydrochloride chernozem, which have a high calcium carbonate content, average pH of more than 8, and strong alkalinity, as well as a low content of organic matter and micronutrient elements that influence the performance of soil function [79]. The northeast region is located in the Sanjiang Plain, which is characterized by small slopes, flat terrain, and balanced water and heat resources. The soil types in this region primarily consist of meadow soil and black soil, which are thick and fertile.

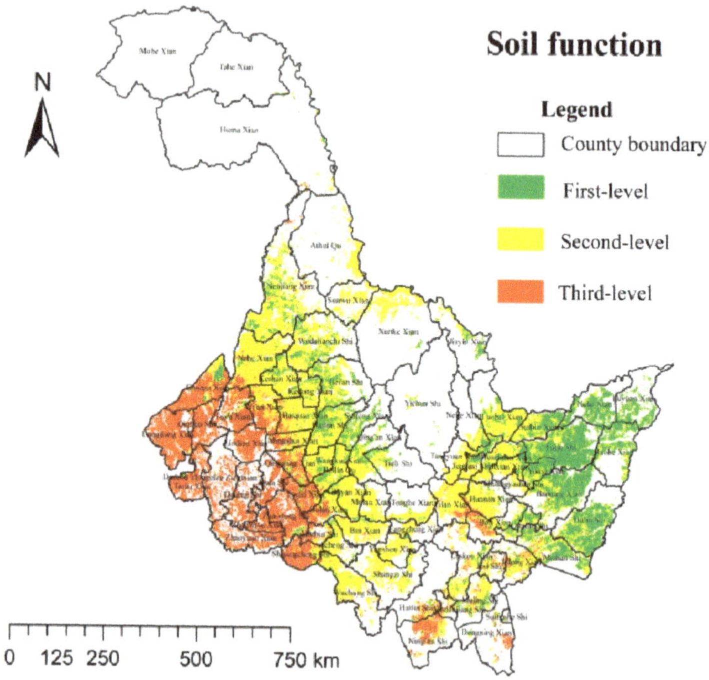

Figure 10. The grade of soil function in Heilongjiang province.

3.2. *Soil Function Supply-Demand Ratio*

Through regional statistics (Figures 11 and 12), the supply and demand of primary productivity function was predominated by functional dissatisfaction, which accounted for 71.32%. This was followed by the functional satisfaction area, which was scattered in the east and occupies 17.00%. The spatial pattern of provision and cycling of nutrient functions showed a gradual increase from southwest to northeast, and the functional satisfaction area is primarily distributed in the northeast, which accounted for 39.39%. The zone of general satisfaction was found to be roughly the same as that of dissatisfaction, accounting for 29.03% and 31.58%, respectively, and this zone was concentrated in the southwest. The functional and intrinsic biodiversity function of 72.53% can satisfy the requirements of human society, while 13.00% scattered in the central and southern can basically meet this demand, and 14.47% was the functional dissatisfaction area, which was mainly distributed in the southwest. The spatial distribution of water purification and regulation functions and carbon sequestration and regulation functions showed the same trend, presenting a spatial pattern of high in the middle and low in the surrounding area. Overall, more than 50% had strong supply capacity to enable a surplus, while most of the southwestern region and a small part of the south had inferior supply capacity that resulted in a state of insufficient supply and weak functionality.

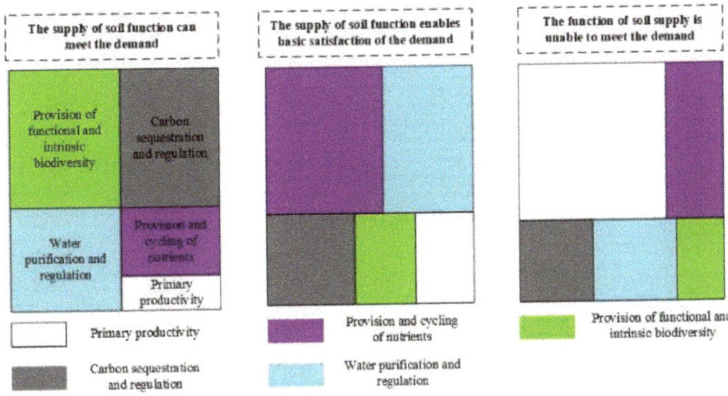

Figure 11. Proportion of supply and demand area of cultivated land soil function.

From a spatial point of view, the supply capacity of the five soil functions in the northeast was relatively strong and in a surplus state, while the supply and demand in the southwest and a small part of the southeast are out of balance and cannot meet the functional requirements. The main reason for this spatial distribution was that the northeast portion is in the Sanjiang Plain, where there is uniform water and heat, flat terrain, less erosion and soil-water loss, and a deep soil layer. Moreover, the region is dominated by a large area of black soil and meadow soil with a balanced acidity-alkalinity, rich nutrient storage, high organic carbon content, and strong water storage and fertilizer retention capacity, which is beneficial to the growth of animals and plants [80]. The main soil types in the southwestern region are aeolian sandy soil and saline-alkali soil. In some low-lying areas, standing water lies in the large area of sodic alkaline soil that bears high salt content and soil alkalinity as well as poor ventilation and water permeability, and is unsuitable for animal and plant growth. Additionally, the border with inner Mongolia is characterized by strong winds throughout the year and little rainfall, making the region susceptible to wind erosion and drought, prominent soil sandification problems, and poor water and fertility retention. As a mountainous area, the southeastern region is characterized by large slopes, serious soil erosion, and water-soil loss, resulting in thin cultivated soil layers, loss of soil nutrients, low soil fertility, susceptibility to drought and waterlogging, and inferior functionality.

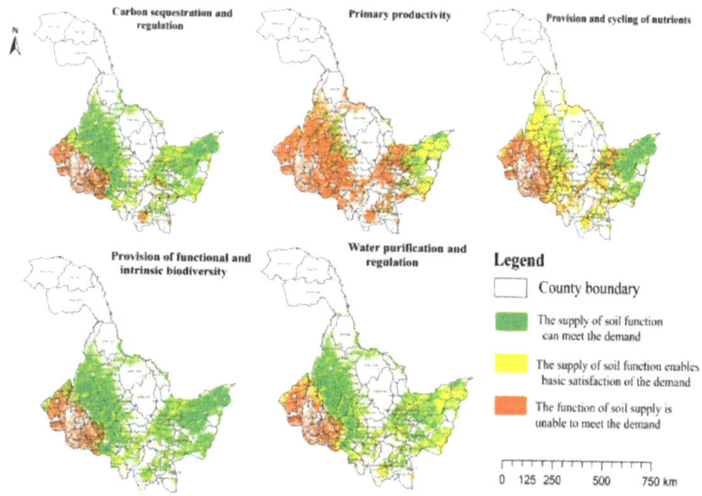

Figure 12. Distribution map of supply-demand ratio of cultivated land soil function.

3.3. *Identification of Dominant Soil Function of Cultivated Land*

Soil has a variety of functional attributes and states, which are usually characterized by a combination of functions. Therefore, we determined the best soil function in the evaluation unit as the dominant function (Figure 13). The largest area of dominant function was water purification and regulation, with 6.11 million ha, or 34.89% of the total area. This was mainly distributed in Suibin County and Luobei County in the northwest and northeast. Soil has a strong ability to absorb and store water, and can therefore effectively prevent natural disasters such as droughts and floods in the region. The second largest area of dominant function was the carbon sequestration and regulation function area, which accounted for about 29.57% of the total and was mainly distributed in the southwest. This area was characterized by a high capacity to store carbon in an unstable form, which can effectively slow the release of carbon dioxide into the atmosphere and adjust the field microclimate. The provision and cycling of nutrients and provision of functional and intrinsic biodiversity were slightly less distributed, while the dominant function with the smallest area was the primary productivity, which had an area of only 0.24 million ha, or 1.35% of the total. This area was mainly distributed in Fujin in the east of the study area and Hailun in the center of the area.

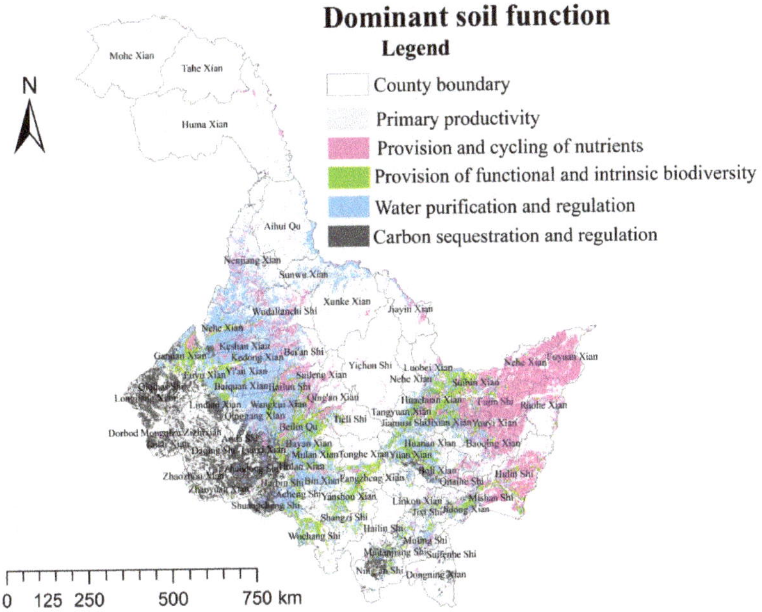

Figure 13. Dominant soil functions of cultivated land in Heilongjiang province.

3.4. Cultivated Land Use Zoning Based on Soil Functions

To clarify the dominant soil functions and functionalities and combine these with the evaluation results for the five soil functions while ensuring the spatial continuity and integrity of the division, the county and city are set as the basic units and the soil function supply and demand evaluation results are superimposed on the basis of the dominant function of the soil to divide the cultivated land into use function zones (Figure 14).

(1) Functional balance region

Most areas in Heilongjiang province were functionally balanced. The five soil functions were in a balanced development, and the compound value of the soil was relatively high. Based on the favorable soil functional conditions, it is necessary to strictly protect and expand the high-yield and stable cultivated land. While maintaining the soil function of the existing cultivated land, deep ploughing is used to break the bottom of the soil plow and increase the permeability of the soil, improve the water storage capacity and water retention capacity of the soil, and promote increased crop production [81]. In the functional balance region, we suggest the characteristics of agricultural production be highlighted with special attention given to various functions such as farmland production ecology. Additionally, the layout of farmland should be optimized and the development of ecological agriculture promoted. To avoid destroying the soil, the requirements of "hiding grain in the ground" should be followed to develop ecological fertile land. The compound functions of farmland soil agricultural production, water conservation, and biodiversity should be considered. Additionally, construction of an agro-ecological leisure tourism complex to guide the rational and compound use of farmland space and further enhance its versatility is suggested [82].

(2) Food supply region

Food supply regions were dominated by the functions of primary productivity and nutrient cycling and supply. The terrain in these areas is mainly valley plains, hilly plains, and slightly inclined high plains, and these areas are characterized by abundant surface water, fertile black soil, high natural fertility, rich mineral nutrients, moderate texture, good agricultural utilization conditions, large and concentrated cultivated land, and high productivity. However, this region was densely populated with a limited amount of cultivated land. The region was also affected by restrictions such as urban construction and poor ecological soil functions. The food supply region could benefit from increased road construction to increase the accessibility of field roads, as well as improved cultivated land irrigation and drainage equipment to ensure the water conditions required for crop growth are met. Additionally, farming conditions should be improved through measures such as land consolidation and land development. This area could also be enhanced by adopting agricultural modern management measures [72]. Food supply regions are considered key areas for high-standard farmland construction. In such regions, it is important to promote measures to return straw to the field, increase the number of soil aggregates, increase soil porosity, and reduce soil compactness [83]. At the same time, these areas can adopt agricultural modernization management measures to enhance the overall utilization efficiency of farming, promote agricultural production mechanization and industrialization, improve farming techniques, strengthen agricultural training and guidance for farmers, and introduce characteristic cash crops based on regional characteristics. It is also possible to mobilize farmers' production enthusiasm through the implementation of agricultural subsidies to ensure high and stable grain production in the region [84]. Based on the original land, we suggest implementation of cultivated land utilization and protection measures, promotion of biological control measures, optimization of farming methods, and promotion of cultivated land rotation while avoiding long-term overload cultivation. Finally, soil can be improved through various utilization measures such as rotation, allowing fields to lay fallow, and alternative planting ecological quality [85].

(3) Ecological agriculture region

The north-central and southeastern parts of Heilongjiang province are typical ecological agricultural regions. The terrain in the north-central part is characterized by hills and undulating mountains. The complex terrain is not conducive to agricultural farming, and there are serious water and soil erosion phenomena in the region [86]. This is mainly because part of the cultivated land was woodland and grassland before reclamation, and the soil was fertile. However, because large areas have been subject to land reclamation for many years, the original forest belt was cut down by farmers and the vegetation coverage rate declined, causing gully erosion to become more serious and soil fertility to decrease annually. As a result, the functions of primary productivity and nutrient supply and circulation are weak [87]. However, the climate is suitable and rainfall is sufficient. For cultivated land with a larger slope, the farming method can be improved. Alternatively, some fruit trees with higher economic benefits can be selected according to local climatic conditions, and a combination of trees should be adopted by fruit farmers. The planting mode takes advantage of the developed root system of fruit trees to reduce soil erosion, consolidate soil, store water, and improve the ecological environment. For some difficult-to-use hills and sloping lands, it is possible to return farmland to grassland, rotate crops and pastures, or plant multiple crops, all of which can improve soil fertility and increase the economic benefits of farmers [88]. Because of the characteristics of the regional topography and the advantages of mountain landscapes, biodiversity can be protected and mountain tourism and agricultural areas can be developed.

(4) Functionally vulnerable region

Functionally vulnerable regions were mainly distributed in counties and cities in the southwest, which is characterized by low-lying, poorly drained, and poor climatic conditions. These regions border Inner Mongolia and are affected by the Siberian dry monsoon, which is characterized by little rainfall and high evaporation. These conditions eventually lead to soil water storage. In these regions, poor capacity and serious soil salinization have led to soil fertility loss, soil stickiness, soil compaction, and difficulty in utilization. These areas have also long been subject to erroneous production activities such as blind land reclamation, improper irrigation and drainage, and over-grazing, which have caused serious soil desertification, fragile ecological functions, and restricted use of soil functions. Focusing on soil improvement, we suggest advanced irrigation and drainage technology be introduced, and that reasonable irrigation and drainage be used to promote soil desalination. Additionally, the groundwater level should be controlled to a critical depth to effectively prevent salt reverse. It is also important to choose salt-tolerant crops and promote the planting of salt-tolerant or salt-resistant crops, as well as to encourage the return of straw to the field, increase the application of organic fertilizers and soil nutrients, improve soil physical and chemical properties, reduce soil water evaporation, promote soil aggregation, and stabilize soil in these regions [89]. Integrating fragmented land and developing innovative agricultural planting technology will help break through the limitations of climate and water resources. For soil desertification areas, ecological conservation can be adopted, starting with protection of the land ecosystem, as well as adjusting the planting structure and layout [90].

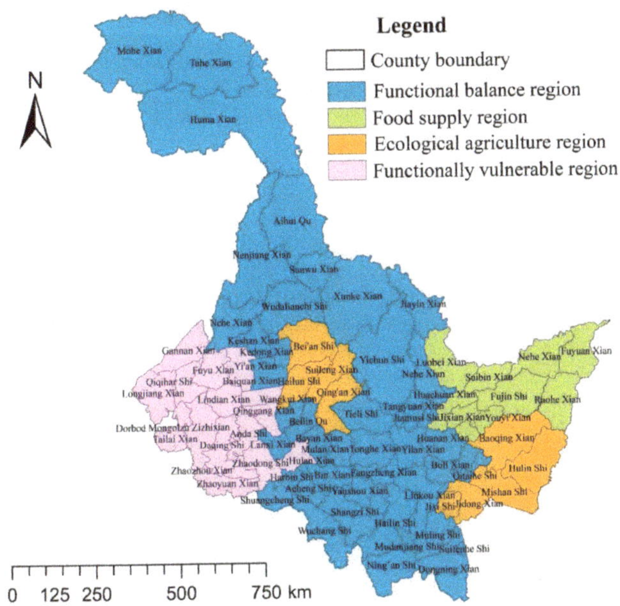

Figure 14. Cultivated land use zones in Heilongjiang province.

4. Discussion and Prospects

The establishment of macro-scale soil function evaluation theories and methods is a bottleneck that needs to be broken through in the field of natural resource management in China [3]. This article describes an evaluation method that is pertinent for the study area. However, the method described herein would be difficult to apply to other areas, especially while large-scale soil biodiversity surveys in China are still in their infancy. Large-scale evaluation of cultivated land soil function is a complex and comprehensive process. The availability of indicators and data quality are essential to soil function evaluation. Here, we discuss some limitations of our study dataset and then look in depth at how the evidence supports our research hypotheses. The protection of national food security is inseparable from the management and protection of soil health. It is important to re-evaluate the properties and functions of soil, and to plan and expand the mission of soil science. Based on this, the present article discusses the current key research directions, provides a reference for future studies, and provides guidance for agricultural production.

4.1. Soil Function Evaluation Scale and Indicator

Soil function assessments have evolved over time with changes in the objectives, tools, methods, and indicators of these assessments [91]. For example, the main objective of assessments before the 1970s was determining the suitability of soils for crop growth, while after 2010 the objective changed to evaluation of multi-functionality, ecosystem services, resistance, and resilience of soils. This change in objectives has resulted in more advanced methods and novel indicators for determination of soil function being developed (Table 3).

Table 3. Evolution of soil function assessment over time in terms of objectives, tools, methods, and indicators.

	Before 1970	1970–1990	1990–2010	2010 Onwards
Objectives	suitability for crop growth	productivity	productivity, environment, animal/human health	multi-functionality, ecosystem services, resistance and resilience
Tools	visual/analytical/digital			
Methods	soil assessment based on color, structure, macrofauna		soil quality test kits, and (bio)chemistry, multivariate statistics	high-throughput methods, add microbiology
Indicator trends	few indicators	many indicators	minimum data sets	novel indicators

Soil is an ecosystem that has different states under different spatial scopes, different time frames, and different management methods [92]. Conducting analyses on different scales solves different problems. Additionally, soil is the result of the combined effects of soil-forming factors at different scales [93]. At the national scale, policy makers need to analyze the overall quality and trends of soil resources to ensure national food security. At the watershed scale, the general public hopes to maintain a healthy production and living environment in the region. At the field scale, farmland managers are concerned about the productivity and sustainability of the soil. The existing literature mainly focuses on the evaluation of soil quality at small and medium spatial scales under specific land use or agricultural land conditions. However, there is almost no related research on cross-spatial scales or multi-functional dimensions [94], which makes it difficult to satisfy different levels of soil health protection decision-making and management behavior needs. Therefore, soil protection needs to be controlled at the macro level, but also implemented at the micro level. China's land resource management and utilization decision-making levels include townships, counties, cities, provinces, and countries. The indicators, units, methods, and applications of soil function evaluations at different scales have their own characteristics [3]. The soil function evaluation indicators and importance of different scales are not fixed, but should be adjusted according to the temporal and spatial change in characteristics of indicators and the purpose of soil management. Data conditions are the main factor restricting the selection of indicators; however, from the perspective of scale differentiation, the availability of data and the selection of indicators are consistent. The large spatial scale mainly focuses on the endogenous potential and evolution trends of soil functions, and static indicators that can reflect the inherent properties of the soil should be used and be relatively consistent with the results of wide-area digital soil mapping. The small spatial scale mainly focuses on the performance of soil functions in the current state. It is necessary to adopt dynamic indicators that can respond to farmland management measures and be obtained through sampling tests [95,96]. Because China has a vast territory and large geographical differences, it is possible to select representative plots under different soil types, land use types, ecological environments, and soil management scenarios to reveal the law of spatial differentiation characteristics of soil functional supply capacity at different scales, and to establish a minimum indicator set for soil function evaluation in different regions [97].

4.2. Classification and Evaluation of Soil Ecosystem Service Function

The soil system is a living system that performs soil functions that provide various ecosystem services [98]. Soils provide a wide range of goods and services that are important for human well-being and sustainable socio-economic development, which are collectively known as ecosystem services [26,99]. However, in recent years, soil function has declined sharply due to the influence of climate change and human activities, seriously threatening the survival and development of human beings [100]. Soil ecosystem services are terminal services formed by soil ecosystems under the effect of human value orientation and the direct contributions of ecosystems to human benefits [101]; accordingly, each soil ecosystem service reflects different soil functions [102]. The various divisions of soil functions have made people realize the multiple uses of soil, as well as the effects that environmental interference and anthropogenic activities have on soil functions. Currently, one of the greatest challenges is determining how to quantify soil ecosystem services [103]. Some unsustainable management measures such as traditional farming, unreasonable fertilization, and blind use of herbicides and pesticides have led to the devaluation and degradation of soil ecosystem services. Conversely, sustainable soil management measures such as conservation tillage, organic agriculture, cover crops, and crop diversification have positive effects on soil ecosystem services [104]. Soil function evaluation has shifted from focusing only on soil production functions in the early days to comprehensive soil multi-function evaluation. Comprehensive coordination and weighing of the diversity of soil ecosystem services are conducted through comprehensive evaluation of soil functions to achieve sustainable soil management and use [105]. However, the comprehensive evaluation of soil functions cannot be simply additive, as soil is suitable for plant habitat and crop production functions simultaneously. Therefore, when conducting a comprehensive evaluation, attention should be paid to the correlation between various soil functions [106]. The soil system integrates physical, chemical, and biological factors and processes, soil properties that are key to providing a particular soil ecosystem service are identified and related to context-specific environmental variables. Thus, soil ecosystem services assessments can benefit from research into dynamic spatio-temporal modelling of soil properties and processes [107].

4.3. Pay More Attention to Soil Biodiversity

Biodiversity not only provides humans with abundant food resources, but also plays an essential role in maintenance of water and soil, climate regulation, water conservation, and air purification [108,109]. Ecologists often pay more attention to above-ground biodiversity than soil biodiversity [110]. However, these two types of diversity are closely related, and soil biodiversity significantly affects the function of ecosystems [111,112]. In addition, Cameron et al. found that areas of mismatch between aboveground and soil biodiversity cover 27% of the terrestrial surface of the Earth [113]. The soil acts as a composite living entity [114], and a more diverse soil biome is conducive to increasing the nutrients needed to produce high-yield and high-quality crops, as well as to protecting crops from pests, pathogens, and weeds. Long-term application of organic fertilizers and chemical fertilizers can be properly combined to increase the community diversity of key soil groups [115]. However, the input of a large number of chemicals and unreasonable management during intensive agricultural production can lead to decreased soil biodiversity, an imbalance in microbial flora, simplification of soil food webs, increased occurrence of pests and diseases, and severe reductions in crop yields [116]. Indeed, the quantity of soil organisms and their activities can sensitively reflect changes in soil health and human management, and can therefore be used as an ecological indicator of soil changes. Soil biodiversity indicators mainly include microbial biomass, soil microorganisms, pathogens, large and medium-sized soil animals, soil enzymes and plants. The EcoFinders (Ecological Function and Biodiversity Indicators in European Soils) project of the European Union has been trying to identify indicators that can comprehensively reflect soil biodiversity, including soil biological community structure and functional gene expression [117]. However, this project is challenged by the following: (1) measurement of most biological indicators takes a long time and the technical requirements and costs are high [118]; (2) there are many types, individuals, and rapid changes among soil organism populations, as well as a certain amount of functional redundancy [99]; (3) a high-resolution, quantitative understanding of the abundance or functional composition of active soil organisms has not been developed to date [119]. Soil is a challenging habitat, and finding clear and unambiguous relationships between soil characteristics and the overall soil biodiversity is very difficult. At a large scale, it is possible to indirectly characterize soil biological diversity from the perspective of suitability of the soil biological habitat and the suitable soil conditions for biological survival. The most important drivers of soil biodiversity on a global scale have been found to be aridity, mean annual temperature, plant richness and cover, soil pH, C content, and clay percentage [120].

4.4. Utilization and Protection of Black Soil Cultivated Land

Black soil is the most fertile and productive soil, which makes it suitable for farming. Strengthening the protection of black soil has become a great concern to all sectors of society [121]. The Great Plains of Ukraine, the Mississippi River Basin in North America, the Black Soil Region of the Northeast Plain of China, and the Pampas Prairie of South America are the four largest black soil regions in the world. Heilongjiang province is located at the core of the Black Soil Region of Northeast China. Since the reclamation of black soil in China, there have been several periods of rotation and fallow, low-intensity utilization by humans and livestock, and high-intensity utilization of mechanization. The natural fertility of black soil has declined annually, the soil has degraded, the organic matter content has been sharply reduced, the cultivation layer has become thin, and the plow bottom has thickened [45]. Because of long-term and unreasonable soil management techniques such as soil plowing, allowing cultivated layers to remain bare, and wanton land reclamation, the original stable micro-ecosystem of black soil in this region has been broken, resulting in soil biodiversity, nutrient maintenance, carbon storage, water purification and water regulation degradation [122,123]. Protection of black soil requires engineering, agronomic, biological and other measures to create a deep and fertile cultivated layer that stabilizes and increases crop yields. After several years of cultivation, fertilization plans need to be developed to maintain and improve soil fertility [6]. Soil fertilization measures include organic fertilizer application and crop rotation [120]. Returning more than 80% of agricultural production waste to the field after the harmless treatment can ensure that the soil organic matter does not decrease, and may even lead to its slowly increasing. Additionally, straw deep return protection technology can be used to maintain and increase the organic matter content of the 0–20 cm black soil layer [124]. It is worth noting that although plowing is conducive to the release of soil microbial activities and soil nutrients, as well as the mineralization of soil materials, plowing also loosens the topsoil, which accelerates the wind and water erosion of the soil and accelerates the decomposition rate of soil organic matter. No-tillage straw mulch is a typical protective tillage measure, but it is still not clear if no-tillage, no mulch, or less tillage is the best protective tillage measure [125].

5. Conclusions

Here, an evaluation system for the five functions of primary productivity (provision and cycling of nutrients, provision of functional and intrinsic biodiversity, water purification and regulation, and carbon sequestration and regulation) was developed based on soil attribute data. Using the soil function discriminant matrix method, soil functions based on supply-demand ratios were constructed to evaluate the current status of supply and demand of soil functions. The comprehensive evaluation results of soil functions in Heilongjiang province demonstrated a distribution pattern of high grade in the northeast and low grade in the southwest, mostly in second-level areas. The paddy fields and dry land showed similar primary productivity functions as well as provision and cycling of nutrient functions, while the intrinsic biodiversity, water purification and regulation, and carbon sequestration and regulation functions of paddy fields are better than those of dry land. The actual supply of primary productivity functions in 71.32% of the region cannot meet the current needs of life. The dominant function of soil in 34.89% of the area is water purification and regulation, and most of the cultivated land belongs to the functional balance region. Cultivated land use zoning and optimization research ideas based on soil multifunctionality make up for the lack of research on large-scale agricultural soil health management and protection in China, and are of great importance to the targeted protection and utilization of black soil.

Author Contributions: Conceptualization, K.W. and R.Z.; methodology, K.W. and R.Z.; validation, J.L.; formal analysis, R.Z. and J.L.; investigation, K.W.; resources, K.W.; data curation, R.Z. and J.L.; writing—original draft preparation, R.Z.; writing—review and editing, K.W. and R.Z.; visualization, R.Z. and L.K.; supervision, K.W.; project administration, K.W.; funding acquisition, K.W. All authors have read and agreed to the published version of the manuscript.

Funding: This study was supported by the National Key R&D Program of China (No. 2018YFE0107000).

Institutional Review Board Statement: Not applicable.

Informed Consent Statement: Not applicable.

Data Availability Statement: The datasets used and/or analyzed during the current study are available from the corresponding author upon reasonable request.

Acknowledgments: Thanks for research assistance from Ganlin Zhang, Fengrong Zhang, Weidong Shan, Baiming Chen, and Wenju Yun. The insightful and constructive comments of the anonymous reviewers are appreciated.

Conflicts of Interest: The authors declare that they have no conflicts of interest.

References

1. Robinson, D.A.; Fraser, I.; Dominati, E.; Davíðsdóttir, B.; Jónsson, J.O.G.; Jones, L.; Jones, S.B.; Tuller, M.; Lebron, I.; Bristow, K.L.; et al. On the Value of Soil Resources in the Context of Natural Capital and Ecosystem Service Delivery. *Soil Sci. Soc. Am. J.* **2014**, *78*, 685–700. [CrossRef]
2. Brevik, E.C.; Sauer, T.J. The past, present, and future of soils and human health studies. *Soil* **2015**, *1*, 35–46. [CrossRef]
3. Wu, K.N.; Yang, Q.J.; Zhao, R. A discussion on soil health assessment of arable land in China. *Acta Pedol. Sin.* **2021**, *58*, 537–544. [CrossRef]
4. FAO; ITPS. *Status of the World's Soil Resources (SWSR)—Main Report*; Food and Agriculture Organization of the United Nations and Intergovernmental Technical Panel on Soils: Rome, Italy, 2015. Available online: http://www.fao.org/3/a-i5199e.pdf (accessed on 31 January 2015).
5. Zhao, Q.G.; Sun, B. Soil quality and sustainable environment: The definition and evaluation method of I. Soil quality. *Soils* **1997**, *29*, 113–120. [CrossRef]
6. Han, X.Z.; Zhou, W.X. Research perspectives and footprint of utilization and protection of black soil in northeast China. *Acta Pedol. Sin.* **2021**. Available online: https://kns.cnki.net/kcms/detail/32.1119.P.20210416.1456.002.html (accessed on 19 April 2021).
7. Wang, J.K.; Xu, X.R.; Pei, J.B.; Li, S.Y. Current situations of black soil quality and facing opportunities and challenges in northeast China. *Chin. J. Soil Sci.* **2021**, 1–7. [CrossRef]
8. Zhao, R.; Wu, K.N.; Yang, Q.J.; Feng, Z.; Zhao, H.F.; Zhang, Z. Farmland Soil Health Evaluation Method Based on Soil Function and Soil Threat. *Trans. Chin. Soc. Agric. Mach.* **2021**. Available online: https://kns.cnki.net/kcms/detail/11.1964.s.20210408.1602.023.html (accessed on 9 April 2021).
9. Paul, C.; Kuhn, K.; Steinhoff-Knopp, B.; Weißhuhn, P.; Helming, K. Towards a standardization of soil-related ecosystem service assessments. *Eur. J. Soil Sci.* **2020**, 1–16. [CrossRef]
10. Banwart, S. Save our soils. *Nat. Cell Biol.* **2011**, *474*, 151–152. [CrossRef]
11. Boyd, J.; Banzhaf, S. What are ecosystem services? The need for standardized environmental accounting units. *Ecol. Econ.* **2007**, *63*, 616–626. [CrossRef]
12. Milne, E.; Banwart, S.A.; Noellemeyer, E.; Abson, D.J.; Ballabio, C.; Bampa, F.; Bationo, A.; Batjes, N.H.; Bernoux, M.; Bhattacharyya, T.; et al. Soil carbon, multiple benefits. *Environ. Dev.* **2015**, *13*, 33–38. [CrossRef]
13. Goss, M.J. Sustaining Soil Productivity in Response to Global Climate Change: Science, Policy, and Ethics. *Can. J. Soil Sci.* **2013**, *93*, 393–395. [CrossRef]

14. Galloway, J.N.; ATownsend, A.R.; Erisman, J.W.; Bekunda, M.; Cai, Z.; Freney, J.R.; Martinelli, L.A.; Seitzinger, S.P.; Sutton, M.A. Transformation of the Nitrogen Cycle: Recent Trends, Questions, and Potential Solutions. *Science* **2008**, *320*, 889–892. [CrossRef] [PubMed]
15. Adhikari, K.; Hartemink, A.E. Linking soils to ecosystem services—A global review. *Geoderma* **2016**, *262*, 101–111. [CrossRef]
16. Sandifer, P.A.; Sutton-Grier, A.E.; Ward, B.P. Exploring connections among nature, biodiversity, ecosystem services, and human health and well-being: Opportunities to enhance health and biodiversity conservation. *Ecosyst. Serv.* **2015**, *12*, 1–15. [CrossRef]
17. Dangi, S.R. Soil Ecology and Ecosystem Services. *Soil Sci. Soc. Am. J.* **2014**, *78*, 335. [CrossRef]
18. Delgado-Baquerizo, M.; Maestre, F.T.; Reich, P.B.; Jeffries, T.; Gaitan, J.; Encinar, D.; Berdugo, M.; Campbell, C.D.; Singh, B. Microbial diversity drives multifunctionality in terrestrial ecosystems. *Nat. Commun.* **2016**, *7*, 10541. [CrossRef]
19. Garland, G.; Banerjee, S.; Edlinger, A.; Oliveira, E.M.; Herzog, C.; Wittwer, R.; Philippot, L.; Maestre, F.T.; van der Heijden, M.G. A closer look at the functions behind ecosystem multifunctionality: A review. *J. Ecol.* **2021**, *109*, 600–613. [CrossRef]
20. Manning, P.; Van Der Plas, F.; Soliveres, S.; Allan, E.; Maestre, F.T.; Mace, G.; Whittingham, M.J.; Fischer, M. Publisher Correction: Redefining ecosystem multifunctionality. *Nat. Ecol. Evol.* **2018**, *2*, 1515. [CrossRef]
21. Xu, W.; Ma, Z.; Jing, X.; He, J.-S. Biodiversity and ecosystem multifunctionality: Advances and perspectives. *Biodivers. Sci.* **2016**, *24*, 55–71. [CrossRef]
22. Greiner, L.; Keller, A.; Grêt-Regamey, A.; Papritz, A. Soil function assessment: Review of methods for quantifying the contributions of soils to ecosystem services. *Land Use Policy* **2017**, *69*, 224–237. [CrossRef]
23. Guerra, C.A.; Heintz-Buschart, A.; Sikorski, J.; Chatzinotas, A.; Guerrero-Ramírez, N.; Cesarz, S.; Beaumelle, L.; Rillig, M.C. Maestre, F.T.; Delgado-Baquerizo, M.; et al. Blind spots in global soil biodiversity and ecosystem function research. *Nat. Commun* **2020**, *11*, 3870. [CrossRef]
24. Schulte, R.P.O.; Ebampa, F.; Ebardy, M.; Ecoyle, C.; Creamer, R.E.; Efealy, R.; Egardi, C.; Ghaley, B.B.; Ejordan, P.; Elaudon, H.; et al. Making the Most of Our Land: Managing Soil Functions from Local to Continental Scale. *Front. Environ. Sci.* **2015**, *3*, 85–91 [CrossRef]
25. Jiang, M.; Lv, X.G.; Yang, Q. Wetland soil and its system of environment function assessment. *Wetl. Sci.* **2006**, *3*, 168–173. (In Chinese) [CrossRef]
26. Costanza, R.; d'Arge, R.; de Groot, R.; Farber, S.; Grasso, M.; Hannon, B.; Limburg, K.; Naeem, S.; O'Neill, R.V.; Paruelo, J.; et al. The value of the world's ecosystem services and natural capital. *Nature* **1997**, *387*, 253–260. [CrossRef]
27. Tzilivakis, J.; Lewis, K.; Williamson, A. A prototype framework for assessing risks to soil functions. *Environ. Impact Assess. Rev.* **2005**, *25*, 181–195. [CrossRef]
28. Burauel, P.; Bassmann, F. Soils as filter and buffer for pesticides—Experimental concepts to understand soil functions. *Environ. Pollut.* **2005**, *133*, 11–16. [CrossRef] [PubMed]
29. Rabot, E.; Wiesmeier, M.; Schlüter, S.; Vogel, H.-J. Soil structure as an indicator of soil functions: A review. *Geoderma* **2018**, *314*, 122–137. [CrossRef]
30. Jost, E.; Schnhart, M.; Skalsk, R.; Balkovi, J.; Mitter, H. Dynamic soil functions assessment employing land use and climate scenarios at regional scale. *J Environ Manag.* **2021**, *287*, 112318. [CrossRef]
31. Thoumazeau, A.; Bessou, C.; Renevier, M.-S.; Trap, J.; Marichal, R.; Mareschal, L.; Decaëns, T.; Bottinelli, N.; Jaillard, B.; Chevallier, T.; et al. Biofunctool®: A new framework to assess the impact of land management on soil quality. Part A: Concept and validation of the set of indicators. *Ecol. Indic.* **2019**, *97*, 100–110. [CrossRef]
32. Thoumazeau, A.; Bessou, C.; Renevier, M.-S.; Panklang, P.; Puttaso, P.; Peerawat, M.; Heepngoen, P.; Polwong, P.; Koonklang, N.; Sdoodee, S.; et al. Biofunctool®: A new framework to assess the impact of land management on soil quality. Part B: Investigating the impact of land management of rubber plantations on soil quality with the Biofunctool® index. *Ecol. Indic.* **2019**, *97*, 429–437. [CrossRef]
33. Schulte, R.; Creamer, R.; Donnellan, T.; Farrelly, N.; Fealy, R.; O'Donoghue, C.; O'Huallachain, D. Functional land management: A framework for managing soil-based ecosystem services for the sustainable intensification of agriculture. *Environ. Sci. Policy* **2014**, *38*, 45–58. [CrossRef]
34. Wu, K.N.; Yang, Q.J.; Chen, J.Y.; Wu, J.F. Soil quality and functions: From science to experiences—A reviewof the wageningen soil conference 2019. *Chin. J. Soil Sci.* **2020**, *51*, 241–244. [CrossRef]
35. Angelini, M.E.; Heuvelink, G.B.M.; Kempen, B. Multivariate mapping of soil with structural equation modelling. *Eur. J. Soil Sci.* **2017**, *68*, 575–591. [CrossRef]
36. Mahajan, G.; Das, B.; Morajkar, S.; Desai, A.; Murgaokar, D.; Kulkarni, R.; Sale, R.; Patel, K. Soil quality assessment of coastal salt-affected acid soils of India. *Environ. Sci. Pollut. Res.* **2020**, *27*, 26221–26238. [CrossRef]
37. Obade, V.D.P.; Lal, R. Towards a standard technique for soil quality assessment. *Geoderma* **2016**, *265*, 96–102. [CrossRef]
38. Chen, Q.-L.; Ding, J.; Zhu, D.; Hu, H.-W.; Delgado-Baquerizo, M.; Ma, Y.-B.; He, J.-Z.; Zhu, Y.-G. Rare microbial taxa as the major drivers of ecosystem multifunctionality in long-term fertilized soils. *Soil Biol. Biochem.* **2020**, *141*, 107686. [CrossRef]
39. Luo, G.; Rensing, C.; Chen, H.; Liu, M.; Wang, M.; Guo, S.; Ling, N.; Shen, Q. Deciphering the associations between soil microbial diversity and ecosystem multifunctionality driven by long-term fertilization management. *Funct. Ecol.* **2018**, *32*, 1103–1116. [CrossRef]
40. Li, J.; Delgado-Baquerizo, M.; Wang, J.-T.; Hu, H.-W.; Cai, Z.-J.; Zhu, Y.-N.; Singh, B. Fungal richness contributes to multifunctionality in boreal forest soil. *Soil Biol. Biochem.* **2019**, *136*, 107526. [CrossRef]

41. Guo, Y.; Luo, H.; Wang, L.; Xu, M.; Wan, Y.; Chou, M.; Shi, P.; Wei, G. Multifunctionality and microbial communities in agricultural soils regulate the dynamics of a soil-borne pathogen. *Plant Soil* **2021**, *461*, 309–322. [CrossRef]
42. Cui, H.; Sun, W.; Delgado-Baquerizo, M.; Song, W.; Ma, J.-Y.; Wang, K.; Ling, X. Phosphorus addition regulates the responses of soil multifunctionality to nitrogen over-fertilization in a temperate grassland. *Plant Soil* **2020**, 1–15. [CrossRef]
43. Qiu, L.; Zhang, Q.; Zhu, H.; Reich, P.B.; Banerjee, S.; van der Heijden, M.G.A.; Sadowsky, M.J.; Ishii, S.; Jia, X.; Shao, M.; et al. Erosion reduces soil microbial diversity, network complexity and multifunctionality. *ISME J.* **2021**, 1–16. [CrossRef]
44. Su, Y.; Liu, J.; Zhang, Y.; Huang, G. More drought leads to a greater significance of biocrusts to soil multifunctionality. *Funct. Ecol.* **2021**, *35*, 989–1000. [CrossRef]
45. Han, X.Z.; Li, N. Research progress of black soil in northeast China. *Sci. Geogr. Sin.* **2018**, *38*, 1032–1041.
46. Xin, J.S.; Wang, J.K.; Xue, Y.D. *Dongbei Heituqu Gengdi Zhiliang Pingjia*; China Agriculture Press: Beijing, China, 2017; pp. 8–37.
47. Liu, B.Y. Issue discussion of sustainable utilization and soil degradation in typical black soil region. *Soil Water Conserv. China* **2003**, *12*, 31–32. [CrossRef]
48. Mueller, L.; Schindler, U.; Mirschel, W.; Shepherd, T.G.; Ball, B.C.; Helming, K.; Rogasik, J.; Eulenstein, F.; Wiggering, H. Assessing the productivity function of soils. A review. *Agron. Sustain. Dev.* **2010**, *30*, 601–614. [CrossRef]
49. Skaalsveen, K.; Ingram, J.; Clarke, L.E. The effect of no-till farming on the soil functions of water purification and retention in north-western Europe: A literature review. *Soil Tillage Res.* **2019**, *189*, 98–109. [CrossRef]
50. Lin, D.; McCulley, R.L.; Nelson, J.A.; Jacobsen, K.L.; Zhang, D. Time in pasture rotation alters soil microbial community composition and function and increases carbon sequestration potential in a temperate agroecosystem. *Sci. Total Environ.* **2020**, *698*, 134233. [CrossRef] [PubMed]
51. Tibbett, M.; Fraser, T.D.; Duddigan, S. Identifying potential threats to soil biodiversity. *PeerJ* **2020**, *8*, e9271. [CrossRef]
52. Chen, Q.; Liu, Z.; Zhou, J.; Xu, X.; Zhu, Y. Long-term straw mulching with nitrogen fertilization increases nutrient and microbial determinants of soil quality in a maize–wheat rotation on China's Loess Plateau. *Sci. Total Environ.* **2021**, *775*, 145930. [CrossRef]
53. Sandén, T.; Trajanov, A.; Spiegel, H.; Kuzmanovski, V.; Saby, N.P.A.; Picaud, C.; Henriksen, C.B.; Debeljak, M. Development of an Agricultural Primary Productivity Decision Support Model: A Case Study in France. *Front. Environ. Sci.* **2019**, *7*, 58. [CrossRef]
54. Ghaley, B.B.; Rusu, T.; Sandén, T.; Spiegel, H.; Menta, C.; Visioli, G.; O'Sullivan, L.; Gattin, I.T.; Delgado, A.; Liebig, M.A.; et al. Assessment of Benefits of Conservation Agriculture on Soil Functions in Arable Production Systems in Europe. *Sustainability* **2018**, *10*, 794. [CrossRef]
55. Van Leeuwen, J.P.; Saby, N.P.A.; Jones, A.; Louwagie, G.; Micheli, E.; Rutgers, M.; Schulte, R.P.O.; Spiegel, H.; Toth, G.; Creamer, R.E. Gap assessment in current soil monitoring networks across Europe for measuring soil functions. *Environ. Res. Lett.* **2017**, *12*, 124007. [CrossRef]
56. Glæsner, N.; Helming, K.; De Vries, W. Do Current European Policies Prevent Soil Threats and Support Soil Functions? *Sustainability* **2014**, *6*, 9538–9563. [CrossRef]
57. Velasquez, E.; Lavelle, P.; Andrade, M. GISQ, a multifunctional indicator of soil quality. *Soil Biol. Biochem.* **2007**, *39*, 3066–3080. [CrossRef]
58. Van den Putte, A.; Govers, G.; Diels, J.; Gillijns, K.; Demuzere, M. Assessing the effect of soil tillage on crop growth: A meta-regression analysis on European crop yields under conservation agriculture. *Eur. J. Agron.* **2010**, *33*, 231–241. [CrossRef]
59. Qi, X.; Fu, Y.; Wang, R.Y.; Ng, C.N.; Dang, H.; He, Y. Improving the sustainability of agricultural land use: An integrated framework for the conflict between food security and environmental deterioration. *Appl. Geogr.* **2018**, *90*, 214–223. [CrossRef]
60. Hoffland, E.; Kuyper, T.W.; Comans, R.N.J.; Creamer, R.E. Eco-functionality of organic matter in soils. *Plant Soil* **2020**, *455*, 1–22. [CrossRef]
61. Obalum, S.; Chibuike, G.; Peth, S.; Ouyang, Y. Soil organic matter as sole indicator of soil degradation. *Environ. Monit. Assess.* **2017**, *189*, 176. [CrossRef] [PubMed]
62. Oldfield, E.E.; Wood, S.A.; Bradford, M. Direct effects of soil organic matter on productivity mirror those observed with organic amendments. *Plant Soil* **2018**, *423*, 363–373. [CrossRef]
63. Schiefer, J.; Lair, G.J.; Blum, W.E. Indicators for the definition of land quality as a basis for the sustainable intensification of agricultural production. *Int. Soil Water Conserv. Res.* **2015**, *3*, 42–49. [CrossRef]
64. Shirato, Y. Use of models to evaluate carbon sequestration in agricultural soils. *Soil Sci. Plant Nutr.* **2019**, *66*, 21–27. [CrossRef]
65. Zwetsloot, M.J.; van Leeuwen, J.; Hemerik, L.; Martens, H.; Simó Josa, I.; Van de Broek, M.; Debeljak, M.; Rutgers, M.; Sandén, T.; Wall, D.P.; et al. Soil multifunctionality: Synergies and trade-offs across European climatic zones and land uses. *Eur. J. Soil Sci.* **2020**, 1–15. [CrossRef]
66. Ellis, E.A.; Baerenklau, K.A.; Marcos-Martinez, R.; Chávez, E. Land use/land cover change dynamics and drivers in a low-grade marginal coffee growing region of Veracruz, Mexico. *Agrofor. Syst.* **2010**, *80*, 61–84. [CrossRef]
67. Pheap, S.; Lefèvre, C.; Thoumazeau, A.; Leng, V.; Boulakia, S.; Koy, R.; Hok, L.; Lienhard, P.; Brauman, A.; Tivet, F. Multi-functional assessment of soil health under Conservation Agriculture in Cambodia. *Soil Tillage Res.* **2019**, *194*, 104349. [CrossRef]
68. Al-Shammary, A.A.G.; Kouzani, A.Z.; Kaynak, A.; Khoo, S.Y.; Norton, M.; Gates, W. Soil Bulk Density Estimation Methods: A Review. *Pedosphere* **2018**, *28*, 581–596. [CrossRef]
69. Liu, L.; Zhou, D.; Chang, X.; Lin, Z. A new grading system for evaluating China's cultivated land quality. *Land Degrad. Dev.* **2020**, *31*, 1482–1501. [CrossRef]

70. Coyle, C.; Creamer, R.; Schulte, R.; O'Sullivan, L.; Jordan, P. A Functional Land Management conceptual framework under sc drainage and land use scenarios. *Environ. Sci. Policy* **2016**, *56*, 39–48. [CrossRef]
71. Burkhard, B.; Kroll, F.; Nedkov, S.; Müller, F. Mapping ecosystem service supply, demand and budgets. *Ecol. Indic.* **2012**, *21*, 17–2 [CrossRef]
72. Han, X.Z.; Zou, W.X. Effects and suggestions of black soil protection and soil fertility increase in northeast China. *Bull. Chin. Aca Sci.* **2018**, *33*, 206–212.
73. Fan, H.M.; Cai, Q.G.; Wang, H.S. Condition of soil erosion in phaeozem region of northeast China. *J. Soil Water Conserv.* **2004**, *1* 66–70.
74. Yang, D.W.; Zhang, M.K.; Zhang, P.Q.; Yang, Y.D. Evolution of soil in microbiology after reclamation of paddy into orchard. *Act Pedol. Sin.* **2018**, *55*, 182–193. (In Chinese) [CrossRef]
75. Yan, B.; Zhang, Y.; Zang, S.; Chen, Q.; Sun, L. Distributions of Particle Sizes in Black Soil and Their Environmental Significance i Northeast China. *Sustainability* **2021**, *13*, 3706. [CrossRef]
76. Yuan, Z.; Jin, X.; Guan, Q.; Meshack, A.O. Converting cropland to plantation decreases soil organic carbon stock and liabl fractions in the fertile alluvial plain of eastern China. *Geoderma Reg.* **2021**, *24*, e00356. [CrossRef]
77. Liu, H.Y.; Zhou, P.; Zhu, H.H.; Wu, J.S.; Zhou, D.S. Effect of land use change on topsoil organic carbon storage of paddy soil in hilly landscape of Red Earth Region. *Agric. Mod. Res.* **2012**, *33*, 359–362. (In Chinese) [CrossRef]
78. Huang, J.F.; Cao, Z.H.; Shi, Y.P.; Ni, X.W. Agricultural and forestry carbon sequestration function analysis and its affection b land use change in Jiaxing Plain, China. *Agric. Environ. Dev.* **2013**, *4*, 23–27. (In Chinese) [CrossRef]
79. Li, M.; Han, X.; Du, S.; Li, L.-J. Profile stock of soil organic carbon and distribution in croplands of Northeast China. *Catena* **201** *174*, 285–292. [CrossRef]
80. Liu, Y.S.; Zhang, Z.W.; Wang, J.Y. Regional differentiation and comprehensive regionalization scheme of modern agriculture ir China. *Acta Geogr. Sin.* **2018**, *73*, 203–218. (In Chinese) [CrossRef]
81. Schneider, F.; Don, A.; Hennings, I.; Schmittmann, O.; Seidel, S.J. The effect of deep tillage on crop yield—What do we reall know? *Soil Tillage Res.* **2017**, *174*, 193–204. [CrossRef]
82. Zhang, C.; Qiao, M.; Yun, W.J.; Liu, J.J.; Zhu, D.H.; Yang, J.Y. Trinity comprehensive regulatory system about quantity, quality and ecology of cultivated land. *Trans. Chin. Soc. Agric. Mach.* **2017**, *48*, 1–6. [CrossRef]
83. Tian, P.; Lian, H.; Wang, Z.; Jiang, Y.; Li, C.; Sui, P.; Qi, H. Effects of Deep and Shallow Tillage with Straw Incorporation on Soi Organic Carbon, Total Nitrogen and Enzyme Activities in Northeast China. *Sustainability* **2020**, *12*, 8679. [CrossRef]
84. Liu, J.; Lu, Y.F.; Xu, Q. Agricultural subsidy, old-age security and farmland transfer. *J. Agrotech. Econ.* **2020**, *12*, 23–37. (In Chinese [CrossRef]
85. Obour, A.; Holman, J.; Simon, L.; Schlegel, A. Strategic Tillage Effects on Crop Yields, Soil Properties, and Weeds in Dryland No-Tillage Systems. *Agronomy* **2021**, *11*, 662. [CrossRef]
86. Sun, J.H.; He, J.M. Hilly field erosion regulation in low mountain region of northern Liaoning. *Res. Soil Water Conserv.* **1997**, *4*, 65–74. (In Chinese)
87. Han, X.Z.; Wang, S.Y.; Song, C.Y.; Qiao, Y.F. Effects of land use and cover change on ecological environment in black soil region *Sci. Geogr. Sin.* **2005**, *25*, 203–208. (In Chinese)
88. Mao, Y.-T.; Hu, W.; Chau, H.W.; Lei, B.-K.; Di, H.-J.; Chen, A.-Q.; Hou, M.-T.; Whitley, S. Combined Cultivation Pattern Reduces Soil Erosion and Nutrient Loss from Sloping Farmland on Red Soil in Southwestern China. *Agronomy* **2020**, *10*, 1071. [CrossRef]
89. Severini, S.; Castellari, M.; Cavalli, D.; Pecetti, L. Economic Sustainability and Riskiness of Cover Crop Adoption for Organic Production of Corn and Soybean in Northern Italy. *Agronomy* **2021**, *11*, 766. [CrossRef]
90. Han, C.; Chen, S.; Yu, Y.; Xu, Z.; Zhu, B.; Xu, X.; Wang, Z. Evaluation of Agricultural Land Suitability Based on RS, AHP, and MEA: A Case Study in Jilin Province, China. *Agriculture* **2021**, *11*, 370. [CrossRef]
91. Bünemann, E.K.; Bongiorno, G.; Bai, Z.; Creamer, R.; De Deyn, G.; Goede, R.; Fleskens, L.; Geissen, V.; Kuyper, T.W.; Mäder, P.; et al. Soil quality—A critical review. *Soil Biol. Biochem.* **2018**, *120*, 105–125. [CrossRef]
92. Powlson, D.S. Soil health—Useful terminology for communication or meaningless concept? Or both? *Front. Agric. Sci. Eng.* **2020**, *7*, 246–250. [CrossRef]
93. Karlen, D.; Andrews, S.; Doran, J. Soil quality: Current concepts and applications. *Adv. Agron.* **2001**, *74*, 1–40. [CrossRef]
94. Grunwald, S.; Thompson, J.A.; Boettinger, J.L. Digital soil mapping and modeling at continental scales: Finding solutions for global issues. *Soil Sci. Soc. Am. J.* **2011**, *75*, 1201–1213. [CrossRef]
95. Vogel, H.-J.; Eberhardt, E.; Franko, U.; Lang, B.; Ließ, M.; Weller, U.; Wiesmeier, M.; Wollschläger, U. Quantitative Evaluation of Soil Functions: Potential and State. *Front. Environ. Sci.* **2019**, *7*, 164. [CrossRef]
96. Craheix, D.; Angevin, F.; Doré, T.; de Tourdonnet, S. Using a multicriteria assessment model to evaluate the sustainability of conservation agriculture at the cropping system level in France. *Eur. J. Agron.* **2016**, *76*, 75–86. [CrossRef]
97. Obade, V.D.P. Integrating management information with soil quality dynamics to monitor agricultural productivity. *Sci. Total Environ.* **2019**, *651*, 2036–2043. [CrossRef]
98. Glenk, K.; McVittie, A.; Moran, D. Deliverable D3.1: Soil and Soil Organic Carbon within an Ecosystem Service Approach Linking Biophysical and Economic Data. Report for EU FP7 SmartSOIL. 2012. Available online: http://smartsoil.eu/ (accessed on 8 October 2012).

9. Zhang, G.L.; Wu, H.Y. From "problems" to "solutions": Soil functions for realization of sustainable development goals. *Bull. Chin. Acad. Sci.* **2018**, *33*, 124–134. (In Chinese) [CrossRef]
00. Ellili-Bargaoui, Y.; Walter, C.; Lemercier, B.; Michot, D. Assessment of six soil ecosystem services by coupling simulation modelling and field measurement of soil properties. *Ecol. Indic.* **2021**, *121*, 107211. [CrossRef]
01. Rinot, O.; Levy, G.J.; Steinberger, Y.; Svoray, T.; Eshel, G. Soil health assessment: A critical review of current methodologies and a proposed new approach. *Sci. Total Environ.* **2019**, *648*, 1484–1491. [CrossRef]
102. Zhao, R.; Wu, K.N.; Liu, Y.N.; Feng, Z. Soil Health Evaluation at a County level Based on Ecosystem Service Function. *Chin. J. Soil Sci.* **2020**, *51*, 269–279. [CrossRef]
103. Hauck, J.; Albert, C.; Fürst, C.; Geneletti, D.; La Rosa, D.; Lorz, C.; Spyra, M. Developing and applying ecosystem service indicators in decision-support at various scales. *Ecol. Indic.* **2016**, *61*, 1–5. [CrossRef]
104. Kragt, M.E.; Robertson, M. Quantifying ecosystem services trade-offs from agricultural practices. *Ecol. Econ.* **2014**, *102*, 147–157. [CrossRef]
105. Pereira, P.; Bogunovic, I.; Muñoz-Rojas, M.; Brevik, E.C. Soil ecosystem services, sustainability, valuation and management. *Curr. Opin. Environ. Sci. Health* **2018**, *5*, 7–13. [CrossRef]
106. Yang, Q.J.; Wu, K.N.; Feng, Z.; Zhao, R.; Zhang, X.D.; Li, X.L. Soil quality assessment on large spatial scales: Dvancement and revelation. *Acta Pedol. Sin.* **2020**, *57*, 565–578. [CrossRef]
107. Calzolari, C.; Ungaro, F.; Filippi, N.; Guermandi, M.; Malucelli, F.; Marchi, N.; Staffilani, F.; Tarocco, P. A methodological framework to assess the multiple contributions of soils to ecosystem services delivery at regional scale. *Geoderma* **2016**, *261*, 190–203. [CrossRef]
108. Yang, X.; Zheng, Q. Progress of agricultural biodiversity conservation in China. *Biodivers. Sci.* **2021**, *29*, 167–176. [CrossRef]
109. Benton, T.G.; Vickery, J.A.; Wilson, J.D. Farmland biodiversity: Is habitat heterogeneity the key? *Trends Ecol. Evol.* **2003**, *18*, 182–188. [CrossRef]
110. Zhang, J.L.; Zhang, J.Z.; Shen, J.B.; Tian, J.; Jin, K.; Zhang, F.S. Soil health and agriculture green development: Opportunities and challenge. *Acta Pedol. Sin.* **2020**, *57*, 783–796. [CrossRef]
111. De Deyn, G.B.; van der Putten, W. Linking aboveground and belowground diversity. *Trends Ecol. Evol.* **2005**, *20*, 625–633. [CrossRef]
112. Bardgett, R.D.; van der Putten, W. Belowground biodiversity and ecosystem functioning. *Nat. Cell Biol.* **2014**, *515*, 505–511. [CrossRef]
113. Cameron, E.K.; Martins, I.S.; Lavelle, P.; Mathieu, J.; Tedersoo, L.; Bahram, M.; Gottschall, F.; Guerra, C.A.; Hines, J.; Patoine, G.; et al. Global mismatches in aboveground and belowground biodiversity. *Conserv. Biol.* **2019**, *33*, 1187–1192. [CrossRef]
114. Lehman, R.M.; Cambardella, C.A.; Stott, D.E.; Acosta-Martinez, V.; Manter, D.K.; Buyer, J.S.; Maul, J.E.; Smith, J.L.; Collins, H.P.; Halvorson, J.J.; et al. Understanding and Enhancing Soil Biological Health: The Solution for Reversing Soil Degradation. *Sustainability* **2015**, *7*, 988–1027. [CrossRef]
115. Fan, K.; Delgado-Baquerizo, M.; Guo, X.; Wang, D.; Zhu, Y.-G.; Chu, H. Biodiversity of key-stone phylotypes determines crop production in a 4-decade fertilization experiment. *ISME J.* **2021**, *15*, 550–561. [CrossRef] [PubMed]
116. Wall, D.H.; Nielsen, U.N.; Six, J. Soil biodiversity and human health. *Nat. Cell Biol.* **2015**, *528*, 69–76. [CrossRef] [PubMed]
117. Pulleman, M.; Creamer, R.; Hamer, U.; Helder, J.; Pelosi, C.; Pérès, G.; Rutgers, M. Soil biodiversity, biological indicators and soil ecosystem services—An overview of European approaches. *Curr. Opin. Environ. Sustain.* **2012**, *4*, 529–538. [CrossRef]
118. Wagg, C.; Bender, S.F.; Widmer, F.; van der Heijden, M.G.A. Soil biodiversity and soil community composition determine ecosystem multifunctionality. *Proc. Natl. Acad. Sci. USA* **2014**, *111*, 5266–5270. [CrossRef] [PubMed]
119. Van den Hoogen, J.; Geisen, S.; Routh, D.; Ferris, H.; Traunspurger, W.; Wardle, D.A.; De Goede, R.G.M.; Adams, B.J.; Ahmad, W.; Andriuzzi, W.S.; et al. Soil nematode abundance and functional group composition at a global scale. *Nat. Cell Biol.* **2019**, *572*, 194–198. [CrossRef] [PubMed]
120. Delgado-Baquerizo, M.; Reich, P.B.; Trivedi, C.; Eldridge, D.J.; Abades, S.; Alfaro, F.D.; Bastida, F.; Berhe, A.A.; Cutler, N.A.; Gallardo, A.; et al. Multiple elements of soil biodiversity drive ecosystem functions across biomes. *Nat. Ecol. Evol.* **2020**, *4*, 210–220. [CrossRef]
121. Sorokin, A.; Owens, P.; Láng, V.; Jiang, Z.-D.; Michéli, E.; Krasilnikov, P. "Black soils" in the Russian Soil Classification system, the US Soil Taxonomy and the WRB: Quantitative correlation and implications for pedodiversity assessment. *Catena* **2021**, *196*, 104824. [CrossRef]
122. Xie, Y.; Lin, H.; Ye, Y.; Ren, X. Changes in soil erosion in cropland in northeastern China over the past 300 years. *Catena* **2019**, *176*, 410–418. [CrossRef]
123. Yang, X.M.; Zhang, X.P.; Deng, W.; Fang, H.J. Black soil degradation by rainfall erosion in Jilin, China. *Land Degrad. Dev.* **2003**, *14*, 409–420. [CrossRef]
124. Ding, R.X.; Liu, S.T. A study on the fertility of black soil after reclamation. *Acta Pedol. Sin.* **1980**, *17*, 20–32.
125. Zhang, H.L.; Gao, W.S.; Chen, F.; Zhu, W.S. Prospects and present situation of conservation tillage. *J. China Agric. Univ.* **2005**, *10*, 16–20. [CrossRef]

MDPI
St. Alban-Anlage 66
4052 Basel
Switzerland
Tel. +41 61 683 77 34
Fax +41 61 302 89 18
www.mdpi.com

Land Editorial Office
E-mail: land@mdpi.com
www.mdpi.com/journal/land

www.ingramcontent.com/pod-product-compliance
Lightning Source LLC
Chambersburg PA
CBHW061613220926
51850CB00005B/497
9783036523231